NOKIA® MOBILE DE

Developing Series 60 Applications

NOKIA MOBILE DEVELOPER SERIES

Developed by Nokia and Addison-Wesley Professional Publishing, the Nokia Mobile Developer Series focuses on application development using Nokia's feature-rich software platforms for mobile phones. The series presents "best practices" guidelines for mobile application development through the presentation of real-world examples that have been reviewed and tested by Nokia's subject matter experts. The Nokia Mobile Developer Series, published and made available around the world by Addison-Wesley, is authorized and approved by Nokia.

NOKIA® MOBILE DEVELOPER SERIES

Developing Series 60 Applications:

A Guide for Symbian OS C++ Developers

Leigh Edwards, Richard Barker
and the Staff of EMCC Software Ltd.

✦Addison-Wesley

Boston • San Francisco • New York • Toronto • Montreal
London • Munich • Paris • Madrid
Capetown • Sydney • Tokyo • Singapore • Mexico City

Many of the designations used by manufacturers and sellers to distinguish their products are claimed as trademarks. Where those designations appear in this book, and Addison-Wesley was aware of a trademark claim, the designations have been printed with initial capital letters or in all capitals.

The authors and publisher have taken care in the preparation of this book, but make no expressed or implied warranty of any kind and assume no responsibility for errors or omissions. No liability is assumed for incidental or consequential damages in connection with or arising out of the use of the information or programs contained herein.

The publisher offers discounts on this book when ordered in quantity for bulk purchases and special sales. For more information, please contact: U.S. Corporate and Government Sales, (800) 382-3419, corpsales@pearsontechgroup.com

For sales outside of the U.S., please contact: International Sales, (317) 581-3793, international@pearsontechgroup.com

Visit Addison-Wesley on the Web: www.awprofessional.com

Library of Congress Cataloging-in-Publication Data

A CIP catalog record for this book can be obtained from the Library of Congress.

Copyright © 2004 by Nokia Corporation and EMCC Software, Ltd. All rights reserved.

Nokia is a registered trademark of Nokia Corporation. EMCC Software and the EMCC logo are trademarks of EMCC Software, Ltd. Other product and company names mentioned herein may be trademarks or trade names of their respective owners.

The information in this book is provided "as is," with no warranties whatsoever, including any warranty of merchantability, fitness for any particular purpose or any warranty otherwise arising out of any proposal, specification, or sample. This document is provided for informational purposes only. Nokia Corporation disclaims all liability, including liability for infringement of any proprietary rights, relating to implementation of information presented in this document. Nokia Corporation does not warrant or represent that such use will not infringe such rights. Nokia Corporation retains the right to make changes to this specification at any time, without notice.

For more information on this book and other titles in the Nokia Mobile Developers Series, please go to: www.forum.nokia.com/books.

All rights reserved. No part of this publication may be reproduced, stored in a retrieval system, or transmitted, in any form, or by any means, electronic, mechanical, photocopying, recording, or otherwise, without the prior consent of the publisher. Printed in the United States of America. Published simultaneously in Canada.

For information on obtaining permission for use of material from this work, please submit a written request to: Pearson Education, Inc., Rights and Contracts Department, 75 Arlington Street, Suite 300, Boston, MA 02116, Fax: (617) 848-7047

ISBN: 0-321-22722-0

Text printed on recycled paper

Second printing

Contents

Foreword by Nokia   xvii
Foreword by EMCC Software Ltd.   xix
Preface   xxi
Introduction to Symbian OS and Series 60 Platform   xxv
Acknowledgements   xliii
Authors and Contributors   xlv

1 Getting Started   3
Series 60 C++ Software Development Kits (SDKs)   5
    Using Multiple SDKs   6
Development Process Overview   6
    Using an IDE versus Command-Line Tools   8
Series 60 Emulators   10
Building for the Emulator   12
    Building from the Command Line   12
    Building from an IDE   13
Running the Emulator   15
    Emulator Executable Locations   15
    Emulator Debug Mode   17
    Running the Emulator from a Command Prompt   17

v

vi Contents

    Running the Emulator from the Visual C++ IDE   18
    Running the Emulator from the Borland C++Builder 6 and C++BuilderX IDEs   18
    Running the Emulator from the CodeWarrior IDE   18
    Locating and Running the Application   18
    Debugging the Application   19
    Further IDE Help   19
Building for a Target Series 60 Device   19
Deploying on a Target Device   21
    Building a SIS Install File   21
    SIS File Installation   22
    Running on a Target Device   22
Summary   23

2 Development Reference   25

SDK Versions and Selection   27
    Series 60 Version 1.x SDKs   28
    Series 60 Version 2.x SDKs   29
The HelloWorld GUI Application   30
    HelloWorld bld.inf   30
    HelloWorld.mmp   31
    Building and Running   35
    HelloWorld GUI Executable and Runtime Files   36
    HelloWorld Project Files and Locations   38
    HelloWorld GUI Source Files   39
    Resource Compiler   41
    Localization of Applications and Resources   42
    AIF Files   43
Console Applications   46
    The Hello World Console Application   47
    Build and Run the Console Application   47
    HelloWorldCon.mmp   48
    HelloWorldCon Emulator Executable Files   49
    HelloWorldCon Target Executable Files   49
Symbian Installation System   50
    SIS File Build Tools   51
    Format of .pkg Files   52
    Building a SIS File   58

Additional Development Tools 58
  Multi-Bitmaps and Bmconv the Bitmap Converter 58
  Series 60 Application Wizards 60
  Miscellaneous SDK Tools 62
  More Tools and Utilities 63
Installation Tips for Series 60 SDKs and IDEs 64
  Microsoft Visual Studio .NET 64
  Emulator Configuration 64
  Application Panics in the Emulator 67
Advanced Application Deployment and Build Guide 67
  Platform UIDs 67
  Device Identification UIDs 68
  Resource File Versions and Compression 70
  Building for ARM Targets 70
Summary 73

3 Symbian OS Fundamentals 75
Naming Conventions 77
  T Classes 78
  S Classes 79
  C Classes 79
  R Classes 79
  M Classes 80
  Namespaces 81
Basic Types 82
Exception Handling and Resource Management 84
  Exceptions, Leaves, Panics and Traps 84
  Leaving Issues and the Cleanup Stack 88
  Two-phase Construction 91
  Symbian OS Construction Methods 96
  Advanced Use of the Cleanup Stack 96
Descriptors 97
  Hierarchy 98
  Nonmodifiable API 99
  Modifiable API 100
  Literals 101
  Using Descriptors 102

viii Contents

    Descriptors as Arguments and Return Types   106
    Package Descriptors   108
Collection Classes   109
    RArray and RPointerArray Types   110
    CArray Types   117
Using Asynchronous Services with Active Objects   126
    The Active Scheduler   128
    Active Objects   129
    Implementing an Active Object   131
    A Practical Example   132
    Common Active Object Pitfalls   142
Files, Streams and Stores   143
    Files   143
    RFs API   144
    RFile API   147
    Streams   151
    Stores   157
Client/Server Architecture   165
    Server Sessions   165
    Server Sessions and Inter-Process Communication   166
    Server Review   169
    Subsessions   169
Summary   171

4 Application Design   173
Application Framework   175
Application Architecture   178
    Core Application Classes   178
    Application Initialization   179
    Important AppUi Methods   182
    Designing an Application UI   183
    Traditional Symbian OS Control-Based Architecture   184
    Dialog-Based Architecture   188
    Avkon View-Switching Architecture   191
    Choosing the Appropriate Application Architecture   197
    File Handling   200
Splitting the UI and the Engine   201

ECom   206
    ECom Conceptual Overview   208
    ECom Interface   208
    ECom DLL   211

Internationalization   213
    General Guidelines for Developers   213
    OS Support for Localization   216

Good Application Behavior   217
    Adopt a Skeptical and Critical Approach to Development   217
    Handle Window Server-Generated Events   218
    Always Exit Gracefully   218
    Check Disk Space before Saving Data   219
    Other Hints and Tips   219

Summary   220

5 Application UI Components   223

Controls   225
    Controls and Windows   225
    Simple and Compound Controls   225
    Window Ownership   228
    Creating a Simple Control   229
    Creating a Compound Control   232
    Establishing Relationships between Controls   236

Skins   237
    Compulsory Skin-Providing Controls   238
    Optionally Skin-Providing Controls   238
    Skin-Observing Controls   238
    Non-Skin-Aware Controls   239
    Defining Skin-Aware Controls   239

Event Handling   239
    Key Events   239
    Redraw Events   243
    Observers   244

Resource Files   245
    Resource File Syntax   245
    Resource File Structure   247

Menus   254
    Submenus   *255*
    Menu Basics   *256*
    Dynamic Menus   *261*
    Context-Sensitive Menus   *262*

Panes   264
    Status Pane   *264*
    Title Pane   *268*
    Context Pane   *272*
    Navigation Pane   *274*
    Main Pane   *286*
    Soft Key Pane   *286*

Summary   286

6 Dialogs   289

Common Dialog Characteristics   291

Standard Dialogs   291
    Creating a Simple Dialog   *292*
    Multipage Dialogs   *298*
    Defining a Menu for Your Dialog   *300*
    Custom Controls in Dialogs   *300*

Forms   302
    Form Lines   *303*
    Form Soft Keys   *304*
    Creating a Form in an Application   *306*

Notes   310
    Wrapped Notes   *310*
    Customized Notes   *313*
    Wait Notes   *316*
    Progress Notes   *319*
    Global Notes   *325*

Queries   326
    Data Queries   *328*
    List Queries   *330*
    Using Global Queries   *332*

List Dialogs   335
    Markable List Dialogs   *338*

Summary   339

Contents

7 Lists   341

List Basics   343

Vertical Lists   344

    Selection Lists   344
    Menu Lists   345
    Markable Lists   346
    Multiselection Lists   346
    List Items and Fields   346
    Finding Items in a List   347

Using Vertical Lists   347

    Basic Lists   348
    Dynamic Lists   355
    Markable Lists   357
    Pop-up Menu Lists   361

Grids   364

    Monthly Calendar Grid   365
    Pin-Board Grid   366
    GMS Grid   367

Using Grids   367

    Grid Basics   368
    Markable Grids   375

Settings Lists   377

Using Settings Lists   378

    Settings List Basics   379

Summary   393

8 Editors   395

Text Editors   397

    Dimensions and Input Capacity   399
    Filtering Keypad Input   400
    Providing Mappings to Additional Characters   402
    Properties   404
    Configuring a Plain Text Editor   404
    Configuring a Rich Text Editor   410
    Using Styles   416

Numeric Editors   417

    Configuring Numeric Editors   418

Secret Editors 422
Using a Secret Editor 422
Multi-Field Numeric Editors 423
IP Address Editor 424
Number Editor 424
Range Editor 424
Time Editor 425
Date Editor 425
Time and Date Editor 425
Duration Editor 425
Time Offset Editor 425
Using an MFNE 425
Summary 427

9 Communications Fundamentals 429
Serial Communication 431
Using Serial Communication 432
Sockets 438
Sockets on Series 60 438
Client and Server 439
Connectionless and Connected Sockets 439
Connected Sockets 440
Secure Sockets 451
TCP/IP 457
IPv6 458
TCP/IP Programming for Series 60 458
CommDB 458
Multihoming 459
Infrared 465
IrDA Stack 466
Programming Infrared on a Series 60 Device 467
Bluetooth 470
Bluetooth Overview 471
Example Bluetooth Application 473
Service Advertisement 474
Bluetooth Security 477
Device and Service Discovery 479
Bluetooth Socket Communication 484
Summary 487

10 Advanced Communication Technologies — 489

HTTP   491
- Example Application   493

WAP   502
- WAP Architecture   502
- Series 60 Implementation   503

Messaging   508
- Key Messaging Concepts   508
- Key Messaging Classes and Data Types   511
- The Messaging APIs   512
- Using the Client MTM API   514
- Using the Send-As API   526
- Using the CSendAppUi Class   531
- Watching for Incoming Messages   539

Telephony   543
- Using the ETel API   544
- Getting Started   544
- Making a Call   546
- Receiving a Call   548
- Retrieving the Last Calling Number   550

Summary   551

11 Multimedia: Graphics and Audio — 553

Overview of Series 60 Graphics Architecture   556
- Window Server   556
- Font and Bitmap Server   558
- The Window Server and the Font and Bitmap Server   558
- Multi Media Server   558

Basic Drawing   560
- Screen Coordinates and Geometry   562
- Graphics Devices and Graphics Contexts   562
- Color and Display Modes   565
- Pens and Brushes   567
- The View from a Window and the Relationship with CCoeControl   567

Fonts and Text   569
- Text and Font Measurements   570
- Key Font Classes and Functions   571

Contents

    Using the Key Font Classes to Enumerate Through All Available Fonts   572
    Text Effects   573

Shapes   **574**
    Rectangles   575
    Ellipses   575
    Arcs and Pies   576
    Polygons   576

Bitmaps   **578**
    Generating Bitmaps for Application Use   578
    Loading and Drawing Bitmaps   580
    Bitmap Masking   580
    Bitmap Functions   581

Animation   **582**
    The Animation Architecture   582
    Off-Screen Bitmaps and Double Buffering   585
    The Client-Side Approach to Animation   586

Direct Screen Access   **588**
    Architectural Overview   589
    Key Classes for Direct Screen Access   590
    Implementation Considerations   593

Image Manipulation   **594**
    Image Conversion   594
    Image Rotation   601
    Image Scaling   602

Audio   **604**
    Recording   605
    Tones   609
    Audio Data   610
    Streaming   612

Summary   **614**

12 Using Application Views, Engines and Key System APIs   617

Using Standard Application Views   **619**
    Phonebook View Switching   620
    Calendar View Switching   621
    Camera View Switching   622

Contents   xv

   Photo Album View Switching 622
   Profiles View Switching 623
   Messaging View Switching 623
   Nonswitchable Applications 623
  Application Engines **625**
   Log Engine 625
   Camera APIs 627
   Phonebook Engine 632
   Compact Business Cards and vCards 638
   Calendar Engine Access 639
   Photo Album Engine 646
  Accessing System Capabilities **649**
   Hardware Abstraction Layer 649
   System Agent 651
   Vibration API Support 653
  Summary **654**

13 Testing and Debugging 657
  Quality Assurance **659**
   Coding Standards 659
   Defensive Programming 661
  Testing **664**
   Strategies for Testing 666
   Tools and Techniques for Testing 669
   Differences of Testing on Target vs. Emulator 677
   Test Harnesses 681
  Debugging **686**
   Debugging an Application on the Emulator 687
   Debugging an Application on Target 692
  Summary **696**

Appendix Emulator Shortcut Keys 699

Glossary 705

References 721
Example Applications 721
Symbian OS Books 722
Other Useful Books 722
SDKs 723
IDEs 724
Other Web Sites 724

Index 725

About EMCC Software Ltd. 747
Company Overview 747
World Leading Expertise 747
- Messaging and Communications 747
- User Interface Creation 748
- Java 748

Services 748
Benefits of Working with EMCC Software 749

Foreword
by Nokia

The era of mobility is upon us. Mobile phone sales are approaching a half billion units a year and climbing. The smartphone is already playing a key role in helping consumers manage their personal and professional lives.

An increasing number of mobile devices provide consumers with rich media and data services as well as voice communications. An equally impressive collection of enablers, such as built-in cameras, high-resolution color displays, streaming media and a level of memory capacity once reserved for PDAs, give developers the ability to build increasingly more sophisticated wireless applications. Mobile versions of Java and C++ provide a rich development environment for creating innovative consumer and business applications.

Programming for mobile devices has traditionally been a time and labor intensive exercise because the technology in each device is different, and device lifecycles are difficult to predict. The Series 60 Platform changes the economies of scale of mobile development by providing a standards-based development environment for a broad array of smartphones.

The Series 60 Platform provides a common screen size, a consistent user interface, a Web browser, media player, calendar, SMS, MMS and common APIs for Java MIDP and C++ programmers. It is also based on the Symbian OS, the only operating system built from the ground up for the special needs of smart mobile devices.

The Series 60 Platform was designed from the outset to be a powerful and robust platform for third-party applications. There are already 2,000 Symbian applications on the market—from 3D action games for the N-Gage mobile game deck, to photo editing software, to utilities for reading books on the go. These applications appeal to consumers and business users alike in imaginative and innovative ways.

The Series 60 Platform is smartphone software licensed by terminal manufacturers such as Matsushita (Panasonic), Nokia, Samsung, Sendo and Siemens. Combined, these players constitute 60% of the worldwide mobile handset market. The Series 60 Platform defines most of the features and functionality in Series 60 devices developed by Nokia and other Series 60 licensees, but some devices may include additional features to appeal to specific consumer segments.

Foreword by Nokia

This platform lets developers shine with a complete, flexible programming environment that is capable of serving the diverse needs of application programmers. The near-universal Java MIDP platform supports portable applications deployable to a wide range of devices. The native Series 60 C++ programming environment is the perfect platform for optimized applications, providing a rich programming API and large address space. And access to tools and resources could not be easier: Nokia and its licensees provide free access to the programming SDKs and documentation.

The Series 60 *Developer* Platform defines the common, relevant and visible interface to the technology enablers and APIs, and details how they are implemented in the Series 60 Platform. The Series 60 Developer Platform covers multiple devices (even if new Series 60 Platform releases with added features have been introduced), thereby resulting in a large installed base of devices with a common set of enablers and APIs.

The mobile market and its leading-edge technologies offer great opportunities for Series 60 developers, but getting started can seem daunting for developers who have spent years working with other operating systems. Forum Nokia has been assisting developers in learning about Symbian OS and Series 60 development since its inception. *Developing Series 60 Applications*, part of the Nokia Mobile Developer Series, is a further commitment by Nokia to support the Series 60 developer.

Developing Series 60 Applications is the most comprehensive reference on Series 60 Platform ever written, walking readers through their very first steps, and then building upon that knowledge to help them create more sophisticated applications. Every aspect of the Series 60 Platform is covered, from unique and powerful user interface components, through communication APIs and multimedia capabilities. The fundamentals of the Series 60 application engine are detailed, as well as best practices for testing and debugging Series 60 applications. Last but not least, over fifty complete projects, with C++ source code and installation scripts are included to help developers jump-start their own projects.

Developing Series 60 Applications draws upon the expertise of EMCC Software Ltd., a Series 60 Competence Center. Authors Leigh Edwards and Richard Barker and the staff at EMCC bring many years of experience and a wealth of expertise in working with the Series 60 Platform and the Symbian operating system.

Using this book as a guide, developers will be armed with the knowledge they need to create world-class applications for todayís most exciting applications market.

Good reading and happy coding.

Pertti Korhonen
Executive Vice President,
Nokia Mobile Software

Foreword
by EMCC Software Ltd.

Symbian OS is a highly sophisticated, powerful and very reliable Operating System, written mainly in C++. Series 60 Platform, from Nokia Mobile Software, builds on the Operating System providing a carefully crafted highly optimized user interface, application framework and much more besides.

EMCC Software specializes in development for Symbian OS and the key smartphone user interface platforms such as Series 60. We have been working with the OS since before the Symbian alliance was created.

I believe strongly in the excellent technical foundations of Symbian OS and in the clear logic behind licensing of Series 60 to other handset manufacturers. A common Operating System and user interface both assist in achieving a high degree standardization and interoperability across a wide range of smartphones.

Writing this book was a major task that we undertook for a variety of reasons. The level of C++ and object orientated design employed throughout Symbian OS and Series 60 leads to a significant learning curve for developers, compared to other environments. Writing this book is an attempt to shorten the learning process. Even after five years there is still a surprising lack of good material available to engineers working with Symbian OS—especially since developing for Symbian OS is quite challenging. As an evangelist and enthusiast for Symbian OS I found this situation slightly irritating, especially since it is the leading player in the market—by far. Additional influences came from interactions with our customers, business partners and the numerous delegates who have attended our training courses. The decision to proceed was also heavily influenced by our business relationship with Nokia—in particular our role in producing much of the Series 60 documentation available via Forum Nokia.

Foreword by EMCC Software Ltd.

A key objective of writing this book was to encourage developers to work with Symbian OS and in particular with Series 60—a genuine attempt to add to the total sum of what is known about Series 60 development. Hopefully, the majority of readers will be pleasantly surprised by the amount of information provided in this work. We firmly believe that helping to increase the number of Series 60 developers out there will benefit the whole ecosystem and will indirectly benefit our company.

EMCC Software works with Symbian, Nokia and a number of Series 60 Licensees on products, on advancing the OS and the platform itself. Typically we work with software and products that are one to two years away from commercial availability. However, what is provided here is necessarily focused on the current products and platform SDKs. So this detailed account of application development aims to support the production of good quality Series 60 software for devices that are currently in the market place or are about to become available very soon.

Leigh Edwards,
EMCC Software Ltd.,
January 2004

Preface

This book is for anyone who is considering or is currently involved in creating software for Series 60 using C++. For software engineers, designers and project managers, it is an in-depth practical guide to Series 60 development. Engineers from a wide range of organizations—independent software vendors, licensees, competence centers, network operators, content providers and so on—should benefit from this work. This book provides an in-depth practical guide to Series 60 software development in C++. We do not attempt to teach C++ or object-oriented design; these are essential prerequisites to getting the best from this book.

We assume at the outset that you have located and installed a suitable Series 60 SDK and a chosen development environment from Borland, Metrowerks or Microsoft. Help on acquiring these necessary materials is provided in the References section at the back of the book.

Generous amounts of documentation, information and example projects are included with the SDKs and tools, so where possible we have avoided duplicating this material. From time to time we refer to sections of the standard documentation and examples where you can find more details. Many other sources of information are available to assist engineers to acquire entry-level Symbian OS development skills, and links to such resources are provided in the References section.

Some basics of Symbian development are provided here to aid the complete beginner, but to avoid too much duplication we focus mainly on the specifics of Series 60 Platform development.

Around sixty separate buildable projects are provided, together with full source code and installation scripts. See the References section for instructions on correct installation of the project materials. Links to the projects materials, updates and errata are available online from:

- http://www.emccsoft.com/devzone/
- http://www.forum.nokia.com/books/
- http://www.awprofessional.com/nokia/

Series 60 is a complete smartphone reference design, including a host of wireless applications, based on Symbian OS. It represents a rich open environment for developers to create their own innovative applications. However, it is a rapidly developing platform, and so this book covers development for versions 1.x and 2.x of Series 60 Platform.

Application developers can choose from Java^{TM} MIDP or C++ as their development language. This book covers only C++, since it currently offers significantly greater capability in terms of performance and access to a huge set of APIs (Application Programming Interfaces).

Guide to Readers

The first part of this book provides an overview of the development process, the essentials of Symbian development and the key structural elements of a Series 60 application. Therefore, if you are new to Series 60 development, we urge you to read Chapters 1–4 completely before reading anything else. The rest of the book can be used as a reference work on Series 60 development.

A brief outline of each chapter is provided here to guide readers of varying levels of previous experience, from novice to expert, on how to proceed.

- **Chapter 1 — Getting Started**

  Introduces the essentials of a Series 60 project, plus building, deploying and running a simple example application.

- **Chapter 2 — Development Reference**

  Builds on Chapter 1 by providing a detailed description of all the essential components of two Series 60 projects, plus the use of other key development tools to build, deploy and run the example applications.

- **Chapter 3 — Symbian OS Fundamentals**

  The essential characteristics of Symbian OS upon which Series 60 Platform is based.

- **Chapter 4—Application Design**

  Examination of the framework architecture behind every Series 60 GUI application and of key elements of application design—this is the first time the source code of a GUI application is examined.

- **Chapter 5—Application UI Components**

  The basics of creating UI controls, plus the essential Series 60 UI controls such as menus, status panes, control panes and so on.

- **Chapter 6—Dialogs**

  Use of Series 60 dialogs for interaction with users and displaying information or editing data.

- **Chapter 7—Lists**

  User interface controls for displaying collections of items for information and user interaction.

- **Chapter 8—Editors**

  Application of user interface components for entering, displaying and editing data.

- **Chapter 9—Communications Fundamentals**

  Basic communication APIs for Series 60 developers.

- **Chapter 10—Advanced Communication Technologies**

  Sophisticated communication APIs for Series 60 developers.

- **Chapter 11—Multimedia, Graphics and Audio**

  Series 60 Graphics Architecture, drawing, fonts, bitmaps, animation and audio.

- **Chapter 12—Using Application Views, Engines and Key System APIs**

  How to invoke the published standard application views, use many of the key application engines and accessing several useful system functions from within applications.

- **Chapter 13—Testing and Debugging**

  Quality assurance, testing techniques and common debugging methods and techniques.

Introduction to Symbian OS and Series 60 Platform

Series 60 Platform is based on Symbian OS, a mobile Operating System from Symbian Ltd. It is an open, highly robust Operating System for data-enabled mobile phones. Symbian OS (formerly called EPOC) is a 32-bit preemptive multitasking Operating System that is central to the success of Series 60 and other user interface platforms such as Series 80 and Series 90, the communicator platforms from Nokia, and UIQ from UIQ Technology AB, a division of Symbian.

Series 60 Platform is a complete smartphone reference design, developed by Nokia Mobile Software, and is currently being licensed by several of the worlds key handset manufacturers.

Symbian OS Structure

In the real world, events often happen simultaneously and with timing that is unpredictable—usually termed asynchronous behavior. Series 60 applications are designed to behave reliably, to interact smoothly with other applications and with the numerous asynchronous services provided by both Symbian OS and Series 60 Platform. For example, a phone call may interrupt a user composing an email message, a user may switch from Messaging to a Calendar application in the middle of a telephone conversation or an incoming SMS may cause the user to access the Contacts database and forward the SMS onward. By complying with the platform architecture and software design guidelines, application designers can routinely manage such occurrences in the daily lives of smartphone users.

From the outset, Symbian OS was designed for use in small battery-powered devices with extensive communications capabilities. Its key design features include:

- Performance—Designed to maximize battery life through careful device-specific power management.
- Multitasking—Telephony, messaging and communications are fundamental components. All applications are designed to work seamlessly in parallel.
- Standards—The use of technologies based on industry standards is a basic principle of Symbian OS, ensuring that applications are interoperable with solutions from other platform vendors.
- Object-oriented software and highly modular architecture.
- Memory management optimized for embedded software environment—very small executable sizes and ROM-based code that executes in place.
- Runtime memory requirements are minimized.
- Security mechanisms for enabling secure communications and safe data storage.
- Application support for an international environment, with built-in Unicode character sets and ease of localization.

Figure I-1 shows a representation of the Symbian OS generic technology (GT) components.

The system kernel, File Server, memory management and device drivers are located in the "Base" Operating System layer. The kernel manages system resources such as memory and is responsible for time-slicing the applications and system tasks. Device drivers provide the control and interface to specific items of hardware—the keyboard, display, infrared port and so on.

The developer interface to most of the base Operating System functionality is through the `EUser` Library, via a huge range of static function calls beginning with `User::`—for example, `User::After()`, which causes the current thread of execution to be suspended until a specified time interval has expired.

The upper layers of the system provide communication and extensive computing services, such as TCP/IP, IMAP4, SMS and database management. Symbian OS components provide data management, communications, graphics, multimedia, security, personal information management (PIM) application engines, messaging engine, Bluetooth, browser engines and support for data synchronization and internationalization.

Symbian C++ APIs enable extremely efficient multitasking and memory management. Memory-intensive operations such as context switching are minimized. Symbian OS applications are primarily event-driven rather than multithreaded. Multithreading is possible and is used with the Operating System, but it is generally avoided in applications, because it potentially creates several kilobytes of overhead per thread. Conversely, a primarily event-driven

Series 60 Structure

approach does not need any context switching and can have an overhead as low as a few tens of bytes. Special design attention has also been given to ensure that Symbian OS is robust and reliable.

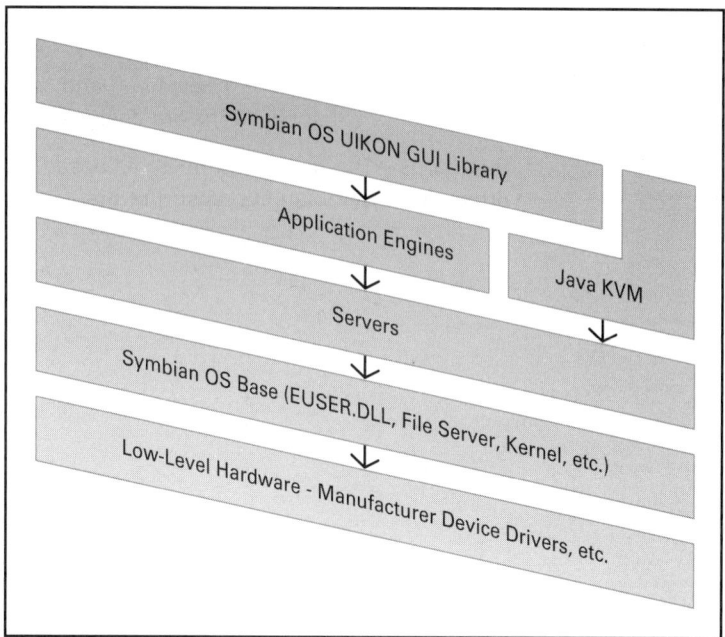

Figure I-1 Symbian OS generic technology structure.

Client/Server architecture is a key design feature of Symbian OS. User applications and system processes are clients that use the resources of a wide variety of system servers. In Symbian OS, servers can be accessed only by their clients via well-defined interfaces. Virtually all servers run with a high priority, but without system privileges, to ensure a timely response to all of their clients while controlling access to the resources of the system.

Some core application engines, written as servers, enable software developers to create their own user interfaces to the application data and databases. Examples include Contacts, Calendar, Multimedia Services (decoding and rendering of image formats) and Messaging.

Data synchronization is provided through a SyncML engine and external connectivity, such as infrared, Bluetooth and a PC Connectivity suite.

Series 60 Structure

Series 60 Platform builds on the Operating System from Symbian, complementing it with a configurable graphical user interface library and a comprehensive suite of applications plus other general-purpose engines. Series 60 is a complete smartphone reference design.

Introduction to Symbian OS and Series 60 Platform

A set of robust components and APIs are provided for developers in Series 60 SDKs. The APIs provided are widely used by the suite of "standard" applications that are an integral part of Series 60 Platform. However, the extensive APIs were designed for use by third-party application developers as well.

The core of Series 60 Platform is Symbian OS GT (Generic Technology) layers—see Figure I–1. Series 60 adds the extensive Avkon UI layer, a full suite of applications based on the Avkon and Uikon libraries plus a number of key application engines—see Figure I–2. Series 60 Platform contains the majority of the user interface and framework APIs used by third-party GUI applications.

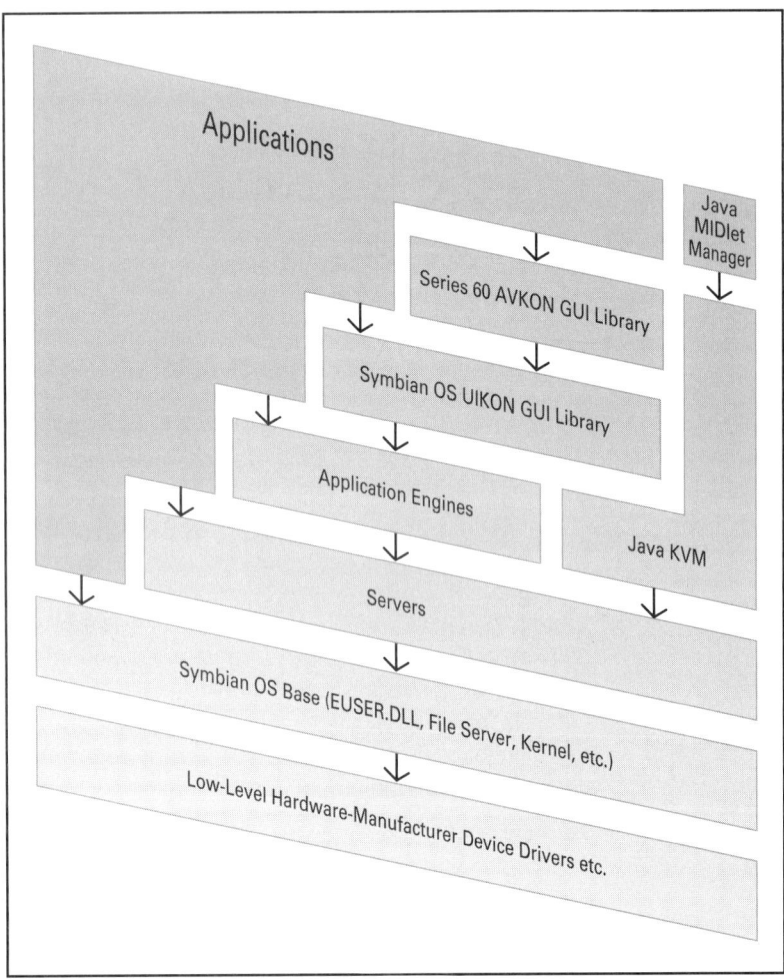

Figure I–2 Series 60 structure.

There are also a number of platform-specific dynamic link libraries, executables and device drivers—for example, to control the specific keyboard, display, real-time clock (RTC), Bluetooth, IrDA and persistent storage devices. Symbian OS communicates with the device's core cellular software through a well-defined interface (ETel), based on a Client/Server architecture.

Porting Series 60 to a new target hardware platform will involve production of some low-level hardware-specific code such as device drivers. This can be a large specialist task and is beyond the intended scope of this book.

A number of the key Series 60 applications or system facilities provide shared access to their data or functionality by making their engines available through public APIs. Examples include Event Logging, Photo Album/Image Gallery, Browser engine and CommDB (a database containing communications and connection settings). These engines use system services from lower layers. For example, the WAP Loader uses the WAP Stack to fetch data using the Wireless Application Protocol. The WAP Stack, in turn, uses the Socket engine in the Communication layer for network access, which in turn uses the ETel engine for hardware-specific telephony data access. In general, engines may be layered upon each other.

The Avkon library defines many user interface components and application framework components. Some key examples are:

- **Status pane**—framework and contents
- **Main pane**—listbox, form, options menu, grid, query, note, soft notification and settings page
- **Control pane**—soft keys and scrolling indicator

Avkon builds on and extends the framework and controls provided in the generic Symbian Uikon library and a version of the Standard Eikon library (from Symbian) modified for Series 60 by Nokia.

Each GUI application is then based on framework classes provided as part of the Avkon library and on the UI layers below that—for example, in Uikon. The majority of Series 60 user interface elements are based on standardized controls provided by the UI libraries or as specific custom controls provided by the developer. The Series 60 GUI library determines the rendering of all GUI elements, so the individual applications share a common look and feel.

As much as 80 percent of the code for an application may be contained in the application engine, which, if it is well designed, should contain no UI code. The UI may account for only 20 percent of an application's code base, and it can share components and services from all Symbian generic and Avkon elements.

History of Series 60

Series 60 Platform is a smartphone software package developed by Nokia for use in its own smartphone products, but also to be licensed by other device manufacturers for use in their own smartphone designs. An integrated suite of applications, built upon personal information management, multimedia, rich communication, messaging, downloading and browsing, enhances the complete software platform.

Series 60 software can be used as supplied, or, more commonly, the "look and feel" is customized to a manufacturer's required specifications. A graphical interface is used throughout the system and by the wide range of wireless applications provided.

The success of Short Messaging Service (SMS) is a good example of what happens when an easy-to-use application is widely available on terminals from different manufacturers and on several interconnected networks. Its widespread adoption has created a larger unified SMS application market that feeds its own growth.

Series 60 Platform was created for a similar unified market to ensure that a wide variety of applications, all based on a common smartphone platform, are available across multiple terminals from many different manufacturers. A potentially very large application market will result in increasing the adoption of new mobile services, enabling the creation of a new category of smartphones, boosting the market still further. As result a significant new ecosystem is now developing around Series 60, including Independent Software Vendors, Series 60 Licensees, Competence Centers, Contractors, Mobile Network Operators and others.

Developing for Series 60 Platform

Series 60 was originally designed for one-hand-operated smartphones; based around a large color screen and an intuitive user interface (UI). By using standard technologies and open standards it ensures interoperability between different terminal and infrastructure manufacturers. Besides Nokia Mobile Phones, many other major terminal manufacturers are using Series 60 Platform and Symbian OS under license to produce exciting new devices, including Siemens, Sendo, Samsung, and Panasonic (Matsushita).

Series 60 includes ready-to-run applications that hardware manufacturers can incorporate in their Series 60 devices. These applications illustrate the possibilities of the platform, but they also guide developers in designing software that complies with the user interface style. Some of the applications provide public APIs for accessing their services from other applications. For example, the Phonebook application has a service (view switching) for displaying a list of contacts; the Photo Album application has a service for finding images; and the Messages application has a service for sending emails.

Series 60 fully supports installing and running applications designed natively for Symbian OS. The installation file format used is the Symbian-specified ".sis" format. Applications can be downloaded via the browser, email, file transfer by IrDA or Bluetooth, or added via a PC connection through an IrDA or a Bluetooth connection.

The mobile phone device market is opening up for software application developers and content creators. Creating an application, downloading it to a phone, and connecting it to the world provides a new business opportunity for the software industry. Investing in a standards-based platform, which enables deployment of the same software on different types of phones and communicators, is a priority for software companies. Series 60 has been designed to ensure a safe investment in creating applications for this new mobile software market.

Developers have the freedom to create applications for a single platform available on phones from multiple manufacturers in a larger unified application market. More applications lead to greater platform adoption, which in turn leads to more devices, thus enlarging the market for everyone.

Forecasts, from several trusted industry sources, predict that during 2007, around 200 million smartphones will be shipped to customers — and by then the total number of Symbian OS based devices in use will be about 500 million. It is expected that Symbian OS will have by far the largest market share in smartphone Operating Systems.

Developing applications for a platform that originates from mobile technology and Internet standards facilitates interoperability. Freedom to create products and services for mobile phones by using, C++, Java[TM], multimedia messaging, and other popular technologies opens up new markets, such as entertainment and multimedia, and extends existing market opportunities into areas such as handset to corporate back-end system connectivity.

C++ Development

Each version of Series 60 Platform provides its own specific C++ SDK based on the relevant version of a Symbian OS SDK. The APIs that are provided enable third parties to develop Series 60 applications for inclusion in new Series 60 terminals or to be distributed as value-added, after-market applications.

Symbian OS is written largely in C++; the language therefore represents a strong development choice for third parties. The Series 60 SDK provides documentation, tools, and sample code to assist developers, along with a Microsoft Windows-hosted emulator (see Figure I–6). The SDK is essential for developing, testing, and debugging C++ applications.

Although C++ development is typically more complex than Java[TM] MIDP development, the advantages more than outweigh the additional effort required. Almost all of the device's capabilities are directly accessible to C++ developers. Creating native Series 60 applications gives the developer access to the Symbian OS APIs, to the Series 60 UI libraries, and to a number of exported application views and a variety of application engines (for example, Contacts, Calendar, PhotoAlbum). Native applications provide better performance. They can take full advantage of a multitude of functions provided by Symbian OS, including access to functionality such as Bluetooth, Infrared, networking functions, messaging interfaces, graphics libraries, multimedia services, telephony and more.

The Series 60 C++ Software Development Kit (SDK)

The final composition of the SDK will depend on the version of Series 60 Platform and your chosen development environment. Typically an SDK package will include:

- Series 60 Application Programming Interfaces
- Software libraries providing a vast range of programming APIs for developers
- Series 60 emulators for testing and debugging
- Various Symbian specific developer tools—Build tools, Bitmaps, Installation files and so on
- Documentation in compiled HTML (.chm) format
- Application wizards to generate basic graphical user interface (GUI) applications
- Example applications (source code and documentation) for Symbian OS basics and Series 60-specific examples
- The PC-hosted GCC C++ cross-compiler for target devices

A number of C++ integrated development environments (IDEs) are suitable for Series 60 development, including offerings from Borland, Metrowerks and Microsoft. Where possible this book is IDE neutral. In some sections, however, such as the build process and testing sections, all of the key options will be covered.

Sample applications are provided in the SDK to help you produce professional Series 60 applications as quickly and as efficiently as possible. They illustrate how key technologies and features of Series 60 Platform can be employed. Sample projects typically illustrate most of the existing functionality and also reflect any new features added to the respective SDK version.

The SDK documentation provides different types of "how-to" documents that help developers not familiar with Symbian and Series 60 to gain competence and start developing Series 60 applications as quickly as possible. It also includes API reference documents that describe the available APIs and their use for developers. Documentation is divided into two main categories, generic Symbian and Series 60 documentation.

Selected add-on tools are typically relatively small, standalone tools, designed to simplify the development process by providing assistance in specific areas.

For example, some Series 60 SDKs include Series 60 Application Wizards for creating basic application projects (availability depends on the chosen IDE and platform version). The SDKs also include a number of other tools created by Symbian (for example, **AIF Builder** and **Sisar**).

Series 60 Principal Characteristics

Key features of Series 60 Platform are the large color screen (current minimum specification is 176 by 208 pixels, and at least 4096 colors, 64K colors in Series 60 2.x) and various interaction modes (two soft keys, five-way navigator and a number of other dedicated keys).

Though the display resolution is currently 176 by 208, the future will bring new resolutions. It is therefore highly recommended that developers write applications in a resolution-neutral way now, to increase compatibility and ease future porting. Essentially this means getting the screen size and calculating positioning at runtime rather than using fixed coordinates. These hardware features are also current requirements for any manufacturer considering using this platform, thereby providing common features that a software developer can depend upon.

A primary UI design objective was operation using only one hand. This has important implications, as it provides convenience for users on the move. A few exceptions exist for functions that are targeted to power users and require pressing two keys simultaneously, for example selecting text to copy and paste.

Since single applications fill the available screen, an application switcher is available via a long press of the menu button—this greatly enhances productive use of the device. Any user with experience of mobile phones will grasp the workings of this intuitive UI very quickly.

Series 60 UI is intended for use in higher-end mobile phones that feature Personal Information Management (PIM) and multimedia applications such as:

- Calendars
- Contacts
- Text and multimedia messaging
- Email
- WAP or other browsers
- Imaging

Many applications are supplied and some have ready-made views available to application developers—for example, the Contacts (Phonebook) application can be called on to display a list of contacts for selection, as illustrated in Figure I–3.

Introduction to Symbian OS and Series 60 Platform

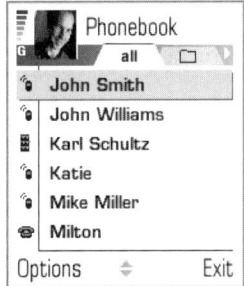 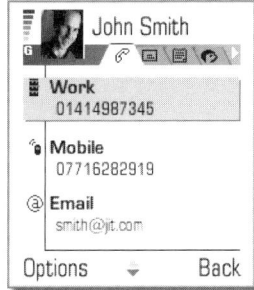 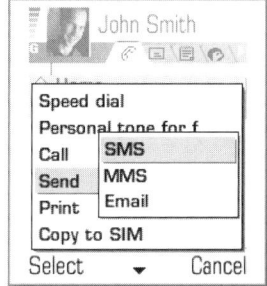

Figure I–3 Use of application views.

Series 60 UI may not be an optimal user interface for very basic phones today, but end-user expectations, even for a basic phone, are increasing. All basic phone functions can be performed with it, but the capabilities of Series 60 UI would probably not be fully used today.

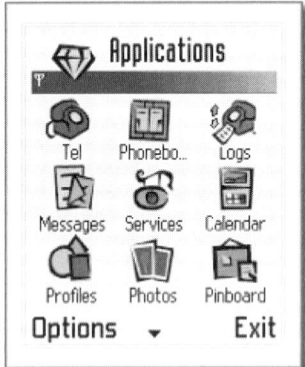

Figure I–4 Applications in a grid view.

The success of smartphone-category devices is highly dependent on the availability of innovative applications and content from third parties—in other words, the growth of mobile services and applications—opportunities which themselves are highly dependent on the availability of an open mobile platform, like Series 60. End users are becoming accustomed to downloading content and many new applications onto their devices. (See Figure I–4 for a scrollable grid view of applications.) The perceived value of terminals will therefore be influenced by the availability of high-quality services, applications and content.

Diversification between handset designs and capabilities has greatly increased. The result is minimal similarity amongst competitive devices in terms of screen size, keypad, browser and other elements of the user interface. Applications, services and other content have to be adapted to these different devices. Series 60 is designed to limit device differentiation to vital elements such as user needs.

Nokia has made Series 60 available for licensing by other handset manufacturers, enabling them to bring phones to market with equivalent and compatible functionality. Common screen size, input methods, APIs, UI libraries and supported technologies allow services and applications to interoperate seamlessly, but still give licensees the freedom to design innovative smartphones.

Standardization and licensing of Series 60 to other handset manufacturers provides an excellent opportunity for third-party developers. They can reach a very much larger market, consisting of users with a variety of handsets—

devices that may be differentiated as products in many exciting respects, but are uniform from a software-development-platform point of view. Software developed for one Series 60 based device will function with little or no change on devices from any other handset manufacturer, providing the version of the platform is the same or is defined to be backward compatible with an earlier version. Developers do need to take care to only use public documented Series 60 Platform APIs. Using licensee-specific API additions, unless done with care, may limit the potential addressable market.

Benefits for Series 60 developers include:

- A larger market—increasing potential revenues
- A market accessible through one common platform—lowering development costs
- Wider availability of applications will fuel demand for terminals, creating positive feedback
- Rich set of APIs and functionality will enable rapid application development

Hardware Requirements

Series 60 UI has specific requirements concerning hardware. The following lists the assumed hardware specification for the first product implementations; it is possible to extend and modify the hardware to some extent for subsequent generations of Series 60 Platform.

Specifications for the current Series 60 UI display (see Figure I–5) are as follows:

Figure I–5 Series 60 UI display.

- Resolution: 176 pixels (width) by 208 pixels (height)
- Square pixels
- Physical size: about 35 mm (width) by 41 mm (height), corresponding to approximately 0.2 mm pixel pitch. Using a significantly smaller pitch risks making some fonts too small to be readable. Using a larger pitch is possible after considering usability issues.
- Color capability (minimum of 4096)

(Note that these specifications are likely to change in future releases of the platform.)

Introduction to Symbian OS and Series 60 Platform

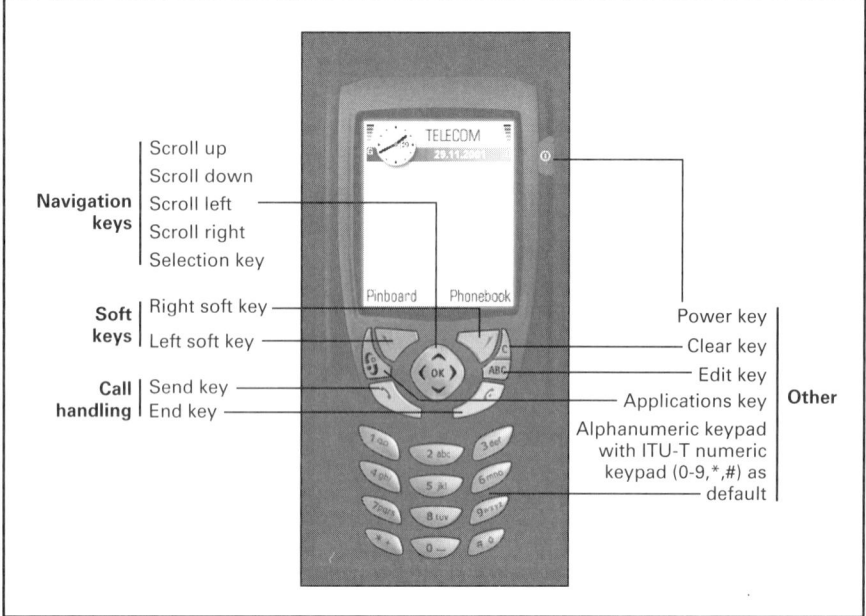

Figure I-6 Series 60 emulator.

Keys

A minimum number of keys events are required for Series 60 UI and they are all illustrated in Figure I-6.

- The navigation keys can be ordinary buttons, or they can be implemented using various control devices—a roller, joystick, jogdial and so on.

- Each soft key has a corresponding textual label on the bottom of the screen.

- The **Edit** key is the only key that can be used simultaneously with another key press—for example, holding down the **Edit** key and using the navigation keys to select text.

Licensee Customization

A smartphone is primarily a mobile phone, a fact reflected in optimization of the user interface for one-handed operation. But a smartphone is also a handheld device for diverse applications and data communication. Fitting these requirements together into a graphical user interface that anyone who can use a mobile phone can master—this distinguishes a smartphone from other handheld devices. This is a crucial distinction when compared to PDAs (Personal Digital Assistants) that require a pen or a keyboard and the use of both hands to operate.

The user interface design of an application for Series 60 typically starts with dividing the structure into browsing elements and detailed views. For example, the browsing view allows the user to select one entry from a list of elements. When the user activates an entry, the respective detailed view displays the data. A wide selection of user interface elements is available for developers, ranging from list boxes, standard dialogs, pop-up menus, check boxes and radio buttons to rich text and color graphics.

While key elements of the UI are mandated to ensure service and application compatibility, there are still numerous possibilities for customization, so that manufacturers can continue to use the different elements that have contributed to their own branding and success. It is possible for terminal vendors to make significant customizations to the Series 60 user interface in order to ensure their own distinct look and feel. However, developers can rely on programming interfaces remaining the same, even when a manufacturer has customized the UI.

Though such customization has always been possible, Series 60 2.0 introduces direct support for "themes," allowing the device to be customized much more easily. Themes (or "skins") may be built into the ROM image by the terminal manufacturer or service provider, or can be downloaded or provided via .sis file. End users may now also change embedded or preset Themes via the Themes application to their personal preference.

Terminal manufacturers can do the following customizations without breaking Series 60 compatibility criteria:

- Replace all provided bitmaps with manufacturer-specific graphics of the same size.
- Embed manufacturer-specific sounds, animations, color schemes and fonts.
- Perform manufacturer-specific internationalization and localization, including terminology.
- Modify, in a straightforward manner, key applications such as the menu system or screen savers.
- Add other specific keys to a product to emphasize or facilitate some functions—for example, to control applications or hardware such as spoken commands, sound recording and audio volume control.
- Customized look and feel—branding for licensees, network operators and potentially for end users.
- Add new applications.

Style Guides

Nokia has produced various style guides for Series 60 UI development, games developers and even for specific products, such as the Nokia OK guide for the 7650. These are typically included in the SDKs—see the References section for details. Terminal manufacturers are very likely to make customizations to their own implementations of Series 60 and therefore may also produce specific style guides for developers working with their products. It is very important to follow closely the general Series 60 style guide and any manufacturer-specific guide; applications will then feel familiar to the user, and this is likely to enhance the sales of your products.

Series 60 Platform Versions

Series 60 1.x is based on Symbian OS 6.1 (with some Nokia modifications), whereas Series 60 2.x is based on Symbian OS 7.0s. Some features in Series 60 2.x come from enhancements to Symbian OS and others come from Series 60 Platform enhancements—the origins of the changes are summarized below.

Symbian OS version 6.1 and 7.0s differences

- On-board camera API
- Multi Media Framework replaces Multimedia Server
- Multiple PDP contexts (3+1 primary)
- HTTP transport framework: HTTP 1.0 and 1.1 client stack
- Language support: Arabic and Hebrew text rendering
- ECom plug-in architecture
- Dual-mode IPv4/v6 Stack
- Multimode ETEL—(GSM, CDMA, and so on)
- W-CDMA 3GPP R99/R4-support
- Java^{TM}: MIDP 2.0 (supports JSR 118 Security)

Series 60 1.x and 2.x Differences

The changes in Series 60 from 1.x to 2.x are summarized here. Many of the platform and Operating System changes will have no effect on developers, but others will—compatibility issues are discussed next.

- Combined Application Manager for both native Symbian applications (.sis) and MIDlet installations
- CLDC 1.0 HotSpot JVM performance is increased
- Open Mobile Alliance (OMA) Client Provisioning: Configuration of WAP Client over the air, with a minimum of user interaction
- SIM Application Toolkit enhancements

Series 60 Platform Versions   xxxix

- Media Player and Media Gallery
- MMS: HTTP support, SMIL enhancements
- PC Suite: Contacts Manager, PIM (Outlook Express, Outlook 2002, Notes 6.0); Windows Bluetooth stack support
- SyncML 1.1 Device Management
- UI for speaker dependent voice commands
- 3GPP streaming
- Embedded download links in Applications
- Wallet 2.0
- Enhanced Camera application with support for two times digital zoom and self-timer
- Theme Support—UI customization for vendors, operators and users (wallpapers, bitmaps, color schemes, icons and so on): new themes can be delivered via a `.sis` (Symbian installation system) file and are downloadable over the air (OTA).

Compatibility Issues

Series 60 Platform 2.x incorporates significant changes due to enhancements in the underlying Symbian OS; others are due to enhancements in Series 60 Platform itself. New features, such as multimode telephony and a Multi Media Framework available in Symbian OS 7.0s, have required the introduction of new APIs and modification of some existing ones. Where possible, compatibility has been maintained between Series 60 Platform 2.x and earlier implementations, but, unfortunately, this has not been possible in all cases. A great deal of effort has been put into maintaining compatibility across the versions. Symbian OS 7.0s implements a number of enhancements to existing (6.1) features, as well as completely new functionality that has no equivalent in the earlier version (for example, IPv6 addressing). In the same way, Series 60 2.x implements new and improved features such as UI themes that have no equivalent in Series 60 1.x. It is therefore possible that a small number of applications built on Series 60 1.x will either fail to build or may exhibit unexpected behavior on Series 60 2.x.

From the developer's point of view, therefore, there are two main issues: application source code compatibility and application binary compatibility.

- **Application Source Code Compatibility** represents that situation where code will execute across versions if it is rebuilt.
- **Application Binary Compatibility (BC)** represents that situation where code will execute across versions without intervention.

Note that in either case, backward compatibility may not be achievable, especially where the implemented features rely on functionality of the newer OS.

In many cases, [forward] compatibility has been maintained by implementing wrapper functions around newer functionality (deprecating the older functions), so that older source code will build/execute without modification. However, where this type of feature could not be implemented, a compatibility break will occur, this is generally manifested by a difference between the library a component was linked against, and the DLL against which the component is being run. In this situation, all client code using this DLL service must be amended; thus, source code compatibility is no guarantee of eventual binary compatibility in these areas.

When designing your application it is advisable to keep the following points in mind:

- If possible, avoid using lower-level API calls where higher-level calls are available.
- Underlying functionality changes provided in modified functions may break binary compatibility.
- If backward compatibility is not an issue, use the new APIs rather than the wrapper functions.
- The use of deprecated functions is not recommended, as these may be removed in future releases.
- Avoid using hard-coded paths. Applications should access media files via interfaces that resolve paths dynamically.

The following are known areas where binary compatibility breaks can or do occur:

- **Comms Framework**—Support for Multihoming has rendered some APIs (Nifman/RGenericAgent) obsolete. CommDB, the database that stores communications-related settings, has been amended, and, although BC has been maintained, data formats have changed.
- **Secure Sockets Layer**—The API for Secure Sockets has been updated and binary compatibility broken.
- **Telephony API**—Multimode telephony has been introduced with a resultant break in BC.
- **SMS Applications**—Changes to the telephony architecture have meant a break in BC by the SMS module (this does not affect the Messaging APIs).
- **Multi Media Framework**—The Multimedia Server has been deprecated. Efforts have been made to retain BC through wrapper functions, but plug-ins to the Multi-Media server will no longer be compatible.

Compatibility Summary

Series 60 Platform 2.x has evolved to improve platform functionality and employ the latest mobile technologies. As with any evolution, a consequence of the progression is that some APIs published in earlier releases are now obsolete.

To develop applications for cross-platform release while maximizing the potential of each requires careful consideration. Given that this (in)compatibility situation exists, the developer has two real choices:

- Generating different source code modules and different builds for each platform release, tailoring the application engine implementation separately for each build and using platform-specific APIs, avoiding the use of deprecated libraries and wrapper functions.

- Using the same source code modules and different builds for each platform. By separating common source code from platform-specific using a preprocessor macro, applications can be tailored to exploit the functionality available in each platform release.

More detailed information on the differences between version 6.1 and 7.0s of Symbian OS can be obtained from the Symbian Web site—see **http://www.symbian.com/technology/whitepapers.html**.

For detailed coverage of compatibility between Series 60 Platform 2.0 and previous versions, search the Forum Nokia Web site (**http://www.forum.nokia.com**) for a document entitled "Series 60 Platform 2.0: Compatibility Issues."

Additionally, see the Series 60 2.x SDK documentation and search for "Deprecated List"—a list of classes and methods that either no longer exist (very rare) or are not recommended for future use, typically because they have been replaced with better alternatives.

Acknowledgements

Here we would like to offer our considerable thanks and gratitude to the many people who provided the benefit of their skills, experience and just plain encouragement during the production of this book.

Contributions and feedback from many organizations and individuals, external to EMCC Software, greatly assisted in the production of this work—helping to ensure technical accuracy and to make the content as clear and relevant as possible.

Apologies are also offered to anyone who we may have overlooked—despite taking great care in keeping records of all who provided input.

Management Team

It is likely that without the persuasive powers of Colleen Romero from Nokia that this book would never have been written—thanks are due for her vision, encouragement and invaluable assistance in achieving this finished work.

Mary Franz of Addison-Wesley, Stacey Johnson of Zen Consulting and Patrick Ames of BookVirtual all played very significant roles in the many stages involved in producing this book. It is a complex process involving many legal, logistical and technically challenging issues. Grateful thanks are due to them all for their excellent contributions in producing this work to such a high standard and under such severe time pressure.

Special thanks must go to Jim Markham of Addison-Wesley, for his attention to detail and for the addition of clarity to every chapter.

Production

The Addison-Wesley production team consisted of Gail Cocker-Bogusz, Patti Guerrieri, Bob Lentz, Meg Van Arsdale—thanks are due for their skills and hard work, delivered under pressure, during a holiday period.

Reviewers and Technical Advisors

Our review and advisory team consisted of Nokia and Siemens subject matter experts and many independent reviewers with relevant technical backgrounds, engaged by Addison-Wesley. This team grounded our manuscript with a real-world balance. We are indebted to them all for the time, care and expertise they put into this project.

Ari Aumala, Petri Backstrom, Richard Bloor, Gerard Bruen, David Carson, Anurag Chandra, Tuomas Eerola, Juergen Fey, Riku Granat, Juha Hietasarka, Pekka Jokela, Juuso Kanner, Jari Karlsberg, Heikki Koivu, Michael Kroll, Ari Kuusisto, Eero Kutola, Vladimir Minenko, Ralph Moore, Seppo Pakarinen, Matti Parnanen, Markus Penttila, Robert van der Pool, Harri Poyhonen, Tero Pulkkinen, Krister Rask, Larry Rudolph, Jussi Ruutu, Joseph Satter, Patrick Stickler, Marko Suoknuuti, Jari Suutarinen, John Szeder, Eero Tunkelo, Mikko Tuokko, Michael Urlich, Charlie White.

Authors and Contributors

Developing Series 60 Applications was produced entirely by the staff of EMCC Software Limited; a United Kingdom based company who has been working with Symbian OS in its various forms for nearly six years. As result of this experience in depth, EMCC Software has become a Symbian Competence Center, a Symbian Training Partner and Platinum Partner. At the launch of the Series 60 product creation community Nokia Mobile Software appointed EMCC Software as a Series 60 Competence Center.

However, despite this experience and expertise, the clarity and quality of this work would have been considerably diminished without the close technical collaboration with expert staff from Nokia, the guidance of Addison-Wesley and significant input from many other individuals external to EMCC Software. Grateful thanks are given to those people in the acknowledgements section above.

Authors

Richard Barker has been developing in C++ for 10 years, with a background in R & D, Artificial Intelligence and real-time training simulators for Defense Systems. He has been with EMCC Software as a Symbian OS and Series 60 developer for three years, working on interface and engine design, client software to deliver dynamic data services and development of Symbian OS itself. He is a regular contributor of Series 60 documentation for Forum Nokia.

Leigh Edwards has been working with, and conducting training on, Symbian OS for over six years, since it was called EPOC32 (Symbian OS version 1.0) and was owned exclusively by Psion Software. He is a co-founder of EMCC Software Ltd., specialists in Symbian OS and the key UI platforms. Leigh firmly believes that resilience, technical excellence plus the unique industry collaboration behind Symbian Ltd., are the keys to the on-going success of Symbian OS and Series 60 Platform.

Mark Whitaker has been working with C++ for six years, five of them in Symbian OS. Much of this time has been spent working with messaging technologies, writing messaging applications and MTMs for a range of Symbian OS phones. Mark wrote the majority of the communications and messaging text for this book, including many of the example projects.

Samantha Richards has been with EMCC Software for over three years. She spent a lot of time developing two-player games and a communications stack that allows the games to be played over infrared, Bluetooth or SMS. She has ported many applications to a number of different Symbian-based user interface platforms, for example, Eikon, the Ericsson R380 UI, Nokia 9210, Series 60 and UIQ-based devises. Working as part of the Symbian Messaging team, Samantha worked on an XML Parser and integrating multihoming technology into the MMS module. She has written several Series 60 documents for Forum Nokia. Samantha has a degree in Mathematics, a Masters in Computing and is a qualified high school teacher.

Sarah Bardill has been a Software Engineer for eight years. She is a qualified Java Professional with over 6 years of Java, OO design and C++ development experience. During her career Sarah was lead designer of a large enterprise Java system. While working at EMCC Software she has concentrated on Symbian OS and Series 60.

Bill Vance has four years of C++ experience, over one year of general Symbian OS experience, plus six months of Series 60 Platform experience. He has a First Class degree in Mechanical Engineering and an MSc in Applied Computer Science. Over the past two years with EMCC Software he has worked on the core Symbian OS applications for a number of Symbian based UI platforms. More recently Bill has focused specifically on development for Series 60 Platform.

Robert Lamb did his University degree dissertation on EPOC (Symbian OS version 3.0) before graduating and joining EMCC Software. He has over two and half years of commercial Symbian OS experience. He has worked on various licensee projects during that time helping to bring Series 60 devices to market, including working at Symbian's Cambridge offices as part of the Series 60 Platform team, and with Siemens in Munich as part of the SX1 team. Additionally, Robert has written a number of white papers for the Series 60 section of Forum Nokia.

Steve Pearce has over three years C++ experience and has considerable experience in Symbian OS, with particular expertise in the Messaging system. He has a degree in Physics and is a qualified high school teacher. He has employed his extensive knowledge of Symbian OS, Series 60 and his considerable skills as a teacher when presenting a wide range of Symbian based training courses on behalf of EMCC Software.

Raphael Espino has three years of C++ development experience, over one year Symbian OS experience and over six months working with Series 60 Platform. He has worked on many communications projects, developing in both Java and C++. Raphael also has GUI development experience with MS Windows and X-Windows. He has a Degree (BSc) and PhD in Computer Science.

Peter Mullarkey started Symbian OS programming in 2000, following the progression from Symbian OS version 5.0 to 7.0s and Series 60 2.0. He has extensive expertise in the Symbian OS Client/Server architecture and in porting 'C' code to Symbian devices. He has a BSc in Applied Mathematics and Computing and a PhD in Engineering, where his research interests lay with the computational modeling of Auxetic materials.

Contributors

In addition to the main authors, many other staff of EMCC Software made valuable direct contributions to this work, including writing sections of the book text, developing example applications, providing technical advice, commenting on and reviewing text, conducting code reviews, plus testing and validating the example projects. Credit and considerable thanks are due to the following people:

Wandeep Basra, Mick Smith, Stuart Smith, Howard Simms, Simon Kelly, Richard Gunn, Stephen Wilkerson, Graham Douglas, Garry Partington, Jonathan Gibson, Chris Woods, Kevin Boyce, Johann Madlberger, Adam Fleming, Katie Evans, Steve Wilkinson, Iain Rosie.

In addition we would like to acknowledge the many other members of staff who provided the benefit of their knowledge and expertise directly or indirectly to the authors and contributors.

chapter 1

Getting Started

Covering the essentials of a Series 60 project, plus building, deploying and running an example application

Chapter 1   Getting Started

This chapter overviews the process of building, deploying and running an example Series 60 application with a Graphical User Interface (**GUI**). If you have already been involved in Series 60 development, you may want to skip this chapter. Chapter 2 builds on the overview, providing a detailed description of the project files and the development tools that are used on a regular basis.

We assume that you have installed both a Series 60 Software Development Kit (**SDK**) and a chosen Integrated Development Environment (**IDE**). Both installations are reasonably easy and are well documented elsewhere. If you do not have the necessary SDK and IDE, see the Preface for details of how to obtain them.

Topics covered are:

- **Series 60 C++ SDKs**—Different versions of the SDKs that are available and how they relate to versions of Series 60 Platform releases, and hence to specific Series 60 products.

- **Development Process Overview**—A high-level description of the process of specifying, building and running a Series 60 project for an emulator or a target device, plus a guide to the many IDE and build options available to developers.

- **Series 60 Emulators**—The features and layout of the Series 60 emulators, both debug and release versions. An overview of the similarities and differences between an emulator and a real Series 60 device.

- **Building for the Emulator**—How to build the **HelloWorld** project for the emulator using IDEs from different vendors and also from a PC command prompt.

- **Running the Emulator**—Each method of starting an emulator, and how to locate and run the example **HelloWorld** application—as both release and debug build variants.

- **Building for a Target Device**—How to build the **HelloWorld** example project for a Series 60 device as an appropriate ARM binary executable.

- **Deploying on a Target Device**—How to package up the various components of an application, options for transferring it to a Series 60 device and then how to locate and execute it.

A step-by-step overview of the development process shows you the essential steps. This is the fast-track guide to using Series 60 C++ build tools and various IDEs. You will see how to build and run an example "Hello World" application with the Series 60 emulator (version 1 and 2), plus how to build, deploy and run the application on a target device. All the information needed to build and run the **HelloWorld** example is provided within this chapter.

All the project files, source files and deployment information associated with the examples in this book are available online as noted in the Preface. If you have not already obtained the source materials, you are advised to download and install them — they will be helpful for reference as you read through this chapter.

Chapter 2 gives detailed explanations of the components that make up a typical project and the key build tools. Chapter 3 covers the fundamentals of Symbian OS and the key concepts you must understand fully to develop efficient, reliable code for Series 60 devices. In Chapter 4 the class structure of a GUI application is described and a detailed examination of typical application code is begun.

Details on debugging and testing applications are provided in Chapter 13.

Series 60 C++ Software Development Kits (SDKs)

Symbian OS is widely used in multiple smartphone platforms, such as Series 60, Series 80 and Series 90, three of the UI platforms from Nokia, and UIQ (the UI platform from UIQ Technology).

Series 60 SDKs are built upon specific versions of Symbian OS C++ SDKs released by Symbian. An SDK contains a wide range of tools, APIs, libraries and documentation to enable you to develop new applications, typically as after-market applications.

As a developer you may need to work with SDKs for more than one version of Series 60 (or even SDKs for different Symbian OS UI platforms). The most important issue is to select the correct SDK version for your chosen Series 60 product.

Nokia or Series 60 Licensees release SDKs that are suitable for development for a specific version of Series 60 Platform. That is to say, each Series 60 product is based on a particular release of the platform, and SDK releases are made to be suitable for development for a platform version. A particular SDK release may also be suitable for use with earlier versions of the platform as well. Such "backward compatibility" will depend on the version of the Symbian OS used as the basis of the release, the APIs used by the application developer and any changes that have occurred in those APIs between platform releases. Particular Licensees also may introduce additional product-specific APIs to allow developers to access the features that differentiate their Series 60 product from

Chapter 1   Getting Started

those from other Licensees. For example, the Siemens SX1 smartphone includes an FM radio, so the relevant SDK may include "add-on" APIs to manipulate the radio. For maximum compatibility across different Licensee products you may want to avoid using such product-specific APIs. You may even choose to limit your use of the general APIs to those that are common and unchanged across a selected range of platform versions.

You will find the key differences between releases of Series 60 Platform described in broad detail in the Introduction to this book.

More extensive variants of Series 60 Development Platforms are available to Licensees, Competence Centers, and other software and technology partners to allow them to develop at the system level, rather than at the application level. This book, however, will focus on the publicly available SDKs.

Using Multiple SDKs

Using a single Symbian OS SDK is very easy, and the installation process will prepare it for immediate use. However, developers often need to work with SDKs for different versions of Series 60, or even SDKs for different user interface (UI) platforms. As described in Chapter 2, you can install multiple SDKs on your development PC, with a few restrictions on where they can be installed and how they are selected for use.

Development Process Overview

PC-based platform emulators are provided as part of the SDKs so that most development and testing can be performed without target hardware. Series 60 project executables can be created as debug or release variants for emulators and for target hardware (although currently some restrictions apply to on-target debugging). Additionally, when building for target devices it is possible to create executable code in various binary formats (for example, `ARMI`, `ARM4` and `Thumb`, explained later in this chapter). The compilation and linking process can be performed using command-line tools or from within a variety of IDEs. The IDEs covered here are Microsoft Visual C++ version 6, Metrowerks CodeWarrior, and Borland C++Builder 6 Mobile Edition and Borland C++BuilderX Mobile.

Symbian devised a method of specifying development projects in a platform-neutral way. Two universal project files can be created (`projectname.mmp` and `bld.inf`), where `projectname` is the name of the component or application to be developed (**HelloWorld** for our example project). These two text files can then be used as a starting point for any of the build options, IDEs and platform variants.

The `bld.inf` file specifies the names of all the project component(s) to be built, with each component specified in its own `.mmp` file. Both types of file are plain text, and often you will simply have a single .mmp file that defines the application you are creating. If the project consists of multiple components, such as the application itself and specific function libraries, then each component would have its own .mmp file. Each of the libraries plus the application would have an .mmp file, and each filename would be listed in the bld.inf file for the project. The syntax of .mmp and bld.inf files is detailed in Chapter 2.

In the **HelloWorld** example, there are only two project specification files: bld.inf and helloworld.mmp. Using these two files, any platform-specific project and command files required can be created.

Typically you employ a Symbian tool called bldmake, using the two project specification files as input, to generate a command file called abld.bat. You can then use abld.bat, from the command prompt, to perform a number of project-related actions. For example, abld.bat can be used to generate platform- and IDE-specific project makefiles. If the project source code exists, and is complete, abld can be used to build the project for one or more platforms.

Since most development projects are built and run from within an IDE, you would usually create the project files suitable for your chosen IDE.

In the case of Microsoft Visual C++, you use abld at the command line to create the HelloWorld.dsp and HelloWorld.dsw project files. The .dsw file is the workspace file to be opened from within the IDE, and it may reference one or more .dsp files.

For Metrowerks CodeWarrior you can either create the project file from within the IDE in the usual way or import the .mmp file directly into the IDE. The import process will create the CodeWarrior specific project (HelloWorld.mcp) file required.

Similarly, the Borland C++ Builder Mobile Edition IDEs can perform an equivalent import task to the Metrowerks IDE, but by importing the bld.inf file instead of the .mmp file. Borland C++BuilderX, for example, will create a project file called HelloWorld.cbx.

Opening the IDE-specific project file will then allow you to develop, build, run and debug the application with full IDE support.

For developers who prefer working at the command-line level, abld can also be used to compile and link from a command prompt.

All of the methods for creating the IDE (or command-line) specific project files outlined are described in detail later in this chapter. In addition, Figure 1–1 illustrates the use of the two generic Symbian OS project files to generate the required platform-specific project files, either via IDE import options or using the Symbian tools.

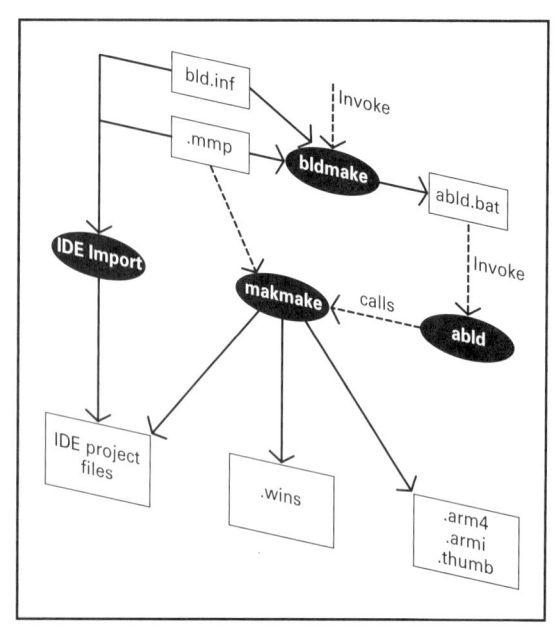

Figure 1-1 Generation of IDE and platform-specific project files from generic Symbian files.

Using an IDE versus Command-Line Tools

The different options currently available to you for working with Series 60 development projects are summarized in Table 1-1. More options are becoming available all the time from Symbian, Nokia and the development tool vendors Borland and Metrowerks. However, it is likely that the Microsoft development tools, though currently viable and still widely used, will no longer be supported in the not-too-distant future.

There are currently considerable differences between the capabilities of the various IDEs from Microsoft, Borland and Metrowerks. Development to enhance the latter two development environments is ongoing. EMCC Software Ltd uses IDEs from all three vendors as the basis of its development activities. We also use the command-line tools every day—not because we want to but because we currently have to. This is primarily for building for target devices and for automating overnight builds. The C++BuilderX and CodeWarrior IDEs can now build for target devices—but not all essential build operations are covered by some variants of the IDEs.

At the time of writing, the easiest and most generic starting point for any new Series 60 project is to define a pair of bld.inf and .mmp project files. At EMCC Software, these files are used for generation of any IDE-specific project files and for use during any command-line builds needed. This approach is taken throughout the book. It is also the approach currently adopted by every single example project provided by both Symbian and Nokia in their SDKs and in the

Chapter 1   Development Process Overview   9

documentation they supply. Using an IDE exclusively is not currently possible on a day-to-day basis. Things will change before very much longer, but these are the current facts of life for Symbian OS and Series 60 developers.

Table 1-1 PC-Based Development Options Summarized

Option	Description
Command line building	Using the Symbian OS tools combined with compilation and linking using the Microsoft Visual C++ compiler/linker for emulator builds invoked from a command line. Emulator can also be invoked from the command line. Source-level debugging on the PC emulator requires working from within an IDE.
	The GNU C++ cross compiler and linker used for ARM target device builds invoked from the command line.
Microsoft Visual C++ IDE	Compilation and linking using the Microsoft Visual C++ compiler/linker for emulator builds from within the IDE.
	The GNU C++ cross compiler and linker used for ARM target device builds invoked from a command prompt.
Borland C++ IDE	Two IDE options are currently available; both use the Borland C++ compiler and linker for emulator builds from within an IDE.
	The GNU C++ cross compiler and linker used for ARM target device builds invoked from the command line.
Metrowerks CodeWarrior C++ IDE	Using the Metrowerks compiler and linker for emulator builds from within the IDE.
	The GNU cross compiler and linker for ARM target device builds, either invoked from the command line or directly from within the CodeWarrior IDE.

Through the rest of this chapter, the **HelloWorld** example project is used to illustrate all the steps involved in the development, debugging, and deployment process for the Series 60 emulator.

All of the other files required for the example application are provided—for example, the header (.h), source (.cpp) and user interface resource (.rss) files.

You can test the application using the Series 60 emulator, either started from within an IDE or run from a command prompt. However, debugging an application on the emulator must be performed from within an IDE.

After developing, running, testing or debugging an application on the emulator you typically will want to build and run it on a target device. So you will then be shown how to build the example application for a target device, how to deploy it and then run it on target Series 60 hardware.

Series 60 Emulators

Development, debugging and initial testing of Series 60 applications is usually carried out on a PC-hosted emulator that provides a Microsoft Windows-based implementation of a Series 60 device. In most cases you will discover that the emulator-based development process closely mimics the operation of an application running on a real device; so that the majority of your development work can take place even before hardware is available. The exact appearance of the emulator may vary from the figures provided and will depend on the target platform you are working with, the version of Series 60 you are using, and the selected IDE. For example, the fascia bitmap may have been changed to closely resemble a particular Series 60 device from a Licensee. In addition, buttons and other interaction elements may be moved or added to emulate the configuration of a manufacturer's actual device. Also, the applications available on the emulator will depend on the platform version and the device manufacturer's preferences.

Some differences between an emulator and a real device cannot be overcome. Real Series 60 devices will have hardware accessories such as cameras and other features such as vibration feedback. Thus, at some point, hardware will be necessary for development and testing. In addition, PC-based emulators do not accurately mimic issues of precise timing, application performance and memory management.

For PC-based development, the edit/compile/build cycle is based on a Microsoft Windows-hosted development toolset. However, instead of linking and building against Win32 or MFC libraries, developers link and build against the headers and PC-format libraries installed by the Series 60 SDK. The resulting Windows-format binary executable is then run under the PC-hosted emulator.

During development, the project file for the specific IDE manages all linking and building details. It also ensures that all outputs from the build, and other required resources, such as application resource files, are built into the appropriate location for running or debugging under the emulator.

Referring to Figure 1–2, the Series 60 display is logically divided into three areas: status pane, main pane and control pane. (See the "Series 60 UI Style Guide" provided with the Series 60 SDK documentation for a comprehensive description of user interface elements and standards.)

The status pane is the graduated (blue, on the emulator) bar near the top of the screen plus the area above it. The main pane is the middle section of the screen between the status pane and the soft key labels at the bottom of the screen. The control pane is the area immediately below the main pane and includes the soft key labels.

The status pane may display information for the current application, as well as general information about the device status, such as the signal strength and battery charge. It is visible in most situations, but sometimes it may be hidden. Many games, for example, will use the whole screen.

Figure 1-2 Series 60 emulator.

The main pane is the principal area of the screen where an application can display its data. Typically, this area, also referred to as the client rectangle, is fully occupied by an application's data display.

The control pane occupies the bottom part of the screen and displays the labels associated with the two soft keys and a scroll indicator when required. Like the status pane, the control pane can also be hidden at times. In such situations it is within Series 60 style guidelines to have the user assume the availability of the **Options** menu (the default label for the left soft key), even though it may not be visible. (See the "Nokia Series 60 Games UI Style Guide" provided with the Series 60 SDK documentation.)

The two buttons below the control pane are the left and right soft keys and are used to select the currently associated **Options** menu or labeled action. The four-way navigation key will scroll up, down, left, right, or will select if pressed (clicked) in the center.

You can interact with the emulator via the PC mouse or cursor keys for navigation around the objects on the display. It is possible to use mouse clicks to select folders and other displayed objects directly rather than via clicking on the four-way navigation key (on the emulator only). For data entry the PC keyboard can be used, or you may click on the twelve-way keypad on the emulator fascia. Therefore the interaction with the emulator is close to, but not exactly the same as, using a real Series 60 device—there is no pen input on a real device, so all movement and selection is through cursor (joystick) navigation.

Building for the Emulator

Since Series 60 applications can be built from a command prompt or from within an IDE, we have detailed both methods here. We'll start by building the project to run under the PC-hosted emulator (that is, for an x86 instruction set) using the C++ compiler supplied with the IDE. We almost always use a debug build, so that symbolic debug information and memory-leak checking are available (checking for memory dynamically allocated on the heap that is not released correctly).

Building from the Command Line

Open a command prompt and change to the drive/folder that contains your Series 60 SDK. Navigate to the folder where the project definition (`helloworld.mmp`) and component description (`bld.inf`) files are located— for example:

`\Symbian\Series602_0\EMCCSoft\HelloWorld\group`

for a Series 60 2.x project

`or`

`\Symbian\6.1\Series60\EMCCSoft\HelloWorld\group`

for a Series 60 1.x project

and type:

`bldmake bldfiles`

After a second or two this command completes without any visual output. It uses the `bld.inf` and `helloworld.mmp` files to generate a new file: `abld.bat`. This command file is always generated in place, as required. Unlike the `bld.inf` and `.mmp` files, `abld.bat` is not portable between different IDEs and should never be edited by hand.

To compile and link the project, type:

```
abld build wins udeb           —for Visual C++
abld build winsb udeb          —for Borland C++
abld build winscw udeb         —for CodeWarrior
```

The `abld` command will build the project (in other words, compile and link) for the Series 60 emulator (the `wins`, `winscw` or `winsb` variant) with debugging (`udeb`—Unicode debug) information included in the binary executable.

Building from an IDE

Projects, such as our example **HelloWorld** application, normally are built and run from within an IDE, so we need to create IDE-specific project files from the `bld.inf` and `HelloWorld.mmp` files. For Visual C++ this must be performed from the command line, using tools supplied by Symbian. For Borland and CodeWarrior this is optional, since both IDEs can import either the `bld.inf` or `.mmp` file, respectively, to create the IDE project files.

When working from the command prompt it may be necessary to create the `abld.bat` file if it does not exist already, or recreate it if the `.mmp` file or `bld.inf` file has changed. At a command prompt you create the `abld` command file by typing:

```
bldmake bldfiles
```

Building Using Microsoft Visual C++ IDE

Open a command prompt and navigate to the drive/project folder for the **HelloWorld** project and then type:

```
abld makefile vc6
```

This will create project and workspace files (`helloworld.dsp` and `.dsw` files) suitable for Microsoft Visual C++. They will be located under the `\Epoc32\Build` sub-folder structure; the complete path will depend on the location of your SDK, for example:

```
\Epoc32\Build\EMCCSoft\HelloWorld\HelloWorld\Wins
```

By opening the workspace file (`helloworld.dsw`) in Visual C++, you can compile or link the application, either by pressing **F7** or via the IDE menu option, **Build|Build HelloWorld.app**.

Building Using Borland C++IDE Builder 6

If you are using Borland C++Builder 6 Mobile Edition, which is based on C++Builder 6 Personal Edition with the Mobile plug-in, you can simply import the `bld.inf` file for the **HelloWorld** project into the IDE. Use the **File|New|Other** menu option. Then select the **Mobile** tab in the resulting dialog: **Import Mobile Application**. Browse to the location of the component description (`bld.inf`) file and open it.

Use **Ctrl+F9**, or **Project|Make** from the menu, to build. To build and run, use **F9** or **Run|Run** from the menu. Note that **F9** or **Run|Run** will cause a project rebuild each time! To just run the emulator, use **Tools|Mobile Build Tools|Run Emulator**. You may be prompted to save a number of project-related files, for example, Borland project (`.bpr`), Borland project group (`.bpg`) files. These files will be saved in the same folder as your `bld.inf` file.

It is also possible to execute individual **abld** commands and run other SDK tools such as **aifbuilder** and **sisar** from the **Tools|Mobile Build Tools** menu. These tools are described in Chapter 2.

Building Using Borland C++BuilderX

When using any of the commercial C++BuilderX products, you can simply import the `bld.inf` file for the **HelloWorld** project into the IDE.

Use **File|New** and click the **Mobile C++** tab in the Object Gallery. Select **Import Symbian C++ Project**. Select the correct Series 60 SDK from the drop-down list, browse to the location of the project `bld.inf` file. Press the **Next** tab, give the project a name, press the **Finish** tab and the project will open. Press **Ctrl+F9** or select **Project|Make Project**. To run, Press **F9**, or select **Run|Run Project** or use the toolbar tab to **Make and Run** the project.

Building Using CodeWarrior IDE

If you are using Metrowerks CodeWarrior for Symbian (Personal v2.5, other editions may vary slightly), you can simply import the `HelloWorld.mmp` file using the IDE menu option, **File|Import Project from .mmp File**.

This runs a project conversion wizard. Select the SDK to use with this project, select (or browse to) the `.mmp` file, and select a platform of `WINSCW` (or all by leaving it blank). The build variant will default to `UDEB`. Use **F7** or **Project|Make** from the menu to build the project.

The CodeWarrior Project files (`.mcp`, `.xml`, `.resources` and `.pref` files) are created automatically in the same directory as the `HelloWorld.mmp` file.

Alternatively you can create a CodeWarrior IDE project from the command line. To do this, run `bldmake bldfiles` as described; then, to generate a CodeWarrior IDE project, use:

```
abld makefile cw_ide
```

This creates an importable project file `HelloWorld.xml` in the directory:

```
\Epoc32\Build\EMCCSoft\HelloWorld\HelloWorld\Winscw
```

You can now use CodeWarrior to import this file and to generate a native project (`.mcp`) file. Choose the **File|Import Project** menu option, select the `HelloWorld.xml` file, and choose a name for the project (such as **HelloWorld** again). CodeWarrior will now generate and load the project, which you can build, run, debug and so on, using the normal IDE commands.

Running the Emulator

In a Series 60 SDK two versions of the emulator executable are available: a version built containing symbolic debugging information, and another built as a release variant. The release emulator is limited to evaluation and demonstration of applications—it starts up considerably quicker because of the absence of the debugging information.

Both versions are called `epoc.exe`, but they are located in their own subdirectories. The name `epoc` is historical—it was the name of the operating system prior to Symbian OS.

In normal development activities, it is usual to use the debug variant of the emulator. Depending on your choice of IDE, you may be able to run the debug version normally or in "debug mode." To be able to run the same "debug emulator" in two modes may seem a little confusing at first.

Sometimes you may want to start the emulator, locate the application and run it (as described later) simply for testing purposes. If a serious error occurs, the emulator and application will shut down in a controlled way.

Other times you may want to put a breakpoint in your code at a specific point where you think a problem exists, and then have the IDE run the emulator in "debug mode." You then locate the application and run it as before. Suitable interaction with your application will cause the breakpoint in the code to be reached. At that point, the source code will be displayed in the IDE along with all of the symbolic debug information associated with the application. Then you can use the debugging features of the IDE to step through sections of code, in a controlled manner. All the while you are able to view the application source code, data, call stack and other debug-related information displayed by the IDE to assist you in tracking down errors in code or logic.

The appearance of a typical Series 60 emulator is shown in Figure 1–3. Debugging an application under an emulator using the Microsoft Visual C++ IDE is illustrated in Figure 1–4.

Emulator Executable Locations

For a Series 60 1.2 SDK the release build emulator is typically located under the following:

For Visual C++

`\Symbian\6.1\Series60\Epoc32\Release\wins\urel\epoc.exe`

For Borland C++

`\Symbian\6.1\Series60\Epoc32\Release\winsb\urel\epoc.exe`

For CodeWarrior

`\Symbian\6.1\Series60\Epoc32\Release\winscw\urel\epoc.exe`

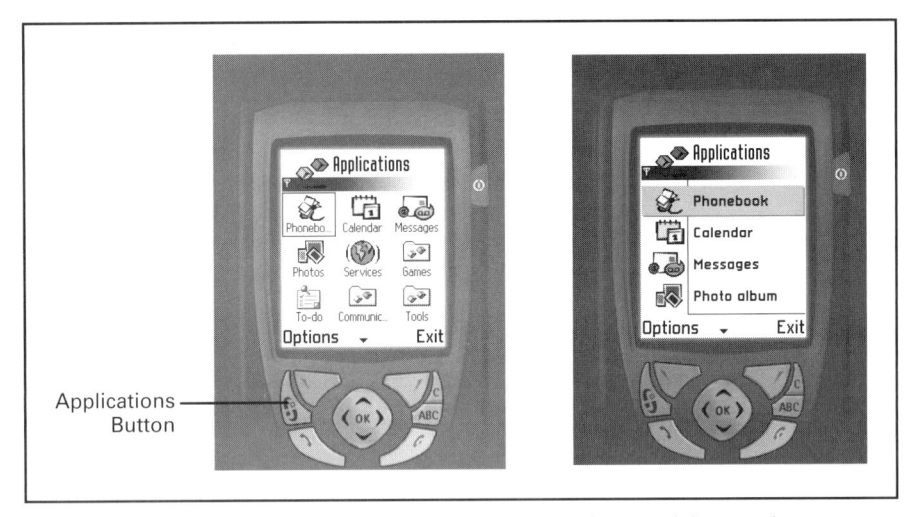

Figure 1–3 Applications grid and list views of Series 60 Platform 1.2 debug emulator.

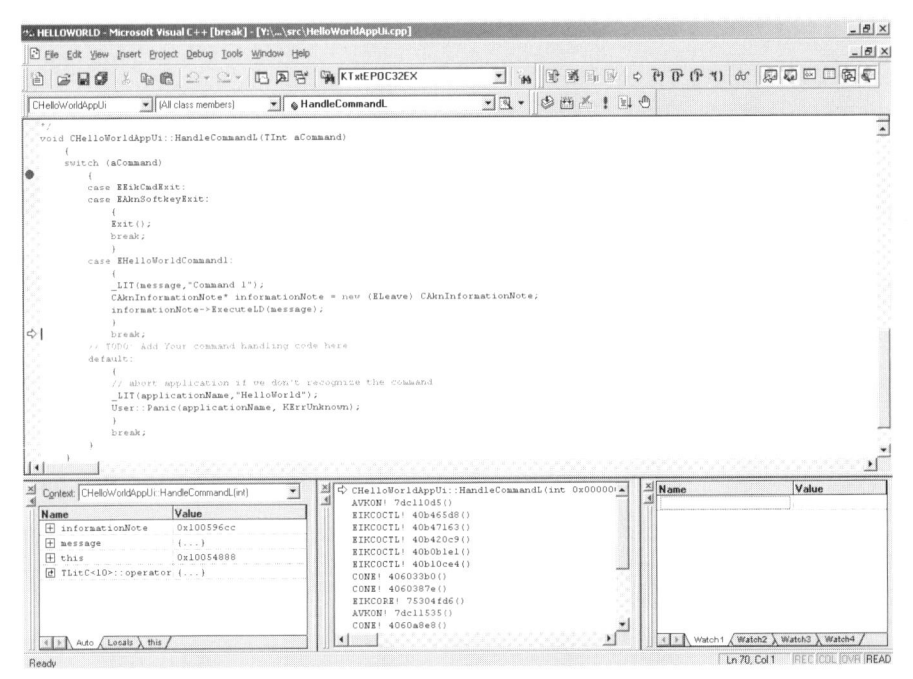

Figure 1–4 The HelloWorld application on the emulator in debug mode under the Microsoft Visual C++ IDE.

The debug build emulator is typically located under:

For Visual C++

`\Symbian\6.1\Series60\Epoc32\Release\wins\udeb\epoc.exe`

For Borland C++

`\Symbian\6.1\Series60\Epoc32\Release\winsb\udeb\epoc.exe`

For CodeWarrior

`\Symbian\6.1\Series60\Epoc32\Release\winscw\udeb\epoc.exe`

The exact paths will depend on the options you choose during the installation of the SDK. In the case of the Series 60 2.x SDK the paths to the emulator will be very similar—for example:
`\Symbian\Series602_0\Epoc32\release\wins\udeb\epoc.exe`.

Figure 1–3 shows the Series 60 1.2 emulator. It starts up showing the Applications main menu as either a grid view or a list view.

The Series 60 2.0 emulator starts with a mock-up of the phone application, and you have to navigate to the applications menu by pressing the applications button shown in Figure 1–3. Whatever version of Series 60 Platform you are using, always specify the debug (`udeb`) version of the emulator executable as the default for development projects—for example:
`\Epoc32\Release\wins\udeb\epoc.exe`.

Emulator Debug Mode

When running the application in debug mode under the emulator, the source code, function call stack, variable information and so on are shown as soon as the breakpoint in the code is reached, as shown in Figure 1–4. The emulator window itself may disappear (it is minimized) if the application code is not at a point where user input is required.

Running the Emulator from a Command Prompt

To run the debug emulator from the command line, open a command prompt, change to the folder in your Series 60 SDK where the `epoc.exe` application is located (for example, `\Symbian\6.1\Series60\Epoc32\Release\wins\udeb`) and type the following:

`epoc`

This will start the debug emulator and you can then locate and run your application, but not in debug mode. To debug an application you need to run the emulator in debugging mode and this can only be done from within an IDE. To run the release emulator enter the following:

`epoc -rel`

Running the Emulator from the Visual C++ IDE

From within the Visual C++ IDE you can start the debug version of the emulator by pressing **Ctrl+F5**, or from the menu use **Build|Execute Epoc.exe**. This will run the emulator in non-debug mode. Alternatively, you can use **F5** or select **Build|Start Debug|Go** from the menu to run the emulator in debug mode.

The first time you run the emulator for a Visual C++ project, a dialog will appear asking you to supply the name of the executable. Navigate to `epoc.exe` in the folder `\Epoc32\Release\wins\udeb` in the root of your SDK.

Running the Emulator from the Borland C++Builder 6 and C++BuilderX IDEs

You can start the debug version of the emulator using the **Tools|Mobile Build Tools** menu option, then select **Run Emulator**. Alternatively you can use the **Run|Run** menu option (**F9**), but it will cause a project rebuild each time — this can be a lengthy and time-consuming process! Using this option, you will need to cancel the build dialog ("Compiling") before the emulator will start up.

If you are running the emulator from C++BuilderX, use the **Run|Debug Project** menu option or press **Shift+F9**. If you wish to rebuild the project and start the emulator, select **Run|Run Project** or press **F9**.

Running the Emulator from the CodeWarrior IDE

Select the **Project|Run** menu option or press **Ctrl+F5** to run the emulator. Press **F5** or use the **Project|Debug** menu option to run the emulator in debug mode.

Locating and Running the Application

Navigate to, and select the **HelloWorld** application by clicking on the image of the cursor keys on the emulator fascia bitmap, or by using the PC keyboard cursor (arrow) keys. Click on the **Selection** button (in the middle of the cursor controls) to start the application.

> **TIP** Applications that do not have their own specific icon (as specified in an `.aif` file, detailed in Chapter 2) will be given a default icon, which looks like a piece of a jigsaw puzzle, by the system.

Under some SDK/IDEs (for example, versions of Borland and CodeWarrior), the application you have built may be located in a folder called "**Other**" rather than on the main desktop. If so, navigate to and select the **Other** folder and then open it by clicking on the Selection button. Navigate to and select the **HelloWorld** application and click on the **Selection** button to invoke the application.

The **HelloWorld** application will run and should appear as shown in Figure 1–5.

Debugging the Application

Alternatively, you could run the application on the emulator from within an IDE in debugging mode—the procedure will vary depending on the IDE in use. Typically you would first set a breakpoint at an appropriate point in the source code.

Start the emulator in debugging mode. Since it is the application (essentially a dynamic link library—**DLL**) that will be debugged, not the emulator itself, navigate to and run the application as described earlier in "Locating and Running the Application." The application will start up, and then execution will stop at the breakpoint you set earlier. You can the use the facilities of your chosen IDE to step through the execution of the application source code.

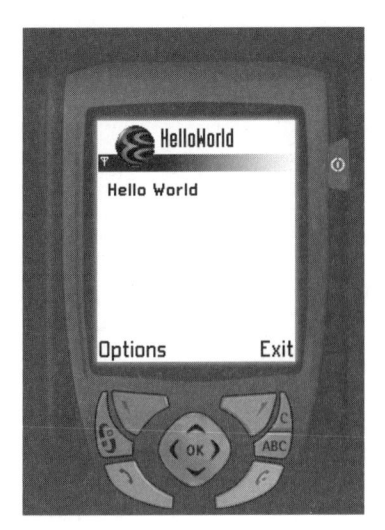

Figure 1–5 The "Hello World" application.

Further IDE Help

Further explanation of the various IDE functions is beyond the scope of this chapter, so for more details refer to the IDE help information, available through the **Help** menu option.

For Microsoft Visual C++ this is accessed through the **Help|Contents** menu option, provided you installed MSDN with your IDE.

For Borland C++Builder 6 this is accessed through the **Help** menu option. You will find a separate **Help|Borland C++ Mobile Edition Help** page as well as the standard Borland help files. For C++BuilderX, select **Help|Help Topics** and choose **Mobile Development**.

For Metrowerks CodeWarrior there is a **Help|Online Manuals** menu option that contains lots of valuable information on working with Symbian OS.

Additional specific IDE information can be obtained online—for example, for Visual C++ information go to **http://msdn.microsoft.com/**, for Borland C++ go to **http://bdn.borland.com/** and for CodeWarrior go to **http://www.metro-werks.com/MW/Develop/Wireless/**.

Building for a Target Series 60 Device

Building for a Series 60 device with the Visual C++ IDE must be performed at the command prompt using the `abld` command, as described next. However, CodeWarrior and C++Builder allow building for a target device from within the IDE.

Chapter 1 Getting Started

Building for a Series 60 device based on an ARM processor on a PC requires the use of a suitable PC-hosted cross compiler (for example, the GNU **gcc** C++ Compiler, as supplied with the Series 60 SDK) to build the executables in a suitable ARM binary format.

When building the project for a Series 60 device you would typically use a release build, since that is what you would do to create a final, deliverable application.

To build for target hardware, open a command prompt window and navigate to the group directory for the GUI **HelloWorld** project, then enter the following commands:

```
bldmake bldfiles
abld build armi urel      — For Visual C++ and CodeWarrior
or
abld build armib urel     — For Borland C++
```

This will cause `abld` to invoke the build (cross-compilation and linking) system to produce an `armi` (ARM Interworking) release (`urel` — Unicode release) build of the application for execution on a target device using the **gcc** tool chain.

There are currently three build variants for ARM based devices — `ARMI`, `ARM4` and `Thumb`. `ARMI` executables will work with the other two build variants. Typically you should build the `ARMI` binary executable variant for compatibility with the maximum number of real devices. `ARM4` builds give maximum performance at the expense of increased code size. `Thumb` builds will reduce the code size at the expense of a slight reduction in execution speed.

The build steps actually include C++ compilation, linking, resource compilation, and production of the application information (`.aif`) file — the file that contains the application icon and other specific details.

When building for a target device a Symbian OS-specific tool called **petran** is automatically invoked behind the scenes. **Petran** translates `HelloWorld.app` into a form suitable for loading at runtime by the Symbian OS executable loader.

If you are using the CodeWarrior IDE to build for a Series 60 device, use **Project|Set Default Target** either to select the required target (for example, `ARMI UREL`) or to choose **Build All**, and then build with **Project|Make**.

With C++Builder X, you need to select the **Project|Properties** menu item and then select the **Symbian settings** tab. From this dialog you can change the **Platform** and **Build** options to `ARMI` and `UREL`, respectively. *Note:* At the time of writing `ARMI` was the only target Platform option available.

Pressing **Ctrl+F9** will make the project, creating the installation package (`.sis`) as part of the project; installation packages are described in the next section.

Chapter 1 Deploying on a Target Device   21

The executable and data files (`HelloWorld.app`, `HelloWorld.rsc` and `HelloWorld.aif`) will now be located in a folder such as `\Epoc32\Release\Armi\Urel`.

For testing on a Series 60 device, all of these files have to be transferred to a device and located in a folder called `\System\Apps\HelloWorld\`.

File transfer to a device can be performed by copying to a memory card, or via a USB cable, if the Series 60 device manufacturer has included appropriate support in their product. Typically though, the file transfer is performed by packaging all the application files into a special installation file. The next section describes how this is achieved.

Deploying on a Target Device

Applications are delivered to target hardware in the form of a **Symbian Installation System** (.sis) file. A .sis file is a single compressed archive file, containing all of the files required for installation, plus optional information about the installation process. The Symbian Installation System provides a simple and consistent user interface for installing applications, data or configuration information onto devices based on Symbian OS. Developers (or end users) install components, packaged in .sis files.

Production of .sis files can be performed using the interactive **sisar** tool provided with the Series 60 SDK. **Sisar** packages all the application files into one .sis file for ease of installation onto target hardware. Alternative methods for producing installation files are described in Chapter 2.

Everything required to make an installation (.sis) file is provided with the example **HelloWorld** project—under the `\install` folder. In this example project we will use a special installation package source file called `HelloWorld.pkg` and a tool from Symbian called `makesis.exe`.

Building a SIS Install File

After building the `armi` release version of the **HelloWorld** application, as described above, you need to package up the application components into an installation package (.sis) file. Open a command prompt and navigate to the SDK folder for the **HelloWorld** project. Change to the `\install` folder, then build the .sis file by typing:

```
makesis helloworld.pkg
```

A successful build will produce an output message such as "`Created helloworld.sis`". The installation package (.sis) file will have been created in the `\install` folder. Now you need to transfer it to the device, as described in the next section.

SIS File Installation

You may choose among three potential installation options, depending on the device you are using, and other facilities available to you—for example, whether you have access to a PC with infrared or Bluetooth capability, or access to appropriate software based on Symbian Connect (Nokia PC Suite, for example, or a branded equivalent provided by the device manufacturer):

- Installation through the invocation of a .sis file located on a PC, with subsequent application installation on to the Series 60 device through an infrared or Bluetooth session between the PC and the target device, established via software such as Symbian Connect.

- Installation by transfer of a .sis file through **OBEX** (OBject EXchange), over infrared or Bluetooth, from another device such as a PC, Symbian OS phone or any OBEX-enabled device. This process will be managed via the **Messaging** application, which intercepts the file attached to the message—when you open up the message, it will automatically start the application installation process on the phone.

- Alternatively, .sis files can be sent as email attachments. Application installation is again managed via the **Messaging** application on the phone. When you open the message, it will automatically start the installer.

The first two options depend on establishing a connection between your development PC and the Series 60 device. The device manufacturer typically supplies suitable communications software, and you will need to refer to the specific instructions supplied with the connection software.

After installation a much-reduced version of the .sis file remains on the Series 60 device to control the uninstallation of the application, if required, using the application "**Manager**." This reduced .sis file contains only the information required to uninstall the application and is typically very much smaller than the original file.

Often the original .sis file may still exist on the device, if it was delivered as a message attachment and the original message has not been deleted from the **Messaging** application's **Inbox** folder.

Running on a Target Device

Transfer the helloworld.sis file provided to the target hardware, using one of the methods described above. After the transfer you will be offered the chance to install the application on the device. To run the application follow the procedure outlined in "Locating and Running the Application" earlier in this chapter. You will be reassured to find that locating and running the application on a target device is identical to the process on the emulator—with one small difference: the application will not be located in an "**Other**" folder.

Summary

This chapter has described the whole process of building, deploying and running a simple GUI application on the Series 60 emulator and on a real target device. The skills you have gained here will be used on a daily basis as you continue to develop for Series 60.

In the next chapter, you will build on this knowledge and look in detail at the creation of a simple Series 60 application. You will learn about the composition of project files, examine application information files and understand how to produce installation package files. Some key additional SDK tools that are used on a regular basis will also be described.

chapter 2

Development Reference

Covering all of the essential components of two Series 60 projects in detail, plus using other key development tools to build, deploy and run the example applications

This chapter focuses on the project structure, files and the required build tools. The chapter has two parts. The first part is a tutorial on the definition of sophisticated projects and the use of the build tools in more depth. The second part, beginning with the **Additional Development Tools** section, is intended mainly for reference. You may want to skim through it, though, because it describes more developer tools, advance project information, and build and application deployment issues. (Chapter 4 discusses the application design, structure and the source code.)

The tutorial is a guide to the more advanced features of the build process—you should refer to Chapter 1 to remind yourself of the basic procedures and syntax. The contents of each of the key files in three example projects are described, with emphasis on the composition and syntax of `.mmp` and `bld.inf` files introduced in Chapter 1.

A GUI application illustrates the details and composition of more complex Series 60 projects. Two variants of this project are described, the second one showing how to create localized application project.

Also described is an example console application. Typically you will use this simple type of executable as the basis for testing other code modules—for example, to test components such as application engines.

You will be introduced to some of the more commonly used SDK tools, such as those used for building application resources and for generating bitmaps and icons, and even wizards for generating the files and code for basic Series 60 applications.

Localization of applications—that is, designing and coding applications for use in multiple locales and languages—is vital to maximize the size of the market for an application. There is excellent support for this in Symbian OS, and the key elements of this part of the development process are described from a project and building perspective.

Application deployment is covered in more depth, and the production of more sophisticated installation packages is addressed, including an installation package for a localized application.

Detailed links to information on SDK and emulator configuration are provided, including an overview of the test and debugging facilities, information about the emulator hot keys and details of how you can use infrared and Bluetooth-enabled devices to add communications capability to the emulator.

Chapter 2 SDK Versions and Selection

Topics covered are:

- **SDK Versions and Selection**—Information on installing and using different versions of the SDKs that are now available. How to set the value of EPOCROOT and use the **devices** command, with tips on setting up the system path correctly.

- **The HelloWorld GUI Application**—A detailed description of the two Symbian OS generic project files. Building and running an application, plus a description of the runtime files produced. Discussion of the structure of the project folders and the source files. Application UI resource file and the resource compiler, application information files, icons, captions.

- Coverage of application localization is demonstrated by a variant of the **HelloWorld** example, called **HelloWorldLoc**. English, French and German variants of the example application are shown.

- **Console Applications**—A detailed description of the two generic project files, and the differences between console and GUI project files. Building and running the application, plus a description of the executable and source files.

- **Symbian Installation System**—How to package up all of the components of an application for installation on a real Series 60 device. Running executables during installation and localized application installation.

- **Additional Development Tools**—Symbian OS bitmap files, the application creation wizard, plus coverage of a wide range of other tools and utilities for developers.

- **Installation Tips for Series 60 SDKs and IDEs**—Information about configuring the emulator and using its many facilities.

- **Advanced Application Deployment and Build Guide**—Series 60 Platform UIDs to avoid installation of software on an incompatible device. Some essential Series 60 Platform differences to be aware of, and where build differences exist between the emulator and target hardware.

All the information needed to build and run the three example projects is provided within this chapter. All the project files, source files and deployment information associated with the example are available online, as noted in the Preface. If you have not already obtained the source materials, you are advised to download and install them—they will be helpful for reference as you read through this chapter.

SDK Versions and Selection

If you install more than one Symbian OS SDK on your development PC system you will need to select the particular SDK you wish to work with before you can use it.

Series 60 Platform SDKs 1.x or earlier are based on Symbian OS 6.1. From Series 60 2.x onward Symbian OS 7.0s was used and the approach to configuring multiple SDKs was modified.

Series 60 Version 1.x SDKs

If you are using only Series 60 Platform 1.x or earlier (based on Symbian OS 6.1), then selecting an SDK simply involves setting the value of a special environment variable called EPOCROOT. During the installation of a Series 60 SDK EPOCROOT is created, and it is set to point to the root directory of the particular SDK. Some of the material installed, such as documentation and tools, may be common to more than one SDK, and so it is placed in shared directories. However, the runtime environment for each platform emulator, associated system header files and libraries are different. Hence there is a separate directory tree per SDK; it is to this directory that the environment variable EPOCROOT must point. For example, it might be set something like:

`\Symbian\6.1\Series60\`

The SDK-specific tools, resources and example projects are under this subtree. So if you need to work with SDKs for multiple versions of Series 60, you may need to ensure that the value of EPOCROOT is set correctly each time you change to a particular SDK. Otherwise you cannot use a particular SDK without getting error messages.

You can change between SDKs from the command prompt by using the set command to change the value of EPOCROOT:

`set EPOCROOT=\Symbian\6.1\Series60\`

Note that EPOCROOT must not contain the (PC file system) drive letter—you specify the folder path only.

Alternatively you can use the **EpocSwitch** utility to switch between different SDK environments (Series 60 1.0, 1.2 and so on). More information about **EpocSwitch** is provided in the *Miscellaneous SDK Tools* subsection later in this chapter.

Additionally, your chosen IDE may have facilities to set the value of EPOCROOT directly.

The Symbian OS tools for C++ development, with the exception of the emulators, are found in two directories:

- Tools developed by Symbian are in the `\epoc32\tools\` directory in the root of your SDK installation.

- The ARM cross complier supplied by Symbian is the GNU **gcc** compiler adapted and supplied by Cygnus. This target cross compiler and its supporting programs are in the `\epoc32\gcc\bin\` directory in the root of

Chapter 2   SDK Versions and Selection   29

your SDK, or possibly in the `\Shared\epoc32\gcc\bin\` directory, at a level higher than the root of your SDK—this will depend on the option chosen during installation of your SDK.

Both of the tools directories should be included in the Windows system path. The SDK installation process normally updates the path and registry settings.

> **TIP** If you are using Microsoft Visual C++, and you want to be able to perform command-line builds for WINS, you must ensure that the environment variables are set up for access to the Visual C++ tools. This is normally the case if the "command-line builds" option was selected when Visual C++ was installed. Alternatively, these can be set using a command file called `vcvars32.bat` provided with the product.

Series 60 Version 2.x SDKs

In Symbian OS 7.0s, the approach to configuring multiple SDKs was modified. So, for Series 60 2.x Platforms (with or without version 1.x SDKs installed) the selection of a specific SDK is done via the **devices** command. The environment variable `EPOCROOT` no longer needs to be set; instead you use the **devices** command to choose the particular Symbian OS device you wish to develop for. A device is specified in the form of `kitname:device`, where `kitname` refers to the SDK. You can see a list of all the available devices by simply opening a command prompt and typing `devices`. To set a device as the default you would type, for example:

```
devices -setdefault @Series60_v20:com.nokia.series60
```

You can then switch to a different device either by changing the default or by using the `set EPOCDEVICE` command.

Note that any SDKs you have installed will not normally be displayed by the **devices** command unless they are based on Symbian OS 7.0s or later. However, you can add them to the list using the `devices -add` command. Further details can be found in the SDK documentation.

Note that the **EnvironmentSwitch** tool, available from Series 60 2.x, can be used to automate the process of changing to different development environments. The **Additional Development Tools** section later in this chapter provides further details.

Use of Devices with Metrowerks CodeWarrior IDE

The CodeWarrior IDE is aware of **devices** information and will query you at appropriate times for the device kit to use—for example, when you use the **Import Project from .mmp File** option from the **File** menu, as described later in this chapter. Note that in order for CodeWarrior to detect the device, the device-name part of the identifier must be of the form `com.name.name`.

The HelloWorld GUI Application

Using exactly the same example application as used in Chapter 1, we look at the two generic project files for the **HelloWorld** example in detail and examine the contents of each line by line. You must have a component definition (`bld.inf`) and a project definition (`.mmp`) file for each development project. The `bld.inf` file for a project may refer to one or more `.mmp` files, one for each component to be built. In simple projects there will be a single component—for example, a single reference to the `.mmp` file for the application itself. In more complex projects, it may involve other components in addition to the application, such as dynamic link libraries (**DLLs**), and there may be multiple `.mmp` files. The component definition and project definition files are used by the tool chain to construct a build command file (**abld.bat**). You can use this file for various purposes, such as creating other project and workspace files for your chosen platform or development environment, or building the project for the emulator or target device in debug or release formats.

HelloWorld bld.inf

The **HelloWorld** component description file (`bld.inf`) refers to a single project definition file (`HelloWorld.mmp`).

```
// HelloWorld bld.inf
PRJ_MMPFILES
// List the .mmp file(s) required for the project
HelloWorld.mmp
```

For a simple application (or single component), such as the **HelloWorld** example, this is typically all there is in a `bld.inf` file. A list of the `.mmp` file(s) in the project is placed after the `PRJ_MMPFILES` statement. In this example this is simply the name of the `.mmp` file for the application. In a more complex project there might be a list of several other `.mmp` files—one for each component part of the application.

The **bldmake** tool processes the component definition file (`bld.inf`) in the current directory and generates the batch file **abld.bat** and several build batch makefiles (`.make`). The makefiles are used by **abld** to carry out the various stages of building the component. To see a general list of the syntax of the **bldmake** command, type:

```
bldmake
```

Then, by typing the following, you will see a general list of the syntax of the other statements valid in a `bld.inf` file:

```
bldmake inf
```

HelloWorld.mmp

A project definition (.mmp) file specifies the properties of a project component in a platform- and compiler-independent way. It may then be used, along with the bld.inf file, by the SDK build tools (**abld.bat**) or an IDE to produce platform-specific makefiles. The HelloWorld.mmp project definition file is shown here, followed by an explanation of each statement in Table 2–1.

```
// HelloWorld.mmp file

TARGET     HelloWorld.app
TARGETTYPE app
UID   0x100039CE 0x101F6148
TARGETPATH \system\apps\HelloWorld

LANG 01

SOURCEPATH ..\src
SOURCE   HelloWorldApplication.cpp
SOURCE   HelloWorldAppUi.cpp
SOURCE   HelloWorldDocument.cpp
SOURCE   HelloWorldContainer.cpp

RESOURCE ..\data\HelloWorld.rss
RESOURCE ..\data\HelloWorld_caption.rss

USERINCLUDE . ..\inc

SYSTEMINCLUDE \epoc32\include

LIBRARY euser.lib apparc.lib cone.lib eikcore.lib
LIBRARY eikcoctl.lib avkon.lib commonengine.lib
AIF HelloWorld.aif ..\aif HelloWorldaif.rss c12
context_pane_icon.bmp context_pane_icon_mask.bmp list_icon.bmp
list_icon_mask.bmp
```

TIP To see a complete list of the valid statements in an .mmp file, open up a command prompt, navigate to your SDK, and type makmake -mmp.

UIDs

A UID is a globally unique identifier consisting of a 32-bit number. In Symbian OS, objects are identified by their **UID type**, which has three component UIDs: UID1, UID2 and UID3. UID1 determines whether a file is a document or executable code. It should be seen as a system-level identifier, distinguishing between executables, DLLs and such like. UID1 is determined by the TARGETTYPE statement in the .mmp file.

Table 2–1 Statements from the HelloWorld.mmp File

.mmp Statement	Function
`TARGET`	The name of the application, which must have the correct filename extension: .exe, .app, .dll and so on.
`TARGETTYPE`	Defined as `app` (this option determines the value of `UID1`), meaning this is a GUI application. There are many other types, including `dll`, `exe`, `tsy`, `csy` and `ldd`.
`UID`	Specifies a unique system identifier (`UID2`) for a GUI application: `0x100039CE` and `0x101F6148` that is absolutely unique to the application itself (`UID3`). `UID2` and `UID3` are not required for `.exe` applications. UIDs are explained in the next section of this chapter.
`TARGETPATH`	The location of the final built application and its components—this is always under `\system\apps\` relative to the root of a device drive or an emulated drive such as `c:` or the emulated ROM (`z:`) drive.
`LANG`	This statement is used if an application is to support different languages. Each language supported has a two-digit code. This is described later in this chapter in the *Localization of Applications and Resources* subsection. It is also covered in more detail in Chapter 4.
`SOURCEPATH`	The location path of the source files for the project.
`SOURCE`	This statement refers to the name(s) of the source file(s) for the project. This statement can occur multiple times and can have multiple filenames on each statement line.
`RESOURCE`	Refers to a source file that defines the majority of the user interface elements, such as menus, dialogs, text strings and so on. Also the source file for the application captions.
`USERINCLUDE`, `SYSTEMINCLUDE`	These statements refer to the location of the application-specific header files for the project with path information for system header files. All Symbian OS projects should specify `\epoc32\include\` for their `SYSTEMINCLUDE` path—relative to the root of the SDK.
`LIBRARY`	Lists the application framework and graphics libraries required for linking to—these are the `.lib` files corresponding to the shared library DLLs whose functions you will be calling at runtime.

Table 2–1 continued

.mmp Statement	Function
`AIF`	The statement refers to an Application Information File (`.aif`) that contains icons and other application properties defined in the application resource file. `.aif` files are explained later in this chapter. The application details and characteristics are specified in the text-based resource file `HelloWorldaif.rss`, and the source bitmaps and masks for the icons are listed as a series of Windows `.bmp` files. The `c12` parameter refers to the color depth (12-bit) to be used when the files are converted from `.bmp` to Symbian OS `.mbm` format as part of the icon creation.

For a document file, `UID2` determines the type of document it is (application data file, or `.aif` file). For an executable code file, the value of `UID2` determines the type of executable it is (for example, `.app`, `.dll` or `.tsy`). For executable code files, `UID2` and `UID3` are specified in the `.mmp` file on the `UID` line.

`UID3` is an application-level identifier; the value for your application must be absolutely unique across all other applications. These values are issued only by Symbian. Unique UIDs for applications and other purposes can only be obtained from Symbian by emailing **uid@symbiandevnet.com**. A range of UIDs: `0x01000000` to `0x0FFFFFFF`, is available for use during development, but be very careful never to release an application with an experimental UID.

Symbian OS uses UIDs to associate documents (data files) and `.aif` files with their respective application; `UID3` is included in the header of each data file, executable file or `.aif` file. However, it is still necessary for executables and their components to have the correct filename extensions, otherwise they will not run correctly, appear in the **Application Launcher** or be associated correctly with their document.

When an application creates a document file, all three UIDs, plus a 32-bit checksum, are automatically incorporated into the first 16 bytes of the file header by the application framework.

> **TIP** Sometimes you will get a link error when you build your project because a required library is missing from the `.mmp` file. To find out which library you need to list in the `.mmp` file, you can check out the SDK documentation or use the **clindex** utility that produces an up-to-date ordered listing—for details see **More Tools and Utilities** later in this chapter.

Stack and Heap Sizes

Automatic (locally scoped) variables, function arguments and so on are stored in an area of memory called the stack. Memory allocated dynamically, through operators such as `new` (or `new (ELeave)` in Symbian OS code), is taken from a pool called the heap. Both a stack and a heap are created by the system for each application or thread of execution. Stack memory allocation and deallocation is controlled by the runtime environment. Memory allocated and deallocated on the heap is under the direct control of the developer. In most cases such allocations are under your control; in some cases they are under the control of the operating system designers. Careful control of allocation, and in particular the deallocation of heap memory, is a vital requirement placed on all Symbian OS developers.

By default the stack size for an application is 8KB, but you may need a larger stack for certain types of application. However, you must be careful not to use this facility to misuse the stack. In Symbian OS code only small objects should be put on the stack; all large objects are normally allocated on the heap. If you do need more than 8KB of stack—for example, in a highly recursive application—you can use the EPOCSTACKSIZE statement in your application's .mmp file to override the default 8KB. The size of the stack (in bytes) can be specified in decimal or hexadecimal format. On target hardware if a stack overflow occurs on ARM, it will cause a page fault; the application will be panicked and will terminate, and an error (KERN-EXEC 3) will be reported. See the SDK documentation for comprehensive error-code information—search for "Panic categories and numbers" or "System panic reference".

The default heap size for an application is 1MB. Again, you can increase (or decrease) this if required. The EPOCHEAPSIZE statement will allow you to specify minimum and maximum heap values—this is not problematic under the emulator, since the maximum heap limit is not enforced.

Depending on the IDE you are using, the EPOCSTACKSIZE and EPOCHEAPSIZE statements may have no effect under the emulator environment. It may be necessary to change project settings directly within your IDE to allocate more (or less) stack or heap space to the application.

WARNING
Trying to use too much stack when building for the emulator under Microsoft Visual C++ may result in the following error:
```
Engtest.obj : error LNK2001: unresolved external
symbol __chkstk
\EPOC32\RELEASE\WINS\DEB/ENGTEST.exe : fatal
error LNK1120: 1 unresolved externals
Error executing link.exe.
```
To fix this for WINS builds in the Visual C++ IDE, select **Project|Setting** from the menu and select the **C++** tab in the dialog. Insert a new line such as:
`/Gs9999 /FD /c`
where `9999` is the new stack size (decimal) in bytes.

Additional Project Components

As a matter of good object-oriented design and to aid maintainability and flexibility, Symbian OS applications are typically split into two main parts: the UI, and an application engine—also known as the model. The UI is typically subdivided into on-screen representation(s) of the data and a handler or controller that determines the overall behavior of the application. Application engines typically encompass the data structures needed to represent the application's data, the algorithms and any data persistence.

Often, engines are created as a separate component—typically as a DLL. If this is the case, the engine DLL is created as an additional component project with its own .mmp file. This file will be listed as a component in the bld.inf file for the main application. The name of the engine DLL must then be included in the LIBRARY statement in the .mmp file for the application.

Application architectures and the key classes in a Symbian OS UI application are discussed in more detail in Chapter 4; where an example project is provided that shows how to create a DLL to encapsulate the engine functionality.

Building and Running

Open a Windows command prompt and navigate to the folder containing the project definition (.mmp) and component definition (bld.inf) files—for example, \EMCCSoft\HelloWorld\group in the root of your SDK.

To create the **abld.bat** file, type:

```
bldmake bldfiles
```

To compile and link the project from the command line, type:

```
abld build wins udeb          —for Visual C++
abld build winsb udeb         —for CodeWarrior
abld build winscw udeb        —for Borland C++
```

This will build the project for the Series 60 debug emulator. Normally projects such as the **HelloWorld** example application are built and run from within an IDE.

TIP The build process in some IDEs may not build all the resources required by a complete application—for example, bitmap (.mbm) and icon (.aif) files. Command-line builds, using **abld**, will do a complete build, including creating all the resource and components specified in the project. Type **abld** to see the general syntax of the command file, or type **abld help commands** to list all of the command options.

To run the application you need to first start up the Series 60 emulator. At the command prompt type:

epoc

Generally this is all that is required to start the emulator from a command prompt. If you have more than one SDK installed, you may need to ensure the correct version is selected—for example, by using the **devices** command. Normally you could create the IDE project files and build and run the emulator from within your chosen IDE, as described in Chapter 1. For application debugging, running your application from within an IDE in debug mode is a requirement, not an option.

Once you have started up the Series 60 debug emulator, navigate to and select the **HelloWorld** application, then click on the **Selection** button (in the center of the navigation control) to invoke the application. The application will run and should appear as shown in Figure 2–1.

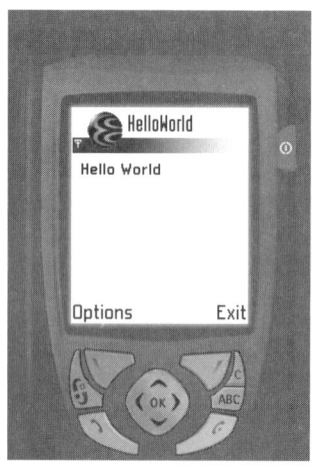

Figure 2–1 The GUI application build process summary.

HelloWorld GUI Executable and Runtime Files

This subsection looks at the files produced by the build process, showing the type and purpose of each file, and its location on an emulator and a target device. The whole build process in terms of the input source plus header files and the output runtime files is summarized in Figure 2–2.

At least two files make up a minimum GUI application, the .app and .rsc files (for example, HelloWorld.app and HelloWorld.rsc). The HelloWorld.app file is the application executable—in other words, the output from the compilation and linking process. The HelloWorld.rsc file is the binary resource file produced by the resource compiler.

In an application intended for release, an application information (.aif) file must also be supplied—it is not essential during development. An .aif file contains icons, optionally captions in the supported languages, and specifies certain application properties such as whether embedding is supported. An application will still function if the .aif file is not present. However, it will have a default icon provided by the system, and other application characteristics will be set by the system to defaults. If a caption resource file (for example, HelloWorld_caption.rss) was specified, an additional file for installation HelloWorld_caption.rsc will be generated. See "Captions" in the *AIF Files* subsection later.

In Symbian OS, a GUI application is a special form of DLL that always has an .app filename extension. A specific framework is provided by the system to

Chapter 2 The HelloWorld GUI Application

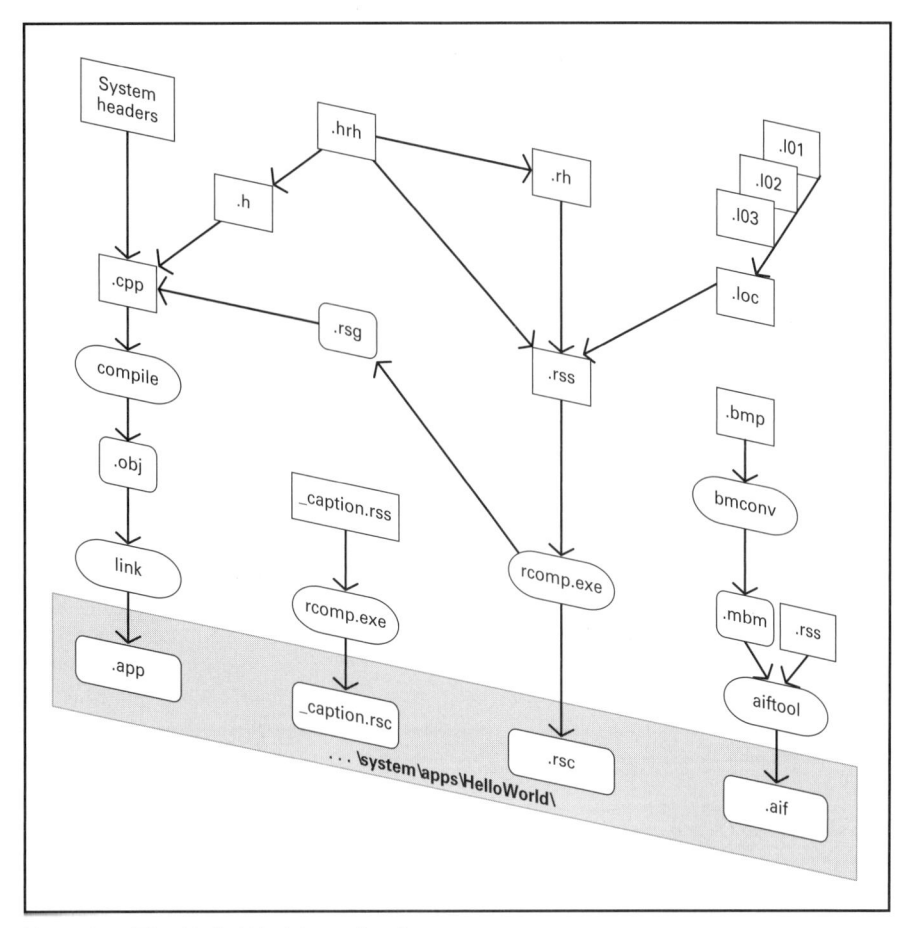

Figure 2-2 The HelloWorld application.

load and correctly initialize GUI applications. As a developer, you must conform to the requirements of this framework and will have to provide implementations of a number of pure virtual functions in order that your application will start up correctly. Optionally, you may also provide implementations of a number of other virtual framework functions in order to achieve additional specific behavior that you may require. The application framework and the associated virtual and pure virtual functions are discussed in detail in Chapter 4.

In order to be automatically recognized by the system, all GUI applications must follow a location convention—they must be under a folder such as: \system\apps\appname\. So in the case of the **HelloWorld** example, the files described earlier (HelloWorld.app, HelloWorld.rsc, HelloWorld_caption.rsc, and HelloWorld.aif) should be located in a folder called \system\apps\HelloWorld\. This folder has to be at the root level of a target device drive or an emulated drive such as c: or z: (the emulated ROM drive).

WARNING Keep the name of your application to 16 characters or less (case is not significant), otherwise your application may not appear under the Series 60 emulator, even though the build process did not report any errors!

A number of other files may be essential for the correct functioning of a particular application. For example, there may be need for a compressed multibitmap (.mbm) file, application-specific data files and so on. It is usual to put such essential files in the same location as the key application files—in other words: \system\apps\appname\.

TIP Sometimes you will get an infuriating "Not found" error message when you try to run your application on target or under the emulator—something is missing or incorrect, but the system does not give you any further clues! Check that your application UID is correct—you may have forgotten to update all the UIDs in your code (including your .mmp and .pkg files). Also a DLL may be missing—this is more common when you install to a target device and forget to move all DLLs across. One further common problem is that another file is missing, such as the .mbm, resource file or another required application data file.

HelloWorld Project Files and Locations

You will find a number of related folders located under \EMCCSoft\HelloWorld\, and they are set out in Table 2–2.

Table 2–2 The Source Folders in the HelloWorld Project

Folder	Contents
\aif	HelloWorld.aif and the source bitmaps (.bmp) for the .aif file.
\data	HelloWorld.rss and HelloWorld_caption.rss—the application resource files to generate HelloWorld.rsc and HelloWorld_caption.rsc.
\group	HelloWorld.mmp and bld.inf platform-neutral project files.
\inc	HelloWorld header (.h) files and other #included files such as Helloworld.loc and Helloworld.l01.
\install	HelloWorld.pkg package file for creating the installation (.sis) file via makesis.exe. HelloworldFR.pkg is the "run application on install" version. The HelloWorld.sis files are placed here when built.
\src	HelloWorld source (.cpp) files.

This subfolder and file organization for the project—under the project name folder—will reflect the structure defined in the `.mmp` file. In some cases the `data` and `install` folders may not exist, and the associated files may be located in the `group` folder instead. Virtually all files generated during a build and the intermediate build files are placed outside of this folder structure, except for:

- The **abld.bat** file—This is not portable between SDKs and must be regenerated when the `bld.inf` or `.mpp` file changes.
- On building, the `HelloWorld.sis` file is placed in the folder containing the `.pkg` file.

HelloWorld GUI Source Files

This subsection outlines the contents and the function of each of the source files that form the example project, but the detailed discussion of Series 60 application design principles is left until Chapter 4.

Source and Header Files

The files listed in Table 2–3 make up the body of the C++ application source, header files and class modules.

Table 2–3 Source Files in the HelloWorld Project

File Name	Contents	Class Module	Comments
`HelloWorldApp.h`	Class and function declarations	The application	Application creation/initialization
`HelloWorldApp.cpp`	Class and function definitions		
`HelloWorldAppUi.h`	Class and function declarations	The application UI	Command handler/controller
`HelloWorldAppUi.cpp`	Class and function definitions		
`HelloWorldDocument.h`	Class and function declarations	The document	Owns the data or application model
`HelloWorldDocument.cpp`	Class and function definitions		
`HelloWorldContainer.h`	Class and function declarations	The application's data view	Displays the application's data
`HelloWorldContainer.cpp`	Class and function definitions		

Once you have the **HelloWorld** project open in your IDE, you will probably discover another source file called `HelloWorld.uid.cpp`. This is the UID source file; the build tools automatically generate it under a folder such as: `\epoc32\Build\Symbian\6.1\Series60\HelloWorld\group\HelloWorld\wins` in the root of your SDK. There should be no need for you to edit this file. It simply defines the application's UIDs as a data segment and is required by some types of DLL running under the emulator (WINS) environment.

Application Resource Files

Resource Source Script (`.rss`) files (for example, `HelloWorld.rss`) are used by applications in Symbian OS to define the way a GUI application will look on screen. Much of the information that defines the appearance, behavior and functionality of an application is stored in a resource file. Created external to the main body of the executable, everything from the status pane, menus, general user interface controls, through to individual dialog boxes can be defined in the resource file. Individual resources are loaded very efficiently as required at runtime, and hence the memory requirements are minimized.

Application resources are compiled (and may be compressed) at build time into binary files that are used at runtime (by default with `.rsc` extensions). During the build steps the resource compiler processes this file. It is typically compiled automatically before the compilation of the source files (but only if the resource file has been updated since the last running of the resource compiler). The output of the resource compiler is a binary file that is used at runtime to supply the resource information as required. The default output name has the same name as the application, but with an `.rsc` filename extension.

For the **HelloWorld** example application, the name will be `HelloWorld.rsc`. In addition, a header file, with an `.rsg` filename extension, is generated by the resource compilation stages. This file contains a series of symbolic name/numeric constant pairs (as `#define`s) that identify each of the user interface elements. This file must be `#include`d in the `.cpp` files wherever the resources are referenced or used.

Resource files and their internal structure may initially appear complex, but taken step by step they are quite straightforward. Comprehensive explanations and examples of application resource files are provided throughout this book, and many other examples are available in the Series 60 SDK example projects.

Optionally another resource file, `HelloWorld_caption.rss`, may be created to define captions for the application—in other words, the application name for display in the **Application Launcher**. Two caption sizes are quoted to accommodate the display options of the **Application Launcher** grid and list views. The resource compiler also processes this file during a full project build.

Chapter 2   The HelloWorld GUI Application   41

Other Header Files

Note that `avkon.hrh` (located in `\epoc32\include`) contains many enumerated constants defined by the Series 60 user interface and application framework. This is a very important file for Series 60 applications. For example, one of the constants defined is `EAknCmdExit`; an application's menu option **Exit** should always generate this command. The standard value passed by the framework when an application should close down (for example, under low memory conditions or for a system backup) is `EEikCmdExit` as defined in `Uikon.hrh`.

A special header file, called `HelloWorld.hrh`, is part of most projects and contains definitions of any enumerated constants that may be used in both source files (`.cpp`) and resource files (`.rss`). These constants are very specific to an application and are typically used for menu commands, key handling, view identifiers, or for command handling in dialogs.

The structure of the UI resources and controls that are used in the application's resource file are defined in files with `.rh` extensions. It is possible to define your own resource structures for custom user interface components that you produce yourself. Specific SDKs provide numerous resource structures, and the `.rh` files that define them are located in the `..\epoc32\include\` folder under your SDK folder tree. Resource files and resource header (`.rh`) files are covered in more detail in Chapter 5.

Resource Compiler

Virtually all user interface elements and controls can be defined in the application resource file. Individual resources are loaded as required at runtime from the compiled resource file (by default with `rsc` extensions), and hence the memory requirements are minimized.

Resource files can be localized without recompiling the main program. For ease of localization, all user interface text is usually separated out from the main resource (`.rss`) file into separate header files that are conditionally `#included` into the main resource file at build time. It is the included language text files, not the main resource file, that are later translated into the different supported languages. A version of the application resource file is produced for each language. Localization is discussed in more detail later in this section.

Application caption resources may be used to provide captions (the application's name) in two different lengths and in multiple languages. The captions are displayed with the application's icon in the **Application Launcher**. The resource compiler also processes the caption resource file(s) during a full build. See the *AIF Files* subsection later in this chapter for more details.

The resource compiler (`rcomp.exe`) is usually invoked automatically from within an IDE. Alternatively it can be run from the command prompt; typically this is performed through a command file (`epocrc.bat`) that combines the actions of passing a resource file through the C++ preprocessor, and then compiling it with `rcomp.exe`. Compilation of resource files generates a binary output file plus an `.rsg` header file that is `#included` into C++ code to identify specific user interface resources such as text strings, dialogs and menu components.

Localization of Applications and Resources

If your application is to be distributed outside of your own locale, then localization needs should be considered from the outset of a project. Good support for localization is provided in Symbian OS and the SDK build system and tools.

For Latin character sets, adapting an application for a different language text should only require changing the resource file, since user visible text should never be embedded in C++ source files. However, you will have to adopt a flexible attitude toward layout, as text translated into different languages will obviously have different lengths.

Each supported language has a two-digit identifier code; for example, English is `01`, French is `02` and German is `03`. These language codes are all defined in `\Epoc32\Include\e32std.h`. For each language you will typically create a specific text file by having the default language file translated. Each language file can then be conditionally included in the main application resource file at build time. The process of localizing an application in this way is demonstrated by a variant of the **HelloWorld** example used in this chapter, called **HelloWorldLoc**. Each specific text file typically has the two-digit language code forming the last two characters of the filename extension. So, in the **HelloWorldLoc** variant of this example project you will see there are three variants of the included text file with extensions of `.l01`, `.l02` and `.l03` for the English, French and German variants.

There are also a number of support classes for other locale-sensitive information and formatting. For instance, the `TLocale` and related classes are defined in `e32std.h` and allow dates, times, currency and many other locale-sensitive data types to be correctly formatted for the current system locale. You need to become aware of the data types that are locale sensitive, the formatting classes available and be sure to use them whenever you are presenting data on screen.

A number of approaches are possible for providing language variants and installation files that give the user options at application install time.

Resource files compiled for specific languages will typically have filename extensions such as `.r01`, `.r02`, `.r03` and so on, and the user may be given a choice of language at install time. The default compiled resource file has the extension `.rsc`; any language variant can be renamed to this extension during installation. A Series 60 project file (.mmp) can list the supported languages via the LANG statement.

LANG 01 02 03

When a full project build is performed, the system will compile the English (01), French (02) and German (03) resources for the application. More information on the installation of multilingual applications is provided later in this chapter, and localization of dates is described in Chapter 4.

With non-Latin languages, the localization of text is much more complex. Symbian OS and Series 60 do provide some support for producing devices using languages such as Chinese, Japanese, Korean, Hebrew, Arabic and Thai. However, writing applications that support these languages is a specialist subject and is beyond the scope of this book.

AIF Files

Application information (.aif) files are used at runtime and store data concerning an application, including:

- Icons in two sizes (defined later in this chapter), are used by the system to represent your application.

- Capabilities, such as document embedding, new-file creation, whether the application is hidden or not and MIME-type support priorities.

- Optionally, application captions (names) in the supported languages— although better support for captions is provided via the caption resource file system described later in this section.

Without an .aif file an application will use a default icon, and the application name as a caption (without the filename extension). It can be created independently of the application by the **aifbuilder** tool provided with the Series 60 SDK. The **aifbuilder** tool allows you to define the properties of the .aif file and saves the details of a project's definition in a file with an .aifb filename extension. Note that **aifbuilder** depends on a Java™ 2 Runtime, which is available from **http://java.sun.com**.

You can specify the composition of .aif files manually through text-based resource files with an .rss filename extension (for example, HelloWorldaif.rss). This is the option used in all of the example GUI applications described in this chapter. This resource file is compiled by the **aiftool** utility—it can be performed manually at a command prompt, or automatically during a full project build. The .aif build process also converts the bitmap files that form the icon (with **bmconv**, as described later in this chapter) and then creates the .aif file. The **aiftool** utility and the syntax of .aif files are documented in the Series 60 SDK.

Icons

Icons are used to represent applications and their associated files/documents, both when they are embedded and when they are shown on the application shell. Series 60 icons are usually provided in two sizes. The most appropriate size for the current container "zoom state" will be displayed, if an icon of the specific size is not available—for example, the **Application Launcher** grid and list views use icons 42 pixels wide by 29 pixels high, whereas the application context pane of the status pane (shown at the top of the screen when an application has focus) uses icons 44 by 44 pixels in size. As illustrated in Figure 2–3, in the application view the **Messages** icon is 42 by 29 pixels, whereas the **Messages** icon at the top of the screen in the status pane is 44 by 44 pixels.

Figure 2–3 Series 60 icon sizes.

Supplying a variety of icon sizes helps to ensure that the system can use the most appropriate size. Supplying only a single icon will result in dynamic scaling when it is drawn at a particular size—scaling small bitmaps generally results in a marked loss of quality.

Aifbuilder can launch an icon designer facility that will generate the bitmaps and masks that make up the icon. AIF icon designer helps you produce such icons in the required Symbian OS-specific bitmap file format, called the multibitmap file format (.mbm).

AIF File Localization

If your application design requires different icons for different languages, you can achieve this by producing multiple copies of the .aif file, each containing the appropriate bitmaps, using the **aiftool** utility. Each localized .aif file produced is usually saved with the extension .aXX, where XX is the two-digit language code associated with the appropriate locale. The application framework software (Apparc) has been modified to attempt to load an .aif file associated with the current system language setting.

MIME Support

Native Symbian OS files—those created by Symbian compliant applications—are recognized by the system through the UID system described earlier, in

Chapter 2   The HelloWorld GUI Application

particular through `UID3`. For nonnative files, MIME type recognition is an excellent alternative. Multipurpose Internet Mail Extensions (**MIME**s) define a file format for transferring nontextual data, such as graphics, audio and fax over the Internet. See **http://www.iana.org/assignments/media-types/** for more details on MIME types.

A Symbian OS application may specify in its `.aif` file any MIME types that it supports and the priority of support that each type is given. For example, you may wish to have your application support the display and editing of plain text files. When a file is to be opened, Symbian OS launches the application that has the highest-priority support for the selected file type.

There are four priority levels (defined by enumeration in `\epoc32\include\aiftool.rh`), of which only `EDataTypePriorityNormal` or `EDataTypePriorityLow` should normally be used.

For example, your text editor is clearly going to be good at editing text/plain files, and hence should be given a priority of `EDataTypePriorityNormal` for that file type. A Web browser is less suited to handling text files and would be assigned the lower priority `EDataTypePriorityLow`. Hence either application can be launched to handle a text document. If both applications are present, however, the text editor is launched by preference. Given two applications with the same MIME type priority, Symbian OS will arbitrarily launch one of the applications.

For more information on `.aif` files and MIME type support, see the Series 60 SDK documentation.

Captions

Avkon, the Series 60 UI and application framework, provides alternative capability to associate captions with an application—functionality that should be used in preference to that available through the caption normally supplied via an `.aif` file. By default, the system will use the caption supplied in the `.aif` file. However, it is possible for application authors to generate a separate caption file, containing both a normal length caption and a short caption—these are provided for use in either the list or grid view of the **Application Launcher**. The short caption is used in application grid view, whereas the normal caption is used in application list view. The caption file is generated in the same way as a normal GUI resource file. For the **HelloWorld** project, the caption source file is `HelloWorld_caption.rss`, and the binary output file, used at runtime, is called `HelloWorld_caption.rsc`. This is placed in the `\system\apps\HelloWorld\` folder with the other application files—for example, `.app`, `.rsc`, `.aif` and so on.

If required, you can generate a separate caption file for each language supported by your application. The convention for naming the caption resource files is as follows:

`NAME_OF_APP_caption.rss`, where `NAME_OF_APP` is the name of the application—in the **HelloWorld** example that would be

`HelloWorld_caption.rss`. The caption resource can be built as part of the normal build process by adding the additional line to an application's `.mmp` file:

`RESOURCE <NAME OF APP>_caption.rss`

The compiled resource file will then appear in:

`\system\apps\appname\appname_caption.rXX`, where `XX` is the two-digit language code specified in the `.mmp` file.

Console Applications

Often during the development of system components, communications modules, application engines and other self-contained functional units you may want a simple test application. Such "test harnesses" allow you to run and exercise your code as it develops.

GUI applications are often not the best option for this purpose; they can be too heavyweight just for testing, particularly so for regression testing (reproducible functional testing). Console applications are often a better option, and they have one key advantage. Since they are `.exe`'s rather than `.app`'s, and therefore do not need the GUI libraries, the emulator and application start up very much faster.

Console applications differ from GUI applications in the following ways:

- They provide text output only—there are no menus, icons, dialogs and so on.

- Console applications are not suitable for deployment onto an end users device—typically there is no way of locating or launching them.

- They must provide their own essential runtime objects, such as a Cleanup Stack and Active Scheduler—UI applications have these provided by the framework classes from which they derive. Note that by examining the code in the console application example (**HelloWorldCon**) you can see how these objects are created.

- Console applications also do not have application information (`.aif`) files or resource (`.rsc`) files.

So, if like many developers you prefer to create and test your code incrementally, the fast start-up speed of a console application is an important consideration.

A key disadvantage of console applications (apart from the simple text interface) is that only one application can be running in the emulator at once. This is because the PC-based Series 60 emulator (**epoc.exe**) is single process,

Chapter 2 Console Applications

and Symbian OS `.exe` applications must run in their own process. Therefore console `.exe`'s are linked with `eexe.obj`, and this causes the emulator to be launched and to run the application as the main process.

The Hello World Console Application

An example console application, **HelloWorldCon**, is provided, and when built, it consists of a single executable file with a filename extension of `.exe`. In Symbian OS, such executables are used for two main purposes; as servers with no user interface or as test harnesses with very simple text-based interfaces.

Note that the project code in the `HelloWorldCon.cpp` file is quite brief and relatively simple. It implements a single function called `doExampleL()`, called from `callExampleL()`, a function defined in a header file `CommonFramework.h`. This header file is provided as part of the Series 60 SDK and is ideal as the basis for any other console-based test harnesses used during general development work.

The main entry point for `.exe` programs is `E32Main()`. This function sets up a memory leak macro, Cleanup Stack, and trap harness for handling any exceptions occurring in the `callExample()` function. This function creates a text console, a console class object derived from `CConsoleBase`, and calls `doExampleL()`. You should place or call your test code from within the `doExampleL()` function. Concepts such as `TRAP` harnesses and the Cleanup Stack are explained in Chapter 3.

Build and Run the Console Application

Open a command prompt and navigate to the folder that contains the project code—for example, `\EMCCSoft\HelloWorldCon`—in the root of your SDK. This folder will contain three files:

- `HelloWorldCon.cpp`   –the source file.
- `HelloWorldCon.mmp`   –the project definition file.
- `bld.inf`   –the component description file.

This project also requires the `CommonFramework.h` header file, provided with the SDK. Depending on your SDK, this will be located in a folder like:

`...\Epoc32Ex\Base\CommonFramework`

or

`...\Epoc32Ex\Basics\CommonFramework.`

You can build and run this example either from the command prompt or from within an IDE. To create **abld.bat**, type:

```
bldmake bldfiles
```

To compile and link the project for the debug emulator, enter:

```
abld build wins udeb      — for Visual C++
abld build winsb udeb     — for Borland C++
abld build winscw udeb    — for CodeWarrior
```

Then, to run the **HelloWorldCon** application, navigate to the folder that contains it—for example, `...\Epoc32\release\wins\udeb` and at the command prompt, type:
`HelloWorldCon`

Alternatively, create the IDE project file (as described in Chapter 1), load and build the project from within the IDE. To run the console application you may have to specify the executable as `HelloWorldCon.exe`. The Series 60 emulator starts up, and the application will appear immediately as shown in Figure 2–4.

Figure 2–4 The console emulator.

Note that **HelloWorldCon** starts up as the only application available. It displays "Hello World" and simply waits for any key press on the PC. Pressing any key will complete the application and close the emulator.

HelloWorldCon.mmp

The project definition (`.mmp`) file for console applications has a few key differences from the equivalent GUI project file as outlined here.

```
TARGET              HelloWorldCon.exe
TARGETTYPE          exe
UID                 0

SOURCEPATH          .
SOURCE              HelloWorldCon.cpp

USERINCLUDE         .
USERINCLUDE         \Epoc32ex\Basics\CommonFramework
SYSTEMINCLUDE       \Epoc32\include


LIBRARY             euser.lib
```

Chapter 2   Console Applications   49

Note the name of the application executable and in particular its extension (`.exe`) as specified in the `TARGET` statement is `HelloWorldCon.exe`. The `TARGETTYPE` is `exe` and the `UID` line is set to zero, because UIDs are not required for `.exe` applications. A value of zero on the `UID` line will suppress a build tools warning. However, at install time supplying a nonzero UID has another very important effect—it prevents one `.exe` replacing a previously installed one that also had a UID of zero—see the section entitled **Format of .pkg Files** later in this chapter for more details.

None of the GUI libraries are required in the `LIBRARY` statement; only `euser.lib` is needed in this very simple example.

HelloWorldCon Emulator Executable Files

Under the Series 60 emulator the **HelloWorldCon** console application consists of a single executable file: `HelloWorldCon.exe`. It is generated in the `\epoc32\release\wins\udeb` folder for debug emulator builds and the `\epoc32\release\wins\urel` folder for release builds (emulator release builds are rarely created in practice). Under the emulator, these locations are essential for the correct execution of `.exe` files.

HelloWorldCon Target Executable Files

The **HelloWorldCon** console application built for a target hardware device consists of a single executable file: `HelloWorldCon.exe`. It is generated, on the PC, in the `\epoc32\release\armi\urel` for an `armi` release build (or `\epoc32\release\thumb\urel` for a `thumb` release build)—the ARM build options for target devices are explained in Chapter 1.

On a target device, `.exe` executables would be located under a `\system\programs` folder by convention, but unlike `.app` programs, they can be run from any location—in theory. The problem is that getting the `.exe` file onto a device, and then locating and launching this type of application on a Series 60 device, may not be straightforward. On most systems the target device's **Messaging** Inbox will recognize the Symbian OS executable (`.exe`) file format in an incoming message attachment, and may refuse to open the message and its attachments for security reasons.

It is not usual to supply `.exe` applications for installation by end users, but as a developer you may have legitimate reasons for wanting to do this. You could supply the `.exe` in a `.sis` file, but Series 60 does not automatically recognize and display an icon for `.exe` applications, so you still have the problem of launching the console application. There are two options:

- Install the executable, and select and launch it via a special utility application.

- Install the executable with an automatic launch by the installation process.

Chapter 2 Development Reference

For the first option a GUI utility application called **Exelauncher** is provided to assist with this problem—it is available as part of the example materials supplied online, as specified in the Preface.

You should package up the `HelloWorldCon.exe` application in an installation (`.sis`) file and then transfer it to the device from your PC via infrared or Bluetooth, followed by the `.sis` file for the **Exelauncher** utility. Read both messages and install both applications. Locate the **Exelauncher** icon and run it. You will then be presented with the option to select and launch the **HelloWorldCon** application. In order for **Exelauncher** to find the `HelloWorldCon.exe` file, it must be installed in `c:\EMCC\Exes\` on the target device.

The second option has the disadvantage that it is a one-time operation, and the installation has to be repeated every time the application is to be run. All that is required to have the application run during the installation process is to specify one additional parameter (`FR`) after the name of the application file (`.exe`) in the `.pkg` installation package file. This process is shown later in this chapter under *Running Executables During Installation*, and the composition and use of `.sis` files is described in detail in the next section.

Everything required for building and running the **HelloWorldCon** project is supplied; including all of the source code, the installation system files, and the **Exelauncher** utility.

Symbian Installation System

The Symbian Installation System (SIS) provides a simple and consistent approach to installing applications, data or configuration information to Symbian OS-based devices. This system is not designed for software installation on an emulator. Developers and end users install components packaged in (`.sis`) files either from a PC, using the installer, or from a Series 60 device, using the Application Controller utility.

Using digital certificates, the Symbian OS software installation system allows users to identify the software vendor, prior to installing software, and hence verify that the installation file has not been tampered with since it was created. This functionality is particularly important within an environment in which there is easy access to a wide range of freely downloadable software—software that might be infected with viruses. The **Certificate Generator** (`makekeys.exe`) can create a public/private-key pair. You may then choose to have the **Installation File Generator** (`makesis.exe`) create a digitally signed installation file. This process is described in detail in the SDK documents—see the section on **Tools and Utilities** and refer to the **Installation Guide** and **Installation Reference** sections.

An installation file can incorporate other embedded installation files. This feature might be used for packaging files into logical components that are

Chapter 2   Symbian Installation System

installed and removed as a complete set. For example, a shared library might be put in a separate installation file from the application files that use it.

A number of options may be offered to the user so that during installation the system will perform different actions depending on user selections. It is possible to install a different set of files, display text or to run custom programs, during the installation and/or removal of an application. The most common use of this facility is to display a license agreement during installation. Users may be prompted about installation options such as where to install the components, to accept a security certificates that verifies the creator or to choose a particular language.

After the installation process completes, a small "stub" .sis file remains on the device to control the uninstallation of the application if required.

Example installation project files are provided with the examples in this chapter, and with the majority of projects associated with this book, to allow the example applications to be installed on a Series 60 device.

SIS File Build Tools

Two tools are provided with the Series 60 SDK to enable developers to build installation files.

- The Installation File Generator (makesis.exe) creates the installation (.sis) files from a specification provided in a package source (.pkg) file. A .pkg file is a text file containing installation information for applications or files. This is the option used in all of the example GUI applications described in this chapter.

- Alternatively, you can produce .sis files using the sisar tool provided with the Series 60 SDK. **Sisar** packages all the application files into one .sis file for ease of installation onto target hardware. It stores the installation project details in a file with a .sisar filename extension. Manually created .pkg text files can be imported into the **sisar** utility, but they cannot be exported again. In addition, **sisar** can import .aifb files generated by the **aifbuilder** tool described earlier.

TIP   **Sisar** is a convenient and easy-to-use tool for creating simple .sis files. However, using a .pkg source file with the makesis.exe utility is, at the time of writing, more flexible. Most importantly, the **makesis** method permits the use of the Series 60 Platform UID code (explained later in this chapter), and **sisar** currently does not. **Makesis** also allows for fine-tuning of the installation options and process.

Another SDK tool, the Certificate Generator (makekeys.exe), can create private/public-key pairs, which you may optionally use, via the Installation File

Generator (`makesis.exe`), to digitally sign installation files. During installation the users will be warned that they are about to install an unsigned application.

Both **makesis** and **makekeys** can be run from the command line or may be invoked from within **sisar**. **Sisar** itself can also be invoked from the command line to build a `.sis` file.

Text source (`.pkg`) files specify the executable code files and other resources needed to install an application onto a device. It is usual to create the required `.pkg` files by hand with a text editor or within an IDE, and then build the `.sis` file using the **makesis** tool.

Using the **sisar** tool is basically self-explanatory, but has its own internal help mechanism and is separately documented in the Series 60 SDK. Note that **sisar** depends on a Java™ 2 Runtime, which is available from **http://java.sun.com**.

Format of `.pkg` Files

There is another version of the **HelloWorld** example project, called **HelloWorldLoc**, which is a localized alternative to show how you can supply an application installation package suitable for several different locales. During installation, appropriate language versions of the resource file, caption file and even the application information file will be used. Hence all the text of the application, the name of the application and even the icons can be tailored to the needs of each supported locale. This subsection looks at the syntax of the nonlocalized `.pkg` file and then points out the essential differences in the localized version.

```
; Languages - Only English so the next line is optional
&EN

; Name/caption needed for each language, caption file used if
supplied
; Name, the app UID, version, minor version, build & package
type
#{"HelloWorld"},(0x101F6148),1,0,0,TYPE=SISAPP

; Platform UID code required for recognition by a compatible
; device at installation time
(0x101F6F88),0,0,0, {"Series60ProductID"}

; Four files to install
"\Epoc32\release\armi\urel\HelloWorld.aif" -
"!:\system\apps\HelloWorld\HelloWorld.aif"

"\Epoc32\release\armi\urel\HelloWorld.r01" -
"!:\system\apps\HelloWorld\HelloWorld.rsc"

"\Epoc32\release\armi\urel\HelloWorld_caption.r01" -
"!:\system\apps\HelloWorld\HelloWorld_caption.rsc"
```

Chapter 2 Symbian Installation System

```
"\Epoc32\release\armi\urel\HelloWorld.app" -
"!:\system\apps\HelloWorld\HelloWorld.app"

; End of File
```

The **HelloWorld** example is in English only. If you wish to support other languages, you must specify each one at the top of the file, and then the syntax of the rest of the file must provide options for each language/locale—this is illustrated later using the **HelloWorldLoc** example.

```
; Languages - Only English so the next line is optional
&EN
```

The package header contains the component name (application name in this case), the UID of the application or component, the major and minor version numbers, the build number and the package options (SISAPP)—whether this is an application installation, upgrade, patch, system option and so on.

```
#{"HelloWorld"},(0x101F6148),1,0,0,TYPE=SISAPP
```

TIP Note that .exe applications do not require a value for the application UID (UID3) in their .mmp and .pkg files—for a test or experimental .exe, the application UID is typically set to zero. However, if you are going to release an .exe (say as a server) then you will need to give it a UID so that it is truly unique. If the same application UID (0) is specified in two different .pkg files, then a previously installed .exe will be removed when installing a new one.

Note that within the package file, it is a requirement that a Series 60 Platform identification code be included as follows.

```
(0x101F6F88),0,0,0, {"Series60ProductID"}
```

The platform identification code (**Platform UID**) enables a built-in system mechanism to issue a warning if a user attempts to install incompatible software onto a Series 60 device. The Platform UID is a 32-bit number assigned to identify a specific version of Series 60 or a particular device—it has no relationship to the three UIDs associated with executables or application documents.

All installation packages of Series 60 applications should carry the sequence in order to facilitate smooth installation of software. If the Platform UID sequence is not found, or if the UID refers to a platform release newer than in the terminal, the user will get a notification about the potential conflict. Depending upon the product, the installation process may be automatically aborted. If it is allowed to continue, there is obviously a risk of application functionality failure.

The identifier for applications that are based on Series 60 Platform first release (0.9) is 0x101F6F88 (for example, early Nokia 7650 devices were based upon this first release of Series 60). Application installation packages specifying this value are compatible with Series 60 Platform 0.9 and potentially with later releases of Series 60 Platform.

A new Series 60 Platform UID was introduced for Series 60 2.0, namely 0x101f7960. You should use this Platform UID in applications that rely on Series 60 2.0 features and APIs. Earlier version Platform UIDs are still supported and they must be used in applications intended to run on Series 60 Platform version 0.9, 1.x, 2.x and so on.

Multiple Series 60 Platform UIDs can be defined in a .pkg file to signify all platforms and devices that an application is able to run on.

SEE ALSO See the **Advanced Application Deployment and Build Guide** section later in this chapter for a more complete list of Platform UIDs.

Finally, each of the application components to be installed is listed—in this case: HelloWorld.aif, HelloWorld.r01, HelloWorld_caption.rsc and HelloWorld.app. Note that the format of each line is the source filename followed by the location and name of the destination file.

Also note that the file locations of the application components (resulting from the build process) on the development PC will vary depending on the version of Series 60 Platform you are working with.

Version 1.x of the SDK places the .app and all of the other files in the relative location: \Epoc32\release\armi\urel.

A version 2.x SDK build (and all CodeWarrior builds) will only put the .app file in that location; any other files (.r01, .aif, .mbm) are placed in the relative location: \Epoc32\data\z\system\apps\appname\.

Using the **AnsPhone** example project as an illustration, the version 1.x .pkg file would be:

```
"\epoc32\release\armi\urel\AnsPhone.aif" -
"!:\system\apps\AnsPhone\AnsPhone.aif"
"\epoc32\release\armi\urel\AnsPhone.app" -
"!:\system\apps\AnsPhone\AnsPhone.app"
"\epoc32\release\armi\urel\AnsPhone.r01" -
"!:\system\apps\AnsPhone\AnsPhone.rsc"
"\epoc32\release\armi\urel\AnsPhone.mbm"  -
"!:\system\apps\AnsPhone\AnsPhone.mbm"
```

An equivalent version 2.x (or CodeWarrior) .pkg file would contain:

```
"\epoc32\data\z\system\apps\AnsPhone\AnsPhone.aif" -
"!:\system\apps\AnsPhone\AnsPhone.aif"
```

Chapter 2 Symbian Installation System

```
"\epoc32\release\armi\urel\AnsPhone.app" -
"!:\system\apps\AnsPhone\AnsPhone.app"
"\epoc32\data\z\system\apps\AnsPhone\AnsPhone.r01" -
"!:\system\apps\AnsPhone\AnsPhone.rsc"
"\epoc32\data\z\system\apps\AnsPhone\AnsPhone.mbm" -
"!:\system\apps\AnsPhone\AnsPhone.mbm"
```

Note that the Platform UID will need to be changed between versions 1.x and 2.x of the `.pkg` file.

Also note that the filename extensions of the source file and the destination file in the `.pkg` script are not always the same. The reason is that there may be a choice of alternative source files, and this is illustrated in the **HelloWorldLoc** localized example later in this section.

To give the user the choice of installing the application on any drive on the device (for example, on a removable memory card instead of flash memory), the drive letter is specified by a "!" character, as illustrated in the example given here. During installation the user will be prompted with a list of alternative drives for storing the application.

> **TIP** To ensure that the application always appears with the correct icon, arrange the order of the files as shown in the example provided—specifically, place the name of the `.aif` file before the `.app` file.

Running Executables During Installation

To have the **HelloWorld** GUI application run during the installation process an additional parameter is required immediately following the application filename as follows:

```
"\Epoc32\release\armi\urel\HelloWorld.app" -
"!:\system\apps\HelloWorld\HelloWorld.app", FR
```

Or in the case of the **HelloWorldCon** console application:

```
"\Epoc32\release\thumb\urel\HelloWorldCon.exe"
-"c:\EMCC\Exes\HelloWorldCon.exe", FR
```

Make sure that all of the other application component files are installed before the application in the `.pkg` file.

The "`FR`" parameter means "File Run". During the installation process, do not be fooled into thinking that the application has shut down again—the application is left running; it is simply that the installation process starts the application up, then tasks back to itself to complete the rest of the installation process. Example "file run" installation package (`.pkg`) files are provided for each of the example projects. They are located in the `\install` folder, under the main project folder, with the filename ending in `FR`—for example, `HelloWorldConFR.pkg`.

Chapter 2 Development Reference

Multi-Locale Installation

The **HelloWorldLoc** version of the example application supports three alternative locales: English, French and German. You will find the project located under: `\EMCCSoft\HelloWorldLoc\` — relative to the root level of the Series 60 SDK.

Here are sections from the `HelloWorldLoc.pkg` file that are relevant to the localization of the application installation:

```
; Languages - English, French and German
&EN,FR,GE
```

Note here how the supported languages are specified—as `EN`, `FR` and `GE`, unlike the method used in the `LANG` statement in the project `.MMP` file—as `01`, `02` and `03`. It seems that this discrepancy in identifying the supported languages is simply due to a historical oddity. The other language codes for `.pkg` files are documented in the SDK documentation under "Package file format".

The package header must contain the component name in each of the supported languages:

```
; Name/caption for each language, caption file is used if
supplied
#{"HelloWorldLoc","BonjourMondeLoc","HalloWeltLoc"},(0x101F614
9),1,0,0,TYPE=SISAPP
```

The package (`.pkg`) file syntax forces you to duplicate the `Series60ProductID` string for each language:

```
(0x101F6F88), 0, 0, 0,
{"Series60ProductID","Series60ProductID","Series60ProductID"}
```

Next, a list of three locale-specific `.aif` files is supplied, of which only one will be installed:

```
{
"\Epoc32\release\armi\urel\HelloWorldLoc.a01"
"\Epoc32\release\armi\urel\HelloWorldLoc.a02"
"\Epoc32\release\armi\urel\HelloWorldLoc.a03"
}-"!:\system\apps\HelloWorldLoc\HelloWorldLoc.aif"
```

Now the name of the application to be installed is specified. Other application components can each have their own line, or components can be installed using embedded installation (`.sis`) files. See *Multicomponent Installation* later for more details.

```
"\Epoc32\release\armi\urel\HelloWorldLoc.app"        -
"!:\system\apps\HelloWorldLoc\HelloWorldLoc.app"
```

Next, a list of three locale-specific *resource* files are supplied, of which only one will be installed:

Chapter 2 Symbian Installation System

```
{
"\Epoc32\release\armi\urel\HelloWorldLoc.r01"
"\Epoc32\release\armi\urel\HelloWorldLoc.r02"
"\Epoc32\release\armi\urel\HelloWorldLoc.r03"
}-"!:\system\apps\HelloWorldLoc\HelloWorldLoc.rsc"
```

Finally, a list of three locale-specific *caption resource* files are supplied, of which, again, only one will be installed:

```
{
"\Epoc32\release\armi\urel\HelloWorldLoc_caption.r01"
"\Epoc32\release\armi\urel\HelloWorldLoc_caption.r02"
"\Epoc32\release\armi\urel\HelloWorldLoc_caption.r03"
}-"!:\system\apps\HelloWorldLoc\HelloWorldLoc_caption.rsc"
; End of File
```

Multicomponent Installation

Sometimes it is better to handle the installation of components needed by an application using embedded installation (`.sis`) files. This is easily achieved as follows:

```
; Embedded SIS file for components e.g. DLLs
@"..\..\Engine\group\Engine.sis", (0x101F6150)
@"..\..\Controls\group\Emccctrls.sis", (0x101F6151)
```

Note that each line that specifies the name of an embedded `.sis` file begins with an @ symbol. The path and filename for the `.sis` file is provided—relative to the current application location—inside quotation marks. Following the filename and path, is the `UID3` for the application or component that is to be included as part of this main installation package.

Each component will either be a separate project or will form part of the main project, and each will require a `.pkg` (or `.sisar`) file of its own to generate the corresponding embedded `.sis` file. If one or more embedded `.sis` files are specified, to be included as part of the installation of the application, the components will be removed when the application or component within which they were embedded is removed (provided no other component has a dependency upon them).

Conditional Component Installation

Occasionally it may be necessary to decide during the installation process itself which components are needed or are appropriate. Conditional statements can be incorporated into an installation script file and how to do this is illustrated later in the chapter—see the subsection titled *Device Identification at Install Time* under the **Advanced Application Deployment and Build Guide** section.

Building a SIS File

To build the `HelloWorld.sis` file, open a command prompt, navigate to the folder for the **HelloWorld** project and build the project as follows:

```
bldmake bldfiles
abld build armi urel
```

Then change to the `\install` folder. To build the `.sis` file enter:

```
makesis helloworld.pkg
```

Or to build the auto-running version:

```
makesis helloworldfr.pkg
```

Note that the Borland C++BuilderX IDE can generate the `.sis` file during the build process if a suitable `.pkg` file is included in the project. Select the **Project|Properties** menu item, then select the **Symbian settings** tab. Set the platform to a suitable target device and the build to `urel`, and the `.sis` file will be generated when the project is built.

To run the application on a target device, transfer the `.sis` file to the device using an infrared or Bluetooth connection.

Additional Development Tools

Symbian OS SDKs include a number of tools and utilities; Series 60 SDKs provide even more tools. EMCC Software Ltd has also placed additional tools and utilities into the public domain. These extra tools are listed here because they are used frequently, are particularly helpful, are important or are not very well documented elsewhere. References on how to obtain the additional tools are also provided.

Multi-Bitmaps and Bmconv the Bitmap Converter

Bitmaps provide the pixel patterns used by pictures, icons and masks, sprites and brush styles for filling areas of the display. To optimize bitmap performance, Symbian OS uses files containing multiple bitmaps in its own highly compressed format. A tool is provided (**bmconv.exe**) with the Series 60 SDK that takes one or more Windows bitmap (`.bmp`) files as input and generates a single Symbian OS multi-bitmap file (.mbm), optimized for efficient runtime loading.

TIP To see the **bmconv** syntax and options simply type **bmconv** at a command prompt.

The **bmconv** tool also generates a header file (.mbg) with symbolic definitions (identifiers) for each bitmap in the file. This generated header file is placed in the \epoc32\include folder of the SDK. It is intended for inclusion (#include) in your C++ code to allow you to reference any of the individual bitmaps as required.

Two types of .mbm files are possible, file store bitmap files that are loaded into RAM and ROM image bitmap files that do not use any RAM when being accessed — only system developers normally create ROM-based .mbm files.

During the conversion process it is possible to specify the number of bits per pixel for the converted bitmaps and whether they should be color or grayscale. The program can also split Symbian OS multi-bitmap (.mbm) files back into component bitmap files in Windows bitmap (.bmp) format.

You do not always have to use the **bmconv** tool directly to produce bitmap files that are dedicated to a particular application. Building the required bitmaps can be performed as part of the standard **abld** project building process. A list of .bmp files can be specified in the project definition (.mmp) file. The next section shows an example of the START BITMAP statement in the project definition (.mmp) file syntax. See the Series 60 SDK documentation for further details of the START BITMAP syntax.

Multi-Bitmap Files, Colors and Palette Support

The hardware color capabilities of Series 60 devices may differ from product to product. Typically, the display is capable of producing a minimum of 4096 colors (12-bit) for Series 60 1.x, and up to 64K colors (16-bit) for Series 60 2.x. All of these color bits may be used to provide the best possible reproduction, for example, when displaying a full-color JPEG photograph.

It is common to specify the generation of a Symbian OS multi-bitmap (.mbm) file and set the target color modes in the application's .mmp file as illustrated below:

```
START BITMAP myapp.mbm
HEADER
TARGETPATH   \system\apps\MyApp
SOURCEPATH   ..\Data
SOURCE c8    MyPict.bmp         // Color256
SOURCE c8    MyOtherPict.bmp    // Color256
SOURCE 1     MyOtherPict_Mask.bmp // Black&White mask
END
```

When the project is rebuilt completely, typically from the command prompt, **bmconv** will generate the .mbm file. It can be invoked directly from the command line if preferred.

It is possible to specify 12-bit color (4096 or Color4k) or 16-bit color (Color64k), however, there is a palette with 256 color (8-bit) entries that is used for drawing user interface graphics, such as icons and window borders. Using this palette saves memory space while also ensuring a consistent look for the graphics from product to product.

For Series 60 SDKs the **bmconv** utility has been changed so that the caller can optionally specify a target palette file as parameter for use in the conversion. Bitmaps with a target color mode other than `Color256` are not affected. Bitmaps with a target color mode of `Color256` are converted using the Series 60 palette—in other words, the 216 colors in the color cube, and 10 gray shades. The palette file used as a **bmconv** parameter can be found under the root level of your SDK at: `\epoc32\include\ThirdPartyBitmap.pal`.

The color indices in the palette are divided into the following sections:

- **Static colors**—This section contains 216 colors commonly known as web-safe colors, and an additional 10 grays, for a total of 226 indices. These colors are constant, so they stay the same regardless of the color scheme in use.

- **Scheme colors**—Twenty-three indices are reserved for use in color schemes. The RGB values in this part of the palette can be switched so that the user can change the overall UI color scheme, for example, from blue to green. These indices are used in those parts of the UI graphics that should follow the color scheme.

- **Notification colors**—These four colors are to be used in the respective notes (confirmation, information, warning and error notes).

- **Series 60-specific coding identification colors**—It is useful to recognize a bitmap that is intended for use with the Series 60 color palette and that supports the color-schemes feature. This can be achieved by checking the RGB values of certain palette indices. If the values match the predefined values, the image is identified as Series 60 specific, and the palette index values in the product will be used to render the image instead of the image's own palette values. If the palette entries of the identification color indices do not match (or if a different bitmap format than indexed noncompressed Windows bitmap is used), the RGB values of the image will be used.

- **Masking color**—One index (255) is reserved for masking color. The pixels of this index value are rendered as transparent in the mask.

Series 60 Application Wizards

A number of example projects have been provided throughout the book, and these can be used as the basis of any of your own new projects and applications.

Alternatively, you can use the wizard facilities provided by the various IDEs to create a project for a basic Series 60 application.

Microsoft Visual C++ 6.0

If you are using Microsoft Visual C++ as your IDE, you can create new projects via the Series 60 Application Wizard provided as part of the SDK. It offers you a simple and convenient way to generate a basic application framework from within the Visual C++ IDE. The wizard will generate a project based on one of three alternative UI display designs: the dialog-based, Avkon view-switching or traditional Symbian OS control-based application architectures. It will create:

- Skeleton code and declarations for the four basic classes (App, AppUI, Document and View) associated with most Symbian OS applications.
- All the build and project files necessary to build the project.
- Simple resource files and bitmaps that are used as default application icons.
- The files needed to install the application onto a device.

Using the wizard is very easy and is well documented in the Series 60 SDK, so its use is not covered here in great detail. However, it does present you with options to create projects conforming to one of three basic application designs. Each of these is covered in Chapter 4, so it is recommended that you read that chapter before using the wizard.

Borland C++BuilderX

As described in Chapter 1, you can open an existing Series 60 project file by importing its `bld.inf` file into the IDE. This is done through the **Object Gallery**, invoked by selecting **File|New** from the menu and then choosing the **Import Symbian C++ Project** option. Other alternatives available at the time of writing are:

- New Symbian GUI Application
- New Symbian Empty Project
- New Symbian DLL
- New Symbian C++ File
- New Symbian .H File
- New Symbian Resource File
- New Symbian AIF wizard

Selecting **New Symbian GUI Application** will create the basic framework for a Symbian project. This, however, is currently based on the Uikon `CEik-` base classes, not the `CAkn-` base classes required by a Series 60 GUI application.

Metrowerks CodeWarrior

CodeWarrior can create new projects based on supplied template projects (called stationery). To use this, choose **File|New...** and select **Symbian Stationery Wizard** from the **Project** tab. Give the project a name and set the location—use the **Set** option to browse. Choose a vendor and the specific SDK version you are working with—press **Next** and select the specific type of project you want from a drop-down tree under the C++ check box. A wizard then will create a basic Series 60 project for you with source and header files.

It is possible to create your own stationery. It is also expected that some Licensees may provide device-specific stationery as part of Licensee Development Kits and SDKs.

Miscellaneous SDK Tools

Series 60 SDKs provide a number of additional tools and utilities as a convenience to developers. Look in the \Series60Tools folder under the root of your SDK for up-to-the-minute availability, details of installation and correct usage.

EPOCSwitch

EpocSwitch sits in your PC system tray and allows you to select the SDK you want to use (by right-clicking the icon). You can also double-click the icon or select a command prompt from the menu to give you a command prompt with the environment set up correctly. Unfortunately it is not compatible with the new approach to SDK configuration management adopted by Symbian OS 7.0s—see **EnvironmentSwitch**.

EnvironmentSwitch

EnvironmentSwitch is available from Series 60 2.0. It replaces **EpocSwitch**, offering the same functionality but being fully compatible with both the latest SDKs and earlier versions. As with **EpocSwitch**, it sits in the system tray and is activated by right-clicking.

EPOCToolbar

This makes available a Symbian OS-specific toolbar—but only in the Microsoft Visual C++ IDE. The six icons give access to six convenient utilities.

MBMViewer

This allows you conveniently to browse the bitmaps/icons embedded in Symbian OS .mbm files on your PC.

MMPClick

Allows you to right-click on an `.mmp` file and choose from a context menu the target platform you want to build for (`armi urel`, `wins udeb`, `wins urel`), or else create the Microsoft Visual C++ workspace and project files.

SMS Inbox

This is an emulator application that you can leave running in the background while you are testing the process of sending/receiving SMS messages from within your application. When an SMS is sent, it is copied to a `\smsout` folder. Received SMS messages are copied into a `\smsin` folder. The usual SMS viewer and editor can be used for editing the SMS messages.

More Tools and Utilities

Customizing Microsoft Visual C++ – Syntax Highlighting

Syntax highlighting for Symbian OS and other user-defined keywords can be enabled using file called `usertype.dat`. User-defined keywords can be set to a color different from that of C++ keywords, making Symbian OS keywords easier to spot. It works for Visual C++ 6.0—see **clindex** below for how to keep the keywords file right up to date. `Usertype.dat` and instructions on its use are available from **http://www.emccsoft.com**.

Customizing Microsoft Visual C++ – Symbian OS Variable Expansion

During a debugging session it is useful to be able to view the contents of a variable in a Watch window. However, Visual C++ is not always able to usefully display user-defined types. Fortunately, you can customize Visual C++, using a file called `autoexp.dat`, so that complex new data types are properly expanded in Watch windows and datatips. `Autoexp.dat` and instructions on its use are available from **http://www.emccsoft.com**.

Clindex

To find out which objects are supplied by which library (and hence list in your `.mmp` file) you can check out the SDK documentation or use the **clindex** utility from EMCC Software Ltd. Run over your Microsoft Visual C++ SDK installation, it produces an up-to-date ordered listing of all objects and their corresponding libraries. In addition **clindex** has an option to merge all found class names into your `usertype.dat` file. **clindex** and instructions on its use are available from **http://www.emccsoft.com**.

Installation Tips for Series 60 SDKs and IDEs

From Series 60 Platform 2.0, SDK installation is managed through a dedicated Package Manager application. Installation and use of multiple SDKs is easier via new tools. EPOCROOT is set automatically, based on the default device setting and any specified overrides. To maintain backward compatibility EPOCROOT can be set manually by the user, and it will not be overridden by the **devices** command.

Microsoft Visual Studio .NET

The Visual C++ 6.0 IDE is no longer supported by Microsoft. If you wish to use Microsoft Visual Studio .NET with Series 60 SDKs, check out **FAQ-0835** on the Symbian Developer Network Knowledge base (**http://www3.symbian.com/faq.nsf**) for assistance with setting it up correctly—use the search option and enter **FAQ-0835** as the search subject.

Emulator Configuration

The Series 60 emulator is used to test applications under development as well as to interact with application data and other content in the Series 60 environment. The emulator provides the functionality and generic look and feel of Series 60 in a PC environment, so testing the application logic, behavior and complete user experience can be easily performed while developing the software, without the need to have real physical Series 60 devices.

The emulator provides a full target environment, which runs under Microsoft Windows on a PC. This implementation of the platform is known as the WINS (WINdows Single process) platform. Applications run as separate threads within a single process under the PC emulator, whereas on target hardware each application runs in its own process, typically protected from one another by memory management hardware.

The emulator enables development to be substantially PC based. It has excellent debugging support; when a thread is panicked, the debugger provides good information about the panic. Only the final development stages focus on the target hardware. Typical exceptions to this rule are for low-level programming, where target hardware must be directly accessed, such as programming for a physical device driver and where the connectivity/communications requirements are not available via an emulator alone.

The emulator is very flexible and highly configurable, but the options available will vary slightly depending on the version of the platform you are using. The options are summarized in Table 2–4. The information resources referred to in the right-hand column are listed immediately after the table.

Chapter 2 Installation Tips for Series 60 SDKs and IDEs

Table 2–4 Emulator Configuration Options and Information Sources

Configuration Option	Information Resources
Mapping of the Symbian OS file system onto the PC including: Symbian OS drive mapping to PC folder assignments, location of executable programs and data files, emulation of additional (virtual) drives—for example, to simulate a Media card acting as a removable drive.	SDK01
Use the PC's mouse, screen and sound facilities as if they were the real input and output devices of a target machine.	SDK01
Emulating drives using the configuration file `epoc.ini`.	SDK02, FN01
Emulator screen size, grayscale/color depth, fascia bitmaps—configuration using initialization (`epoc.ini`) scripts to specify screen size.	SDK02, SDK03, FN01
Configuring emulator memory capacity. The maximum total heap size can be set in the initialization file `epoc.ini`. This allows you to emulate target machines with different memory capacities.	SDK02, SDK03, FN01
Configuring the emulator the key map. The `KeyMap` keyword maps real keys on the device to the PC keyboard and areas of the facia bitmap—as defined in `epoc.ini`.	SDK02, SDK03, FN01
Virtual keys are represented as rectangular regions of the screen. The `VirtualKey` keyword allows developers to emulate target machine keys drawn on the fascia bitmap.	SDL02, SDK03
Specifying aliases for device keys—`DefineKeyName` defines aliases for device key codes. This allows device-specific key names to be used in the `epoc.ini` configuration file. The key codes used are defined in `\epoc32\include\e32keys.h` file in the root of your SDK.	SDK02
In debug builds, the emulator GUI provides special key combinations for resource checking and redraw testing, which can help you to trap memory leaks early in the development cycle. All debug keys are accessed using a **Ctrl+Alt+Shift** prefix.	SDK01

Chapter 2 Development Reference

Table 2-4 continued

Configuration Option	Information Resources
Internet applications can be tested on the emulator by accessing peripheral devices attached to the PC. On a real device, the Internet applications use dial-up networking or GPRS to connect to an Internet Service Provider (ISP). To use the Internet programs on the emulator, you need a PC with a modem, an ISP account and SMTP/POP3 email accounts.	SDK01, FN01
Internet access is possible using a data-enabled mobile phone with an infrared port and IrDA COM-port serial adapter. The recommended adapter is the Extended Systems, Inc. Jeteye ESI-9680.	SDK01, FN01
Integration with the Windows runtime environment providing support for Comms, Internet, IrDA and Bluetooth through the host PC.	FN01
CommDB—systemwide storage for communications-related settings.	SDK05, FN01
RAS (Remote Access Server) settings.	FN01, SDK04
Which telephony hardware (phones) can be used for testing the IP connectivity of the OS? Recommended phones are listed here.	SDN01
EPOCROOT—how to configure the environment variable.	SDK01
How to configure emulator communications settings. Settings syntax for `setupcomms.bat` when used to configure modem and Internet connection settings.	SDK01, SDK03
Speeding up the Series 60 emulator.	SDN02, SND03

Resources

When searching the SDK documentation (the `.chm` files), a literal search means to put the search string inside quotation marks and exactly as specified. Depending on the version of the SDK documentation you have, the search string may be slightly different.

To find the information referenced in the Symbian Developer Network Knowledge Base use the FAQ number in the Knowledge Base search facility—it can be found at **http://www3.symbian.com/faq.nsf**.

Chapter 2 Advanced Application Deployment and Build Guide 67

SDK01—Series 60 SDK documentation—do a literal search for "Emulator Guide".

SDK02—Series 60 SDK documentation—do a literal search for "Configuring the Emulator" or "Emulator configuration".

SDK03—Series 60 SDK documentation—do a literal search for "Comms settings syntax".

SDK04—Series 60 SDK documentation—do a literal search for "Enabling RAS for the emulator".

SDK05—Series 60 SDK documentation—do a literal search for "CommDb Overview".

FN01—"Series 60 Emulator Configuration" in the Series 60 section of the Forum Nokia Web site.

SDN01—Symbian Developer Network Knowledge Base FAQ-0893.

SDN02—Symbian Developer Network Knowledge Base FAQ-0867.

SDN03—Symbian Developer Network Knowledge Base FAQ-0713.

Application Panics in the Emulator

Symbian OS uses the term "panic" for the abnormal termination of an application or system thread. Panics may be created by the system or by the application itself when an unrecoverable error has occurred. From Series 60 SDK 1.2 onward the emulator only displays the message `Program Closed` when an application panics without the usual panic number and information text. To display the details such as the panic number and a text context message as per earlier versions of the platform, you need to create an empty file called `ErrRd` in the folder `\epoc32\wins\c\system\bootdata`. If you do this, any panics that occur will be handled in the normal way.

Advanced Application Deployment and Build Guide

A miscellaneous set of more advanced build and deployment guidelines is provided here. Some are relevant to working with the emulator, some are for target devices (including some crucial differences between emulator and target hardware). Some information is more relevant to software installation and distribution.

Platform UIDs

Platform UIDs are an essential part of the definition of a Symbian Installation System (`.sis`) package file (`.pkg`)—they must be included before a Series 60 application is readied for external distribution. Table 2–5 provides a list of current Platform UIDs you can use in your `.pkg` file to specify the target Series 60 Platforms that are compatible with your application:

Table 2–5 Platform UIDs for Series 60 Platform Versions and Devices

Platform or Product	Platform UID
Series 60 0.9	0x101F6F88
Series 60 1.0	0x101F795F
Series 60 1.2	0x101F8202
Series 60 2.0	0x101F7960
Nokia 7650	0x101F6F87
Nokia 3650	0x101F7962
Nokia N-Gage	0x101F8A64
Nokia 6600	0x101F7963
Siemens SX1	0x101F9071

It is generally preferable to use the Series 60 Platform UIDs (rather than target device-specific ones) unless your application is only targeted at a very specific phone.

Multiple Series 60 Platform UIDs can be defined in a `.pkg` file to signify all platforms and devices that an application is able to run on. Use the widest possible number of Platform UIDs for maximum compatibility, but use discretion. Do not use the Series 60 2.x Platform UID unless you have built with the Series 60 2.x SDK.

For a fuller explanation of this requirement, and for up-to-the-minute implementation details, see **http://www.forum.nokia.com**—the relevant document is entitled **Series 60 Platform Identification Code**.

Device Identification UIDs

All Series 60 devices can return a unique identifier UID assigned by the manufacturer. Note that this is not the same as the Platform UID.

Occasionally it may be useful to determine the exact identity of the device your application is being installed on during the process, or at runtime. Both requirements are possible and are described in the next two sections.

A partial list of device UIDs is provided in Table 2–6; however, you may need to refer to the licensee for the UID of a specific device.

Chapter 2 Advanced Application Deployment and Build Guide

Table 2–6 Machine UIDs

Product	Machine UID
Nokia 7650	0x101F4FC3
Nokia 3650	0x101F466A
Nokia N-Gage	0x101F8C19
Nokia 6600	0x101FB3DD
Siemens SX1	0x101F9071
Sendo X	0x101FA031
Series 60 1.2 emulator	0x10005F62

Device Identification at Install Time

Determining the exact identity of the device your application is being installed on during the process has a number of potential benefits. For example, it allows decisions about which variants of particular components or files are needed for the specific device. Here is an example section from a fictitious installation package script (.pkg) file that shows how you could achieve this conditional installation processing.

```
IF MachineUID=0x101F4FC3; install Nokia 7650 specific files
"\Epoc32\release\armi\urel\N7650.dat" -
"!:\system\apps\ExampleApp\N7650.dat"

ELSEIF MachineUID=0x101F466A ; install Nokia 3650 specific
files
"\Epoc32\release\armi\urel\N3650.dat" -
"!:\system\apps\ExampleApp\N3650.dat"

ELSEIF MachineUID=0x101FB3DD ; install Nokia 6600 specific
files
"\Epoc32\release\armi\urel\N6600.dat" -
"!:\system\apps\ExampleApp\N6600.dat"

ELSE ; install option for all other devices
"\Epoc32\release\armi\urel\Default.dat" -
"!:\system\apps\ExampleApp\Default.dat"
ENDIF
```

Device Identification During Execution

It may be useful to get the exact identity of the device an application is executing on at runtime—typically you might want to do this early on during the application's initialization. This is possible, and a Symbian OS utility API can provide the device UID—see Chapter 12 and the section titled **Hardware Abstraction Layer** for some example code.

Resource File Versions and Compression

Symbian OS 7.0s introduced a resource file compression algorithm that was not present in generic Symbian OS 6.1. However, Series 60 versions prior to version 2.0 implemented its own resource file compression scheme—a different scheme from that in Symbian OS 7.0s.

Series 60 2.x compressed resource files are not backward compatible with Series 60 1.x.

However, to maintain a degree of forward compatibility, Series 60 1.x resource, .aif, .sis, and bitmap (.mbm) files can be used with devices based on Series 60 2.x.

Building for ARM Targets

Building for ARM will in general be more difficult than building for the emulator (WINS), and it is normal to find additional compiler errors and warnings from **gcc** on the first attempt. This is because **gcc** is, in general, stricter than, say the Microsoft compiler, and also it has some subtle differences that will come out the first time an ARM build is attempted. The following covers a few of the most common pitfalls.

Function Exports

The **gcc** tool-chain is stricter than WINS builds when it comes to specifying exported functions. The correct way to export a function from a DLL is as follows:

In the header (.h) file:

```
class CMyClass : public CBase
    {
IMPORT_C void Function();
    }
```

and then in the source (.cpp) file:

```
EXPORT_C void CMyClass::Function()
    {
    }
```

The `WINS` tool chain does not mind if the `EXPORT_C` is excluded from the `CPP` file; it exports the function anyway. However, the **gcc** tool chain requires the `IMPORT_C` and `EXPORT_C` to be perfectly matched. If they are not, the function will not be exported from the DLL, which will eventually lead to errors such as "`Cannot Find Function`" when attempting to link to this DLL.

Writable Static Data in DLLs

Other compilation differences include errors such as the "`The MyDll.DLL has (un)initialized data`" error arising from the build tool **petran**. The petran tool strips a PE format file (Win32 Portable Executable file format) of its irrelevant symbol information for an ARM target; thus making DLLs much smaller. As a consequence, ARM targets only support linking by ordinal. **Petran** is also responsible for adding UID information into the headers of executable files.

The Symbian OS architecture does not allow DLLs to have a data segment (static data, either initialized or uninitialized). There are fundamental problems in deciding whether a data segment should be used:

- Do all users of the DLL share it?
- Should it be copied for each process that attaches to the DLL?
- There are significant runtime overheads in implementing any of the possible answers.

However, as the `WINS` emulator uses the underlying Windows DLL mechanisms, it can provide per-process DLL data using "copy-on-write" semantics. This is why the problem goes undetected until the code is built for an ARM-based Symbian OS device.

The associated `.map` file generated during the build process contains information that helps to track down the source file involved. Look in `..\epoc32\release\armi\urel\dllname.map` and search for "`.data`" or "`.bss`". See "Coding Idioms for Series 60" on the Series 60 section of Forum Nokia for more detailed coverage of this topic.

Building and Freezing DLLs

When a DLL is loaded, it supplies a table of addresses, one for each exported symbol and one for the entry point of each exported function. This is the DLL's public interface, and DLLs should freeze their exports before release, so as to ensure the backward compatibility of new releases of a library. This is termed maintaining binary compatibility (**BC**), and the index of each export must remain constant from one release to another.

While you are developing a DLL, you can use the EXPORTUNFROZEN keyword in the .mmp file for the project to tell the build process that exports are not yet frozen. When you are ready to freeze, you must remove the EXPORTUNFROZEN keyword from the .mmp and supply a .def file listing the exports.

Symbian OS uses export definitions (.def) files to manage this requirement. Each exported symbol is listed in the exports section of the file with its ordinal number—ordinals start at 1.

The first time a build is done, a warning may be generated to say that the frozen .def file does not yet exist. Once the project has been completed, and has been built in its release form, you can freeze it with abld by using:

```
abld freeze
```

This method will create the frozen .def file containing the project's exported functions. Then, to build an application against the newly frozen interface, enter the following command:

```
abld build wins udeb          —for Visual C++
abld build winsb udeb         —for Borland C++
abld build winscw udeb        —for CodeWarrior
```

To maintain BC, every export defined in an earlier release must be defined in the new release. Any ordinals for new exports introduced in a new release must come after those defined in earlier releases.

For DLL builds, the command-line tools automatically create the .def file within the build tree for the specified target. Once these have been generated for a build, they can be archived with the project source and used in future builds to freeze the exports against change. This is done by copying the .def files into a default location and including the directive:

```
DEFFILE projectname.def
```

into the project .mmp file.

Normally only command-line builds should be released—note that some of the IDEs now support freezing of exports. In any subsequent command-line build of the project the exports will be guaranteed compatible with the current version.

If new exports are added, the new .def files should be copied from the build directory and archived with the new release.

Note that all ARM platforms share a common .def file, but WINS/WINSCW/WINSB have different .def files.

Summary

This chapter has looked in detail at how to select and use a specific SDK and how to specify a Series 60 project in a manner that allows it to be built for an emulator or a target device. It has given details on the process of using project files to enable application development using various IDEs. Also, it has described how to build a project from a command prompt, for both the emulator and target hardware.

A detailed description was provided of the source and other files that make up a typical project—acting as the inputs to build tools—and also of the files that form the outputs of that process. The build process and output files were demonstrated using a GUI application, a localized GUI application and also a console application example. The chapter also defined application resources and covered the process of defining and building localized applications.

Deploying applications on Series 60 devices was described through a detailed coverage of the Symbian Installation System, including how to have applications start on installation.

By now you should have gained a good knowledge of all of the main build and deployment tools that are used regularly during Series 60 development. In addition, we have examined some of the less-common tools supplied with Series 60 SDKs, and the numerous configuration options available for the Series 60 emulator.

chapter 3

Symbian OS Fundamentals

Covering the essential aspects of Symbian OS, upon which Series 60 is based

Chapter 3 Symbian OS Fundamentals

This chapter introduces the basic concepts, APIs, and data structures that you *need* to *understand* in order to develop for Symbian OS and therefore Series 60. All of the topics discussed in this chapter will be used throughout the book, usually without further explanation. It is therefore *essential* that you understand them fully before proceeding further.

Symbian OS is written almost entirely in C++, but developing applications for it is not trivial, and so this chapter requires the reader to have a solid background in advanced use of C++ and object-oriented design. Readers with previous Symbian OS development experience may wish to skip this chapter, as no Series 60-specific elements will be covered here. However, it may still prove to be a useful refresher course.

Although Symbian OS provides part of the User Interface (**UI**) implementation of the platform, Series 60 provides many of the concrete UI controls—so UI features will not be covered here. In fact the contents of this chapter could easily apply to any platform that is based on Symbian OS.

Topics covered are:

- **Naming Conventions**—Symbian OS uses naming conventions to increase readability and consistency. Following these conventions allows other Symbian OS developers to more easily follow your code and also allows for some automatic verification of code.

- **Basic Types**—How Symbian OS manages basic variable types such as integers and floating-point numbers.

- **Exception Handling and Resource Management**—Resource management is a key element when developing applications for devices that have small amounts of available memory and are infrequently rebooted. Symbian OS has its own efficient paradigms for handling resource allocation and deallocation. This also impacts object construction and exception handling.

- **Descriptors**—Symbian OS defines its own efficient types for handling text strings, in both Unicode and narrow (8-bit) formats. Descriptors can also be used for storing arbitrary binary data.

- **Collection Classes**—Symbian OS does not support the Standard Template Library (STL). Instead, it provides a set of small, efficient collection classes, the most commonly used of which are covered here.

Chapter 3 Naming Conventions

- **Using Asynchronous Services with Active Objects**—it is uncommon for Symbian OS applications to use multiple threads. Instead, a cooperative multitasking paradigm is used through the **Active Object** API. This section will cover why and how Active Objects are used to interact with the asynchronous services provided by Symbian OS servers.

- **Files, Streams and Stores**—This section covers how to use the Symbian OS file system and how to read and write to files. **Streams** and **Stores** provide an elegant and efficient way to persist any data.

- **Client/Server Architecture**—Symbian OS uses Client/Server architecture throughout in order to provide safe, efficient, extensible and typically asynchronous services to multiple client applications. This section will focus on client-side interaction with Symbian OS Servers.

Each of these fundamental topics builds on information given earlier in the chapter, so you are advised to approach each topic in turn: The naming conventions and basic types are used throughout; descriptors and collection classes require resource management; file handling and Client/Server architecture use Active Objects, and so on.

All of the topics covered in this chapter are illustrated using code snippets, mostly taken from an example application called **Elements**, and the details of how to download the full buildable source of this, along with the other example applications, are given in the Preface.

For the sake of simplicity, this example is implemented as a console application—in other words it has no Graphical User Interface (**GUI**). Using a console application allows the focus to be placed on the underlying principles of the topics being explained, rather than having to examine the Series 60 UI framework at this early stage.

The **Elements** application is a contrived example that stores and manipulates lists of chemical elements. The elements can be loaded in from a standard comma-separated value (**CSV**) text file, or from a Symbian OS stream-based binary file (this will be explained further in the section **Files, Streams and Stores**). Once loaded, the list can then be sorted, searched and written out to the console display, or back to a file.

Note that lines of code are sometimes omitted for clarity in the book listings, and while it is unusual for a console application to be run on a real Series 60 device, details are provided with the full source to allow you to run the application on target hardware.

Naming Conventions

Naming conventions are used throughout Symbian OS for a number of reasons. Most significantly, they provide important clues to the developer as to how a class, function or variable should be used. Consistent use of naming conventions also allows the use of automatic verification tools to check for coding errors in your code. Such tools will be covered further in Chapter 13.

This section introduces naming conventions that relate to classes and data. Other naming conventions that relate to methods will be introduced as you meet them later in the chapter, such as those in the **Exception Handling and Resource Management** section. Further information on all of the naming conventions used within Symbian OS can be found in the SDK documentation.

Class names consist of a single letter prefixed to the class name: `T`, `S`, `C`, `R`, `M` or `N`. The remainder of the class name should provide a clear indication of what the class does, with words delimited by capitalizing the first letter of each—for example, `CActiveScheduler`, `TRect`, and `RArray`. Note that classes that consist solely of static member functions have no prefix letter, such as `User`, `StringLoader`, and `SysUtil`.

One other important programming concept that is used throughout this chapter is the difference between the **stack** and the **heap**. This is not a Symbian OS-specific concept, and so to refresh your memory:

- The (program) stack is a fixed-size, last-in, first-out (LIFO) area of system memory, reserved for each application. This area is where temporary data, such as local variables, are stored. At the end of each function call, all of the local variables created in that function are destroyed as the stack unwinds.

- The heap is an area of memory that is dynamically allocated by the system, as needed by an application. Heap memory can be allocated or deallocated at any point during program execution—its lifetime is controlled by the application, not the system. Heap memory is allocated (in general) using a system call such as `malloc()`, or in C++ using the `new` operator. As you will see in the section **Exception Handling and Resource Management**, Symbian OS requires the use of more complicated methods for allocating heap memory.

T Classes

Any class prefixed with a "`T`" denotes a simple class that owns no heap-allocated (dynamic) memory or other resources, and so does not require an explicit destructor. Such a class may be instantiated on either the heap or stack, but bear in mind that "simple" need not mean "small." Some of these `T`-classes can be quite large, and it is easily possible to exceed the stack size, for instance, by instantiating a handful of `TFileNames`. Stack overflow is an unrecoverable error that is caught by the system and results in your application being shut down.

Also note that these classes are not merely data structures—they can have extensive APIs. For example, `TRect`, which is used to represent rectangular regions of the display, has a number of methods such as `Width()`, `Height()`, `Shrink()` and so on.

T-classes do not have a common base class, and you must ensure that each is fully constructed before use.

S Classes

C++ `struct`s may also occasionally be used, and these are given the prefix "S". The use of these is quite rare, however, and for the purpose of this book they can be treated as T-classes. If they are used, then such structures should not have any member functions—a (T-type) `class` should be used instead.

C Classes

A "C" prefix denotes a class that derives from `CBase`. These classes are designed to be constructed on the heap and should not be constructed on the stack—many have private (or protected) constructors to prevent this. The prefixed "C" denotes "Cleanup", as the implementation of `CBase` (particularly its virtual destructor) is integral to the memory management scheme of Symbian OS—this will be covered in much more detail in the **Exception Handling and Resource Management** section of this chapter. In addition to a virtual destructor (all C-class destructors are virtual), `CBase` has an overloaded new operator (`new (ELeave)`) that zero-initializes all member data. Therefore it is never necessary to zero-initialize any member data for a `CBase`-derived object. (*This applies only to C-classes.*) Examples of C-classes include `CActive` and `CArray`.

Note that any C-classes you create *must* derive from `CBase` (although not necessarily directly—deriving from another C-class will do). Symbian OS memory management will work correctly only if all C-classes derive from `CBase`.

R Classes

An "R" prefixed class denotes a **client-side handle** to a resource. This resource is not actually owned by an application, but typically by a Symbian OS Server elsewhere on the device. The client can use these handles to access the resource being managed by the Server and to request functionality of it.

In spite of some common functionality between these classes, they have no common base class. Generally they are instantiated on the stack or nested within C-classes, and then "opened" in some way (usually by a call to a method such as `Open()` or `Connect()`). When they are finished with, it is essential to use the appropriate functionality to dispose of the class—usually a `Close()` function. Failure to do so will result in memory or other resources being leaked by the server that the R-class connected to.

For example, the `CElementsEngine` class connects its file system session (`iFs`, an `RFs` instance) when the class is constructed, and closes it in its destructor:

```
void CElementsEngine::ConstructL()
  {
  User::LeaveIfError(iFs.Connect());
  }
```

```
CElementsEngine::~CElementsEngine()
    {
    iFs.Close();
    }
```

The file system will be discussed in the **Files, Streams and Stores** section later in this chapter. `User::LeaveIfError()` will be explained in the **Exception Handling and Resource Management** section.

> **NOTE** Symbian OS has two main conventions for naming data: all class attributes (member data) are prefixed with a lower-case "`i`" (for "instance data"), as in `iFs`, and all function parameters are prefixed with a lower-case "`a`" (for "argument"). There are no conventions for automatic variables.

M Classes

Classes prefixed with "`M`" are **mixins**. They define an interface but do not provide an implementation. Such classes are **Abstract Base Classes** and consist of nothing but pure virtual functions. As such, they have no member data and cannot be instantiated. `M`-classes provide the only form of multiple inheritance used in Symbian OS—you never derive from more than one `C`-class, but you can derive from as many `M`-classes as necessary.

`M`-classes are used to define a variety of interfaces throughout Symbian OS, with one of the most common uses being to implement the **Observer Design Pattern**, a standard way of allowing objects to communicate without tight coupling.

> **SEE ALSO**
> Design Patterns
> Gamma, Helm, Johnson and Vlissides
> Addison-Wesley
> ISBN: 0201633612

The following example is taken from the **Elements** application. The `M`-class, `MCsvFileLoaderObserver`, is defined with one pure virtual function:

```
class MCsvFileLoaderObserver
    {
    public:
    virtual void NotifyElementLoaded(TInt aNewElementIndex) = 0;
    };
```

In this example (with numerous functions omitted for clarity), `CElementsEngine` implements the `MCsvFileLoaderObserver` interface. It achieves this by inheriting (publicly) from `MCsvFileLoaderObserver`.

```
class CElementsEngine : public CBase, public
MCsvFileLoaderObserver
    {
private:
    // From MCsvFileLoaderObserver
    void NotifyElementLoaded(TInt aNewElementIndex);
    };
```

Next is a code snippet that shows how the `CCsvFileLoader` class uses a pointer to a `MCsvFileLoaderObserver` object to notify it that an element has been successfully loaded into its element list:

```
    // Report the element that's just been loaded to the observer
    iObserver.NotifyElementLoaded(lastElementLoadedIndex);
```

By using the `MCsvFileLoaderObserver` mixin, the `CCsvFileLoader` does not have to know anything about the `CElementsEngine` that is observing the loading process. All the loader needs to know about is the `MCsvFileLoaderObserver` interface, which the engine implements.

Namespaces

C++ namespaces allow the packaging of components (encapsulated software modules that equate to UML component packages) to control their scope and visibility and help avoid name clashes. They are fairly new to Symbian OS, since they have only (relatively) recently been fully ratified and supported by all compilers, so you may not encounter them all that often. However, Series 60 coding standards state that namespaces should start with an "N". Examples include `NEikonEnvironment` and `NBitMapMethods`. Generally, they are used in place of static classes, to wrap up functions that do not belong anywhere else, without polluting the global namespace. An example of a namespace declaration is shown here:

```
namespace NElementComparisonFunctions
    {
    // Equality functions
    // (True if aElement1 == aElement2)
    TBool ElementsHaveSameName(const CChemicalElement&
aElement1, const CChemicalElement& aElement2);
    TBool ElementsHaveSameSymbol(const CChemicalElement&
aElement1, const CChemicalElement& aElement2);
    TBool ElementsHaveSameAtomicNumber(const CChemicalElement&
aElement1, const CChemicalElement& aElement2);
    };
```

Before the advent of namespaces, utility classes were used to achieve the same goal. These were classes that consisted of nothing but static member functions, and they were given names such as `User` and `Math`—without prefixed letters. You will still find these classes used throughout Symbian OS code.

Basic Types

The basic types used throughout Symbian OS are shown in Table 3-1. They are mainly defined in the system include file e32defs.h (which is often #included implicitly, as part of e32base.h), and they all have a T-class interface, so they do not require complex construction or explicit destructors. Many of them are actually implemented as compiler-specific typedefs of standard C++ types, providing guaranteed platform independence. Some are implemented as actual simple C++ classes. Note that the Symbian OS types should always be used in preference to built-in C++ types. For example, you should always use TInt instead of int.

Note that the implementation of TBool seems very wasteful, considering that 32 bits are used to store one of two possible values. There are several reasons for this, including:

- Ease of passing as an argument, since individual bits cannot be passed.
- Lack of ratified native boolean type when Symbian OS was developed.
- Necessity of a direct mapping onto int for use in conditional expressions.

However, note that you should never compare a TBool directly to ETrue, as any nonzero value is also considered "true." For example:

```
TInt b = 2;

if (b == ETrue)
    {
    // This statement is "false"!
    }

if (b) // Will return "true" as required.
    {
    // Use this form instead.
    }
```

Similarly, you should not compare TBool values to EFalse—even though the code will work, it is inefficient and inconsistent. Instead use the following form:

```
if (!b)
    {
    // b is "false".
    }
```

If your class uses a lot of Boolean attributes, and if minimum memory usage is a goal, then you might consider using bit packing. However, this is beyond the scope of this book.

Table 3–1 The Basic Types in Symbian OS

Name	Type	Size (bytes)	Comments
TInt	Integer	At least 4	Signed and unsigned integers, guaranteed to give at least 32 bits of precision, independent of the compiler.
TUInt	Integer	At least 4	
TInt64	Integer	8	64-bit integer, implemented as two 32-bit integers.
TInt8	Integer	1	Explicitly sized signed integers.
TInt16	Integer	2	
TInt32	Integer	4	
TUInt8	Integer	1	Explicitly sized unsigned integers.
TUInt16	Integer	2	
TUInt32	Integer	4	
TReal	Floating point	8	Double-precision floating-point value.
TReal32	Floating point	4	Explicitly sized floating-point values.
TReal64	Floating point	8	
TRealX	Floating point	12	Extended-precision floating-point values, with a dynamic range from as small as ~ $\pm 1 \times 10^{-9863}$ to as large as ~ $\pm 1 \times 10^{9863}$.
TText8	Character	1	Unsigned char for narrow (8-bit) character storage.
TText	Character	2	Unsigned short int for Unicode character storage. Note these cannot be used for storing characters in the extended Unicode range.
TText16	Character	2	
TChar	Character	4	A 32-bit character that can be used for storing extended Unicode characters. Used in a host of character-related helper functions.
TBool	Boolean	4	Enumerated values of ETrue and EFalse are defined, but naturally any nonzero value is considered true.
TAny	Void	N/A	Used exclusively as TAny* (implying a pointer to *anything*), which is more meaningful than void* (implying a pointer to *nothing*).

Exception Handling and Resource Management

This crucial section describes the exception handling methodology used by Symbian OS. C++ programmers may well be familiar with `try`, `catch` and `throw` for exception handling, and they may wonder why Symbian OS does not make use of this well-understood paradigm. The reason is mainly that the Symbian OS implementation is far more lightweight and appropriate to the constraints of small devices, but also that C++ exception handling was not fully supported by the target (GNU) complier when Symbian OS was first developed.

The Symbian OS implementation does have similarities. For example, it employs a **trap harness**, `TRAP`, that can be compared to the try and catch elements of traditional C++, while **leaving**, a call to `User::Leave()`, is the equivalent of throw. These concepts, and other exception handling techniques, will be explained in detail in this section.

Rigorous exception handling is central to the stability of Symbian OS—it is closely integrated into resource management and also affects function naming conventions. It is incredibly important that applications do not leak memory, and so you must take steps to ensure that any application resources stored on the heap are cleaned up when an exception occurs. The **Cleanup Stack** is central to this methodology, as it offers a way to delete data that may otherwise be orphaned—more about this later. Also of importance in this regard is the Symbian OS approach to object construction. The traditional C++ `new()` operator is generally not used. An overloaded version is provided for simple object construction, and a two-phase approach is taken when complex objects are involved. These techniques are employed to ensure that constructors do not leave.

An unfortunate consequence of the Symbian OS exception handling methodology is that it can sometimes be intimidating to programmers who are new to the platform. It forms such an integral part of the language that even writing a simple "Hello World" application can seem like a major task. However, you should not be discouraged, as most of the rules will soon become second nature and you will be happy in the knowledge that you are writing good code.

Exceptions, Leaves, Panics and Traps

In simple terms, an exception is a runtime error that is not the programmer's fault. As a programmer, you may like to think that no errors are your fault! However, common problems, such as array out of bounds errors, or the use of dangling pointers, are certainly the responsibility of the programmer. Other errors, such as a lack of memory, the inability to open a socket because of a dropped Internet connection, or the failure of a nonexistent file to open, cannot necessarily be attributed to bad programming—they are due to conditions of the system that were not expected at the time of execution.

In other words, these errors occur because the system is in an unusual state. Sufficient memory *should* be available—the Internet connection *should not* have failed—the user should specify a file that *does* exist. The failures of these conditions are the *exception* rather than the rule, but they cannot be ignored just because they are unusual. So, although you cannot predict *when* any of these problems may occur, it is important that you consider the possibility that they *could* happen at some time and that you cater for them. Good software should be written to cope with all such exceptional circumstances, and cope gracefully, *without compromising the user's data* if it cannot proceed.

Contrast this with a **panic**. In Symbian OS, a panic is a programmatic error, which *is* the programmer's fault, such as an out-of-bounds array. Often the system will invoke a panic to indicate that something has gone seriously wrong and immediately terminate the offending application with a brief message, as shown in Figure 3–1.

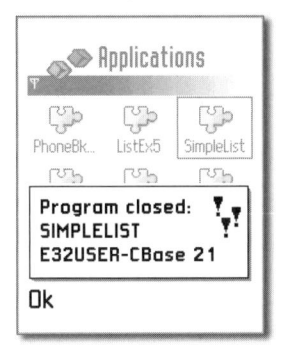

Figure 3–1 A sample panic message.

Needless to say, thorough testing of your application should highlight any programmatic errors; therefore such panics should not occur in a released version of your application and so should *never* be visible to the end user!

Writing code to cope with every single exceptional circumstance that may occur would quickly become tiresome, and so the technique is to "throw" any such exception up the call stack to some centralized exception handler. In Symbian OS this is referred to as a "leave." Execution of a function continues under normal circumstances, but if anything untoward happens, then execution *leaves* the function and jumps back to the exception handler. This handler is known as a **trap**, or **trap harness**. Such traps should be few and far between—trapping every possible leave where it occurs defeats the object of centralized exception handling and consumes additional system resources.

Here is part of a function containing a trap harness:

```
void DoMainL()
   {
   ...
   TRAPD(error, DoExampleL()); // Perform example function
   if (error)
      {
      console->Printf(KFormatTxtFailed, error);
      }
   else
      {
      console->Printf(KTxtOK);
      }
   ...
   }
```

The job of this function is simply to invoke the `DoExampleL()` function (shown later) within a trap harness, then handle any errors that may have occurred (by informing the user via the console). Note the form of the trap harness: `TRAPD`. The variable error is declared by the macro to be a `TInt` (and so should not have already been declared) that will contain the error code. The state of this error can be checked later. If all is well, a reassuring OK message can be displayed. Otherwise, the leave code is embedded in a warning message and output to the user in this example. It is naturally up to the programmer to decide what is to be done with errors at each trap harness, which may include displaying a warning message, attempting to take some remedial action or simply "throwing" the leave up to the next nested trap.

There is another form of trap harness: `TRAP` (without the trailing "D"). This is almost identical but does not declare the error variable. Instead, it depends on the variable having been declared previously as a `TInt`, perhaps explicitly, or perhaps by a previous `TRAPD` macro. Remember, the "D" here stands for "declare."

NOTE Traps can be nested—in other words, one TRAP may enclose a function that itself contains other traps. It is also common for the error-handling code following one TRAP to deal with just a few possible errors—any others that occur can be thrown up to the next TRAP harness using User::Leave().

So how do the leaves occur in the first place? There are three basic ways:

- Use of an explicit leave, in other words: `User::Leave()`, `User::LeaveIfError()`, `User::LeaveNoMemory()`, or `User::LeaveIfNull()`.
- Use of the overloaded `new (ELeave)` operator.
- Calling a leaving function.

The first of these is quite straightforward (similar to `throw` in regular C++ exception handling). `User::Leave()` simply leaves at that point. A single `TInt` argument is provided, which is the leave code (that would be passed to the trap harness via the error variable in the previous example). `User::LeaveIfError()` leaves if the value passed in is negative—all error codes are less than zero—otherwise it does nothing. The value returned to the trap harness will simply be the leave error code as before. This is a way of causing a function that *returns* an error code to leave instead—simply place the non-leaving function inside a `User::LeaveIfError()` wrapper. The code below shows this:

```
TInt foundAt = iElementArray.Find(example, aRelation);
// Leave if an error (i.e. KErrNotFound)
User::LeaveIfError(foundAt);
```

Chapter 3 Exception Handling and Resource Management

If the `Find()` function returns a negative value (the only likely one is `KErrNotFound`, which equals -1), then `User::LeaveIfError()` will leave with this value—otherwise the flow of control continues as normal onto the next line of code.

`User::LeaveNoMemory()` takes no argument and is effectively shorthand for `User::Leave(KErrNoMemory)`. Finally, `User::LeaveIfNull()` leaves with `KErrNoMemory` if the pointer it is passed is `NULL`.

The next two leaving scenarios are illustrated in the `DoExampleL()` function, shown here. This is the function that the `DoMainL()` function (shown earlier) calls within its trap harness:

```
void DoExampleL()
    {
    // Construct and install the Active Scheduler
    CActiveScheduler* scheduler = new (ELeave) CActiveScheduler;
    CleanupStack::PushL(scheduler);
    CActiveScheduler::Install(scheduler);

    // Construct the new element engine
    CElementsEngine* elementEngine = CElementsEngine::NewLC(*console);

    // Issue the request...
    elementEngine->LoadFromCsvFilesL();
    // ...then start the scheduler
    CActiveScheduler::Start();
    CleanupStack::PopAndDestroy(2, scheduler);
    }
```

The first executable line shows the use of `new (ELeave)` to allocate memory for a `CActiveScheduler` instance. In Symbian OS, heap (dynamic memory) allocation of new simple objects should almost always be performed this way. `new` on its own is used only in certain very specific circumstances—for example, when an instance of a new application is created before the framework is fully constructed. `new` on its own does not take advantage of the leaving mechanism and requires an explicit check of success each time.

The next line, and one or two others, show the third way of potentially causing a leave—calling a potentially leaving function. *All functions that could leave must end with a trailing* L. `CleanupStack::PushL()` and `LoadFromCsvFileL()` are both leaving functions, as is `DoExampleL()` itself.

When writing a function, how can you tell if it is a leaving function? Simple: if any line can leave and that leave is not trapped locally, then it is a leaving function. So all functions that contain calls to `User::Leave()`, `new (ELeave)` or calls to other leaving functions are themselves leaving functions. Their name *must* have a trailing L.

Chapter 3 Symbian OS Fundamentals

Why the need for the trailing L? It is to ensure that any caller of your function knows that a leave may occur. Partly this is to give them the option of trapping the leave should they wish to. More important are the issues associated with jumping out of a function should a leave happen, as discussed in the next subsection.

> **WARNING** The compiler will not detect if a function without a trailing L could leave. However, the **Leavescan** and **EpocCheck** tools described in Chapter 13 can be used to eliminate this problem.

Leaving Issues and the Cleanup Stack

This subsection focuses on the precautions you need to take to ensure memory is not leaked when a function leaves. It introduces the Cleanup Stack and describes the essential role it plays in Symbian OS resource management.

To understand the problems that can be caused by a function leaving, consider the following (poorly implemented) function—a slightly modified version of the DoExampleL() function seen previously:

```
void IncorrectDoExampleL()
    {
    // Construct and install the Active Scheduler
    CActiveScheduler* scheduler = new (ELeave) CActiveScheduler;
    CActiveScheduler::Install(scheduler);

    // Construct the new element engine
    CElementsEngine* elementsEngine =
CElementsEngine::NewL(*console);

    // Issue the request...
    elementsEngine->LoadFromCsvFilesL();
    // ...then start the scheduler
    CActiveScheduler::Start();

    delete elementsEngine;
    delete scheduler;
    }
```

The differences are subtle, but crucial. The references to the Cleanup Stack have been removed, and if you are very observant you will have noticed the missing "C" from the CElementsEngine::NewL() call. Two deletes have been added that were not there previously—cleaning up the heap-allocated objects. On first inspection, all seems well, but consider what would happen if the LoadFromCsvFilesL() function were to leave:

As mentioned previously, the trap harness would catch this leave, but what does this *mean* in real terms (or processor instructions)? It means (among other

Chapter 3 Exception Handling and Resource Management   89

things) that a long jump (`longjmp`) will be executed back to the `TRAPD` macro. As part of this, the stack will be "unwound," and all automatic variables (`scheduler` and `elementsEngine`) will be destroyed.

The vital point is while these *variables* will be destroyed, the objects which they point at will not be. The runtime environment is not clever enough to realize that these are locally scoped pointers—when execution leaves *before* the two `delete` operations, the pointers drop out of scope and their contents are lost. There is now no way to deallocate the two heap-based objects, and a memory leak occurs.

"So what?" you may think. "It is only a few hundred bytes, and that can be reclaimed next time the system reboots."

Such leaks may be tolerable on machines with multi-gigabytes of virtual memory that are rebooted daily, but remember that Symbian OS runs on devices that may have only a couple of megabytes of RAM and may go months, or even years, between reboots. All leaks must be anticipated and avoided by design.

What is needed is a way of making some kind of "backup copy" of any locally scoped pointers, just in case a leave should occur. If a leave does occur, then all of these pointers should be automatically deleted to ensure that the heap cells they point to are not leaked. This is precisely what the Cleanup Stack is for.

The Cleanup Stack

The Cleanup Stack (`CleanupStack` is defined in `e32base.h`) is a special stack that is crucial to Symbian OS resource mangement. It is essentially a way of making sure that if a leave occurs, then all resources are cleaned up, so there are no memory leaks—for example, heap cells that were pointed to by locally scoped variables are not orphaned.

The basic idea is that before you call a potentially leaving function, you use one of the `PushL()` methods to add any locally scoped pointers to the top of the stack. If the function leaves, then all pointers to objects placed on the stack since the last `TRAP` will be removed and the objects they point to will be deleted. If the function does not leave, then you use one of the `Pop()` methods to remove the pointer from the stack yourself. As you would expect, the Cleanup Stack operates on a last-on, first-off basis.

You will see some practical examples of how to use the stack throughout this section, and the crucial points ("Resource Handling Rules") will be highlighted.

> **Resource Handling Rule 1**
>
> Any locally scoped pointer to a heap-allocated object must be pushed onto the Cleanup Stack if there is a risk of a leave occurring and there is no other reference to the object elsewhere.

Note that if there is no chance of a leave, then there is no reason to push such a pointer onto the Cleanup Stack—there is no danger of the pointer going out of scope before it is deleted.

Also, and perhaps less obviously, objects owned elsewhere must *not* be pushed onto the Cleanup Stack.

> ### Resource Handling Rule 2
>
> Instance data (data owned by an instance of a class) must never be pushed onto the Cleanup Stack.

If a leave were to occur with the instance data on the Cleanup Stack, then the instance data would be deleted as the Cleanup Stack was unwound. This is fine until you consider that the *owning object* must also be on the Cleanup Stack somewhere—when the *owner* gets deleted (as the Cleanup Stack unwinds further), its destructor will try to delete the same piece of instance data, leading to a double-deletion.

Remember that instance data in Symbian OS always has a prefixed "i". So the following is *always wrong*:

```
// This is always wrong!
CleanupStack::PushL(iMemberPointer);
// Never push instance data!
```

Instance data is (and must always be) deleted in the destructor of the class that *owns* it. (Note that a class may maintain a pointer to data without actually owning it—in this case deleting a pointer in a non-owning class would also cause a double-deletion! However, this is not a Symbian OS-specific issue.)

Once the leaving code has safely completed, there is no danger of the local pointer being lost, so it can safely be popped off the stack using `CleanupStack::Pop()`. The same is true if ownership is being transferred to another object, which is itself on the Cleanup Stack somewhere, as in this example:

```
// Append new element to the element array
AppendL(newElement);

// Remove the element pointer from the cleanup stack,
// since ownership has successfully passed to the array
CleanupStack::Pop(newElement);    // Now owned by array
```

Since the array in question is owned elsewhere, once the `newElement` has been appended, it is the responsibility of the array's owner to dispose of it. So if the `AppendL()` function was successful (in other words, it did not leave), then the new element must be popped off the Cleanup Stack to avoid the same double-deletion problem as highlighted previously. Notice that the `newElement` pointer is passed into the `Pop()` function as an argument. (Cleanup) Stack pop operations *always* remove items from the top, so technically the argument is not necessary. However, supplying the pointer is *strongly encouraged*, since, for a debug build, the system will check it against the address of the object being popped, and panic if they do not match—remember that a panic indicates a programmer's error. This allows any Cleanup Stack imbalance to be tracked down quickly!

If you have more that one object on the Cleanup Stack that you wish to pop, you can use an overloaded version of the Pop() function. Pop(TInt account, TAny* aLastExpectedItem) takes an integer value to indicate how many objects should be popped, and a pointer to the last item you want to be popped.

Often you will wish to dispose of an object at the end of the function. You could Pop() followed by delete, but this is unnecessary, since Symbian OS provides PopAndDestroy() to do both operations atomically. You can see its use at the end of the original DoExample() function—the 2 denotes the number of objects to be popped, and scheduler is the address of the last item to be popped. See the SDK documentation for details of the overloads for both Pop() and PopAndDestroy(). Here is the code again:

```
void DoExampleL()
   {
   // Construct and install the Active Scheduler
   CActiveScheduler* scheduler = new (ELeave) CActiveScheduler;
   CleanupStack::PushL(scheduler);
   CActiveScheduler::Install(scheduler);

   // Construct the new element engine
   CElementsEngine* elementEngine = CElementsEngine::NewLC(*console);

   // Issue the request...
   elementEngine->LoadFromCsvFilesL();
   // ...then start the scheduler
   CActiveScheduler::Start();
   CleanupStack::PopAndDestroy(2, scheduler);
   }
```

How come there are two objects to be popped, when there appears to be only one PushL()? Remember the trailing "C" that disappeared in the second implementation? The next subsection will explain its significance and address another pivotal Symbian OS concept: two-phase construction.

Two-phase Construction

So far you have learnt about the clean-up of objects owned by locally scoped pointers, which must always be pushed onto the Cleanup Stack whenever there is danger of a leave occurring. What about complex objects, when one object owns another via a heap pointer? As mentioned previously, such objects are usually `CBase`-derived. Consider, by way of an example, the engine class from the **Elements** application (the example used throughout this chapter) where the engine owns an array of elements. The declaration of the relevant data members looks like this:

```
class CElementsEngine : public CBase
    {
private:
    CElementList*          iElementList;
    CConsoleBase&          iConsole;
    };
```

`CElementList` is a `CBase`-derived class that represents a list of chemical elements. The destructor for the class looks like this:

```
CElementsEngine::~CElementsEngine()
    {
    delete iElementList;
    }
```

What about the constructor? You might think this would suffice:

```
CElementsEngine::CElementsEngine(CConsoleBase& aConsole)
: iConsole (aConsole)
    {
    iElementList = new (ELeave) CElementList();
    }
```

However, such constructors are not used in Symbian OS, and for good reason. To see why, you need to consider what would happen if the allocation of the new `CElementList` were to fail, causing a leave. Well, presumably this constructor would have been invoked elsewhere something like this:

```
CElementsEngine* myEngine = new (ELeave) CElementsEngine(*console);
```

Remember that this line does two important things: First, the `new (ELeave)` operator allocates memory for the new `CElementsEngine` instance (and all its nested data). Assume for the moment that this allocation is successful. Next it calls the C++ constructor shown above. If the allocation of the `CElementList` in the `CElementsEngine` constructor were to fail, say, due to running out of memory, there would be a problem. When this leave occurs, there is no pointer pointing to the area of memory successfully allocated for the `CElementsEngine` object, so this memory will be orphaned, and a memory leak will occur. Because of this possibility, in Symbian OS a C++ constructor must not leave. Also note that destructors must never leave and must never assume that full construction has occurred.

> ### Resource Handling Rule 3:
> Constructors and destructors must not leave, and destructors must not assume full construction.

Chapter 3 Exception Handling and Resource Management

So how can complex objects be constructed? The solution is to perform the construction *in two phases*.

The first phase is to do the normal, non-allocating construction in a normal C++ constructor. The *real* `CElementsEngine` constructor is shown below—just as you might expect but with the construction of `iElementList` missing:

```
CElementsEngine::CElementsEngine(CConsoleBase& aConsole)
 : iConsole(aConsole)
    {
    }
```

The second, potentially leaving phase of construction is performed by the function `ConstructL()`, the name that is always given to such second-phase constructors (except those in abstract base classes, which are often called `BaseConstructL()`). The relevant part of `CElementsEngine::ConstructL()` is shown below:

```
void CElementsEngine::ConstructL()
    {
    iElementList = new (ELeave) CElementRArray;
    }
```

Note that `CElementRArray` is a specialization of `CElementList`.

So, is it always necessary to use two lines of code to perform the construction of a complex object in Symbian OS? No, because such classes will usually provide a static `NewL()` function. Here is a possible `CElementsEngine::NewL()`:

```
CElementsEngine* CElementsEngine::NewL(CConsoleBase& aConsole)
    {
    CElementsEngine* self = new (ELeave) CElementsEngine(aConsole);
    CleanupStack::PushL(self);
    self->ConstructL();
    CleanupStack::Pop(self);
    return self;
    }
```

Consider what each line does. Firstly, the overloaded `new` operator allocates memory for a new `CElementsEngine` instance. Should this allocation fail, a leave will occur (and, as always, this will be caught by the last `TRAP`, which will unwind the Cleanup Stack). If not, the non-leaving C++ constructor will do some initialization, *but no allocation*. The newly allocated and initialized `CChemicalElement` is assigned to the local pointer `self`.

Next you need to push this local pointer onto the Cleanup Stack, because you are about to call a leaving function. Note that `CleanupStack::PushL()` may itself leave, since it must allocate memory for a new stack cell. In fact it always has a "spare" stack cell, to which it assigns the pointer being pushed *before* allocating the new stack cell. This means that even if the call to `PushL()` leaves, the pointer being pushed is guaranteed to be on the stack before it leaves, and so is guaranteed to be cleaned up.

Once the pointer is safely on the stack, you call the `ConstructL()` to perform the potentially leaving allocation. If this does leave, the pointer to the new `CElementsEngine` is on the Cleanup Stack and so it will be deleted as the Cleanup Stack unwinds. This is why destructors must not assume full construction—they may be invoked because `ConstructL()` left before construction was complete. However, because the overloaded `new (ELeave)` operator zero-initializes all member data, it is safe to `delete` this object, since the member pointer `iElementList` will have been set to `NULL`, and `delete NULL` does nothing.

Assuming `ConstructL()` did not leave, you now have a fully constructed `CElementsEngine` instance and can safely pop the pointer to it from the Cleanup Stack. Finally, you return the pointer to this new object.

This whole process works in much the same way as a normal C++ constructor but encapsulates the two-phase construction process in a leave-safe manner. Since `NewL()` is static, it can be invoked like this:

```
CElementeEngine* myEngine = CElementsEngine::NewL(*console);
```

However, you are now left with a locally scoped pointer (`myElement`) to some heap-allocated data. If you are going to call some leaving functions before deleting it, you should push it onto the Cleanup Stack for safety. Since this is such a common thing to do, most complex classes provide a `NewLC()` function in addition to the `NewL()`. The only difference is that the `NewLC()` function leaves a pointer to the newly constructed object on the Cleanup Stack. This is another Symbian OS naming convention—a trailing "C" after a function name denotes that function leaves a pointer to the object it returns on the Cleanup Stack. You will see other examples of this later.

```
CElementsEngine* CElementsEngine::NewLC(CConsoleBase&
aConsole)
   {
   CElementsEngine* self = new (ELeave)
CElementsEngine(aConsole);
   CleanupStack::PushL(self);
   self->ConstructL();
   return self;
   }
```

If both functions are being implemented, the `NewL()` is best written in terms of `NewLC()`:

```
CElementsEngine* CElementsEngine::NewL(CConsoleBase& aConsole)
   {
   CElementsEngine* self = CElementsEngine::NewLC(aConsole);
```

```
CleanupStack::Pop(self);
return self;
}
```

Note that if you are providing `NewL()` and/or `NewLC()`, it is usually good practice to make the corresponding C++ constructor and `ConstructL()` private (or protected). This prevents the `ConstructL()` from being called more than once and overwriting pointers to previously allocated member data. Also, a private (or protected) C++ constructor prevents construction on the stack and therefore removes the potential use of an object that is only partially constructed.

Private constructors prevent derivation of your class—if you wish to allow derivation, then make the constructors protected instead.

One final rule:

> **Resource Handling Rule 4:**
>
> After deleting a member pointer, always zero it before reallocation.

For instance, consider the `CChemicalElement::SetNameL()` function in the **Elements** application. This function is used to assign a name to a particular element. The element name is stored in an `HBufC` descriptor, `iName`. You will learn more about the `HBufC` descriptor later in the section of this chapter, but for now, all you need to know is that it is a heap-based class for holding a text string.

Here is the implementation of `SetNameL()`:

```
void CChemicalElement::SetNameL(const TDesC& aName)
    {
    // First delete the old name
    delete iName;
    iName = NULL;    // In case the next line leaves!
    iName = aName.AllocL();
    }
```

Since the `iName` member is a pointer to some heap-allocated data, the function must do two things:

- Delete the old name.

- Allocate memory for the new one.

Consider what might happen if the `AllocL()` failed. You have successfully deleted `iName`, but if you did not set it to `NULL`, it remains pointing to the old, deallocated `iName`. If the `AllocL()` leaves, the `CChemicalElement` instance will be destroyed, and its destructor will try to delete the nonzero `iName` pointer, resulting in a fatal double-deletion!

Symbian OS Construction Methods

So, in Symbian OS there are four potential ways of constructing a new heap-allocated instance of a class: `new`, `new (ELeave)`, `NewL()` and `NewLC()`. Following are some summary guidelines for when to use each:

- `new`: Almost never used. One common exception is when constructing a new instance of an application, as the Cleanup Stack does not exist at that point. Creating an application is covered in Chapter 4.

- `new (ELeave)`: Used when constructing a heap-allocated instance of a simple class, such as a `T`-class (normally stack allocated) or a `C`-class that does not own any heap-allocated data. Note that heap allocation of anything other than a `C`-class may require advanced use of the Cleanup Stack.

- `NewL()`: Used when constructing a heap-allocated instance of a compound class and either the instance is assigned directly to a member pointer of another class or there is no danger of a leave before the object is deleted. (Use of the Cleanup Stack is not needed.)

- `NewLC()`: Used when constructing a heap-allocated instance of a compound class that is to be assigned to a locally scoped pointer, where leaving functions will be called before the pointer is deleted. (Use of the Cleanup Stack *is* needed.)

Advanced Use of the Cleanup Stack

There are a few situations when conventional cleanup is not sufficient. The most common of these relate to heap allocation of non-`CBase`-derived objects, and the use of `R`-classes.

Non-CBase-Derived Heap-Allocated Objects

`CleanupStack::PushL()` has an overload that takes a `TAny*` (in other words, an untyped pointer), and any such object pushed onto the Cleanup Stack will be cleaned up by invoking `User::Free()`. While this will deallocate memory assigned to the object, it *will not invoke the object's destructor*. This is sufficient for `T`-classes, since they must not require destructors, but it may not suffice for other arbitrary objects. Such objects must make use of `TCleanupItem`—see the SDK documentation for further details of how to use this class to clean up after arbitrary objects.

R Classes

It is often assumed that the Cleanup Stack is purely concerned with memory. While cleaning up memory is its most common use, it can in fact clean up *any* resource. Since resources are usually owned by `R`-class handles, which do not have a common base class, special functions are provided, depending on the cleanup function required. The functions `CleanupClosePushL()`, `CleanupDeletePushL()` and `CleanupReleasePushL()` push resources onto

Chapter 3 Descriptors

the Cleanup Stack and invoke a corresponding `Close()`, `Delete()` or `Release()` method on the given resource object when `CleanupStack::PopAndDestroy()` is called for that object. Further information can be found in the SDK documentation – a simple example is shown below:

```
RFs myFileSystemHandle;

// Connect to the file system
TInt error = myFileSystemHandle.Connect();

CleanupClosePushL(myFileSystemHandle);   // In case of a leave

// Some leaving code

// Clean up myFileSysteHandle by calling Close()
CleanupStack::PopAndDestroy();
```

The function `CleanupClosePushL()` creates a `TCleanupItem` object that automatically invokes `Close()` on the `myFileSystemHandle` object when `CleanupStack::PopAndDestroy()` is called. Note that, since you do not have the address of this `TCleanupItem`, it is not possible to pass an address into `PopAndDestroy()` for checking.

Descriptors

Descriptors are essential objects that are used throughout Symbian OS to store arbitrary data. While they are most commonly used for storing text, they can also store binary data and even serialized compound objects, since *they do not rely on zero termination*. They are designed for maximum efficiency, and they avoid the use of virtual functions with their associated memory and instruction overhead.

Because of these issues, they have an API and derivation structure that takes a little getting used to, but they do provide:

- Runtime bounds checking.

- An extensive and standard API for data and text management throughout Symbian OS.

- Tight integration with the Symbian OS resource management paradigm.

NOTE Developers new to Symbian OS sometimes want to avoid using descriptors, and instead try to develop their own alternatives. This often leads to dangerously unsafe code (particularly regarding leave-safety), and since many Symbian OS APIs use descriptors, you cannot avoid using them at some point!

Hierarchy

Figure 3-2 illustrates the derivation structure of the nine main descriptor classes. Note that the only instantiable classes are the five shown on the bottom.

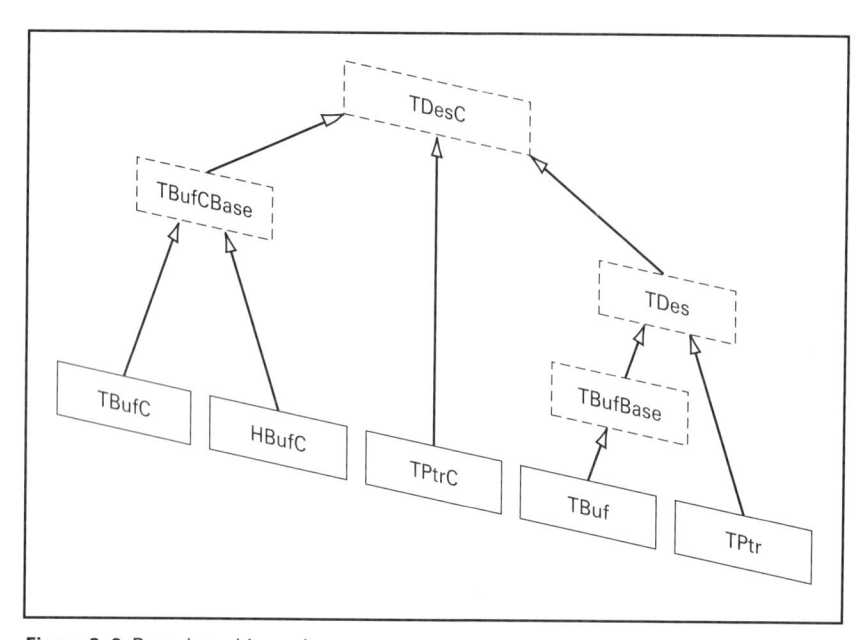

Figure 3-2 Descriptor hierarchy.

Table 3-2 describes each of these classes in a little more detail. It is worth noting that all of the descriptors detailed here actually come in two sizes: 8-bit (or narrow) and 16-bit (or Unicode). These descriptors may be explicitly sized by specifiying 8 or 16 after their name—for example, TDesC8 or TDesC16.

Since all modern Symbian OS platforms (including Series 60) are exclusively Unicode, TDesC may be assumed synonymous with the explicitly sized TDesC16 variant, but this second form may be used for clarity, if required.

TDesC8-derived classes should be specified when manipulating binary data.

This table divides the descriptors into two natural groups: modifiable and nonmodifiable descriptor types. If a descriptor has a modifiable API, it means its contents can be altered; for example, a character could be appended. A nonmodifiable API does not allow the descriptor content to be modified, although it can be reset. The Symbian OS convention is that nonmodifiable types should append a C (for "constant") to their basic name—for example, TBufC.

All of the descriptors have an iLength member that stores the current length of the descriptor. This is how descriptors can function without the need for null termination. Modifiable API descriptors also have an iMaxLength member, since their length needs to be bounded.

Chapter 3 Descriptors

Table 3-2 The Descriptor Classes

Name	Instantiable	Modifiable API	Comments
TDesC	No	No	Base class of all descriptors, used for nonmodifiable argument passing and return type (as const TDesC&).
TBufCBase	No	No	Rarely used intermediate class. Listed here for completeness only.
TDes	No	Yes	Base class for modifiable descriptors, used for modifiable argument passing (as TDes&).
TBufBase	No	Yes	Rarely used intermediate class. Listed here for completeness only.
TBufC	Yes	No	Nonmodifiable buffer descriptor with templated size. Size and data buffer stored in a single allocation cell.
HBufC	Yes	No	Nonmodifiable heap descriptor. Never instantiated on the stack. Size, maximum size and data buffer stored in a single heap cell.
TPtrC	Yes	No	Nonmodifiable pointer descriptor. Size and data buffer pointer are stored separately to the data, which is owned elsewhere.
TBuf	Yes	Yes	Modifiable buffer descriptor with templated maximum size. Size, maximum size and data buffer stored in a single allocation cell.
TPtr	Yes	Yes	Modifiable pointer descriptor. Size, maximum size and data buffer pointer are stored separately to the data, which is owned elsewhere.

Note that although they are "nonmodifiable," the contents of the instantiable classes HBufC and TBufC can actually be modified through use of a pointer descriptor. You will see how this is achieved later in this section.

Nonmodifiable API

Since all descriptor types are derived from the abstract base class TDesC, they share a nonmodifiable API. Some of the most useful members are mentioned here (for full details of the API, see TDesC16 in the SDK documentation):

- Length(): Simply returns the number of characters stored in the descriptor.
- Size(): The number of bytes occupied by the descriptor. For a 16-bit (Unicode) descriptor the size will be twice the length; for an 8-bit descriptor they will be the same.
- Ptr(): Returns a const TUint* (for a 16-bit descriptor or a const TUint8* for an 8-bit descriptor) to the descriptor's data buffer.

- `Alloc()`: Allocates and returns a new `HBufC` heap-based descriptor, containing a copy of the data in this descriptor. If the allocation fails, it returns `NULL`.

- `AllocL()`: Performs the same task, but leaves with `KErrNoMemory` if the allocation should fail.

- `AllocLC()`: Identical to `AllocL()` except that it leaves a pointer to the newly allocated object on the Cleanup Stack.

NOTE The use of `Alloc()`, `AllocL()`, and `AllocLC()` can have an impact on the use of the Cleanup Stack if they are used in conjunction with items on the Cleanup Stack—see the entry on `HBufC` later.

- `Left()`, `Mid()` and `Right()`: Standard string-slicing functions that return a `TPtrC` to the relevant substring of the descriptor. Note they do not affect the descriptor itself, but simply return a pointer to the relevant substring *in place*.

- `Find()`: Search for a descriptor's contents within another, returning the zero-based offset of the first occurrence, or `KErrNotFound` if not found. Note that searching always starts from the beginning of the descriptor; use `Right()` to get the remainder of the descriptor, then repeat the find for further occurrences if necessary.

- `operator<()`, `operator>()`, `operator==()` and `operator!=()`: These operators are overloaded for text, according to the standard text comparison rules. They may, of course, be used on descriptors containing binary data, but the results may be unpredictable.

- `operator[]()`: Allows the inspection of individual characters or data items at a zero-based index.

- `operator=()`: It may come as a surprise that the assignment operator is available for supposedly nonmodifiable descriptors. One may argue that reassignment is not modification, since it is simply *replacing* the entire data with some more. `TPtrC` is an exception—it has no overloaded assignment operator and may be considered genuinely immutable.

WARNING Any attempt to assign data that is longer than the original data length will result in an immediate panic.

Modifiable API

Remember that these APIs are *in addition to* the nonmodifiable APIs. A common feature of all modifiable APIs is that any attempt to increase the length of the descriptor beyond the maximum length will result in an immediate panic.

As before, the following list is not a full API description, but it covers those APIs considered the most useful.

- `MaxLength()`: The maximum number of data items that can be stored in the descriptor. Any operation that would cause the length of the descriptor to exceed this value will result in a panic.
- `MaxSize()`: The maximum number of bytes that the data buffer will occupy.
- `SetLength()`, `SetMax()` and `Zero()`: Adjusts the length of a descriptor to the specified length, the value of `MaxLength()` and zero, respectively.

> **NOTE** These functions do *not* result in changes to the data buffer. In particular this may mean that uninitialized data will be present if the length is *increased*.

- `Append()`: Appends data in the given descriptor to the end of the current descriptor. Note this function cannot leave, since the data buffer is already allocated for the descriptor. It will panic, though, if the data appended causes the maximum length to be exceeded.
- `Insert()`: Inserts the given descriptor at the specified location within the original descriptor.
- `Delete()`: Removes the given number of data items from the given location in the descriptor. Any subsequent data is shuffled down.
- `Format()`: Formats a descriptor in a manner similar (but not identical) to `sprintf`, given a descriptor specifying a format string and then an appropriate number of arguments. Note that the format specifiers are mostly the same as for `sprintf`, but there are some differences. Check in the SDK documentation for further details.
- `Copy()`: Copies data into a descriptor from a variety of sources: raw memory addresses, 8- and 16-bit descriptors. Copying from an 8-bit into a 16-bit descriptor will result in the addition of padding bytes to convert from ASCII to Unicode; the converse is also true, although Unicode values greater than 255 are converted to 1.

Literals

Although not strictly part of the descriptors API, the string literal is a useful tool to have when learning how to use descriptors. String literals are used throughout the example code as an easy method of including printable text in the program. Before introducing this technique, it is vital to stress that in any commercial-quality (UI) application, such text should be provided in resource files for ease of localization. Localization is covered in more detail in Chapter 2 and Chapter 4.

Literals are stored as objects of type `TLitC`. As with descriptors, `TLitC8` and `TLitC16` are available as explicitly sized alternatives.

String literals are constructed using the `_LIT` macro. The macro takes the name of the literal, and a pointer to the null-terminated string it should contain, usually supplied by simply providing the string in quotes. Here is a simple example:

```
_LIT(KTxtMyStringLiteral, "My string");
```

It is important to appreciate that literals are not actually descriptors themselves. However, they can be converted to descriptors in three ways: implicitly, using `operator()`, and using `operator&`.

Implicit conversion takes place courtesy of `operator const TDesC16&() const` (and its 8-bit equivalent), which is defined for literals. Because of this, any function call that will take a `const TDesC&` will also take a literal without needing any modification or explicit casting.

`operator()` is defined for situations when implicit casting is not enough. For example, in the following code it is necessary to know the length of the literal `KTxtLineDelimiter`. While a constant could be defined to contain this value, it is convenient to use `operator()` and then use `TDesC::Length()`:

```
// Increment the file position
iFilePos += elementBuf16->Length() +
    KTxtLineDelimiter().Length();
```

Finally, if a descriptor pointer is required from a literal, `operator &` is overloaded to return a `const TDesC*`:

```
// Get a descriptor pointer from a literal.
const TDesC* ptr = &KTxtMyStringLiteral;
```

> **NOTE** If you look through some of the examples provided with the SDK, you will probably see literals constructed with the `_L` macro. This has now been deprecated, because it is less efficient than `_LIT` and also because there can be problems with scope. Always create literals using `_LIT`!

Using Descriptors

The common APIs for descriptors were introduced earlier in this chapter; this section provides practical examples to illustrate their use and should help you to understand the fundamental techniques necessary for using descriptors.

Using HBufC

`HBufC` descriptors, despite not conforming to standard naming conventions, are allocated on the heap (they are not derived from `CBase` and so have the unique starting letter, H for heap). Heap desciprtors are generally used in preference to all other descriptors in the following situations:

- If the size of the descriptor is not known at compile time, or its size is not bounded by a small maximum value.
- If the size of the descriptor is known to be large.

For example, in the `CChemicalElement` class in the **Elements** example application, there are two descriptors. One is used to store the *name* of the element, the other its *chemical symbol*. The name of the element can be relatively large—a maximum of 13 characters = 26 bytes (remember it is a Unicode System) for the element Rutherfordium, and to use 13 characters for each name would waste an average of about 10 bytes per element—using an `HBufC` is most efficient here. The longest chemical symbol is only three characters in comparison, and most of them are two characters in length—only an average 2 bytes per element is wasted storing these in a `TBuf<3>`.

`HBufC`s are constructed in a number of ways: the simplest is to use `HBufC::New()`, `HBufC::NewL()` or `HBufC::NewLC()` as in the following example, which creates a new `HBufC` as long as `elementBuf8`:

```
HBufC* elementBuf16 = HBufC::NewLC(elementBuf8.Length());
```

`NewLC()` is used here for a locally scoped pointer. `NewL()` could be used if use of the Cleanup Stack were unnecessary (assignment to member data), and `New()` could be used if leaving behavior is not required. `New()` returns `NULL` if allocation fails.

There are no constructors defined for `HBufC`, but a new one can be constructed from an existing descriptor using the `TDesC::AllocL()` function, as in the code below:

```
void CChemicalElement::ConstructL(const TDesC& aName)
    {
    iName = aName.AllocL();
    }
```

Note that `AllocL()` is invoked on the original descriptor and returns a pointer to the new `HBufC`. Since `CChemicalElement::ConstructL()` is private, there is no danger of this function being called more than once (so deleting and assigning to `NULL` is unnecessary).

It may seem surprising that there is no `HBuf` corresponding to `HBufC`. Does this mean that you cannot modify a heap-allocated buffer? Well, by using the `HBufC` API, you cannot (other than complete reassignment). But there is another way of doing it—the method `HBufC::Des()` returns a *modifiable* pointer descriptor (`TPtr`), pointing to the data owned by the `HBufC`. Since this pointer has a modifiable API, then the data owned by the `HBufC` *can* be modified. Here is an example:

Chapter 3 Symbian OS Fundamentals

```
HBufC* elementBuf16 = HBufC::NewLC(elementBuf8.Length());
TPtr elementPtr16 = elementBuf16->Des();
elementPtr16.Copy(elementBuf8);
```

The code segment is used to copy data from a narrow (8-bit) ASCII descriptor into a Unicode HBufC. First it creates a new HBufC elementBuf16 with same *length* (but not *size*) as the original 8-bit descriptor elementBuf8. It then uses the Des() method to get a TPtr, elementPtr16, pointing to the data of elementBuf16. Finally, it uses the Copy() method of TPtr to copy the data from elementBuf8 into elementBuf16 (to which elementPtr16 is pointing).

This particular TDes16 Copy() overload, which takes a TDesC8, converts from 8- to 16-bit by adding zero padding—effectively converting ASCII to Unicode. Note that the TDes8 Copy() overload that takes a TDesC16 does the opposite, stripping every other byte and effectively converting from Unicode to ASCII (so long as the Unicode characters are less than 255, otherwise they are set to 1).

One point worth stressing once more about TPtr objects (which applies to TPtrCs, too)—they *do not own the data that they point at*. In the example above, the data to which elementPtr16 points is **not** deleted when elementPtr16 goes out of scope, nor is there any way of doing this with the TPtr API. The HBufC pointer elementBuf16 must *itself* be deleted. You can have as many TPtrs as you like pointing to the buffer of an HBufC.

HBufC provides another pair of APIs that are extremely useful: ReAlloc() and ReAllocL(). As their name suggests, they are used for reallocating an HBufC.

What does this mean? Basically, it constructs a new HBufC on the heap. Should that allocation fail, ReAlloc() returns zero, and ReAllocL() will leave with KErrNoMemory, but in both cases the original HBufC *remains unchanged*. Only if the allocation is successful will ReAlloc() (or ReAllocL()) then copy the data into the new HBufC() and delete the original one automatically. A pointer to a *new* HBufC() is returned.

This poses a potentially nasty problem. Consider the following code example:

```
// Push onto Cleanup Stack
HBufC* variableBuf = HBufC::NewLC(10);
// Make some (potentially leaving) use of the original
variableBuf...

variableBuf = variableBuf->ReAllocL(20);   // Make it bigger
// The old variableBuf
CleanupStack::Pop();
CleanupStack::Push(variableBuf);           // The new one

// Make some (potentially leaving) use of the new
variableBuf...

CleanupStack::PopAndDestroy(variableBuf);
```

Chapter 3 Descriptors

What would happen if you did not pop the old `variableBuf` off the Cleanup Stack and push the new one on? Well, at very least the `CleanupStack::PopAndDestroy()` would panic, since the pointer passed in (the new value for `variableBuf`) would not match the one on the Cleanup Stack. What would happen if the leaving code after the reallocation were to leave? In that case, as the Cleanup Stack was unwound, a pointer to the *old* `variableBuf` would be deleted. Unfortunately, this has already been deleted as the last `ReAllocL()` step, causing a disastrous double-deletion. Follow the steps above, ensuring that the old `HBufC*` is popped and the new one is pushed, and you will be safe from such catastrophes.

Using TBuf and TBufC

If you refer back to Table 3-2, you will see that the buffer descriptors `TBuf` and `TBufC` are *templated* with a size—maximum size for a `TBuf` and actual size for a `TBufC`. The use of this templated size means that you must know the size of a `TBufC`, or the maximum size of a `TBuf`, at compile time.

The following code snippet shows how to declare a `TBufC` to hold three characters:

```
const TInt KMaxSymbolLength = 3;
...
// Delare a 3-character TBufC
TBufC<KMaxSymbolLength> iSymbol;
```

To demonstrate how you can use a `TBuf`, the `NotifyElementLoaded()` function, declared in `CElementsEngine`, will be discussed. The function is a callback to provide user feedback when an element has been loaded from file (more on files later). It uses a console object that allows descriptors to be printed:

```
_LIT(KTxtLoaded, "Loaded ");
...
void CElementsEngine::NotifyElementLoaded(
   TInt aNewElementIndex)
   {
   // Start with KTxtLoaded...
   TBuf<KMaxFeedbackLen> loadedFeedbackString(KTxtLoaded);

   // ...append the name of the new element...
   loadedFeedbackString.Append(
   (iElementList->At(aNewElementIndex)).Name());

   // ...append a new line...
   loadedFeedbackString.Append(KTxtNewLine);

   // ...and print to the console
   iConsole.Printf(loadedFeedbackString);
   }
```

First, the literal `KTxtLoaded` is defined using the macro `_LIT()`. It is set to contain the string "Loaded ". `KTxtNewLine` simply contains a "\n".

The `TBuf` constructor is templated with the maximum size and can also take a descriptor (or literal via implicit casting in this case) to copy into the new `TBuf`. In this example a `TBuf` of maximum size `KMaxFeedbackLen` is constructed and initialized with the characters "Loaded ".

An element name is then appended to the descriptor buffer using the `Append()` function. The argument passed in may look slightly confusing, but essentially the element at a particular position (`aNewElementIndex`) in the array is located and the name of the element retrieved using `CChemicalElement::Name()`. The new line characters are also added to the buffer. Again implicit casting means you can just pass the string literal as the argument.

`TBufC`s, without a modifiable API, are generally used less frequently. Here is how the `iSymbol` member of `CChemicalElement` is declared:

```
TBufC<KMaxSymbolLength> iSymbol;
```

Remember that only the nonmodifiable APIs listed for `TDesC` are available for `TBufC`, but this includes `operator =` (as used in `CChemicalElement::SetSymbol()` above).

Descriptors as Arguments and Return Types

Descriptors are often passed into functions and returned by them. It is common to use references to the base classes: `TDes` for modifiable descriptors and `TDesC` for nonmodifiable descriptors. This allows the greatest flexibility in implementation, as your API is not limited to one concrete type. Here are some common examples:

```
/**
* Setter function for the element's symbol.
* param aSymbol The element's new symbol.
*/
void SetSymbol(const TDesC& aSymbol);
```

Note that `const TDesC&` is used to pass in a nonmodifiable descriptor. While `TDesC&` might seem appropriate, remember that it is possible to assign to a `TDesC`. Doing this within the body of such a function can mean the value of the descriptor is modified, and so developers are encouraged to use `const TDesC&` in such circumstances.

The implementation of the above function is shown below:

```
void CChemicalElement::SetSymbol(const TDesC& aSymbol)
    {
    iSymbol = aSymbol;
    }
```

If the data to be passed as an argument is stored in an `HBufC`, then programmers new to descriptors sometimes insist on doing this:

```
DescriptorMethod(myHBufC->Des());
```

This is both inefficient and unnecessary *unless* `DescriptorMethod()` *takes a modifiable* `TDes&`. If `DescriptorMethod()` takes a `const TDesC&`, then dereferencing the `HBufC*` is all that is required:

```
DescriptorMethod(*myHBufC);
```

Modifiable descriptors can be used to pass back descriptors as shown below. Note that the Symbian OS convention is to precede such function names with "Get":

```
/**
 * Getter function for the element's symbol
 * @param aSymbol the element's symbol
 */
void GetSymbol(TDes& aSymbol) const;
```

Here is the implementation, but note that this function will panic if `aSymbol` is too small to accommodate `iSymbol`:

```
void CChemicalElement::GetSymbol(TDes& aSymbol) const
    {
    aSymbol = iSymbol;
    }
```

It is usually more convenient to simply return (a reference to) the descriptor. This is most commonly achieved using a `const TDesC&` (note that "Get" should not be used in the function name in this instance):

```
/**
 * Getter function for the element's symbol
 * @return The element's symbol
 */
const TDesC& Symbol() const;
```

And the implementation is as simple as:

```
const TDesC& CChemicalElement::Symbol() const
    {
    return iSymbol;
    }
```

This is only possible if the access function is returning the *whole* member descriptor. Functions that return *part* of a member descriptor must return a pointer descriptor—this allows you to return member data in place, with the efficiency that entails. Copying part of a descriptor to a temporary variable for returning is not only inefficient, but can also be dangerous—do not return a reference to a temporary variable, and always check your compiler warnings!

A classic example of returning a pointer to only part of a member descriptor is `TDesC::Left()`. It returns a nonmodifiable pointer descriptor to represent the leftmost part of the data, as shown in Figure 3–3.

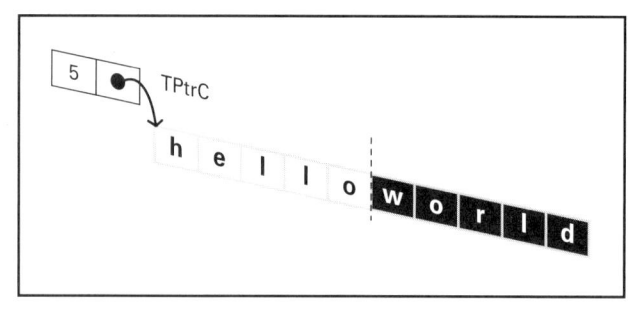

Figure 3-3 Illustration of TDesC::Left().

Note that the pointer descriptor must be returned by value. Also note that `TPtrC` is genuinely immutable—there is *no way* to modify the data to which it points—so it is not necessary to return a *const* `TPtrC`.

> **TIP** When writing your API, it is often worth considering the extra flexibility that returning a `TPtrC` offers. It allows you to return all or part of a member descriptor, and this means you could possibly change the implementation of your function while retaining the same API. Remember, however, that returning a `TPtrC` will be slightly less efficient (it contains a pointer to the data and its size), so consider both facts when making your design choice.

Package Descriptors

Package descriptors are a useful way of packaging complex data into a descriptor (the SDK guide refers to them as package *buffers*, but this is something of a misnomer—only one of them is actually a buffer with its own data). Once packaged, this data can then be passed to any function that takes a `TDesC8` derived class. This technique is used within Symbian OS to pass structured data between clients and servers in a type-safe manner.

Note that package descriptors should not be considered a way of streaming arbitrary data either to memory or to a file. They cannot handle compound data, where one object owns another via a pointer. See the use of streams in the section **Files, Streams and Stores** later in this chapter for a more detailed explanation of how to do this properly.

Three package descriptor types are available in Symbian OS: `TPckg`, `TPckgC` and `TPckgBuf`. Table 3-3 provides some detail about each.

Here is a simple (but rather contrived) example that demonstrates the use of a `TPckg` to contain and modify a `TRect` object:

```
TRect myRect(0, 0, 10, 20);
// Construct the package descriptor
TPckg<TRect> myRectPckg(myRect);

// Note that operator() on myRectPckg will return a
```

Chapter 3 Collection Classes 109

```
// modifiable TRect& to the rectangle
// it points to, i.e. myRect
TInt twenty = myRectPckg().Height();
myRectPckg().SetWidth(30);              // Modify width
TInt thirty = myRect.Width();
```

Table 3–3 Package Descriptor Types

Name	Owns Data	Modifiable API	Comments
TPckg	No	Yes	TPtr8 derived package descriptor. Pointer points to the original data, and allows modification of this data.
TPckgC	No	No	TPtrC8 derived package descriptor. Pointer points to the original data, but no modification of this data is permitted.
TPckgBuf	Yes	Yes	TBuf8 derived package buffer. Buffer contains a copy of the original data, which may be modified. Such modification will have no effect on the original data.

This example may appear somewhat pointless, but remember that package descriptors are really important during Client/Server communication—this is discussed later, in the section **Client/Server Architecture**.

Declaring a package descriptor is simply a case of selecting the appropriate package descriptor variety and then specifying the data type it is to contain as the template parameter. All three package descriptors have constructors which take an instance of the templated data type. Only TPckgBuf has a default constructor that allows an uninitialized buffer to be constructed.

Note the use of operator(). For the modifiable package descriptor types these return modifiable references to the object in the buffer for TPckgBuf, or the object pointed to by TPckg. In the case of TPckgC a nonmodifiable reference is returned, so only const functions can be called on this reference.

The example above also stresses that TPckg objects do not own the data—the original myRect object is modified by the line myRectPckg().SetWidth(30). If a TPckgBuf were used instead, only the package buffer's *copy* of the data would be modified—myRect would remain unchanged. If a TPckgC were used, this line would not compile, since an attempt is being made to call a non-const function (SetWidth()) on a const reference.

Collection Classes

Symbian OS does not support the Standard Template Library (STL) for a number of reasons—primarily because of STL's large footprint. However, Symbian OS does provide a number of templated collection classes, so that developers do not need to write their own arrays, linked lists and so on.

Chapter 3 Symbian OS Fundamentals

This section will concentrate on dynamic arrays, as these are the most commonly used collection classes in Symbian OS. There are two basic types of array: `CArray` and `RArray`. `CArray`s are perhaps the most flexible in some respects, but while `RArray`s do impose some limitations, they are more efficient and often easier to use.

SEE ALSO

See the SDK documentation for a full discussion of other collection classes such as `TSglQue` (singly-linked lists), `TDblQue` (doubly-linked lists), `CCirBuf` (circular buffers) and `TBtree` (balanced trees). Note, however, that these are rarely used as Symbian OS applicatons generally store only a (fairly small) fixed amount of data and so have little need for complex data structures. Usually an array will do.

RArray and RPointerArray Types

Since `RArray`s are easier to use than `CArray`s, they will be discussed first. An `RArray` is a simple array of fixed-length objects, while an `RPointerArray` is an array of pointers to objects. It is worth noting from the outset that `RArray`s impose a limitation of 640 bytes on the size of their elements. Usually this is not an issue, since most objects are likely to be smaller than this limit, and `RPointerArray` can be used to point to arbitrarily large objects anyway. Generally, therefore, owing to their greater efficiency, flexibility and ease of use, `RArray`s are recommended over `CArray` types.

Basic APIs

`RArray`s and `RPointerArray`s tend to be constructed on the stack or simply nested directly into other heap-based objects. They are templated, and therefore require a template argument in their declarations and constructors. In the case of an `RArray`, this template argument should be a simple type or an R-type or T-type object; for `RPointerArray`s it may be any type.

Note that as these arrays are R-classes, they must be closed before they go out of scope.

WARNING

If you create a local (stored on the stack) `RArray`, then you need to use the Cleanup Stack to make sure that there are no memory leaks if a function leaves before the array is closed. Rather than the usual `CleanupStack::PushL()`, you need to use `CleanupClosePushL(myArray)` to push your array onto the stack. This will ensure that if a leave occurs, `Close()` will be called on your array, freeing up all of the memory allocated to the array.

If you are using a local `RPointerArray` that owns the data it points to, then either this data must also be pushed onto the Cleanup Stack, or you should

Chapter 3 Collection Classes

consider making the array a member of the class and calling `RPointerArray::ResetAndDestroy()` in its destructor.

Here is the declaration of the `RPointerArray` used in the **Elements** example application:

```
class CElementRArray : public CElementList
    {
    ...
private:
    RPointerArray<CChemicalElement>       iElementArray;
    };
```

`CElementList` is an abstract base class that defines a common interface for manipulating a list of elements; regardless of what array type is used.

Since the array is nested within the `CElementRArray` class, and no other data is owned on the heap, there is no need for a second-phase constructor. The C++ constructor for the `CElementRArray` class is implemented like this:

```
CElementRArray::CElementRArray() :
iElementArray(KElementListGranularity)
    {
    }
```

This simply invokes the only non-default `RPointerArray` constructor, specifying the granularity of the array—the number of elements that can be appended to the array before reallocation needs to occur (see the subsection on *CArray Types* for a fuller discussion of granularity).

Regardless of the granularity, all newly constructed dynamic arrays have no members. The easiest way to add them is to use the `Append()` function. Here is `CElementRArray::AppendL()`, which simply acts as a proxy and leaves if `Append()` fails (an error code is returned):

```
void CElementRArray::AppendL(const CChemicalElement*
aElement)
    {
    User::LeaveIfError(iElementArray.Append(aElement));
    }
```

Note that, since this is a pointer array, the `Append()` function takes *a pointer* to the templated class. It is also worth considering the issue of ownership—the array is generally considered to take ownership of the object passed by pointer, and so it may be necessary to pop any locally scoped pointer off of the Cleanup Stack once it has been appended:

```
// Construct a new element from the mixed data.
// Remember NewLC leaves element on the Cleanup Stack
CChemicalElement* newElement = CChemicalElement::NewLC(aName,
```

Chapter 3 Symbian OS Fundamentals

```
aSymbol, atomicNumber, relativeAtomicMass, type,
radioactive);

// Append new element to the element array
AppendL(newElement);

// Remove the element pointer from the cleanup stack
CleanupStack::Pop(newElement);    // Now owned by array
```

Failure to do this may result in a double-deletion, should a leave occur.

In addition to append, it is also possible to insert a new element at a specified position. The prototype for this is:

```
TInt Insert(const T* anEntry, TInt aPos);
```

where `T` is the templated class and `aPos` is the insertion position—in other words, the index into the array. The return code indicates whether the insertion was successful or not (perhaps due to lack of memory, for instance).

To find the number of elements stored, simply invoke the `Count()` method as shown below:

```
TInt CElementRArray::NumberOfElements() const
    {
    return iElementArray.Count();
    }
```

Accessing elements is achieved via `operator[]`. Two overloads are provided, one of which returns a `const` pointer, the other a non-`const` pointer. Again, `CElementRArray` provides a proxy for both of these overloads:

```
const CChemicalElement& CElementRArray::At(TInt aPosition)
const
    {
    return *iElementArray[aPosition];    // Const overload
    }

CChemicalElement& CElementRArray::At(TInt aPosition)
    {
    return *iElementArray[aPosition];    // Non-const overload
    }
```

NOTE In Symbian OS, passing an object by pointer generally denotes transfer of ownership. Rather than return a pointer, the method returns a reference, and the pointer is dereferenced to obtain the value pointed to. Ownership is not passed out, but remains with the array.

In common with most arrays, Symbian OS dynamic arrays are zero indexed, and so the value of `aPosition` must not be negative or greater than the

Chapter 3 Collection Classes

number of elements minus one (N - 1). A panic will occur if these conditions are not met.

Once an array is finished with, it must be reset before being allowed to go out of scope (or before the object in which it is nested is deleted). Both RArray and RPointerArray implement a Reset() method that frees all of the memory allocated for storing the elements. RPointerArray also implements an additional method, ResetAndDestroy(). Since the elements are pointers to heap cells, simply calling Reset() would deallocate the pointers *themselves* but not the objects to which they point. That is fine if the array does not own the data, but in the **Elements** example it does. CElementRArray implements a method DeleteAllElements() as a proxy for ResetAndDestroy():

```
void CElementRArray::DeleteAllElements()
    {
    // Deletes all the elements
    iElementArray.ResetAndDestroy();
    }
```

CElementRArray's destructor is then simply:

```
CElementRArray::~CElementRArray()
    {
    DeleteAllElements();
    }
```

Sorting and Finding

RArray provides powerful and flexible finding and sorting APIs. The simpler of these is sorting, so that will be discussed first.

Sorting

The general procedure to sort an RArray (or RPointerArray) is:

- Define a comparison function that returns 0 if the two objects are identical, -1 if the first is smaller or +1 if the first is bigger. This function may be implemented as a *static* member function, a global function or a member of a namespace—you will need a pointer to it for the next step.

- Construct a TLinearOrder object, using a pointer to the desired comparison function in its constructor.

- Invoke Sort() on the array, passing in the TLinearOrder object.

The **Elements** application demonstrates how to use the sorting API. The function CElementRArray::SortByAtomicNumberL() is implemented to sort the elements of iElementArray into an order determined by their atomic number. The comparison function here would simply need to take two integers (atomic numbers) and decide which was the biggest.

Here is the implementation of CElementRArray::SortByAtomicNumberL():

```
using namespace NElementComparisonFunctions;

void CElementRArray::SortByAtomicNumberL()
    {
    // Wrap up a function pointer to the comparison function
    TLinearOrder<CChemicalElement>
order(CompareElementsAtomicNumber);
    iElementArray.Sort(order);
    }
```

The actual sort API is very simple, but it is the `TLinearOrder` object that introduces the power and complexity behind the process.

The constructor of `TLinearOrder` takes, in addition to the template parameter, a function pointer to a comparison function. For the sake of tidiness, all of the comparison functions (a few of which are shown below) have been wrapped up into a namespace:

```
namespace NElementComparisonFunctions
    {
    TInt CompareElementsName(const CChemicalElement& aElement1,
const CChemicalElement& aElement2);
    TInt CompareElementsAtomicNumber(const CChemicalElement&
aElement1, const CChemicalElement& aElement2);
    };
```

Of interest in this example is `NElementComparisonFunctions::Compare ElementsAtomicNumber()`. Here is the implementation:

```
TInt
NElementComparisonFunctions::CompareElementsAtomicNumber(const
CChemicalElement& aElement1, const CChemicalElement&
aElement2)
    {
    if (aElement1.AtomicNumber() == aElement2.AtomicNumber())
        {
        return 0;
        }
    return (aElement1.AtomicNumber() <
aElement2.AtomicNumber()) ? -1: 1;
    }
```

The effect of this code is simply to return zero if the atomic numbers are equal, -1 (less than zero) if `aElement1`'s atomic number is less than `aElement2`'s, and +1 (greater than zero) if `aElement1`'s atomic number is greater than `aElement2`'s.

This function can then be called by the `TLinearOrder` object, as it is used by the `RArray::Sort()` (or `RPointerArray::Sort()`) method to determine if two elements are in the correct order or not. This is the basis of how the sorting algorithm works.

As a matter of interest, here is how the equivalent name comparison function works. It is actually simpler, because `TDesC` provides a `CompareF()` method that does exactly what is required:

```
TInt NElementComparisonFunctions::CompareElementsName(const
CChemicalElement& aElement1, const CChemicalElement&
aElement2)
    {
    return aElement1.Name().CompareF(aElement2.Name());
    }
```

Finding

Finding is slightly more complex, but it builds on the sort process outlined above. The basic steps are:

- Define a comparison function that returns `ETrue` if the objects are identical, `EFalse` otherwise. Again it must be a static, global or function of a namespace.

- Construct a `TIdentityRelation` object, using a pointer to the desired equality function in its constructor.

- Construct a new object, which contains the data you are looking for.

- Invoke `Find()` on the array, passing in the `TIdentityRelation` object and the example object. This function will return the (zero-based) index of the matching element, or `KErrNotFound`.

In the example application, `CElementRArray::FindElementByAtomic NumberL()`, enables an array to be searched for an element with a particular element number. This function creates the `TIdentityRelation` object and then calls `CElementRArray::DoFindL()` where `RArray::Find()` is called.

The `Find()` API takes two arguments:

```
TInt foundAt = iElementArray.Find(example, aRelation);
```

In this case, `example` is an object that contains values matching those being sought, and `aRelation` is a `TIdentityRelation` that is used to equate two objects. Like `TLinearOrder` it is constructed with a pointer to the equality function:

```
TIdentityRelation<CChemicalElement>
relation(ElementsHaveSameAtomicNumber);
```

Here is the `ElementsHaveSameAtomicNumber` function (also defined in the `NElementComparisonFunctions` namespace):

```
TBool
NElementComparisonFunctions::ElementsHaveSameAtomicNumber(
    const CChemicalElement& aElement1,
    const CChemicalElement& aElement2)
```

```
    {
    return aElement1.AtomicNumber() ==
aElement2.AtomicNumber();
    }
```

Some of `CElementRArray`'s find process is factored into one common function that is invoked by `FindElementByAtomicNumberL()`. Both functions are shown below:

```
using namespace NElementComparisonFunctions;

TInt CElementRArray::FindElementByAtomicNumberL(TInt
aAtomicNumber) const
    {
    // Wrap up a function pointer to the identity function
    TIdentityRelation<CChemicalElement>
relation(ElementsHaveSameAtomicNumber);

    // Find the element using the given atomic number only
    return DoFindL(relation, KNullDesC, KNullDesC,
aAtomicNumber);
    }

TInt CElementRArray::DoFindL(const
TIdentityRelation<CChemicalElement>& aRelation, const TDesC&
aName, const TDesC& aSymbol, TInt aAtomicNumber) const
    {
    // A utility function to wrap up the find operation
    CChemicalElement* example = CChemicalElement::NewL(aName,
// The example element to search for
aSymbol, aAtomicNumber, 0);

    TInt foundAt = iElementArray.Find(example, aRelation);
    delete example;          // No longer needed
    User::LeaveIfError(foundAt);
    return foundAt;
    }
```

`FindElementByAtomicNumberL` constructs the `TIdentityRelation` from the appropriate function pointer. It then invokes `DoFindL`, passing dummy values for all but `aAtomicNumber` (since `ElementsHaveSameAtomicNumber()` will ignore them anyway).

`DoFindL()` first of all constructs a new `CChemicalElement` from its arguments. In this example the new `CChemicalElement` will have the atomic number that is being sought. It then invokes `Find()`, passing the example `CChemicalElement` and the `TIdentityRelation` that is to be used for the equality check. Once the find has completed, the `CChemicalElement` may be destroyed. If the value returned by `Find()`, indicates an error (`KErrNotFound`), the function leaves; otherwise the index of the `CChemicalElement` with atomic number `aAtomicNumber` is returned.

It is perhaps worth stressing that both the comparison and equality functions may be as complex as you like for the particular problem in hand. The equality function, for instance, might return `ETrue` if the relative atomic masses are identical to *within a given accuracy* (say, two decimal places).

CArray Types

All of the `CArray` types use buffers to store their data. Buffers (derived from `CBufBase`) provide access to regions of memory and are in some ways similar to descriptors in that respect. However, while descriptors are intended to store data objects whose maximum size is not expected to alter much, buffers are expected to grow and shrink dynamically during the lifetime of the program. There are two buffer types—flat and segmented—and these two types give rise to two basic subtypes of `CArray`.

Flat buffers store their entire data within a single heap cell. Once full, any subsequent append operation requires a new heap cell to be allocated that is large enough to contain the original and new data. Once the allocation has completed, all of the old data is copied to the new cell, and the old cell is released back to the global heap.

Segmented buffers store their data in a doubly-linked list of smaller *segments*, each of which is a separate heap cell of fixed size. Once all of the segments have been allocated, a new segment will be allocated and added into the list, with the old data remaining in place, and without the need for copying. While this can reduce the memory thrashing associated with frequent reallocation, accessing data is less efficient than flat buffers, since the list of segments must be traversed, plus more memory is consumed by the need to store linked list pointers. It can also lead to memory fragmentation.

Note that inserting into the middle of a segmented buffer may also result in the creation and insertion of a new segment into the middle of the segment list. This can also require some moving of data if it causes a segment to be effectively "split."

All `CArray` types have an associated granularity. This specifies how many elements can be added before the array must grow. For flat buffer arrays with a granularity of N, the initial allocation will be large enough to hold N elements. Subsequent expansion will see the buffer size increase by N elements—even to just add a single element. For segmented buffers with a granularity of N, each segment is large enough to hold N elements. If the granularity is too large, then considerable amounts of memory may go unused. If it is too small, then reallocation of the buffer may occur frequently, or there will be a large number of small segments.

An outline of all of the available `CArray` types is provided in Table 3–4.

Figures 3–4 and 3–5 illustrate how each of the different `CArray` types is arranged and stores its data. In each case the granularity shown is 3.

Table 3-4 Summary of CArray Types

Name	Element Size	Buffer Type	Use
`CArrayFixFlat`	Fixed	Flat	Use for fixed-sized `T` or `R` type objects when infrequent allocation is expected
`CArrayVarFlat`	Variable	Flat	Use for variable-sized `T` or `R` type objects when infrequent allocation is expected
`CArrayPtrFlat`	Pointer	Flat	Use for pointers to objects when infrequent allocation is expected
`CArrayPakFlat`	Variable (packed)	Flat	Use to store an array of variable sized `T` or `R` type objects within one heap cell when infrequent allocation is expected
`CArrayFixSeg`	Fixed	Segmented	Use for fixed-sized `T` or `R` type objects when frequent allocation is expected
`CArrayVarSeg`	Variable	Segmented	Use for variable-sized `T` or `R` type objects when frequent allocation is expected
`CArrayPtrSeg`	Pointer	Segmented	Use for pointers to objects when frequent allocation is expected

Arrays of variable sized elements are internally managed as an array of pointers, but with an interface that works with objects by value. Packed arrays are managed by value and contain length information. How the length of a T or R class object varies is not really defined!

Regardless of the of the array type, the APIs remain much the same, although there are, of course, some differences. In the **Elements** example application, the class `CElementCArray` (derived from `CElementList`) uses a `CArrayPtrFlat` array to store the elements. This is a replacement for the `RPointerArray` used by `CElementRArray`, which was covered previously.

Basic APIs

Since `CArray`s are `CBase` derived, they should always be constructed on the heap. Like `RArray`s, they are templated, and pretty much the same rules apply for the type of objects they can contain: simple types for `CArrayFix`, `CArrayVar` or `CArraySeg` arrays, and pointers to objects of any type in the case of `CArrayPtr` arrays, as used in the example application.

Here is a declaration taken from the **Elements** example:

```
class CElementCArray : public CElementList
    {
private:
    CArrayPtrFlat<CChemicalElement>* iElementArray;
    };
```

Chapter 3   Collection Classes

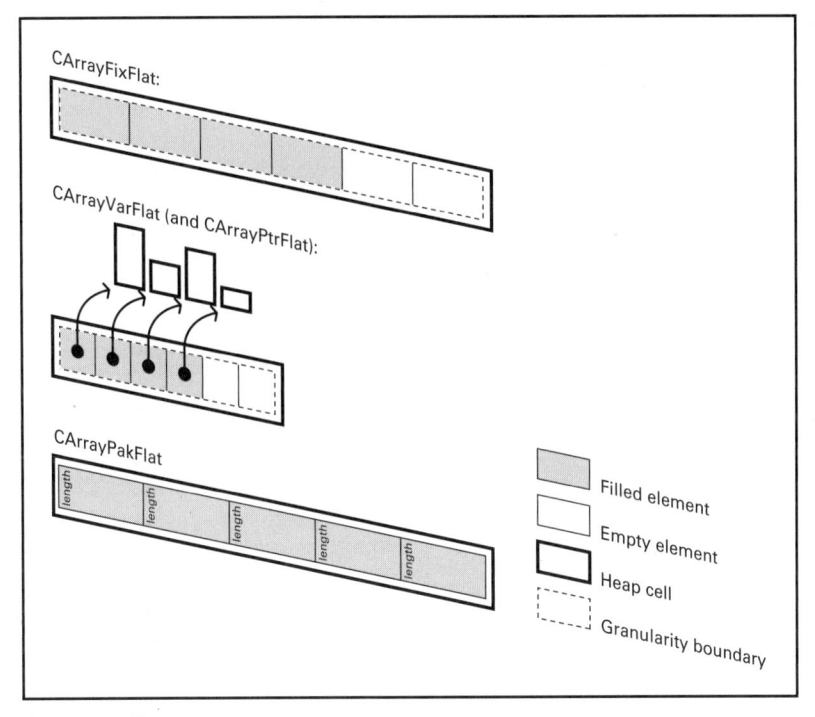

Figure 3–4 Flat arrays.

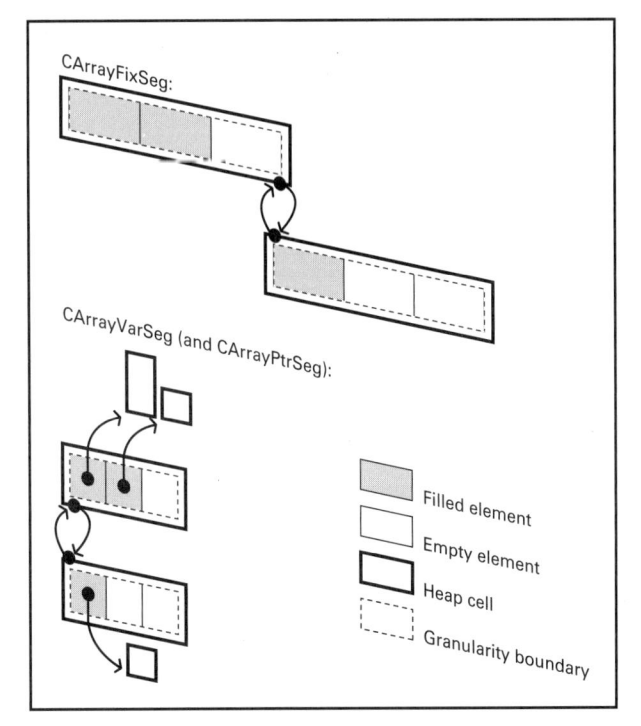

Figure 3–5 Segmented arrays.

Naturally, `CElementCArray` will require two-phase construction because of the array being owned on the heap. Here is the second-phase constructor:

```
void CElementCArray::ConstructL()
    {
    iElementArray = new (ELeave)
CArrayPtrFlat<CChemicalElement>(KElementListGranularity);
    }
```

`CArray`s contain no elements when they are constructed and they implement an `AppendL()` function—note the trailing `L`, it leaves rather than returning an error code. This makes the `CElementCArray::AppendL()` proxy even simpler:

```
void CElementCArray::AppendL(const CChemicalElement*
aElement)
    {
    // Cast away pointer const-ness
    iElementArray->AppendL(
    const_cast<CChemicalElement* const>(aElement));
    }
```

Or at least it would be simpler were it not necessary to cast away the const-ness of `aElement`. This is because `CArrayPtrFlat::AppendL()` requires a `const` pointer to a modifiable object, not a (modifiable) pointer to a `const` object, as would be the case here. A `const*const` pointer (`const` pointer to a `const` object) in the function prototype would not help, either!

TIP It is worth mentioning that `CArrays` also have an `ExtendL()` function. This function returns a non-const reference to a newly constructed object of the templated type. This is often simpler to use than constructing a new object and then appending it—simply use `ExtendL()` and modify the reference it returns directly. Note that this operation uses the class's *default* constructor, so it should be used only with simple types.

`CArray` classes also implement an `InsertL()` method, which operates just like `RArray::Insert()` (other than it leaves on failure), and a `Count()` method. They also have two overloaded `operator[]`s for accessing members, and two `At()` overloads that work just the same as the `operator[]`. It is a matter of preference, but using `At()` with a pointer to a `CArray` can be cleaner than having to dereference the pointer to be able to use `operator[]`:

```
const CChemicalElement& CElementCArray::At(TInt aPosition)
const
    {
    return *(*iElementArray)[aPosition]; // Const overload
    }

CChemicalElement& CElementCArray::At(TInt aPosition)
    {
```

Chapter 3 Collection Classes

```
    return *iElementArray->At(aPosition);// Non-const overload
    }
```

As with `RArray`s, `CArray`s must be `Reset()` before being deleted to deallocate all of their elements. And as with `RPointerArrays`, `CArrayPtrFlat` implements a `ResetAndDestroy()` method to delete all of the objects pointed to as the pointers themselves are deallocated. Again, `CElementCArray::DeleteAllElements()` is called by the destructor in the example application, and is implemented like this:

```
void CElementCArray::DeleteAllElements()
    {
    // Deletes all the elements
    iElementArray->ResetAndDestroy();
    }
```

The destructor must invoke this function and then delete the array *itself*:

```
CElementCArray::~CElementCArray()
    {
    DeleteAllElements();
    delete iElementArray;
    }
```

Sorting and Finding

The sorting and finding APIs provided by `CArray`s are a little less flexible than those of `RArray`s, but they are correspondingly easier to use.

Sorting

The sorting of a `CArray` consists of just two steps:

- Construct an instance of the appropriate key, specifying the field offset, the sort type and (for char arrays) the text length.
- Invoke `Sort()` on the array, passing in the key.

These steps will be explained by examining `CElementCArray::SortByAtomicNumberL()` from the example application. Here is the implementation:

```
void CElementCArray::SortByAtomicNumberL()
    {
    TKeyArrayPtr key(_FOFF(CChemicalElement, iAtomicNumber), ECmpTInt);
    User::LeaveIfError(iElementArray->Sort(key));
    }
```

Notice that, unlike the `CElementRArray` implementation, this method is self-contained. It does not need to refer to other functions to sort the elements into order. All that is needed is a key.

Table 3–5 shows the correct key type for the different `CArray` types.

Table 3–5 Key and Array Types

Name	Key Type	Notes
CArrayFixFlat	TKeyArrayFix	
CArrayVarFlat	TKeyArrayVar	
CArrayPtrFlat	TKeyArrayFix-derived	Requires special implementation (see example code)
CArrayPakFlat	TKeyArrayVar	
CArrayFixSeg	TKeyArrayFix	
CArrayVarSeg	TKeyArrayVar	
CArrayPtrSeg	TKeyArrayFix-derived	Requires special implementation (see example code)

Note the need for a special implementation for pointer arrays. Using a `TKeyArrayFix` instance on a `CArrayPtr` will compile *but will not produce meaningful results*! This is because the key's `At()` implementation expects the array to contain *instances* of the data, not *pointers to instances*. (In effect, you are sorting addresses rather than values.)

The solution is to define a new `TKeyArrayFix`-derived key that overrides the implementation of `TKeyArrayFix::At()`. The implementation of the private nested class `TKeyArrayPtr` in the **Elements** application demonstrates how to do this, and for reference, the overridden function is shown here:

```
TAny* CElementCArray::TKeyArrayPtr::At(TInt aIndex) const
   {
   if (aIndex == KIndexPtr)
      {
      return *(TUint8**)iPtr + iKeyOffset;
      }
   return *(TUint8**)iBase->Ptr(aIndex *
sizeof(TUint8**)).Ptr() + iKeyOffset;
   }
```

You do not need to understand this, but feel free to copy it in your own code should you ever need to sort a `CArrayPtr` class. Naturally it does not necessarily have to be a private nested class—it is merely implemented this way, since it is not needed elsewhere in the **Elements** example.

Once you have chosen the right key class for your array type, the next step is to instantiate it. The constructors for all of the keys are pretty much the same. Here are the ones for `TKeyArrayFix`:

Chapter 3 Collection Classes

```
TKeyArrayFix(TInt aOffset, TKeyCmpText aType);
TKeyArrayFix(TInt aOffset, TKeyCmpText aType, TInt aLength);
TKeyArrayFix(TInt aOffset, TKeyCmpNumeric aType);
```

All of them take an integral offset value, `aOffset`. This is simply the offset of the "field" (or data member) within the class. How can you work this out? The macro `_FOFF(c,f)`, where `c` is the class and `f` is the field, will calculate it for you. The first line of the `CElementCArray::SortByAtomicNumberL()` shows its use:

```
TKeyArrayPtr key(_FOFF(CChemicalElement, iAtomicNumber),
ECmpFolded);
```

Within the key's constructor you must specify the field type. Table 3–6 summarizes those available.

Table 3–6 Key Types

Field Type	Key Types
Integer	`ECmpTInt8`, `ECmpTInt16`, `ECmpTInt32`, `ECmpTInt`, `ECmpTUint8`, `ECmpTUint16`, `ECmpTUint32`, `ECmpTUint`, `ECmpTInt64`
Text	`ECmpNormal`, `ECmpNormal8`, `ECmpNormal16`, `ECmpFolded`, `ECmpFolded8`, `ECmpFolded16`, `ECmpCollated`, `ECmpCollated8`, `ECmpCollated16`

The integer types simply match the integer type of field (for example, use `ECmpTInt` for a `TInt` object). The textual types specify the text "width" (8- or 16-bit), and whether the text should be treated normally, *folded* or *collated*. Folding effectively ignores differences assumed to be insignificant for comparison, such as case differences. Collating removes all such differences and also allows special linguistic characters and character combinations to be compared, such as equating "ss" with "ß" in German, or "Æ" with "ae".

Notice that the length can also be specified for text comparisons. This is not necessary for descriptor types, since the length is implicit. If a length *is* specified, however, the text is assumed to be a simple `char` array.

Here is how the key is defined in `CElementCArray::SortBySymbolL()`:

```
TKeyArrayPtr key(_FOFF(CChemicalElement, iSymbol),
ECmpFolded);
```

`ECmpFolded` specifies that the sorting is done on a text field and that simple case differences should be ignored. Since no size is given, the field `iSymbol` is assumed to be a descriptor with an implicit length.

It is important to realize that the `_FOFF` macro *must be able to access the data members*. This implies that `CElementCArray` must be a friend of

`CChemicalElement` for the sort (and find) methods to be able to calculate the offset of `CChemicalElement`'s private data members. Another alternative would be to make the data public, but that is an undesirable design option.

Once the key has been constructed, sorting is simply a matter of passing the key into the `Sort()` method of the array.

Finding

Finding data in a `CArray` is a little more complex; it involves the following steps:

- Construct an instance of the appropriate key type, specifying the offset of the field to be matched, the comparison type and (for `char` arrays) the text length.
- Construct a new object, which contains the data you are looking for.
- Invoke `Find()` on the array, passing in the key, the example object and a `TInt` to receive the search result. `Find()` will return zero if the item is found and will set the search result to the index the item was found at. If the item is not found, then the search result is set to the size of the array and an error code is returned.

The example application provides `CElementCArray::FindElementByAtomicNumberL()` to allow you to find, within the array, an element with a particular atomic number. This function creates the key and then invokes `DoFindL()` where `Find()` is called.

```
TInt error = iElementArray->Find(example, aKey, foundAt);
```

The first argument, `example`, is an object of the templated type matching the data being sought. `aKey` is the key to use for the search—this specifies *which* data field is to be matched in exactly the same way as with sorting. The final argument, `foundAt`, is a `TInt&` that will contain the index of the matching element if one is found. If no match is found, or some other error occurs, then an error code will be returned by the function—in this case `foundAt` will contain the size of the array.

As with `CElementRArray`, some of the common code within the find functions has been factored out. The implementation of `CElementCArray::FindElementByAtomicNumberL()`, together with `CElementCArray::DoFindL()` is shown below:

```
TInt CElementCArray::FindElementByAtomicNumberL(TInt
aAtomicNumber)
    const
    {
    // Make the key
    TKeyArrayPtr key(_FOFF(CChemicalElement, iAtomicNumber),
ECmpTInt);

    // Find the element using the given atomic number only
```

```
    return DoFindL(key, KNullDesC, KNullDesC, aAtomicNumber);
    }

TInt CElementCArray::DoFindL(TKeyArrayPtr& aKey, const TDesC&
aName,
    const TDesC& aSymbol, TInt aAtomicNumber) const
    {
    // A utility function to wrap up the find operation
    CChemicalElement* example = CChemicalElement::NewL(aName,
    // The example element to search for
aSymbol, aAtomicNumber, 0);
    TInt foundAt = 0;
    TInt error = iElementArray->Find(example, aKey, foundAt);
    delete example;

    User::LeaveIfError(error);

    return foundAt;
    }
```

The `DoFindL()` function has been implemented in such a way that it can find elements based on their name, symbol or atomic number. Since the discussion here is concerned with finding elements of a particular atomic number, a null descriptor, `KNullDesC`, is passed through for both the name and the symbol. Remember that you need to check the value returned from `Find()` to determine whether a match was made. In this example the function, `DoFindL()` will leave if the required element is not found successfully.

Descriptor Arrays

Descriptors commonly need storing in arrays, so Symbian OS provides a number of instantiable descriptor arrays for such situation.

All of the descriptor arrays are derived from `MDesCArray` (which predictably has `MDesC8Array` and `MDesC16Array` variants). `MDesCArray` provides a common interface for accessing the basic functionality of the different descriptor array classes (which include `CDesCArrayFlat`, `CPtrCArray` and numerous others). The (simplified) declaration of `MDesC16Array` is shown below:

```
class MDesC16Array
    {
public:
    virtual TInt MdcaCount() const = 0;
    virtual TPtrC16 MdcaPoint(TInt aIndex) const = 0;
    };
```

`MDescC16Array::MdcaCount()` returns the number of descriptors in the array; `MDescC16Array::MdcaCount()` returns a `TPtrC16` pointing to the descriptor specified by `aIndex`. This API allows any type of descriptor array to be accessed regardless of its actual type.

Full details of creating and using desciptor arrays can be found in the SDK documentation.

Limitations

To finish off the discussion of `CArray` types, it is worth pointing out their limitations, particularly with regard to finding and sorting.

Firstly, it is only possible to search and sort integer or *nested* text fields. It is not possible to search or order floating-point values or *heap-allocated* text (including `HBufC*`).

It is only possible to sort into ascending order.

Searches always start from the beginning of the array, and they will only return the first match — note that this also applies to `RArray`s.

As a final note, it is worth being aware that a number of Symbian OS APIs use descriptor arrays — since these are `CArray` types, it is sometimes impossible to avoid their use!

Array Summary

Symbian OS provides two basic types of dynamic array: `RArray`s and `CArray`s. The array classes are all templated, with the parameter defining the type of object that is to form an array element. Both types of array offer the standard functionality that you would expect from an array class, such as sorting and searching.

Where possible you should favor the use of an `RArray` type, as they offer greater efficiency and ease of use, particularly when using pointers. However, you should be aware that `RArray`s do impose a limit on the size of the elements (640 bytes), and for efficiency, insertions into the array should be infrequent. The main benefit of a `CArray` is its support for segmented storage. Obviously the type of array you use may be dictated by the required parameters of a function you want to call, but, if not, always consider a `RArray` first.

Using Asynchronous Services with Active Objects

Symbian OS is a heavily asynchronous operating system. Virtually all system services are provided through servers, which operate in their own process for high reliability. (A process is a unit of memory protection, and it isolates the application thread from all other processes.) Service provider APIs typically have asynchronous and synchronous versions of their functions available to their client applications, but to avoid blocking the application's user interface you would usually use the asynchronous version. Most time-consuming operations are therefore made as a request to some asynchronous service provider, such as the file system (this will be introduced in the next section). The service request function returns immediately, while the request itself is

Chapter 3 Using Asynchronous Services with Active Objects

processed in the background. This means that the program needs to receive some notification in the future when the request is complete, and in the meantime it needs to be able to get on with other tasks like responding to user input or updating the display.

Such asynchronous systems are prevalent in modern computing, and there are a number of ways of implementing them:

One popular practice is to use multiple threads in a preemptive system: a new execution thread is spawned to handle each asynchronous task, making the request and awaiting completion—often by polling the request provider to see if the request has completed yet. Although this approach is possible in Symbian OS, it is inefficient and strongly discouraged.

SEE ALSO   Further information on implementing multithreading can be found under the `RThread` API reference section of the SDK documentation, but this approach should rarely be used, and further explanation is beyond the scope of this book.

Another alternative, and the one favoured for Symbian OS applications, is to use **cooperative multitasking**. In a preemptive system, the operating system's kernel decides the current thread has had enough use of the processor, and allows another thread to preempt it, usually before it has completed processing. Cooperative (or non-preemptive) multitasking requires that the currently running task gives up use of the processor and allows other tasks to run. All application processing occurs within a single thread, and so the multiple tasks must *cooperate*.

Cooperative multitasking is usually implemented as some kind of wait or message loop, and this is the case with Symbian OS, as shown in Figure 3–6.

The loop starts by waiting—the single application thread is blocked until a service request completes, and the service provider ("server"), which is running in a separate thread (or process), signals that it has completed by incrementing a semaphore. Symbian OS does not poll, as this wastes processor cycles and therefore battery power. Note also that it is important that at least one request is made before the start of the loop, or the wait loop will never be signaled!

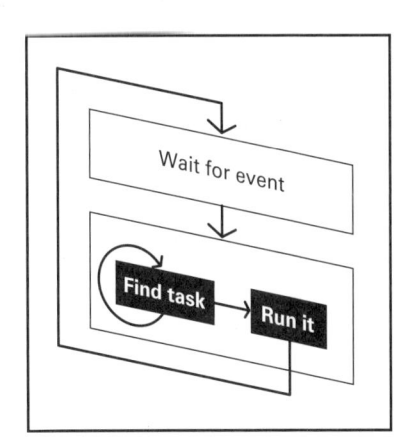

Figure 3–6 A cooperative multitasking wait loop.

Once the wait loop has been signaled, execution can continue. The system searches through each of the outstanding task requests until it finds a completed task and runs its handler function. The loop then waits for the completion of the next request.

If multiple requests complete at around the same time, then the first completed task found may not be the task that signaled. However, as all completed tasks will have incremented the semaphore, this will likely be handled on the next iteration of the loop—event completion signals are queued and handled in sequence, but only one request completion is handled in each iteration of the loop.

To easily allow a Symbian OS application, which typically consists of a single thread within its own process, to issue multiple asynchronous requests and be notified when any of the current requests have completed, a supporting framework has been provided. Cooperative multitasking is achieved through the implementation of two types of object: an Active Scheduler for the wait loop (one per thread—the static functions provided always refer to the current one), and an Active Object for each task—this encapsulates the request and the corresponding handler function (called upon request completion).

Note that all processing in a GUI application takes place in an Active Object, but much of this, along with the creation of the Active Scheduler, is hidden by the Application Framework. GUI applications will be covered in Chapter 4.

In a console application or DLL, you must create and install your own Active Scheduler.

The Active Scheduler

The Active Scheduler is implemented by `CActiveScheduler`, or a class derived from it. It implements the wait loop within a single thread, detecting the completion of asynchronous events and locating the relevant Active Object to handle them.

The Active Scheduler maintains a list, ordered by priority, of *all* Active Objects in the system. The ordering of Active Objects of equal priority is unspecified.

Completed events are detected through a call to `User::WaitForAnyRequest()`. This call suspends the thread until an asynchronous request has completed. When the thread is signaled, the Active Scheduler cycles through its list of Active Objects in priority order to find an Active Object that has completed. Each Active Object is checked to see if it has issued a request (its `iActive` flag is set), and if the request has completed (`iStatus` is set to a value other than `KRequestPending`—note that the request may have returned an error). If both of these conditions are met, the Active Object is first made inactive (its `iActive` flag is unset), so that it will not be found again next time there is a completed request (unless its request has actually been reissued)—remember that the Active Scheduler maintains a list of *all* Active Objects, whether they have currently requested a service or not. Then its handler function is called.

The handler function for an active object is called `RunL()`. Note that since this function may leave, the Active Scheduler calls it from within a trap harness. If a leave occurs, an attempt is made to invoke the Active Object's own exception handler, `RunError()`. If that returns an error, the Active Scheduler's error-handling routine is called, which by default will panic the thread.

Once the completed Active Object is found, the loop waits again. If the Active Scheduler reaches the end of its list without finding a completed Active Object, then this indicates that the Active Object that issued the request was not in its list and results in a stray-event panic.

Active Objects

Active Objects represent asynchronous service requests, encapsulating:

- A data member representing the status of the request (`iStatus`).
- A handle on the asynchronous service provider (usually an `R`-class object).
- Connection to the service provider during construction.
- The function to issue (or reissue) the asynchronous request (user-defined).
- The handler function to be invoked by the Active Scheduler when the request completes (`RunL()`).
- The function to cancel an outstanding request (`Cancel()`).

Here is part of the definition of the abstract base class of all Active Objects, `CActive`:

```
class CActive : public CBase
    {
public:
    ~CActive();
    void Cancel();
    inline TBool IsActive() const;

protected:
    CActive(TInt aPriority);
    void SetActive();
    virtual void DoCancel() = 0;
    virtual void RunL() = 0;
    virtual TInt RunError(TInt aError);

public:
    TRequestStatus iStatus;
    };
```

First, note the public members: a `Cancel()` function that is called when you want to cancel the request, and a function that checks whether the Active Object is active or not (`IsActive()`). The public destructor of your derived Active

Object should always call `Cancel()` to make sure that any outstanding requests are cancelled before the Active Object is destroyed—if there are no outstanding requests, then `Cancel()` does nothing, so there is no need to check this explicitly. Note also that the `iStatus` member is public—this will be covered in more detail later in this section.

It is also worth pointing out that there is no actual function in the base class to issue an asynchronous request—it is up to you to define one in your derived class. This is often called `Start()`.

The protected constructor prevents stack instantiation and takes a priority parameter. This is used to order the Active Objects in the Active Scheduler's list. The higher the priority, the closer to the top of the list the Active Object will be, and the more likely it is to be handled before the others in the event of multiple requests being completed at once. Normally an Active Object is given a standard priority of `EPriorityStandard`—available priorities will be covered later in this section.

> **NOTE** It must be stressed that, regardless of priority, Active Objects cannot pre-empt each other.

The final four methods are the crucial ones:

`SetActive()` must be called once a request has been issued, otherwise the Active Scheduler will ignore it when searching for a completed Active Object, causing an error.

`DoCancel()` is a pure virtual function that must be implemented to provide the functionality necessary to cancel the outstanding request, but note that *it should never be invoked directly*—always use `Cancel()`, which will invoke `DoCancel()` for you and also ensure that the necessary flags are set to indicate the request has completed.

`RunL()` is the asynchronous event-handler function. This is the function called by the Active Scheduler when the outstanding request has completed. What it does will obviously depend entirely upon the nature of the service requested—whatever processing is necessary once the request has completed. You will often see an active object used to implement a state machine—in this case, each time the `RunL()` is called, the next step of a complex series of interdependent asynchronous events is launched.

`RunError()` is called (by the Active Scheduler) if `RunL()` leaves—it gives the Active Object a chance to handle its own errors. If able to handle the error, `RunError()` should return `KErrNone`. Otherwise it should simply return the error it was passed as the argument. In this case, the error will then be passed to the Active Scheduler's `Error()` function, with a default behavior of causing a panic.

Remember that a set of Active Objects belong to a single thread and are cooperating together—the `RunL()` of one Active Object cannot preempt the `RunL()` of another. What this really means is that once the Active Scheduler has invoked one `RunL()`, no others (within the same thread) can run until this one has completed and returned. This means that all `RunL()` functions must be kept as short as possible, and this is most crucial in any GUI application, since all of the code keeping the UI responsive is also running in other Active Objects within the same thread. If your `RunL()` takes ten seconds to complete, your application will appear to freeze for this period of time—no keypresses will be processed, animations will stop and so on. As a general rule of thumb, no `RunL()` should take any longer than one-tenth of a second to complete—otherwise the responsiveness of your application will suffer.

Notice finally the `TRequestStatus iStatus` member. It is basically just an encapsulated integer that is used to represent the status or error code returned by the asynchronous service provider. A reference to the Active Object's `iStatus` member is passed to the service provider (via the Kernel) when the request is made. The service provider's first task is to set it to `KRequestPending`. When the requested service completes, the service provider will set the value of `iStatus` to either `KErrNone` (if the request completes successfully) or an error code.

Implementing an Active Object

The rest of this section covers the practicalities of writing an active object. You are first provided with a checklist summary of the tasks involved and then guided through a full example of how to apply this information to create a simple Active Object-based timer. Finally, common pitfalls that occur when using Active Objects are covered.

Active Object Checklist

Writing an Active Object is actually quite simple. It is summarized in the following steps:

1. Create a class derived from `CActive`.
2. Encapsulate the appropriate asynchronous service provider handle or handles (usually `R`-classes) as member data of the class.
3. Invoke the constructor of `CActive`, specifying the appropriate task priority—usually the default priority.
4. Connect to the service provider in your `ConstructL()` method.
5. Invoke `CActiveScheduler::Add()` in `ConstructL()` (unless there is good reason not to—it must be called somewhere).
6. Implement `NewL()` and `NewLC()` as normal.

7. Implement your asynchronous request function, often called `Start()` or `StartL()`. This should invoke the appropriate asynchronous service function on the service provider handle (`R`-class), specifying `iStatus` as the `TRequestStatus&` argument. Do not forget to call `SetActive()`!

8. Implement your `RunL()` method to handle any necessary work once the request is complete. This may involve processing data, notifying an observer of completion or reissuing the request.

9. Implement `DoCancel()` to handle canceling of the request. This will often simply be a case of invoking the appropriate cancel function on the service provider handle.

10. Optionally override `RunError()` to handle any leaves from `RunL()`. The default implementation will cause the Active Scheduler to panic!

11. Implement the destructor to call `Cancel()` and close the handle(s) on the service provider(s).

Using the Active Object is then simply a case of applying the following steps:

1. Instantiating via `NewL()` or `NewLC()` as appropriate.

2. Calling `Start()`, or whatever function you implemented to make the initial request.

3. If you wish to cancel the request prior to completion, simply call `Cancel()`.

A Practical Example

The **Elements** example application shows the use of Active Objects. It loads lists of element data from three different CSV (Comma-Separated Value) files using an asynchronous interface to the File Server. Active Objects control the reading of data, which is read in a line at a time to break up a potentially long-running task into smaller chunks. This also allows all files to be effectively loaded "simultaneously."

This part of the example has been contrived to demonstrate asynchronous processing, but as Symbian OS is so efficient, a second type of asynchronous request, a timer, has to be introduced to slow the system down in order to illustrate the asynchronous loading process! The file loader class works as a simple state machine: it either loads data or wastes time on a small pseudo-random delay.

Note that Symbian OS already provides a simple timer class, `CTimer`, but it is instructive to look at how this might otherwise be implemented. File operations will be explained properly in the next section.

The file loader class is called `CCsvFileLoader`. Here is part of its class definition; note that it derives from `CActive`:

```
class CCsvFileLoader : public CActive
    {
```

Chapter 3 Using Asynchronous Services with Active Objects

```
public:
   void Start();

private:
   CCsvFileLoader(RFs& aFs, CElementList& aElementList,
      MCsvFileLoaderObserver& aObserver);
   void ConstructL(const TDesC& aFileName);
   // From CActive
   void RunL();
   TInt RunError(TInt aError);
   void DoCancel();

private:
   TFileName iFileName;
   RFile iFile;
   TBool iWastingTime;
   RTimer iTimeWaster;
   };
```

Construction

Two-phase construction is almost always needed for Active Objects, since they usually need to connect to their asynchronous service provider, which may fail. Here is the implementation of the first-phase constructor for the file loader:

```
CCsvFileLoader::CCsvFileLoader(RFs& aFs, CElementList&
aElementList, MCsvFileLoaderObserver& aObserver)
  : CActive(EPriorityStandard),
    ...
    {
    }
```

Note that `CCsvFileLoader()` calls the constructor of `CActive` with a hard-coded priority, in this case `EPriorityStandard`. The standard priorities are defined in `e32base.h` as:

```
enum TPriority
   {
   EPriorityIdle = -100,
   EPriorityLow = -20,
   EPriorityStandard = 0,
   EPriorityUserInput = 10,
   EPriorityHigh = 20,
   };
```

Choose whichever is most appropriate for your Active Object, remembering that all it affects is the order in the Active Scheduler's list. It is rare that an Active Object in application code is given a priority of anything other than `EPriorityStandard`—if you have to worry about specific priorities, then there may be something wrong with your design! For reference, the priorities given to system tasks for UI applications are defined in the class `TActivePriority` in `coemain.h`.

Here is the implementation of the second-phase constructor for the file loader:

```
void CCsvFileLoader::ConstructL(const TDesC& aFileName)
    {
    iFileName = aFileName;
    User::LeaveIfError(iFile.Open(iFs, iFileName, EFileRead));
    User::LeaveIfError(iTimeWaster.CreateLocal());
    // Not done automatically!
    CActiveScheduler::Add(this);
    }
```

It sets the filename and connects to the two asynchronous service providers required via their handles: `RTimer` and `RFile`. If there is insufficient memory for the timer, or the file does not exist, then `ConstructL()` (and so `NewL()` or `NewLC()`) will leave.

Finally, `this` Active Object is added to the Active Scheduler's list. This does not happen automatically, so it is commonly performed as the last line in the second-phase constructor, after construction has been successful.

WARNING Each Active Object must be added to the Active Scheduler once and once only. Failing to add it will result in a stray-request panic.

Note that construction is normally wrapped up in `NewL()` or `NewLC()` methods—these have been omitted here.

Starting the Active Object

Here is the implementation of `Start()`:

```
void CCsvFileLoader::Start()
    {
    // Wait for a randomish amount of time before doing anything else
    TInt delay = (iFileName.Size() % 10) * 100000;
    iTimeWaster.After(iStatus, delay);
    SetActive();
    }
```

This is the function that will be called to start the initial asynchronous request—in this case, all it does is wait for a period of time before the `RunL()` method is called. Remember, though, that the `Start()` function itself will return almost immediately—`RunL()` will be called at some point in the future, once the asynchronous timer request completes.

A "random" delay time is calculated, based on the length of a filename (not very random, but random enough!). The class member `iTimeWaster` is an `RTimer` instance—the handle on an asynchronous service provider. The `After()` method takes two arguments—the time delay in microseconds (millionths of a second) and a `TRequestStatus`.

Chapter 3 Using Asynchronous Services with Active Objects

Note that the presence of a `TRequestStatus` denotes an asynchronous function. All asynchronous functions in Symbian OS take a `TRequestStatus&`, which they use to notify the owning Active Object of completion.

After the given period of time, the asynchronous service provider (running in another thread/process) will signal the thread by doing two things:

- Incrementing the thread's semaphore.
- Setting the given `TRequestStatus` to something other than `KRequestPending` (hopefully `KErrNone` if all is well).

Incrementing the semaphore reawakens the thread, if suspended. (Remember the `WaitForAnyRequest()` at the top of the wait loop suspends the current thread.) Changing the value of the `TRequestStatus` (`iStatus`) will allow the Active Scheduler to tell which Active Object has completed, and this will result in its `RunL()` method being called.

WARNING Failing to call `SetActive()` once a request has been issued will result in a stray-request panic when the request completes.

RunL()

The `RunL()` of this particular Active Object is fairly complex, because it performs a number of tasks:

- It decides what the next iteration should do (load data or waste time).
- It checks the status of the last iteration and reports this to the observer.
- It processes any data loaded.

The `RunL()` of any general Active Object will probably want to do at least one of the jobs suggested above—some may want to do all three.

Here are the first few lines of `RunL()`:

```
void CCsvFileLoader::RunL()
    {
    if (!iHaveTriedToLoad)
        {
        // Fill the read buffer if we haven't yet tried to do so...
        FillReadBufferL();
        return;
        }
```

In other words: if this is the first time `RunL()` has been called, then you need to load some data first. It will become obvious later why this is important.

Chapter 3 Symbian OS Fundamentals

`FillReadBufferL()` looks like this:

```
void CCsvFileLoader::FillReadBufferL()
    {
    iHaveTriedToLoad = ETrue;
    // Seek to current file position
    User::LeaveIfError(iFile.Seek(ESeekStart, iFilePos));
    // Read from the file into the buffer
    iFile.Read(iReadBuffer, iReadBuffer.MaxSize(), iStatus);

    SetActive();
    }
```

Firstly, you need to set `iHaveTriedToLoad` to true, since you have now tried to load something.

Next, seek to the right position in the file. The `iFile` member is an `RFile` object whose API will be discussed in more depth in the next section.

The crucial steps come next. `RFile::Read()` is another asynchronous function. Again, you can tell this by the presence of the `TRequestStatus` object, `iStatus`. And as before, this call will cause `RunL()` to be called once the operation is complete. `SetActive()` is invoked to ensure this.

Here is the next section of the `RunL()`, which handles error notification:

```
if ((iStatus.Int() != KErrNone) || (iReadBuffer.Size() == 0))
    {
    iObserver.NotifyLoadCompleted(iStatus.Int(), *this);
    return;
    }
```

Notice the first test in the `if` statement. `TRequestStatus::Int()` just returns the error value of the `TRequestStatus` as an integer. This error value is returned by the asynchronous service provider as a way of indicating to the Active Object whether the operation was successful or not.

> **NOTE** Generally speaking, you should always check the error value of the `iStatus` member to see if the asynchronous function was a success. It is important to stress that `RunL()` will always be called, whether or not the asynchronous call succeeded.

In the example above, you notify the observing client if the asynchronous call failed, or if no data was read from the file. If the latter is true, it denotes that the end of file was reached. In either case, the completion status is returned to the user, and you return *without reissuing the request*. That is basically the end of the Active Object's useful life (unless `Start()` is called again).

Assuming that there were no errors and some data has been read in, the next step is to decide whether this iteration is to simply waste time. If so, you reissue the time-delay request much like the first time:

Chapter 3 Using Asynchronous Services with Active Objects

```
if (iWastingTime)
    {
    // Just wait a randomish amount of time
    TInt delay = (iFilePos % 10) * 100000;
    iTimeWaster.After(iStatus, delay);
    SetActive();
    // Don't waste time next time around!
    iWastingTime = EFalse;
    return;
    }
```

You can see here that the file position is used to supply the random seed, but other than that it is basically the same as the code in `Start()`. In addition, `iWastingTime` is set to false, preventing the time-wasting step from being executed on the next iteration.

The final section of the `RunL()` processes the data in the buffer and reissues the load request by calling `FillReadBufferL()` again. It also sets `iWastingTime` true, ensuring a time-wasting step on the next iteration to slow things down a bit:

```
// Extract and convert the buffer
TPtrC8 elementBuf8 = ExtractToNewLineL(iReadBuffer);
HBufC* elementBuf16 = HBufC::NewLC(elementBuf8.Length());
TPtr elementPtr16 = elementBuf16->Des();
elementPtr16.Copy(elementBuf8);

// Read the element from the buffer
iElementList.AppendL(*elementBuf16);

// Report the element that's just been loaded to the observer
TInt lastElementLoadedIndex = iElementList.NumberOfElements()
    - 1;
iObserver.NotifyElementLoaded(lastElementLoadedIndex);

// Increment the file position
iFilePos += elementBuf16->Length() +
KTxtLineDelimiter().Length();

// Reissue the request
FillReadBufferL();

// Cleanup
CleanupStack::PopAndDestroy(elementBuf16);

iWastingTime = ETrue;
```

Handling Errors in RunError()

It is perfectly permissible for `RunL()` to leave—the trailing "L" denotes this. If it does, then the error code that it leaves with is passed to `RunError()`:

```
TInt CCsvFileLoader::RunError(TInt aError)
    {
    // Notify the observer of the error
    iObserver.NotifyLoadCompleted(aError, *this);

    return KErrNone;
    }
```

In this instance, that is all there is to it. You simply notify the observer that the load has completed (albeit unsuccessfully, with the appropriate error code), but return `KErrNone` to show that the error has been handled locally. Any nonzero error would be passed to the Active Scheduler's `Error()` function.

Other error handlers may do more complex things such as retrying or the like, but commonly they will do something similar to the example above and just pass on notification of the error.

Canceling an Outstanding Request

All Active Objects must implement a `DoCancel()` method to cancel any outstanding request. Here is the implementation for `CCsvFileLoader`:

```
void CCsvFileLoader::DoCancel()
    {
    if (iWastingTime || !iHaveTriedToLoad)
        {
        iTimeWaster.Cancel();
        }
    }
```

All you need to do is invoke `RTimer::Cancel()` if the timer is running. Since there is no way of canceling an outstanding `RFile` request, there is nothing you need to do in that case.

`DoCancel()` is called by `CActive::Cancel()`. `CActive::Cancel()` itself should never be overridden (it is non-virtual anyway), as it does a number of important things for you:

- It checks to see if the Active Object is actually active—if not, it just returns without doing anything.
- It invokes `DoCancel()`.
- It waits for the request to complete—this must complete as soon as possible. (Note that the original request may actually complete before it is cancelled.)
- It sets `iActive` to false.

Understanding that `Cancel()` waits for completion of the outstanding request is crucial. This means that `RunL()` will not get called as a result of calling `Cancel()` as you might expect—so any cleanup needed as a result of a `Cancel()` must be handled in `DoCancel()`, not in `RunL()`.

WARNING You should never call the `DoCancel()` function of an Active Object explicitly. If you want to cancel a request, then just call `Cancel()` — this will invoke `DoCancel()` and also ensure that `RunL()` is not called.

Destructor

The destructor implementation is shown here:

```
CCsvFileLoader::~CCsvFileLoader()
    {
    Cancel();
    iFile.Close();
    iTimeWaster.Close();
    }
```

The first thing that should happen in any Active Object destructor is a call to `Cancel()` any outstanding requests. If an Active Object is deleted with a request outstanding, a stray-request (`E32USER-CBase 40`) panic will result.

Any handles to the asynchronous service providers must be closed to avoid resource leakage.

The base `CActive` destructor will automatically call `Deque()` to remove the Active Object from the Active Scheduler's list.

Starting the Active Scheduler

The UI framework will automatically create, install and start an Active Scheduler for you, so this can be skipped if you are not going to be writing .exes (console applications or Symbian OS Servers), or DLLs — these need an Active Scheduler to be started explicitly. However, it does provide some additional insight into how Active Objects and the Active Scheduler interact.

Before an Active Scheduler can be started, a number of steps must be undertaken:

- The Active Scheduler must be instantiated.
- It must be installed into the thread.
- An Active Object must be created and added to the Active Scheduler.
- A request must be raised (as the wait loop suspends the thread).

Only then can the scheduler be started.

All of these steps can be seen in the following code example:

```
void DoExampleL()
    {
    CActiveScheduler* scheduler = new (ELeave) CActiveScheduler;
```

Chapter 3 Symbian OS Fundamentals

```
    CleanupStack::PushL(scheduler);
    CActiveScheduler::Install(scheduler);

    CElementsEngine* elementEngine =
CElementsEngine::NewLC(*console);
    elementEngine->LoadFromCsvFilesL();    // Issues requests

    CActiveScheduler::Start();

    CleanupStack::PopAndDestroy(2, scheduler);
    }
```

The first three lines are fairly obvious—they instantiate a new `CActiveScheduler` object, add it to the Cleanup Stack, and then install it as the current Active Scheduler for the thread—only one Active Scheduler can be installed in each thread.

The next two lines create the `CElementsEngine`, which owns Active Objects (three `CCsvFileLoader` objects) and starts them off.

Since there are now some outstanding requests, the Active Scheduler can be started. The call to the `CActiveScheduler::Start()` function does not return until a corresponding call is made to `CActiveScheduler::Stop()` from within the `RunL()` of one of the Active Objects. When it does, the Active Scheduler and engine are deleted, and the thread exits.

There are a number of issues here of profound importance—the `CActiveScheduler::Start()` function pretty much covers the entire lifespan of the thread. To be able to discuss these issues further, consider the lifecycle of the Active Scheduler. This is illustrated in Figure 3-7.

The four "objects" involved (Executable, Active Scheduler, Active Object and Service Provider) are shown as vertical columns. Time runs from top to bottom.

When the executable's main thread is launched, it instantiates and installs the Active Scheduler, as shown previously. Then, at least one Active Object must be created and added to the Active Scheduler. Once done, the Active Object must issue a request. The asynchronous service provider (in another thread and usually another process) starts by setting the Active Object's `iStatus` member to `KRequestPending`. The service provider can then set about servicing the request.

Once there is a request outstanding, it is safe to `Start()` the Active Scheduler. This causes the Active Scheduler to enter its wait loop, starting with a call to `WaitForAnyRequest()`, decrementing the thread's semaphore, and suspending the thread.

This is why it is essential to have an outstanding request before starting the Active Scheduler. Without an outstanding request, there is no way for the thread to be reawakened—nothing will increment the thread's semaphore, and the thread will remain suspended permanently.

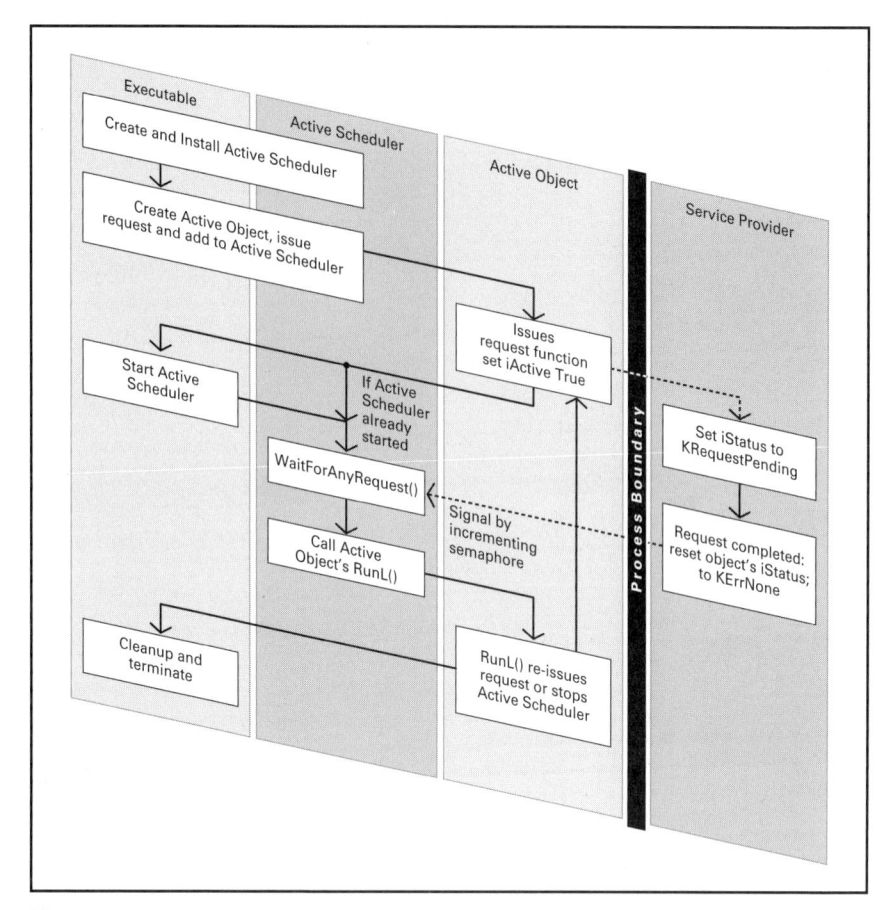

Chapter 3 Using Asynchronous Services with Active Objects 141

Figure 3-7 Lifecycle of the Active Scheduler.

At some point the service provider will complete the request and signal this by incrementing the requesting thread's semaphore and setting the Active Object's `iStatus` member to the appropriate error code. The scheduler then searches through its list to find the Active Object that has just completed (is active and has an `iStatus` not equal to `KRequestPending`) and runs its `RunL()`.

What happens next depends upon the `RunL()`. It will handle the completion of the request and may reissue the request or spawn another Active Object — in these cases the cycle repeats (although the Active Scheduler will not be started again).

Alternatively, the Active Object's `RunL()` may call `CActiveScheduler::Stop()`. This will cause the Active Scheduler to exit its wait loop, returning control from the call to `CActiveScheduler::Start()`. The executable will usually then clean up and exit. Remember that if you are creating a UI application, the Active Scheduler will be created and controlled by the framework and you will not need to call `CActiveScheduler::Stop()` for the main Active Scheduler loop explicitly — in fact you should not do this, just let the application framework handle it.

NOTE Calling `CActiveScheduler::Start()` a second time creates a nested wait loop—this is used for modal processing. For example, modal dialogs need to "pause" the main processing of the application thread but must still allow processing of input and so on. A call to `CActiveScheduler::Stop()` will return control to the initial wait loop. This is an advanced topic, and further details are beyond the scope of this book.

As you can see, the lifespan of the Active Scheduler is pretty much the same as that of the owning executable thread. Certainly this is true of a UI framework application, where the Active Scheduler is created by the framework and populated with a few Active Objects that are responsible for maintaining the UI, responding to user input, and so on.

Again, it is important to reflect that almost all code written by a developer in a UI application is running within the `RunL()` of an Active Object, so it needs to be made as efficient as possible to keep the system responsive. Long-running synchronous tasks will prevent other Active Objects from running, reducing the application's responsiveness.

Common Active Object Pitfalls

The most common problem with an Active Object is a stray-event panic from the Active Scheduler. This is usually caused by one (or more) of the following:

- Forgetting to call `CActiveScheduler::Add()` before starting the Active Object.
- Not calling `SetActive()` after issuing (or reissuing) an asynchronous request.
- Passing the same `iStatus` to two service providers at the same time (thereby having multiple requests outstanding on the same Active Object).

Do not invoke `DoCancel()` directly. It should be private anyway—always call `Cancel()`, and never call `Cancel()` from within `DoCancel()`! Note that `Cancel()` should *always* be called in the destructor of your derived class. If there is an outstanding request when the base `CActive` destructor is called, an `E32USER-CBase 40` panic is raised.

Never forget that Active Objects use cooperative multitasking—remember that no Active Object can preempt another, nor should one `RunL()` take any longer than about one-tenth of a second to complete. Very long-running `RunL()` functions may cause a View Server Time Out panic (`ViewSrv 11`)—this occurs when an application fails to respond to the system within a given time (around ten seconds).

If your Active Object repeatedly reissues requests, then a View Server Time Out panic may also occur if there is insufficient time between iterations—your

Active Object may be "hogging" the Active Scheduler and not allowing lower-priority Active Objects to complete. If you want to have one Active Object called repeatedly, use a technique similar to that outlined in the example, using a timer to give a reasonable delay (at least a few hundredths of a second or so). This delay does not necessarily have to be after every request—in a game application, for example, you may just want to drop an occasional frame to allow other processing to take place.

Note that if you are deriving from, or using `CTimer` to create a delay, you need to `Add()` it to the Active Scheduler—this is not performed by default in the base class! Further details of `CTimer` can be found in the SDK documentation.

Don't forget that if you are writing a console application or DLL, an Active Scheduler is *not* provided by default. Without a valid Active Scheduler installed, any use of Active Objects will result in an `E32USER-CBase 44` panic.

Files, Streams and Stores

The ability to be able to save and retrieve data is a fundamental requirement for most application programmers. Symbian OS offers files, streams and stores to fulfil this need. A stream is basically the representation of an object as a sequence of bytes; read and write streams are responsible for reading in and writing out data. A store is a collection of these streams, with each stream having a unique ID. Files are supported through the `RFile` API, which allows you to create and open files and read and write binary descriptors. In addition, streams can also be externalized to a file, and using the streaming API for saving and recalling data is often the easier option.

Files

The use of asynchronous services has been discussed in some depth in this chapter, so you are now in a position to learn about one of the most important asynchronous service providers, the File Server.

NOTE In order to use any of the functionality outlined in this chapter, you will need to `#include <f32file.h>`. You will also need to add `efsrv.lib` to the `LIBRARY` line of your `mmp` file.

Symbian OS Filenames and Pathnames

Before considering the function of the File Server, it is worth taking a moment to describe the details of Symbian OS filenames and pathnames.

The filing system has been designed to operate in a very similar manner to the DOS/Windows file system with which most developers will already be familiar. Fully qualified filenames generally consist of up to four components:

Chapter 3 Symbian OS Fundamentals

- The drive name: a single letter followed by a colon (":").
- The path: starting from either the root directory (if preceded by "\") or the session's current directory, with each level of the folder hierarchy separated by a backslash ("\").
- The filename.
- The file extension, preceded by a dot (".")

The most important difference between DOS and Symbian OS is that no fully qualified filename (including drive name, path, filename and extension) can exceed 256 characters. That is to say, the fully qualified filename of any file must fit completely within a `TFileName` object.

In common with DOS (but unlike otherwise similar UNIX systems), the file system preserves case but does not differentiate case. In other words, "`c:\Documents\FILE.ext`" is the same as "`C:\dOCUMENTS\FiLe.EXT`" as far as the file system is concerned.

The following characters cannot be used anywhere in the file or path name: < > " / |. Note that spaces can be used. Full details of the rules associated with pathnames and filenames can be found in the SDK documentation.

On most Symbian OS platforms (including Series 60), `c:` is the main system drive (flash filing system) where all user data and system files are stored, `z:` is the ROM drive, and removable drives are usually mapped to `d:` onward. (In the case of Series 60, `d:` is a nonpersistent (temporary) RAM drive and removable drives are mapped from `e:` onward.)

In spite of all that has been said above, Series 60 1.x does not usually expose the file system to the user; however, Series 2.x does partially expose it, providing a file manager application.

RFs API

The File Server is accessed via a handle of type `RFs`, and each `RFs` connected to the File Server demands a certain level of server-side and system resource, so their number should therefore be kept to a minimum. The Control Environment (**CONE**), which forms part of the application framework for UI applications, provides a permanent handle to the File Server to help reduce the number of connected File Server Sessions needed. To access it, invoke `FsSession()` on the `iCoeEnv` member of your View class. The Control Environment and other application framework components will be covered in more detail in Chapter 4.

Connect() and Close()

Alternatively, connecting to the File Server is simply a matter of invoking `Connect()` on a File Server handle (or, to give it the proper name: File Server *session*):

Chapter 3 Files, Streams and Stores

```
void CElementsEngine::ConstructL()
    {
    User::LeaveIfError(iFs.Connect());
    }
```

where `iFs` is simply an `RFs` member of `CElementsEngine`. While it really would be very easy to have each object that needs a session with the File Server create its own, for efficiency's sake a reference to the session is passed down to all objects that need it. For example:

```
void CElementsEngine::LoadFromCsvFilesL()
    {
    iMetalsCsvFileLoader = CCsvFileLoader::NewL(iFs,
*iElementList, *this,
        KMetalsFileName);
```

Before the File Server session is allowed to go out of scope or its owner deleted, it must be closed:

```
CElementsEngine::~CElementsEngine()
    {
    delete iMetalsCsvFileLoader;
    iFs.Close();
    }
```

If the File Server session is locally scoped, and there is the danger of a leave before it is closed, use `CleanupClosePushL()` to ensure it is closed properly, should a leave occur. `CleanupStack::PopAndDestroy()` can then be used to close it explicitly, once done.

Remember that none of other key functions available on a File Server can be called before the session has been connected. All of the following ones return error codes rather than leaving:

- `MkDir()` and `MkDirAll()`: These functions simply take the path of a new directory, relative to the current session path, and create the new directory. For `MkDir()`, all of the preceding directories must already exist. For `MkDirAll()`, any ancestor directories of the new directory will also be created if they do not already exist.

- `RmDir()`: Removes the directory specified as a descriptor. The directory must be empty and cannot be the root.

- `SessionPath()`: Retrieves the current session path, effectively the "current directory."

- `SetSessionPath()`: Takes a descriptor containing the path relative to the current session path to set as the new session path. It must not contain a filename; otherwise an error will result.

Fetching a Directory Listing Using GetDir()

The full function prototype for `RFs::GetDir()` is:

```
TInt GetDir(const TDesC& aName, TUint aEntryAttMask, TUint
aEntrySortKey, CDir*& aEntryList) const;
```

`aName` is simply the name of the directory whose listing is required.

`aEntryAttMask` is a bitmask specifying the entries of interest. `KEntryAttNormal` will specify all entries except directories, hidden and system files, for instance.

`aEntrySortKey` specifies the sort order. The values are specified in `TEntryKey`—`ESortNone` is the default.

Finally, `aEntryList` is a reference to a pointer to the directory listing. This pointer should have no memory allocated to it prior to calling, otherwise the memory will be orphaned. Ownership of the newly created `CDir` object is returned to the caller; it is the caller's responsibility to delete it.

Here is the definition of the `CDir` public API:

```
class CDir : public CBase
   {
public:
   IMPORT_C virtual ~CDir();
   IMPORT_C TInt Count() const;
   IMPORT_C const TEntry& operator[](TInt anIndex) const;
   IMPORT_C TInt Sort(TUint aEntrySortKey);
   };
```

The API is pretty self-explanatory. The `operator[]` can be used to iterate through the `Count()` entries, and return a `TEntry` reference. `TEntry` is defined like this:

```
class TEntry
   {
public:
   IMPORT_C TEntry();
   IMPORT_C TEntry(const TEntry& aEntry);
   IMPORT_C TEntry& operator=(const TEntry& aEntry);
   IMPORT_C TBool IsReadOnly() const;
   IMPORT_C TBool IsHidden() const;
   IMPORT_C TBool IsSystem() const;
   IMPORT_C TBool IsDir() const;
   IMPORT_C TBool IsArchive() const;
   inline const TUid& operator[](TInt anIndex) const;
   inline TBool IsUidPresent(TUid aUid) const;
   inline TBool IsTypeValid() const;
   inline TUid MostDerivedUid() const;
public:
   TUint iAtt;
```

Chapter 3 Files, Streams and Stores

```
TInt iSize;
TTime iModified;
TUidType iType;
TBufC<KMaxFileName> iName;
};
```

This code snippet will return the name of the first (chronological) file in the specified directory:

```
void GetFirstFileNameL(RFs& aFs, const TDesC& aFolderName,
    TDes& aFirstFileName)
    {
    CDir* dir = NULL;
    User::LeaveIfError(aFs.GetDir(aFolderName, KEntryAttNormal,
        ESortByDate, dir));

    if (dir->Count())
        {
        // Some files were found, get the name of the zeroth entry
        aFirstFileName = (*dir)[0].iName;
        }
    else
        {
        // No files found—return zero-length descriptor
        aFirstFileName.Zero();
        }

    delete dir;             // Since we have ownership
    }
```

RFile API

While the File Server session does provide a useful API, there are no functions for writing and reading to and from files.

In order to manipulate individual files, an `RFile` object is needed. This type of handle is known as a subsession, for reasons that will become clearer when Symbian OS Client/Server architecture is discussed later in this chapter.

Only a small number of API methods are discussed here. For full details, see the SDK documentation.

Opening and Closing

As with many other `R`-class objects, `RFile` must be opened before use. Unlike previously encountered objects, `RFile`'s open function takes a number of arguments:

```
Open(RFs& aFs, const TDesC& aName, TUint aFileMode);
```

The first argument is a File Server session (which must itself have been connected previously). Naturally, the name of the file to open needs to be specified. Finally, the mode must be given, from the enumeration `TFileMode`. Options include `EFileRead` and `EFileWrite`.

Here is part of the second-phase constructor of `CCsvFileLoader`, showing its `iFile` member being opened:

```
void CCsvFileLoader::ConstructL(const TDesC& aFileName)
    {
    iFileName = aFileName;
    User::LeaveIfError(iFile.Open(iFs, iFileName, EFileRead));
    }
```

Before being deleted or going out of scope, the `RFile` subsession must be closed:

```
CCsvFileLoader::~CCsvFileLoader()
    {
    iFile.Close();
    }
```

Reading

There are numerous `Read()` overloads, all of which read binary (8-bit) data into a `TDes8`-derived descriptor. Half of the overloads take a `TRequestStatus&`, the other half do not. This means there are basically two types of `Read()` overload—synchronous and asynchronous. In most cases, the asynchronous overload should be used from within an Active Object. Only use the synchronous overload in extreme circumstances (such as a quick test harness), since long file reads will block the thread, affecting responsiveness.

The simplest overload just takes a descriptor reference and reads in data from the start of the file to fill as much of the buffer as possible. The length of the descriptor is naturally set to the number of bytes actually read, which may be less than the maximum length of the descriptor if there is insufficient data in the file to fill it. Further overloads allow you to specify the offset from the beginning of the file and/or the number of bytes to read in.

In all asynchronous cases there is no return type, nor can the function leave. Rather than returning `KErrEOF`, the end of file condition is signaled by simply returning a zero-length descriptor.

Here is the `FillReadBufferL()` function of `CCsvFileLoader` again. It simply fills the buffer with as much data as it can from the file:

```
void CCsvFileLoader::FillReadBufferL()
    {
    User::LeaveIfError(iFile.Seek(ESeekStart, iFilePos));
    iFile.Read(iReadBuffer, iReadBuffer.MaxSize(), iStatus);
    SetActive();
    }
```

Chapter 3 Files, Streams and Stores

The same functionality could be achieved by various other means, but all are variants of the idea shown above.

It is not necessary to check the return value of `Seek()`, even if seeking moves beyond the end of the file. It will be obvious, because `iReadBuffer` will be of zero size when you next enter `RunL()`.

Writing

`Write()` mirrors `Read()` precisely, writing data from a `TDesC8` to a file. Overloads are provided to specify where, within the file, the data is to go, and how much data is to be written. Again, all variants are available as synchronous and asynchronous overloads—asynchronous functions are always preferred.

All file operations use 8-bit descriptors. 16-bit descriptors cannot automatically be converted into 8-bit descriptors, but there are two approaches to achieve this explicitly—one will preserve the length of the descriptor (number of characters), the other the size (number of bytes).

The first method preserves the size of the descriptor, but results in an 8-bit descriptor of double the length. This writes the 16-bit descriptor to file exactly as it appears in memory (endianness issues aside), as shown in Figure 3–8.

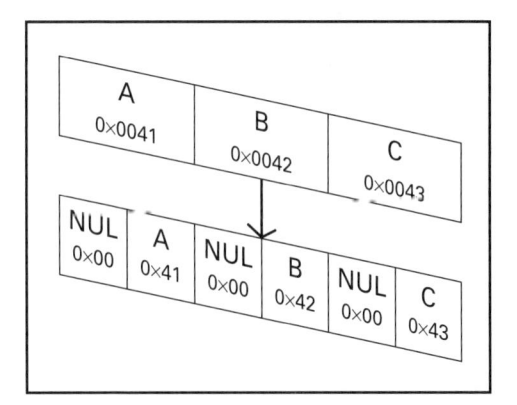

Figure 3–8 Preserving size when writing a 16-bit descriptor to an 8-bit descriptor.

In code, this is written using:

```
TPtrC8 ptrC8((TUint8*)des16.Ptr(), des16.Size());
User::LeaveIfError(iFile.Write(ptrC8, iStatus));
```

The request to read it back in then looks like:

```
delete iBufC8;
iBufC8 = NULL;
TInt size = User::LeaveIfError(iFile.Size());
```

Chapter 3 Symbian OS Fundamentals

```
iBufC8 = HBufC8::NewL(size);
TPtr8 ptr8 = iBufC8->Des();         // To access modifiable API
User::LeaveIfError(iFile.Read(ptr8, iStatus));
```

And, in the `RunL()` method:

```
TPtr16 ptr16((TUint16*)iBufC8.Ptr(), iBufC8.Size() / 2);
iObserver.Notify(ptr16);
```

The observer should be sure to make a copy of the data, using `Alloc()`, since the data buffer is actually owned by the Active Object's `iBufC8` in this case.

The second method preserves the length of the descriptor, effectively converting from Unicode to ASCII, as shown in Figure 3-9.

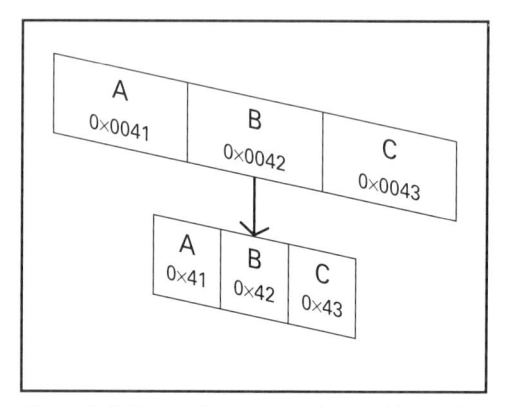

Figure 3-9 Preserving length when writing a 16-bit descriptor to an 8-bit descriptor.

Note that some data may be lost—**all Unicode values greater than 255 (decimal) will be converted to 1**! So, if you are working with a non-Latin character set, then you should probably make use of the first method.

The following code will write the 16-bit descriptor to file, converting it to 8-bit first:

```
delete iBufC8;
iBufC8 = NULL;
iBufC8 = HBufC8::NewL(des16.Length());
TPtr8 ptr8 = iBufC8->Des();         // To access modifiable API
ptr8.Copy(des16);                   // Convert 16-bit into 8-bit
User::LeaveIfError(iFile.Write(ptr8, iStatus));
```

The request to read it back in then looks like:

```
delete iBufC8;
iBufC8 = NULL;
```

Chapter 3 Files, Streams and Stores

```
TInt size = User::LeaveIfError(iFile.Size());
iBufC8 = HBufC8::NewL(size);
TPtr8 ptr8 = iBufC8->Des();          // To access modifiable API
User::LeaveIfError(iFile.Read(iBuf8, iStatus));
```

And, in the `RunL()` method:

```
delete iBufC16;
iBufC16 = NULL;
iBufC16 = HBufC16::NewL(iBufC8.Length());
TPtr16 ptr16 = iBufC16->Des();    // To access modifiable API
ptr16.Copy(iBufC8);               // Convert 8-bit to 16-bit
iObserver.Notify(ptr16);
```

Again, the observer will probably need to make its own copy of the data contained in the Active Object's `iBufC16` (pointed to by the local `ptr16` in the `RunL()`).

Seeking

Moves the current read or write position to the given location within the file. The seek mode can be specified to move the seek position relative to the current location, relative to the start or end or to the absolute address for ROM based files.

```
User::LeaveIfError(iFile.Seek(ESeekStart, iFilePos));
```

The value returned by the function is the current (start-relative) file position. To find the current file position, use:

```
TInt currentPosition = file.Seek(ESeekCurrent, 0);
```

Streams

As you will have gathered from the previous discussion on `RFile::Read()` and `RFile::Write()`, `RFile` is not the easiest way of reading data to and from a file. While a resourceful developer could doubtless find many exotic ways of writing complex data objects into an `RFile`, Symbian OS provides a far more elegant method: streams.

Symbian OS streams provide a simple manner of externalizing and internalizing data, whether it be to file store, memory or any other arbitrary I/O device. Streams take care of converting the data into a suitable external format (handling such issues as endianness and Unicode compression transparently), provided the data is passed to the streams API in a sensible fashion. Clearly, the streams API cannot do much with a pointer — any objects owned by another must be individually streamed out.

First consider the streams API before considering how to stream a complete object to file.

TIP To use streams functionality, you will need to `#include <s32std.h>` at the very least. To use file streams (as described next), you will also need `s32file.h`. You will then need to link against `estor.lib` by adding it to the `LIBRARY` line of your `mmp` file.

RWriteStream

`RWriteStream` is the base class for externalizing data to a stream. Before it can be used, it is necessary to connect the stream to some form of data sink. This depends upon the type of stream to be used, so `RFileWriteStream` will be considered for externalizing to a file.

The code snippet below shows how to create an `RFileWriteStream`. This function assumes that the file `KTxtTestFileName` does not already exist. If it does (an error value of `KErrAlreadyExists` is returned by `Create()`), use `RFileWriteStream::Open()` instead.

```
RFileWriteStream writeStream;
User::LeaveIfError(writeStream.Create(fs, KTxtTestFileName,
    EFileWrite));
writeStream.PushL();
```

Note the use of `RWriteStream::PushL()` to push the stream onto the Cleanup Stack before calling any leaving functions.

The stream can then be used for externalization using either the various write functions (for example, `WriteL()`, `WriteInt32L()`, `WriteReal64L()` and so on), or by using the overloaded `<<` operators defined for explicitly sized simple types. The `ExternalizeL()` function below illustrates the use of both methods:

```
void CChemicalElement::ExternalizeL(RWriteStream& aStream)
const
    {
    aStream << *iName;
    aStream << iSymbol;
    aStream.WriteUint8L(static_cast<TUint8>(iAtomicNumber));
    aStream.WriteUint16L(static_cast<TUint16>(
iRelativeAtomicMass));
    aStream.WriteUint8L(static_cast<TUint8>(iRadioactive));
    aStream.WriteUint8L(static_cast<TUint8>(iType));
    }
```

Note how it is necessary to explicitly specify the size of the integer being written out, and the use of `static_cast` to avoid compiler warnings.

Use of `operator<<` takes care of Unicode compression automatically, and it will also write out the length of the descriptor so that the system knows how much data to read back in at a later point. Any attempt to read a 16-bit descriptor into an 8-bit one will result in a panic. It is also possible to use `WriteL()` to write a descriptor to a stream, but this introduces complexities for reading that data back (since the size and type are not stored for you) and does not automatically compress Unicode data as `operator<<` will.

Note that `operator<<` can also be used for externalizing integers, but it *will not work* for `TInt`. You can only use `operator<<` with explicitly sized integers (or explicitly sized reals).

WARNING It is worth stressing at this point that `operator<<` and `operator>>` can leave, since they invoke the functions `ExternalizeL()` and `InternalizeL()`! Do not use them in a non-leaving function without using a trap.

Once writing has completed, it is necessary to commit the changes made before the stream can be released. This ensures that any data buffered is written to the stream. Once `RWriteStream::Commit()` has been called, it is safe to release the stream:

```
writeStream.CommitL();
writeStream.Pop();
writeStream.Release();
```

RReadStream

`RReadStream` is the conceptual opposite of `RWriteStream`—it is the base class for internalizing data from a stream. It must be connected to a data source prior to reading. The example below shows how the file stream created earlier might be opened.

```
RFileReadStream readStream;
User::LeaveIfError(readStream.Open(fs, KTxtTestFileName,
EFileRead));
readStream.PushL();
```

Once opened, data can be internalized using either `operator>>` or the appropriate `ReadXxxL()` function. Naturally care must be taken to ensure that data members are read in from the stream in precisely the order in which they were written out. In this case, that means the element name followed by its symbol:

```
iName = HBufC::NewL(aStream, KMaxTInt);

TPtr modifiableSymbol = iSymbol.Des();
aStream >> modifiableSymbol;
```

The first line may be a little surprising—but there is an overload for `HBufC::NewL()` which takes a stream. This factory function assumes that the size of the descriptor was streamed out first, followed by the data itself. Luckily, this is exactly what `operator<<` did in the earlier example.

The second argument is the *maximum* amount of data to be read. Should the length of the data in the stream exceed this value, the `NewL()` will leave with `KErrOverflow`. Otherwise a new `HBufC` is created with exactly the size to contain the data in the stream. This is by far the easiest method of reading in a variable-sized descriptor.

If you choose not to use `operator<<` to externalize your descriptor, you must externalize the length of the descriptor explicitly. Your internalizer must read this in and use it to construct an `HBufC` of the necessary size. `RReadStream::ReadL()` can then be used to fill the descriptor.

Returning to the example code, it is not possible to read data directly into `iSymbol`, since `TBufC` has no modifiable API (and so no overloaded `operator<<`). Instead a modifiable pointer descriptor is created using `TBufC::Des()`, and this allows you to access the memory owned by `iSymbol`. (An alternative approach would be to read data into a temporary `TBuf` and then assign this to `iSymbol`.)

Note that any attempt to use `operator<<` with a data type that is not supported (for example, `TBufC`) results in a very obscure error like this:

```
\EPOC32\INCLUDE\s32strm.inl(200) : error C2039:
'InternalizeL' : is not a member of 'TBufC<3>'
\EPOC32\INCLUDE\s32strm.inl(243) : see reference to function
template instantiation 'void __cdecl DoInternalizeL(class
TBufC<3> &,class RReadStream &,class Internalize::Member)'
being compiled
```

It will not even refer to the line of code that caused the error (in this case trying to use `operator>>` with `iSymbol` directly). You would be wise to learn to recognize this type of template error and realize that it is likely to imply the use of `operator>>` with an illegal type.

```
iAtomicNumber = aStream.ReadUint8L();
iRelativeAtomicMass = aStream.ReadUint16L();
```

The final two lines of the code read integers in from the stream. Again, the function names denote the explicit integer size. Use of `operator>>` is possible, but it requires the recipient to be an explicitly sized integer and not a `TInt`. An error similar to that above would result if `operator>>` were to be used with a `TInt`.

ExternalizeL()

Symbian OS convention requires that any object that can be externalized to a stream should provide a publicly accessible `ExternalizeL()` function, declared like this:

```
void ExternalizeL(RWriteStream& aStream) const;
```

Chapter 3 Files, Streams and Stores

Earlier you saw an extract from `CChemicalElement::ExternalizeL()`. Here is the full function:

```
void CChemicalElement::ExternalizeL(RWriteStream& aStream)
const
    {
    aStream << *iName;
    aStream << iSymbol;
    aStream.WriteUint8L(static_cast<TUint8> (iAtomicNumber));
    aStream.WriteUint16L(static_cast<TUint16>
(iRelativeAtomicMass));
    aStream.WriteUint8L(static_cast<TUint8> (iRadioactive));
    aStream.WriteUint8L(static_cast<TUint8> (iType));
    }
```

Such functions are obvious for simple classes like `CChemicalElement` that do not own other complex classes. For those that do, the solution is to invoke the `ExternalizeL()` of the owned class from within the owner's `ExternalizeL()`:

```
void CMyComplexClass::ExternalizeL(RWriteStream& aStream)
const
    {
    aStream << iSimpleMember1;
    aStream << iSimpleMember2;
    iComplexMember1->ExternalizeL(aStream);
    iComplexMember2->ExternalizeL(aStream);
    // ...
    }
```

A more complicated example is `CElementList::ExternalizeL()`:

```
void CElementList::ExternalizeL(RWriteStream& aStream) const
    {
    TInt32 elementCount = NumberOfElements();
    aStream << elementCount;

    for (TInt i = 0; i < elementCount; i++)
        {
        At(i).ExternalizeL(aStream);
        }
    }
```

In such a case, it is simply a matter of externalizing the number of items, `elementCount`, you are about to externalize, and then externalizing the items themselves.

Note that you can also do this using `operator<<` instead of `ExternalizeL()`—any class that has `ExternalizeL()` defined can use `operator<<`:

```
void CElementList::ExternalizeL(RWriteStream& aStream) const
   {
   TInt32 elementCount = NumberOfElements();
   aStream << elementCount;

   for (TInt32 i = 0; i < elementCount; i++)
      {
      aStream << At(i);
      }
   }
```

> **TIP** It is sometimes a good idea to write out a version number (perhaps just a `TInt32`) to your stream before anything else. This might allow future versions of your application with different stream formats to interpret an older version's streams. At the very least it will allow you to inform the user that the stream belongs to an older version, rather than having the inevitable panic or scrambled data that would result otherwise.

InternalizeL()

If an object can be externalized, you will need to implement an `InternalizeL()` to do the converse:

```
void InternalizeL(RReadStream& aStream);
```

Here is `CChemicalElement::InternalizeL()`:

```
void CChemicalElement::InternalizeL(RReadStream& aStream)
   {
   delete iName;    // Just in case this existed previously
   iName = NULL;

   iName = HBufC::NewL(aStream, KMaxTInt);

   TPtr modifiableSymbol = iSymbol.Des();
   aStream >> modifiableSymbol;

   iAtomicNumber = aStream.ReadUint8L();
   iRelativeAtomicMass = aStream.ReadUint16L();
   iRadioactive = static_cast<TBool>(aStream.ReadUint8L());
   iType = static_cast<TElementType>(aStream.ReadUint8L());
   }
```

And `CElementList::InternalizeL()`:

```
void CElementList::InternalizeL(RReadStream& aStream)
   {
   TInt32 elementCount;
   aStream >> elementCount;
```

Chapter 3 Files, Streams and Stores

```
    for (TInt32 i = 0; i < elementCount; i++)
        {
        CChemicalElement* newElement =
CChemicalElement::NewLC(aStream);
        AppendL(newElement);
        CleanupStack::Pop(newElement);     // Now owned by array
        }
    }
```

This is a little different. Rather than constructing a new `CChemicalElement` with dummy values and then internalizing, you can overload the `NewLC()` factory function to take a stream:

Overloading NewL() and NewLC() to take a Read Stream

Providing overloads such as this gives the simplest possible API for constructing a new object from a stream. It does not get much easier than calling `NewL()` with a read stream to get a freshly internalized object instantiated on the heap.

The implementation is pretty simple, although it does require the existence of a (presumably private) default constructor:

```
CChemicalElement* CChemicalElement::NewLC(RReadStream&
aStream)
    {
    CChemicalElement* self = new (ELeave) CChemicalElement;
    CleanupStack::PushL(self);
    self->InternalizeL(aStream);
    return self;
    }

CChemicalElement* CChemicalElement::NewL(RReadStream& aStream)
    {
    CChemicalElement* self = NewLC(aStream);
    CleanupStack::Pop(self);
    return self;
    }
```

These are almost identical to the canonical `NewL()` and `NewLC()` pair, other than `InternalizeL()` replacing `ConstructL()` —a minimal outlay of effort for a considerable increase in ease of use!

Stores

Stores are basically collections of streams. There are a number of different types of store, each catering for a particular need. For example, sensitive data can be stored using `CSecureStore`, which encrypts its data streams. Table 3-7 gives a brief description of the available types of store, and Figure 3-10 shows the class hierarchy of stores.

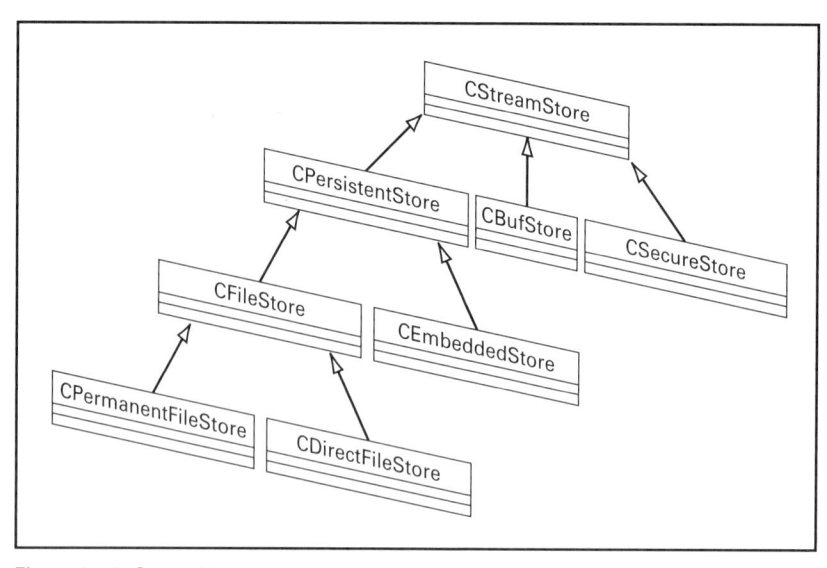

Figure 3-10 Store hierarchy.

Table 3-7 Store Types

Name	Instantiable	Comments
CStreamStore	No	Abstract base class for all stores, providing functionality for streams to be created and manipulated.
CPersistentStore	No	Abstract base class for persistent stores. It provides the behavior for setting and retrieving the root stream ID.
CFileStore	No	Abstract base class for file-based persistent store. Constructors take either an RFs and filename or an open file.
CDirectFileStore	Yes	Derived from CFileStore. File opened during construction. Enables streams to be created and objects externalized to them. Once committed, streams cannot be replaced, deleted, extended or changed in any way.
CPermanentFileStore	Yes	Derived from CFileStore. File opened during construction. Allows full manipulation of store contents. Generally used by database applications.
CEmbeddedStore	Yes	Derived from CPersistentStore. An embedded store is one that is contained with another stream.
CBufStore	Yes	Derived from CStreamStore. Nonpersistent, in-memory store.
CSecureStore	Yes	Derived from CStreamStore. Streams within the store are all encrypted.

Chapter 3 Files, Streams and Stores

TIP To use stores, you need to `#include <s32stor.h>`.

Stores are used in Symbian OS to save documents associated with applications. This aspect of file functionality is tied up with the Application Framework and is covered in Chapter 4. Many of the basic issues, however, can be discussed here with the focus being on file-based stores.

`CFileStore` itself is derived from `CPersistentStore`—its existence can last beyond that of the application that created it. Contrast this with `CBufStore`, which cannot be closed without losing all the data it contains.

There are two basic types of `CFileStore`: `CPermanentFileStore` and `CDirectFileStore`. `CPermanentFileStore` is used by applications that consider the data in the store to be the primary copy of the data. If the application data needs to be modified, it is loaded in from store, amended and then written back. The entire store will never be replaced; individual entries will be inserted, deleted or modified. `CPermanentFileStore` is typically used by database type applications, particularly those that use the DBMS API. Further details about the use of `CPermanentFileStore` can be found in the SDK documentation.

In contrast, `CDirectFileStore` allows no modification of data, once it has been committed. Applications using a direct file store will consider the in-memory version of data to be the primary copy, and modifications will be made to this internal, nonpersistent form of the data. When the data is to be persisted—for example, when the application exits—the original file will be completely replaced. Use of `CDirectFileStore` is explained in detail in the next section.

CDirectFileStore

This is how to create a direct file store, given a connected file server session, a filename and the mode in which to open the store:

```
void CElementsEngine::CreateStoreL()
   {
   CFileStore* store = CDirectFileStore::CreateLC(iFs,
      KTxtStoreTestFileName, EFileWrite);
   store->SetTypeL(TUidType(store->Layout()));
   CleanupStack::Pop();      // Store
   iStore = store;
   }
```

`ReplaceLC()` can also be used instead of `CreateLC()` to overwrite an existing file.

When the store is created, its type should be set with a call to `SetTypeL()`, with the `TUidType` passed in specifying the UID of the store layout, as returned by the `Layout()` method of the concrete type (in other words, identifying it as either a direct or permanent file store). The second and third UID components may be used for the store subtype (typically `KUidAppDllDoc` or `KDatabaseUid`) and the application UID, if required by the application.

Chapter 3 Symbian OS Fundamentals

In the example code shown, only the first UID component (store layout) is used, since .exes do not have documents associated with them. However, in a UI application, the application UID mapping stored in the file can be used by the system to recognize the application that such a file should be opened with.

Once the type has been set, the store is ready for use.

Closing the store is simply a matter of deleting the store object—the destructor takes care of closing the actual file and deallocating any resources. Naturally, since this is a persistent store, deleting the store object does not delete the actual file!

To reopen the store at a later stage—for example, to read it back in—use `OpenL()` or `OpenLC()`:

```
CFileStore* store = CDirectFileStore::OpenLC(iFs,
KTxtStoreTestFileName, EFileRead);
```

If you know the type of store you are dealing with, then you can use the open functions of the required concrete class—in this case, `CDirectFileStore::OpenLC()`. If you were to try and open this store using `CPermanentFileStore::OpenLC()`, then the attempt would leave with KErrNotSupported (-5). Note that if you are unsure of the store type, then the base class (`CFileStore`) implementation can be called to open the store and the type later determined using `Layout()`.

Once the store has been created or opened for writing, the first task is usually to create a new stream within the store. Remember that a store is a collection of streams.

Here is the code to create a new stream and write to it:

```
RStoreWriteStream stream;
TStreamId id = stream.CreateLC(*iStore);
stream << *list;
stream.CommitL();
```

`RStoreWriteStream` is derived from `RWriteStream`. This means, of course, that the `ExternalizeL()` functions developed earlier will take an `RStoreWriteStream` just as they took an `RFileWriteStream`. Once you have created the stream, you can write to it in just the same way as before. In this example, `operator<<` has been used—which, you will remember, is templated to invoke `ExternalizeL()`. Once the writing is complete, you can call `CommitL()` to flush data from the buffer and into the stream.

Notice how the `RStoreWriteStream::CreateLC()` function returns a `TStreamId`. The `TStreamId` object is simply the offset of the stream within the store, but can be considered to be a handle onto the stream, or a "pointer" to it. What do you do with the stream ID? Presumably you need to record it somewhere so that you can refer to it later when you want to read the stream back in—but where is the best place to put it? The answer is: in the **Stream Dictionary**.

The Stream Dictionary

The Stream Dictionary is just another stream that stores a mapping between the other stream IDs and their corresponding (stream) UIDs. These 32-bit unique identifiers are not related in any way to the application or store type UIDs. They only have to be unique within the specific file store, and it is up to the developer to specify the UID of each stream. Once specified, the Stream Dictionary can be used to "look up" the location (stream ID) of each stream from its given UID.

This is best considered by way of an example. To write the list of elements into three streams within the store—one for the metals, one for the semimetals and one for the nonmetals—requires three UIDs, say:

```
const TInt KUidMetalsStreamId = 0x101F613F;
const TInt KUidNonmetalsStreamId = 0x101F6140;
const TInt KUidSemimetalsStreamId = 0x101F6141;
```

The Stream Dictionary will be used to store the stream IDs associated with each UID as shown in Figure 3–11.

So the Stream Dictionary can be used to look up the stream ID of the given stream. However, since the Stream Dictionary is *itself* a stream, how do you find the Stream Dictionary's ID? The answer is to store the Stream Dictionary with the root ID of the store, which is always available for any given store.

In summary, then: the store will return its root stream ID, the root stream will lead to the Stream Dictionary ID, and the Stream Dictionary will allow individual stream IDs to be looked up, given their UIDs (which *you* have set).

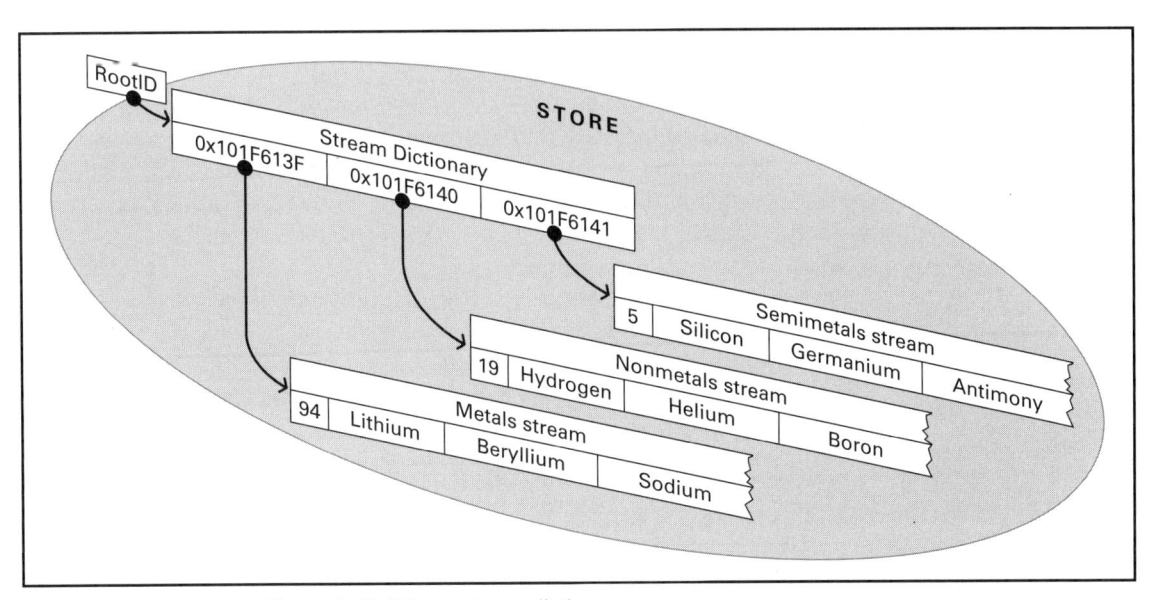

Figure 3–11 Using a stream dictionary.

Here is the code to instantiate a new Stream Dictionary:

```
iRootDictionary = CStreamDictionary::NewL();
```

Each time you create a new stream in the file store, you need to invoke `AssignL()` on the Stream Dictionary:

```
iRootDictionary->AssignL(TUid::Uid(KUidMetalsStreamId),
    metalsStreamId);
iRootDictionary->AssignL(TUid::Uid(KUidNonmetalsStreamId),
    nonmetalsStreamId);
iRootDictionary->AssignL(TUid::Uid(KUidSemimetalsStreamId),
    semimetalsStreamId);
```

passing in the UID for the stream, and the stream ID returned by the function when the stream was created.

Once all the streams have been added, you can then write the Stream Dictionary itself to a stream in the store, and set the store's root ID to point to the Stream Dictionary's stream:

```
RStoreWriteStream root;
TStreamId id = root.CreateLC(*iStore);
root << *iRootDictionary;
root.CommitL();
CleanupStack::PopAndDestroy();    // Root
iStore->SetRootL(id);
iStore->CommitL();
```

Note that with a `CDirectFileStore`, it is not possible to delete or otherwise change the streams in any other way once they are written. This is not the problem it may appear to be—the contents of direct file stores are completely replaced each time a change is made. This type of file store is generally used by applications that consider the in-memory copy of their data to be the primary copy. If your application needs to update a store each time data is amended, and the stored data is considered the primary copy—for example, in a database-style application—then a Permanent File Store should be used. More information on this topic can be found in the SDK documentation.

WARNING   Symbian OS defines a dictionary file store class, `CDictionaryFileStore`. This is not a stream store class; it is traditionally used for handling .ini files, and these are rarely used in Series 60. Do not confuse a direct file store and stream dictionary with a dictionary file store.

Reading from the Store

Here is the complete code to read all of the data back from the three streams plus the Stream Dictionary:

Chapter 3 Files, Streams and Stores

```
void CElementsEngine::ReadElementsFromStoreL()
    {
    // Open the store
    CFileStore* store = CDirectFileStore::OpenLC(iFs,
        KTxtStoreTestFileName, EFileRead);

    // Create and internalize the dictionary from the root stream:
    CStreamDictionary* dictionary = CStreamDictionary::NewLC();
    RStoreReadStream dictionaryStream;
    dictionaryStream.OpenLC(*store, store->Root());
    // Stream in the dictionary
    dictionaryStream >> *dictionary;
    CleanupStack::PopAndDestroy(); // DictionaryStream

    // Now use the dictionary to look up the stream ids
    TStreamId metalsStreamId = dictionary
        ->At(TUid::Uid(KUidMetalsStreamId));
    TStreamId nonmetalsStreamId = dictionary
        ->At(TUid::Uid(KUidNonmetalsStreamId));
    TStreamId semimetalsStreamId = dictionary
        ->At(TUid::Uid(KUidSemimetalsStreamId));
    CleanupStack::PopAndDestroy(dictionary);

    // Read in each stream
    RStoreReadStream stream;
    stream.OpenLC(*store, metalsStreamId);
    stream >> *iElementList;
    CleanupStack::PopAndDestroy(); // stream

    stream.OpenLC(*store, nonmetalsStreamId);
    stream >> *iElementList;
    CleanupStack::PopAndDestroy(); // stream

    stream.OpenLC(*store, semimetalsStreamId);
    stream >> *iElementList;
    CleanupStack::PopAndDestroy(); // stream

    CleanupStack::PopAndDestroy(store);
    }
```

Notice how the `Root()` function on the store is used to obtain the root (Stream Dictionary) stream ID. `CStreamDictionary::At()` returns the ID of the stream from the given UID.

File Stores Overview

To write multiple streams to a direct file store, use the following steps:

1. Specify the UIDs for the streams you wish to define (as part of the design of your application)—these need to be unique within the store.

2. `CreateL()` or `ReplaceL()` the store, specifying a file server session, the filename and `EFileWrite`.

3. Set the type of the store using `SetType()`. Use `Layout()` to get the store layout UID. (This will be `KDirectFileStoreLayout` for a direct file store.)

4. Create a Stream Dictionary object.

5. Create each of the file store write streams you wish to use, and assign each one to the Stream Dictionary with the appropriate UID.

6. Externalize the object data to the correct streams.

7. Commit each store stream once writing is complete.

8. Once all streams have been written, write the Stream Dictionary to the store.

9. Commit the Stream Dictionary.

10. Set the store's root stream to the stream ID of the Stream Dictionary, using `SetRootL()`.

11. Commit the store.

12. Delete the store and Stream Dictionary to free their associated resources.

To read the contents of the store, follow these steps:

1. `OpenL()` the file store, specifying a file server session, the filename and `EFileRead`.

2. Create a Stream Dictionary object and internalize it from the store. Use `Root()` on the file store to specify the Stream Dictionary's stream ID.

3. Retrieve the stream ID of each stream by calling `At()` on the Stream Dictionary, specifying the UID of the stream.

4. Internalize each stream from the store.

5. Delete the store and the Stream Dictionary.

In the example given, there is no clear justification for storing the element list in three separate streams within the file store—it has been done purely for the sake of example. In a real application, each stream may represent different data objects, such as video and audio streams in a media player, or high scores, player data and map information in a game. Streams allow you to keep this data separate; stores allow you to keep the separate streams in one place.

Further information on the other stores mentioned in this section can be found in the SDK documentation.

Client/Server Architecture

Virtually all asynchronous services in Symbian OS are provided by servers through Client/Server Architecture. Clients are simply programs that make use of a particular service provided by a server. A server is the program that services those requests in some way. Client/Server Architecture allows:

- Extensibility, since plug-in modules can be added to service new types of object. Often this can take place at runtime, so that a device can, for instance, support a new type of filing system that has been developed after the File Server that makes use of it.

- Efficiency, since multiple clients can be serviced by the same server.

- Safety, since servers and their clients exist in separate processes (usually) and communicate via message passing. A misbehaving client cannot bring down its server. (Note also that servers are expected to panic misbehaving clients, through a handle to the client's thread.)

- Asynchronicity, since servers use the Active Object framework to notify their clients when their work is done. The use of Active Objects with servers is important—by suspending the client thread, rather than polling the status of a request, Symbian OS reduces the amount of processor cycles required to process that request. This helps to improve power management, which is very important in a mobile device.

Details of how to write a server are beyond the scope of this book—this section simply focuses on how to use a server from its Client-side interface. It is not important to completely understand this when starting out.

One important server has already been introduced during the course of this chapter—the File Server. This uses an `RFs` **session** and an `RFile` **subsession**—these terms will be explained later in this section. Many other servers are used throughout the system: Comms Server, Window Server, Font and Bitmap Server and so on. All of them provide a (relatively) easy to use asynchronous API via an `R`-class server session.

Server Sessions

At the beginning of this chapter, when introducing the different class types, it was noted that `R`-classes do not have a common base class. While this is true, many `R`-classes are in fact server sessions which do have a common base class: `RSessionBase`.

`RSessionBase` provides part of the underlying API for communication between client and server. In the most part this is not used directly, since each derived server session class provides an API that abstracts it, hiding the detail and simplifying the interface. For example, the File Server derives from `RSessionBase` to create `RFs` as its session class. However, it may be informative to learn a little about what is happening "under the hood."

Server Sessions and Inter-Process Communication

The Symbian OS Kernel and Memory Management Unit implement a different memory-mapped address space for each process, preventing a rogue process from being able to overwrite the memory of another. The only process capable of "seeing" the entire physical memory is the Kernel process itself. All threads are clients of the Kernel server, and it is the Kernel that facilitates communication between processes and the threads they contain.

Here is part of the class definition for `RSessionBase`:

```
class RSessionBase : public RHandleBase
   {
protected:
   inline TInt CreateSession(const TDesC& aServer,
       const TVersion& aVersion);
   void SendReceive(TInt aFunction, TAny* aPtr,
       TRequestStatus& aStatus) const;
   TInt SendReceive(TInt aFunction, TAny* aPtr) const;
   };
```

The `CreateSession()` function is used by the server session to connect to its server. The server is specified by name, and the Kernel can use this to establish the connection. As you can see, the function is protected, and methods like `RFs::Connect()` will wrap up this call and present an easy client-side API, so that the caller does not actually have to know the name of the server it is connecting to.

The connection may then be thought of as a "pipe" between the client and server, through which all communication is routed. The communication itself takes the form of messages, passed from the client via the `SendReceive()` methods. Each overload takes a function ID (as a `TInt`) and a pointer to an array of four 32-bit values.

The function ID is used by the server to identify which service is being requested by the client. Depending on the requested service, any one of the four 32-bit values may be treated as:

- An integer.

- A pointer to a descriptor in the client's address space. This may be a package descriptor containing any "flat" data structure—see the discussion on package descriptors earlier.

Chapter 3 Client/Server Architecture

In the latter case the Kernel can, at the server's request, copy the contents of the descriptor into the server's address space. That data may be modified and copied back into the client's address space if required. Note that the server is never granted *direct* access to the client's address space—the Kernel marshals the data back and forth, because it can see the entire extent of the memory as a unified, flat address space. Naturally, copying large amounts of data back and forth can take some time, so clever steps are taken to ensure that servers such as the Media Server and the Font and Bitmap Server do not have to move large quantities of data across the process boundary.

You will notice that the two overloads shown differ by return type and whether or not they take a `TRequestStatus` argument. This is because both synchronous and asynchronous requests are permitted. Naturally, synchronous requests return their completion code directly from the function, whereas asynchronous requests return it via the `TRequestStatus` of the Active Object used to encapsulate the request.

On the server side, the client's message is represented as an `RMessage` object:

```
class RMessage
   {
public:
   void Complete(TInt aReason) const;
   void ReadL(const TAny* aPtr, TDes8& aDes) const;
   void ReadL(const TAny* aPtr, TDes16& aDes) const;
   void WriteL(const TAny* aPtr, const TDesC8& aDes) const;
   void WriteL(const TAny* aPtr, const TDesC16& aDes) const;
   TInt Function() const;
   TInt Int0() const;
   TInt Int1() const;
   TInt Int2() const;
   TInt Int3() const;
   const TAny* Ptr0() const;
   const TAny* Ptr1() const;
   const TAny* Ptr2() const;
   const TAny* Ptr3() const;
   };
```

It is pretty clear from the section of the API shown above how the server can use the `RMessage` class to access the data that has been sent over from the client. `Function()` returns the value of the function ID that the client requested. `ReadL()` is used to read data into the server's address space—`WriteL()` is used to write the data back to the client. `IntX()` and `PtrX()` provide different interpretations of *the same four data items*—the array of four 32-bit values in the `RServerSession`'s `SendReceive()` call.

Once the request has been completed, `Complete()` is invoked with the appropriate error code (hopefully `KErrNone`). In the case of a synchronous function being requested, this will become the return value of `SendReceive()`.

Chapter 3 Symbian OS Fundamentals

If an asynchronous function was requested, then this will be the value of the `TRequestStatus`.

How does the server get the message in the first place? Well, here is some of the class declaration for `CSharableSession`, the base class of the server-side objects that represent the session with the client:

```
class CSharableSession : public CBase
   {
   friend class CServer;
public:
   virtual void CreateL(const CServer& aServer);
   virtual void ServiceL(const RMessage& aMessage) =0;
   const CServer* iServer;
   };
```

The server creates the session when a client initially connects. It calls `CreateL()` to complete construction, passing in a reference to itself to allow the session to access it.

When a request is issued by the client, the server invokes the `ServiceL()` function on the relevant session. It is this method that invokes the appropriate functionality as determined by the function ID of the message, acting on any of the data in the message. The functionality is usually built into the session itself.

For completeness, here is some of the declaration of `CServer`:

```
class CServer : public CActive
   {
public:
   IMPORT_C void StartL(const TDesC& aName);
   IMPORT_C void DoCancel();
   IMPORT_C void RunL();
   };
```

Note that all servers derive from `CActive`—they are all Active Objects. The `StartL()` method adds them to the Active Scheduler and starts them waiting for their first request. It is the implementation of `RunL()` that is actually responsible for invoking `ServiceL()` on the session, and this is executed when the Kernel signals the server's `TRequestStatus` that a request has come in from the client.

How does the server itself come into existence? This depends on the nature of the server. Many of the essential servers are started as the system boots up, and they are always available. Others, particularly servers that are added to the system at a later date, are started explicitly by their client APIs when a request is issued for the first time—a server of this type will usually keep a count on the number of clients, and automatically unload itself when this number falls to zero.

Server Review

The basic process by which a server services an **asynchronous** request is given here. For simplicity's sake you can assume that the server in question has already been started.

1. The client establishes a connection with the server via an `RSession`-derived object which presents the client-side API.
2. The server creates a new session object associated with the client session.
3. The client makes a request via the client-side API.
4. The client API wraps up any data associated with the request into integers, descriptors or package descriptors, depending on the request.
5. The client API invokes `SendReceiveL()`, with the appropriate function ID and argument pointers.
6. The Kernel completes the server's `iStatus` member.
7. The server's `RunL()` is invoked, passing the message received from the client via the Kernel to the session's `ServiceL()` method.
8. The session's `ServiceL()` invokes the appropriate service handler based on the value of the function ID in the message.
9. The service handler copies data from the client's address space, using `ReadL()` as appropriate.
10. The service handler services the request.
11. The service handler copies data to the client's address space, using `WriteL()` as appropriate.
12. The service handler calls `Complete()` on the message with the appropriate error code.
13. The Kernel increments the client thread's semaphore and sets the `TRequestStatus` to the value specified in `Complete()`.
14. The client's `RunL()` is called by the client thread's Active Scheduler.
15. The client's `RunL()` performs any processing on the data returned via the client API.

So long as the client-side API is sufficiently well designed and implemented, the client need be concerned only with steps 1, 3 and 15.

Subsessions

The methodology described above works well, but it can be inefficient. Many of the resources needed in the Kernel to maintain the session have been left out, as they are nontrivial and their coverage is beyond the scope of this book.

However, each time a session is opened, more Kernel resources are consumed. This is why it was suggested earlier that the number of File Server sessions (RFs) should be kept to a minimum.

It was also mentioned that RFile is a *subsession*. A subsession is a lightweight way for a client to communicate with its server without needing individual sessions to represent each client-side object. The client establishes a single session with the server, then creates subsessions for each object. Each subsession is associated with a session through which the actual Inter-Process Communication (**IPC**) takes place.

The subsession base class API looks like this:

```
class RSubSessionBase
   {
protected:
   TInt CreateSubSession(RSessionBase& aSession, TInt aFunction, const TAny* aPtr);
   void CloseSubSession(TInt aFunction);
   void SendReceive(TInt aFunction, const TAny* aPtr, TRequestStatus& aStatus) const;
   TInt SendReceive(TInt aFunction, const TAny* aPtr) const;
private:
   RSessionBase iSession;
   TInt iSubSessionHandle;
   };
```

Notice how the call to CreateSubSession() requires an RSessionBase object, as outlined above (for example, opening an RFile needs an RFs). Also note how a function ID must be specified. This is the function ID which tells the server to create a subsession. Closing a subsession also needs a (different) function ID.

SendReceive() is available just as for an RSessionBase-derived class and is available with both synchronous and asynchronous overloads.

Finally, notice that the subsession maintains a copy of the RSessionBase and also a subsession handle that it uses to identify itself to the server. This is how the server is able to keep track of *which* subsession has made a particular request.

As with RSessionBase-derived classes, RSubSessionBase-derived classes should provide an easy-to-use API that hides the underlying IPC architecture — for example, the File Server derives from RSubSessionBase to create RFile as its subsession class.

Summary

This chapter has covered the essential concepts and techniques needed to begin writing applications for Symbian OS. Some of the more complex concepts, such as Active Objects, you may not need to use right away, but others such as the naming conventions and exception handling are fundamental, and an understanding of them is required for even the simplest application. You are encouraged to spend time getting to grips with the basics, but you will probably need to refer back to this chapter many times as you progress.

None of the topics discussed in this chapter are specific to Series 60; rather they are applicable to Symbian OS in general. The rest of the book will build on the foundations laid in this chapter, introducing techniques that will enable you to write applications that take advantage of the technologies offered by Series 60. The example application used throughout this chapter was console based; later you will learn about how to create Series 60 UI applications, add sound and graphics to them and support various communication protocols.

The next chapter describes the various Series 60 application architectures and frameworks and explains the basic structure of all Series 60 applications.

chapter 4

Application Design

Examining the framework architecture behind every Series 60 UI application and key elements of application design

Chapter 4 Application Design

Every Series 60 device comes with a large number of standard applications preinstalled. This chapter expands on the fundamental idioms detailed in Chapter 3, examining the main design principles used by such applications and explaining how to use Series 60 APIs to create similar GUI-based applications of your own.

The public APIs provided by the Series 60 SDK enable you to quickly develop your own applications. However, developing *good* applications requires a little more knowledge of the APIs, and the underlying Operating System components that they interact with. The explanations and design tips in this chapter will help you to produce high-quality applications, and increase your awareness of the capability of the platform.

Topics covered are:

- **Application Framework**—An introduction to the components, layers and classes that comprise the application framework of a fully fledged Series 60 application. Both Symbian OS and Series 60 components are covered.

- **Application Architecture**—A more detailed description of the core application framework classes—how they are created, and how they interact. The various application architectures available are described, along with the reasons for using each.

- **Splitting the Application UI and the Engine**—This section covers why, and how, to separate the implementation of an application's **Engine** (its main data store, and the definition of how it manipulates that data) from its **User Interface** (how the data is displayed, and how the user interacts with it).

- **ECom**—Introduced in Series 60 2.x, ECom is a framework for delivering additional functionality to an application at runtime through the use of plug-ins. This section discusses ECom and provides a practical example of its use.

- **Internationalization**—How to prepare your application for use in different countries and languages. This is a short guide covering such issues as date, currency and other locale-specific formats, and enabling support for user-visible text in multiple languages.

- **Good Application Behavior**—This section provides guidelines for how a well-behaved Series 60 application should correctly respond to system events and notifications from system watchdogs.

Chapter 4   Application Framework   175

Details of how to use specific UI components will be covered in the chapters that follow.

The frequent code excerpts used throughout this chapter are mainly taken from the following example applications: **ShapeDrawer**, **HelloWorld**, **SimpleDlg** and **EComExample**. Details of how to download the full buildable source for these example applications are given in the Preface.

The **ShapeDrawer** application provides two different views: the first lets the user draw circles and rectangles on the screen; the second represents these drawn items, along with their coordinates, in a list box. From the second view, the user can select and delete items. Upon shutting down and restarting the application, the user will see that their data has been saved from the previous session. This application also demonstrates splitting the engine from the application UI.

The **HelloWorld** and **SimpleDlg** applications have no real functionality—they just serve to demonstrate a particular Series 60 framework architecture. The **EComExample** application is also very simple and just draws a particular shape chosen by the user, demonstrating the ECom framework and how to write ECom plug-ins.

It is recommended that the reader download, build and run these applications, as this will help to provide a more practical view of the information presented here.

Application Framework

Series 60 UI applications rely upon a number of OS components to operate. Many common application requirements, such as screen drawing and the persistence of application data, need not be programmed from scratch into the application by the developer—this would be like reinventing the wheel. Instead, an application's common requirements are satisfied by various servers in the OS. For example, the Window Server provides extensive UI support, and the File Server lends support for persisting data.

The application framework constitutes a set of core classes that are the basis of all applications. These classes form the structure of all applications, and they also encapsulate interplay between the application and required OS servers.

This section provides background information concerning the design of the Series 60 application framework. It shows how Series 60 framework classes derive from, and functionally augment, generic Symbian OS framework classes. This contextual information will help you to better understand and more effectively use these APIs and will prepare you for the class-specific discussions in the **Application Architecture** section.

Chapter 4 Application Design

Primarily, Figure 4–1 shows conceptual groupings of the application framework classes. It also shows real API classes contained within these groupings, and inheritance relationships between them:

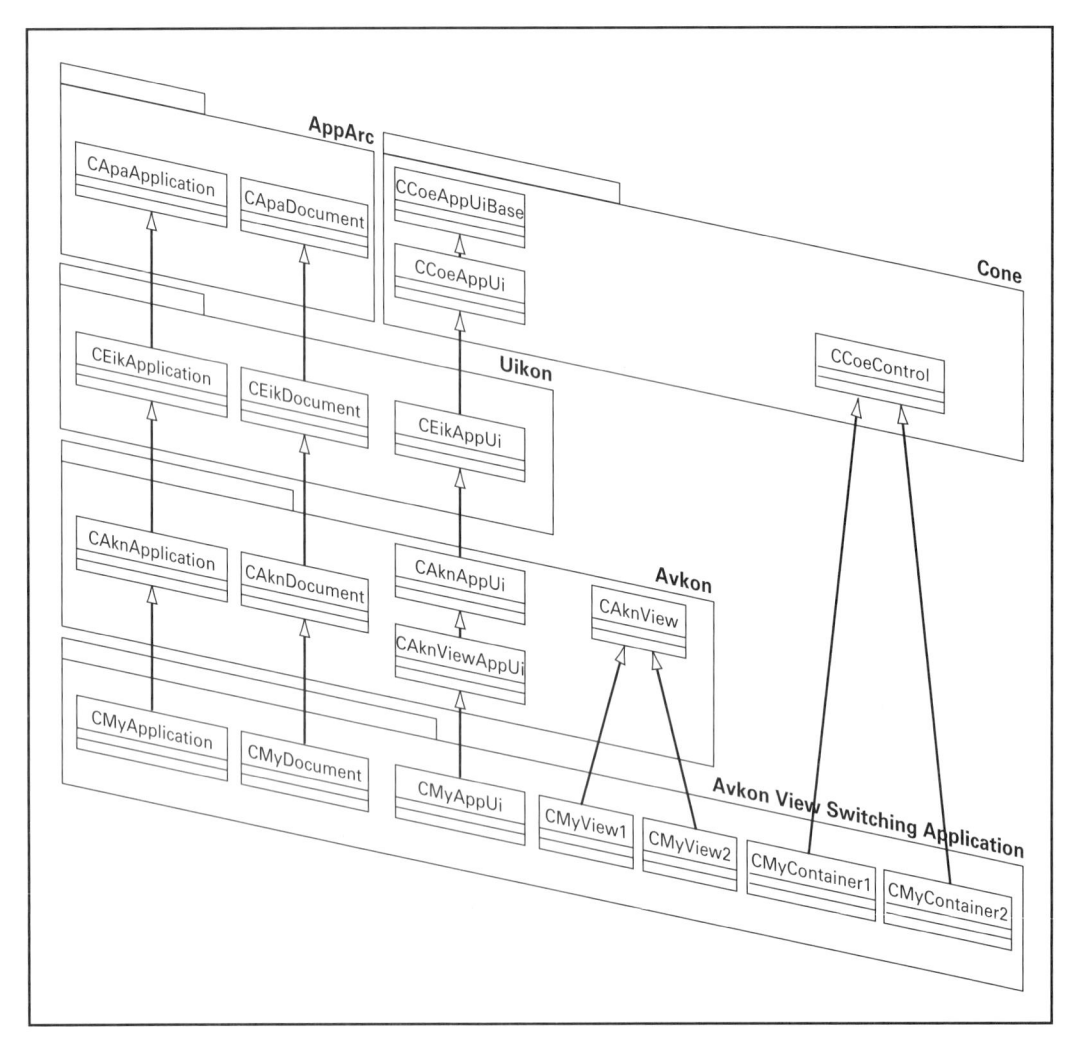

Figure 4–1 The application framework classes for an Avkon View-Switching application.

The architecture used here will be explained in detail in the *Avkon View-Switching Architecture* subsection later. Note that Figure 4–1 is a UML class diagram that shows four class layers, omitting associations, mixin classes and derivations from `CBase` for clarity.

The first layer in the class hierarchy is divided into two fundamental components, namely **AppArc** and **CONE**.

AppArc stands for "Application Architecture." These classes provide the basic application structure, and mechanisms for delivering system information to the application and persisting data using the File Server. This is a Symbian OS category, and classes in it are named with the prefix "*Apa"—for example, CApaApplication.

CONE is short for "CONtrol Environment," and the classes in this component provide mechanisms for handling user input and creating the User Interface—these classes interact primarily with the Window Server. (Briefly, the Window Server is a Symbian OS Server that controls drawing and handles user input.) Again, this is a Symbian OS component, and classes in it are named with the prefix "*Coe"—for example, CCoeControl.

NOTE Remember that a Symbian OS Server is a program, without a UI, that runs in a separate process and manages (possibly shared) resources. A client-side API is provided, with parameters passed over the process boundary.

The second layer of classes fit into the Symbian OS **Uikon** component. This is a generic, device-independent implementation of highly functional, nonabstract framework classes, and provides a UI library layer that is common to all Symbian OS platforms. Some concrete UI controls, for example, list boxes and scroll bars, can be created in this layer and these are sometimes called **Eikon** controls. Note, however, that **Avkon** (Series 60-specific) controls should always be used in preference to Eikon controls, as they provide the correct behavior for Series 60 applications. Classes in the Uikon/Eikon component are named with the prefix "*Eik", for example, CEikApplication.

The third layer consists of the Avkon classes, and these provide core Series 60 UI functionality, such as menu support. This layer is a purely Series 60-specific implementation, and your application should always derive from the Avkon classes in preference to generic Symbian OS classes. The Avkon classes are named with the prefix "*Akn"—for example, CAknApplication.

The fourth layer, the application-specific layer, demonstrates how you would derive from the Series 60-specific Avkon classes in order to design custom applications. Please note that this layer of Figure 4–1 is application architecture-specific and the classes from which you derive may vary, depending upon UI requirements. The **Application Architecture** section will address these issues.

Many of the classes in the first layer are abstract and merely define the interface to the framework APIs. The second layer adds a common Symbian OS implementation, and this will be shared with other Symbian OS UI platforms. The third layer adds a Series 60-specific implementation to the framework, and the fourth layer adds custom application implementation.

The next section will discuss the core application framework classes in more detail and examine their relationship with each other and the system in general.

Application Architecture

The term application architecture refers to the collection of application framework classes that collectively make up an application. Applications can have slightly different architectures, based upon their required UI designs. These variations are covered in detail later, in the *Designing an Application UI* subsection, which describes the merits and drawbacks of each and guides you towards choosing the design that is most applicable for your application, based on clearly defined UI requirements. Regardless of these UI design variations, the functionality of each of the architectures overlaps significantly. Therefore, the first part of this section focuses on the commonality across the various architectures—the core application classes. The next subsection examines the functionality that these classes offer, and also their interplay with the system.

Core Application Classes

All Series 60 UI applications share some basic functionality:

- They provide a User Interface displaying information and allowing user interaction.
- They respond to various user-initiated events (such as the user choosing a menu option).
- They respond to various system-initiated events (such as Window Server events that cause redrawing of the screen).
- They are able to save and restore application data.
- They can uniquely identify themselves to the framework.
- They provide descriptive information to the framework about the application (such as icons, captions and so on).

The application framework classes that provide this functionality fit into the following high level categories: **View**, **Document**, **Application** and **Application UI** (or **AppUi**). Figure 4–2 shows how the classes interrelate to make up an application's basic form.

The **Application** class is notably the least coupled—its role is static. Besides serving as the application's main entry point, it delivers application-related information back to the framework, such as representational icons and captions, as can be seen in the Series 60 Application Launcher menu. The Application class does not involve itself with the application's data and algorithms. Classes in this category derive from `CAknApplication`.

The **Document** class, as its name indicates, provides a context for persisting application data. As documents are typically made to be edited, this class also provides a method for instantiating the AppUi class. Classes in this category derive from `CAknDocument`.

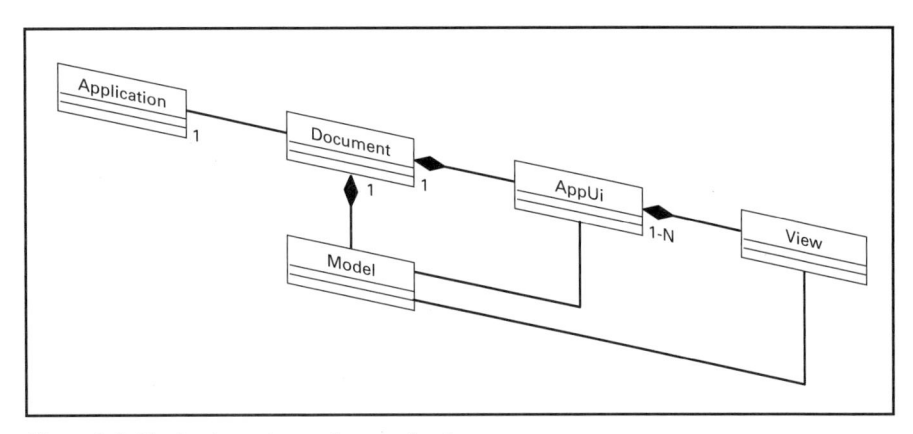

Figure 4–2 The basic anatomy of an application.

The **AppUi** is not, in itself, a drawable control. More accurately, it is a recipient of numerous framework-initiated notifications, such as key presses from the user, or important system events. The AppUi will either handle events itself, or if appropriate, relay them to the Views that it owns for them to handle. Classes in this category derive from `CAknAppUi`.

View is an umbrella term describing any representation of the model's data on the screen. It does not refer to specific UI controls that may be used to make these representations. The actual concrete application framework class used for this category depends on the application architecture used and will be discussed later in the *Designing an Application UI* subsection.

Also present in Figure 4–2 is the **Model**, which doesn't map to a specific Symbian OS or Series 60 class, but encapsulates an application's data and its algorithms. The model is owned by the document and can use the document's facilities for storing data. It is worth noting that numerous classes hold references to the model—this is a visible demonstration of its importance.

Note also, that Series 60 applications have a limitation (by design) of only one open document (file) per application.

Application Initialization

Figure 4–3 demonstrates the framework's role in initializing an application, using the particular case of the **ShapeDrawer** application.

Each of the methods shown here *must* be created in order to provide a minimal Series 60 application.

First, all Series 60 UI applications implement a global function, `E32Dll()`, that is called by the framework in the first step, when the application is launched. It is referred to as the **DLL entry point** and must be present in your application. Remember that every Series 60 UI application is a polymorphic DLL.

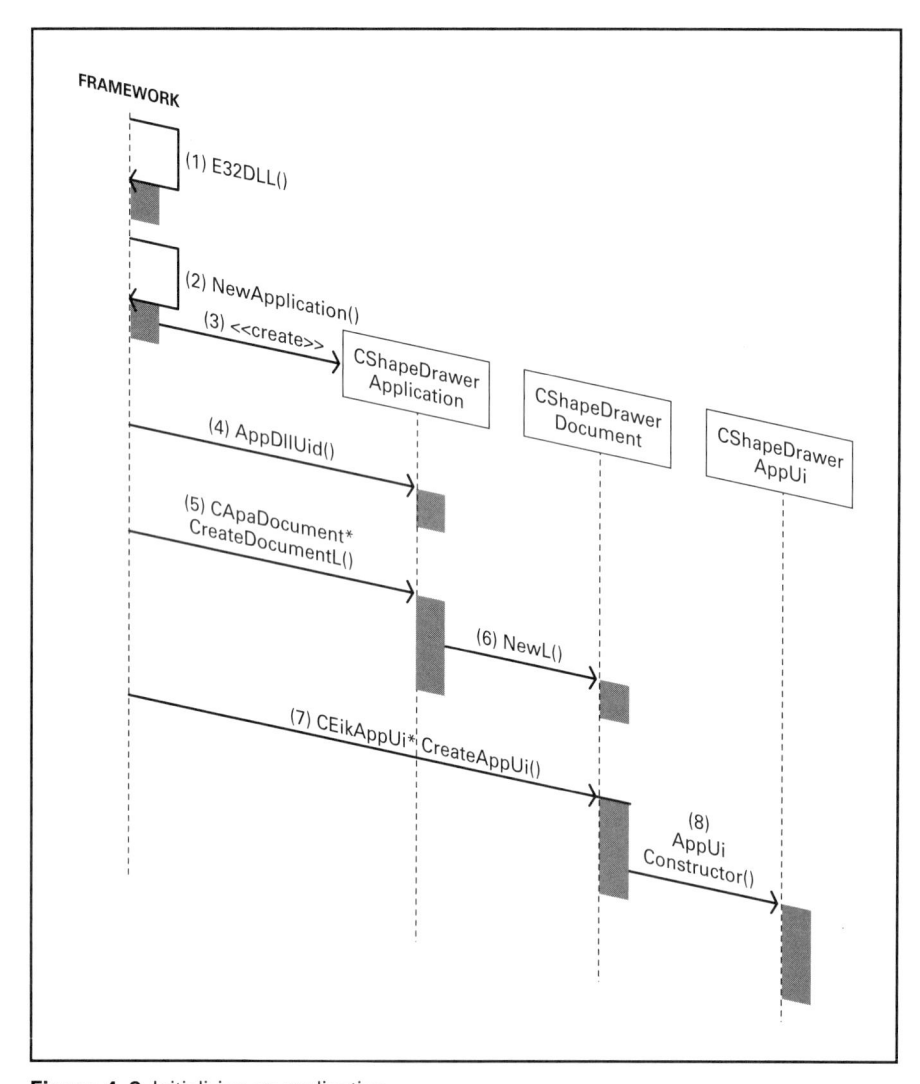

Figure 4-3 Initializing an application.

```
// DLL entry point, return that everything is OK
GLDEF_C TInt E32Dll(TDllReason)
    {
    return KErrNone;
    }
```

The second step in the initialization of an application is for the framework to call NewApplication(), which is the single function exported by the DLL:

```
EXPORT_C CApaApplication* NewApplication()
    {
    return new CShapeDrawerApplication;
    }
```

Chapter 4 Application Architecture

This creates an instance of your application class (step three), in this case `CShapeDrawerApplication`, and returns a pointer to it. The framework subsequently uses this pointer to complete construction of the application.

NOTE As the Cleanup Stack does not yet exist, the overloaded `new` (`ELeave`) operator cannot be used for construction at this stage.

The call to `AppDllUid()` by the framework in step four returns the UID of the application:

```
TUid CShapeDrawerApplication::AppDllUid() const
    {
    // Return the UID for the ShapeDrawer application
    return KUidShapeDrawerApp;
    }
```

This must return the value specified in the application's `.mmp` file and can be used to determine whether an instance of the application is already running.

The application startup process also allows the framework to obtain pointers to the newly created Document and AppUi classes. In the case of the Document, the framework receives a pointer to the abstract `CApaDocument` base class through a call to `CShapeDrawerApplication::CreateDocumentL()` in step five/six:

```
CApaDocument* CShapeDrawerApplication::CreateDocumentL()
    {
    // Create a ShapeDrawer document and return a pointer to it
    CApaDocument* document =
CShapeDrawerDocument::NewL(*this);
    return document;
    }
```

Note that the Document receives a reference to the Application, as it needs to be able to call `AppDllUid()` when handling application files.

In the case of the AppUi, the framework receives a pointer to the `CEikAppUi` base class through a call to `CShapeDrawerDocument::CreateAppUiL()` in step seven/eight:

```
CEikAppUi* CShapeDrawerDocument::CreateAppUiL()
    {
    // Create the application user interface, and return a
pointer to it
    CEikAppUi* appUi = new (ELeave) CShapeDrawerAppUi();
    return appUi;
    }
```

This is how the system framework is able to call methods in these classes when necessary.

Chapter 4 Application Design

These implementations, though brief and straightforward, are enough to create the application framework. What remains is to consider the other important functions needed to provide an executable skeleton application.

Important AppUi Methods

The AppUi exposes many functions that the framework calls in order to inform each application of various events. The following AppUi functions are important to know about:

- `HandleKeyEventL()` —For handling user key presses.

- `HandleForegroundEventL()` —Called when the application switches to or from the foreground. The default implementation handles changes in keyboard focus.

- `HandleSystemEventL()` —Delivers Window Server-generated events.

- `HandleApplicationSpecificEventL()` —Notification of custom events that you can define yourself. The default implementation handles notification of color-scheme changes.

- `HandleCommandL()` —For handling user selection of menu options.

Below is a partial implementation of the **ShapeDrawer** example's `HandleCommandL()` method. Its parameter is set by the system to represent the command issued. Apart from handling menu options, it demonstrates the typical approach toward invoking Avkon View-Switching, and this will be explained later in the *Avkon View-Switching Architecture* subsection:

```
void CShapeDrawerAppUi::HandleCommandL(TInt aCommand)
    {
    switch (aCommand)
        {
        case EShapeDrawerSwitchToListView:
            {
            TUid viewId;
            viewId.iUid = EShapeDrawerListViewId;
            ActivateLocalViewL(viewId);
            break;
            }
        case EShapeDrawerSwitchToGraphicView:
            {
            TUid viewId;
            viewId.iUid = EShapeDrawerGraphicViewId;
            ActivateLocalViewL(viewId);
            break;
            }
        case EAknSoftkeyExit:
        case EEikCmdExit:
            if (CheckDiscSpaceL())
                {
```

Chapter 4 Application Architecture

```
            SaveL();
        }
        Exit();
        break;
        ...
    default:
        User::Panic(KShapeDrawerPanicName,
EShapeDrawerUnknownCommand);
        break;
    }
}
```

It is usual for a `switch` statement to be used to ensure the appropriate action is taken, depending on the command received. Note that some of these commands are custom defined—for example, `EShapeDrawerSwitchView`, as defined in `ShapeDrawer.hrh`—and others, such as `EAknSoftKeyExit`, are defined by the system.

The command `EAknSoftkeyExit` is called when the **Exit** soft key is pressed. The command `EEikCmdExit` is called by the system when the application needs to close—this can be due to user selection of the **Exit** option from the **Options** menu or to system shutdown of the application. These commands and menus in general are covered in more detail in Chapter 5.

Note also that the function will Panic if an unexpected command is given.

Designing an Application UI

The three common application architecture design approaches used for application UIs are:

- Traditional Symbian OS Control-Based Architecture
- Dialog-Based Architecture
- Avkon View-Switching Architecture

This subsection explores these three application architectures, giving example code to show how to implement each option and to clarify the benefits and drawbacks of each approach.

View Terminology

Before exploring these different architectures, you need to be clear on the terminology used in this chapter:

- A "view" is a conceptual term meaning "the representation of the model's data on the screen," and this does not refer to a specific architecture. A view is actually rendered by one or more (`CCoeControl`-derived) UI controls, organized in a hierarchy. The parent control is often referred to as the **Container**, except where a **Dialog** is used to implement the view.

- In the Avkon View-Switching Architecture, the term "**Avkon view**" refers to a class that is registered with the systemwide View Server and controls the view's instantiation and destruction. As you will see, an Avkon view is a `CAknView`-derived class and not an actual control itself—typically it owns a Container control to create its "view."

Each of these architectures offers different approaches to designing application UIs—all architectures offer a means of delivering "views," or visual representations of application data, and a mechanism by which users can interact with it. Each architecture also has features and capabilities that differentiate it from the others. The material in the sections that follow will overview the respective features, demonstrate the programming tasks required in order to create each architecture and provide guidance in choosing the appropriate one.

Before addressing the architectures individually, it is helpful to compare and contrast them in general terms. Dialog-Based and Traditional Symbian OS-Based Architectures, though very different, are conceptually much more similar to one another than to the Avkon View-Switching Architecture. They are conceptually similar because:

- They are characterized solely by the type of UI controls they use to generate views.
- Architecturally, they are nearly identical. That is, in both designs the AppUi class simply "owns" the view controls and is therefore responsible for managing them directly.

The Avkon View-Switching Architecture is fundamentally different from these two approaches. (You will see this immediately when inspecting it from an architectural perspective.) Whereas the other architectures are characterized by the UI controls they specify and the different features these controls offer, Avkon View-Switching does not specify the type of control to be used for rendering a view. Moreover, this architecture facilitates a new way to manage views. Whereas views in the other architectures are managed by the AppUi, the Avkon View-Switching Architecture delegates the management responsibility to a systemwide View Server. Avkon View-Switching is best understood after understanding the more fundamental designs. Therefore, it is presented last in this section.

Traditional Symbian OS Control-Based Architecture

As previously stated, the Traditional Symbian OS Control-Based Architecture is created such that the AppUi owns its view controls directly—these controls always inherit directly from `CCoeControl`. The standard term for a view class that derives directly from `CCoeControl` is "Container."

Although Chapter 5 covers `CCoeControl` in great detail, a brief overview of its characteristics is required here, so that you can understand the capabilities of

Chapter 4   Application Architecture   185

this approach. `CCoeControl` can be thought of as a blank canvas. By inheriting from this class, you can create a wide range of custom controls, the functionality and complexity of which are limited only by your ability and imagination. The only downside to this flexibility is that the control is *literally* like a blank canvas, and a lot of coding effort is required to provide any significant functionality.

In terms of handling view-switches, the AppUi is responsible for handling user-initiated view-switch requests. Subsequently, the AppUi ends up behaving like a giant switch, activating and deactivating Containers according to user or system input.

Figure 4–4 illustrates the framework classes for the Traditional Symbian OS Control-Based Architecture and shows you which classes you need to derive from when creating your own application. The fourth layer is the custom layer that represents the classes that you would write.

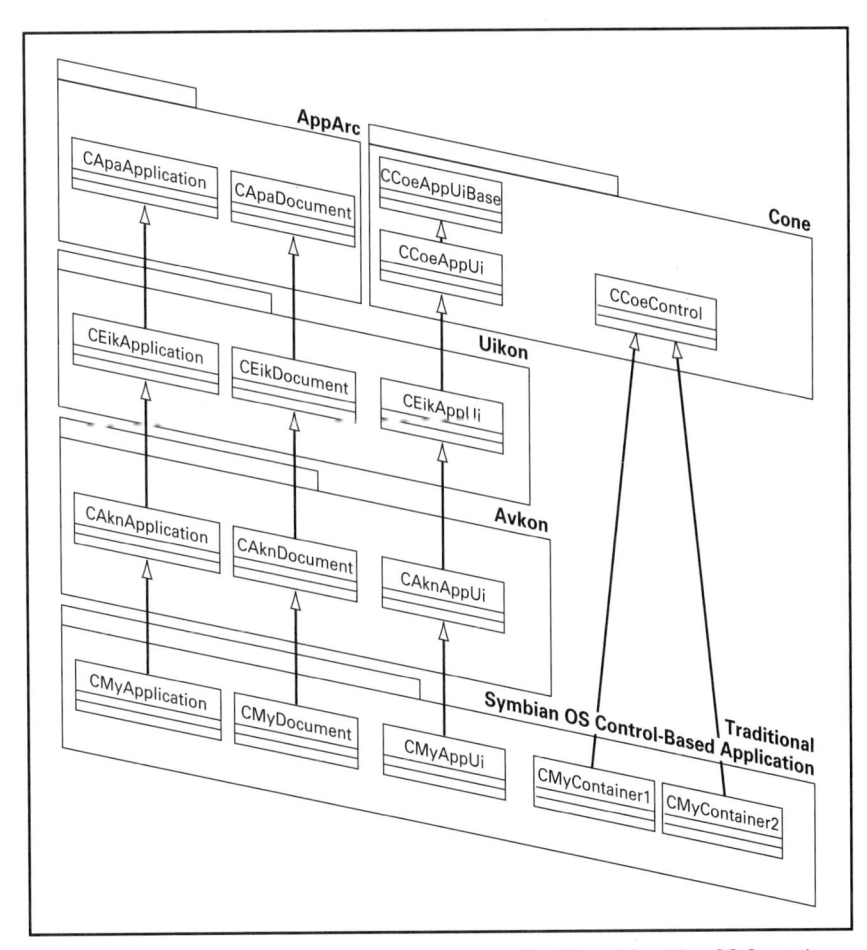

Figure 4–4 The application framework classes for a Traditional Symbian OS Control-Based UI application.

How to Use Traditional Symbian OS Control-Based Architecture

The **HelloWorld** application introduced in Chapter 1 uses the Traditional Symbian OS Control-Based Architecture. It will be used to demonstrate the essential steps in creating an application using this architecture. This application is very simple and only has one UI control, CHelloWorldContainer, but you could supplement the application with additional Containers.

Remember that the different architectures reflect only the implementation of the UI controls, and therefore there are no differences in the application class or document class. The AppUi will create the Container class, and the Container class will display the data. Note that in this very simplistic example there is no model class.

The class declaration for the CHelloWorldContainer class is shown below:

```
class CHelloWorldContainer : public CCoeControl
   {
public: // Constructors and destructor

   static CHelloWorldContainer* NewL(const TRect& aRect);
   static CHelloWorldContainer* NewLC(const TRect& aRect);
   ~CHelloWorldContainer();

private:

   void ConstructL(const TRect& aRect);

private: // from CoeControl

   void SizeChanged();
   TInt CountComponentControls() const;
   CCoeControl* ComponentControl(TInt aIndex) const;
   void Draw(const TRect& aRect) const;

private: //data

   CEikLabel* iLabel;                    // example label
   };
```

The first thing to notice is that the Container class derives from CCoeControl — this is the base class for all controls. The "Hello World" text is displayed in the form of a label — iLabel, the label itself being a control. CHelloWorldContainer implements four methods from CCoeControl — all of them are called by the framework. SizeChanged() allows the control to respond to a change in its size. Draw() is called to draw the control. In this example, Draw() just clears the screen (the label control draws the text, and its Draw() method will also be called by the framework). CountComponentcontrols() returns the number of controls the Container owns — in this case, just one (the label). For each control owned by the

Container, the framework makes a call to `ComponentControl()` to retrieve it. More information on all of these methods can be found in the **Controls** section of Chapter 5.

The Container is constructed in the AppUi class:

```
void CHelloWorldAppUi::ConstructL()
   {
   BaseConstructL();
   iAppContainer = CHelloWorldContainer::NewL(ClientRect());
   iAppContainer->SetMopParent(this);
   AddToStackL(iAppContainer);
   }
```

Calling `SetMopParent()` on the Container is important to establish the child-parent relationship between controls. `AddToStackL()` pushes the Container on top of the control stack so that, for example, it can receive key events. Both of these methods will be examined in more detail in Chapter 5.

The menu is defined in the resource file, `HelloWorld.rss`:

```
RESOURCE MENU_BAR r_helloworld_menubar
   {
   titles =
      {
      MENU_TITLE
         {
         menu_pane = r_helloworld_menu;
         }
      };
   }

RESOURCE MENU_PANE r_helloworld_menu
   {
   items =
      {
      MENU_ITEM
         {
         command = EHelloWorldCommand1;
         txt = COMMAND_ONE;
         },
      MENU_ITEM
         {
         command = EAknCmdExit;
         txt = text_softkey_exit;
         }
      };
   }
```

Resource files and structures are explained in more detail in Chapter 5. Note, however, that the **HelloWorld** example provides the menu option `EHelloWorldCommand1`. If the user selects this menu option, then the framework will call the `HandleCommandL()` method in the AppUi class, which handles each command appropriately:

```
void CHelloWorldAppUi::HandleCommandL(TInt aCommand)
   {
   switch (aCommand)
      {
      ...
      case EHelloWorldCommand1:
      {
      HBufC* message = StringLoader::LoadLC(R_DIALOG_TEXT);// Pushes message onto the Cleanup Stack.

      CAknInformationNote* informationNote = new (ELeave) CAknInformationNote;

      informationNote->ExecuteLD(*message);

      CleanupStack::PopAndDestroy(message);   // Removes message from the Cleanup Stack.
      break;
      }
      ...
   }
```

That covers the basics for an application using the Traditional Symbian OS Control-Based Architecture. The details of this example are not relevant to the architecture discussion here, but the functionality shown will be discussed in Chapters 5 and 6.

You can implement your own view-switching mechanism by using `AddToStackL()` and `RemoveFromStackL()` to swap between Containers (and therefore views) if using this architecture to implement an application with multiple views. These methods will be covered in Chapter 5, but note that if your application has multiple views, then an Avkon View-Switching Architecture may well be more appropriate. Further details on using controls to create fully interactive application UIs are provided in Chapters 5 through 8.

Dialog-Based Architecture

Like the Traditional Symbian OS-Based Architecture just described, the Dialog-Based Architecture similarly establishes the AppUi as the control-owning class. The difference is that the control that it owns inherits directly from one of a family of dialog classes. The idea is to use the built-in features of these classes in order to render data views and to handle switching between them.

Dialogs are covered in greater detail in Chapter 6, but some of their main characteristics are worth mentioning here in order to provide a basic rationale for this approach. Series 60 comes complete with a number of dialog classes that satisfy common requirements for data representation, entry and editing. A primary benefit of dialogs is that they require much less development work

Chapter 4 Application Architecture   189

than controls that derive immediately from `CCoeControl`, as they automatically manage the layout of their child controls. Additionally, their content and layout can be changed in resource files, without rebuilding any C++ code.

A typical use case is where the main view requires a simple layout—for example, a "settings" application. The dialog used for the main view is modeless; in other words, it does not need to keep the focus at all times. (See Chapter 6 for further information on the difference between modal and modeless dialogs.) Multipage dialogs may be used to provide a set of views in conformance with the Series 60 UI Style Guide.

Figure 4–5 illustrates the framework classes for the Dialog-Based Architecture and shows which classes you need to derive from when creating your application. The fourth layer is the custom layer that represents the classes that you would write.

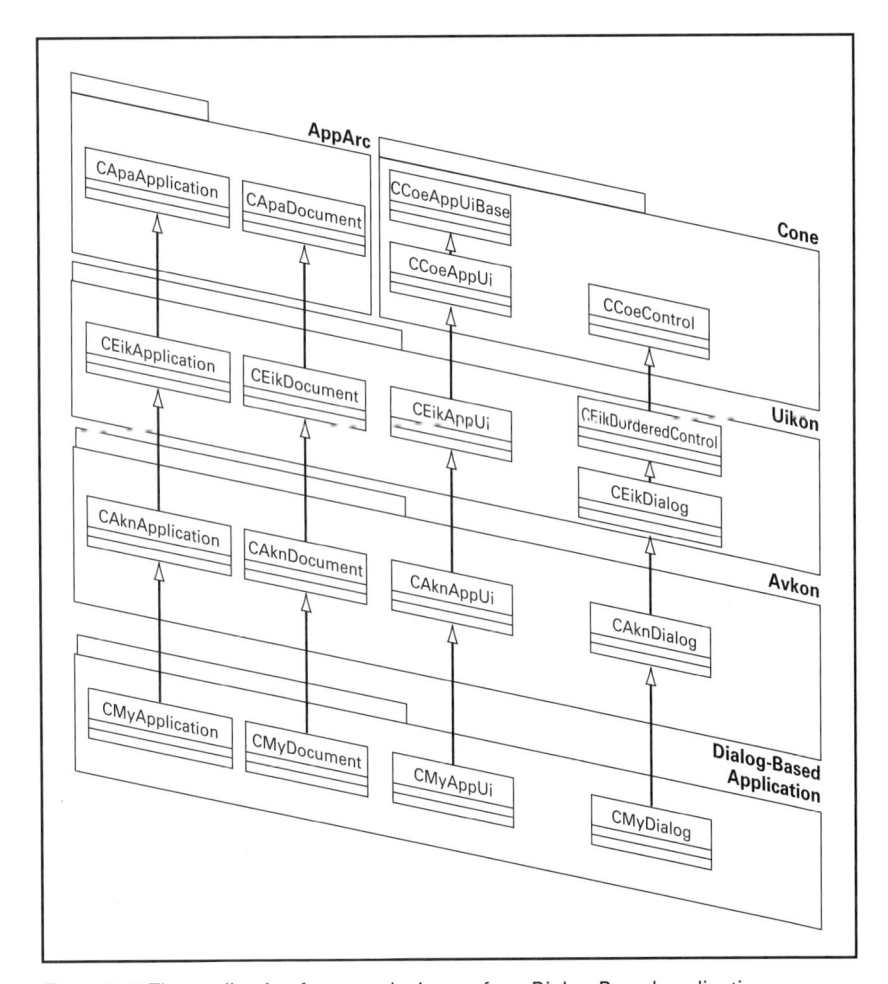

Figure 4–5 The application framework classes for a Dialog-Based application.

How to Use Dialog-Based Architecture

The example, **SimpleDlg**, demonstrates how to create a simple application where the main application view is a dialog. Again, this application can be used to create a skeleton application, if required.

The dialog is defined in the resource file `SimpleDlg.rss`:

```
RESOURCE DIALOG r_simple_dialog
   {
   flags = EEikDialogFlagNoDrag |
      EEikDialogFlagNoTitleBar |
      EEikDialogFlagFillAppClientRect |
      EEikDialogFlagCbaButtons |
      EEikDialogFlagModeless;

   buttons = R_AVKON_SOFTKEYS_OPTIONS_BACK;

   items =
      {
      DLG_LINE
         {
         id = ESimpleDlgCIdGameName;
         type = EEikCtLabel;
         control = LABEL
            {
            txt = GAME_NAME_TEXT;
            };
         }
      };
   }
```

Resource files and structures are covered in more detail in Chapter 5.

The mode of the dialog is set by using the `EEikDialogFlagModeless` flag. Flags are also set to achieve the filling of the client rectangle and to suppress a title bar. The soft keys are defined by `EEikDialogFlagCbaButtons` to be set to **Options** and **Back**. Note that the dialog must be modeless in order for the AppUi to receive commands and events—otherwise all input would be directed straight to the dialog itself.

Constructing and running the dialog is quite straightforward, as you would expect. This is performed in the AppUi class:

```
void CSimpleDlgAppUi::ConstructL()
   {
   BaseConstructL();
   iAppDialog = new (ELeave) CSimpleDlgDialog;
   iAppDialog->SetMopParent(this);
   iAppDialog->ExecuteLD(R_SIMPLEDLG_DIALOG);
   AddToStackL(iAppDialog);
   }
```

Chapter 4 Application Architecture   191

Because the dialog is modeless, `ExecuteLD()` returns immediately after being called. The dialog has to be added to the control stack with the `AddToStackL()` because modeless dialogs do not do this for themselves. The application extends the dialog just as it would normally extend a `CCoeControl`-derived view in order to achieve the desired functionality.

One further point to note is that the dialog must be destroyed in the destructor of the AppUi:

```
CSimpleDlgAppUi::~CSimpleDlgAppUi()
    {
    if (iAppDialog)
        {
        RemoveFromStack(iAppDialog);
        delete iAppDialog;
        }
    }
```

Again, the reason is that the dialog is modeless and so is not destroyed when the `ExecuteLD()` method returns. Note that the dialog must also be first removed from the control stack.

Avkon View-Switching Architecture

Figure 4-1 depicts the basic design for an application using the Avkon View-Switching Architecture. The architecture is clearly more complicated than the previous two. Specifically, another class has been introduced as an intermediary between the AppUi and the Container. This is a `CAknView`-based class, whose purpose will be clarified shortly. Another variation from previous designs is that the AppUi class inherets from `CAknViewAppUi` instead of just `CAknAppUi`. This is because these two classes (`CAknView` and `CAknViewAppUi`) offer complementary functionality—you never use one without the other.

In the other two architectures, the AppUi was directly responsible for handling view switching—it had to manage the instantiation, deletion and display of the view-rendering controls. But the `CAknView`-based class significantly reduces the AppUi's role in this respect.

The AppUi still handles view-switch requests, but now, instead of deleting the old Container and instantiating a new one, the AppUi need only call one of its special view-activation functions—`ActivateViewL()`, for example. These special `CAknViewAppUi` functions make an activation request to the View Server—the View Server then explicitly coordinates the deactivation of the current view and the activation of the requested view by calling activation/deactivation member functions in the relevant `CAknView`-based classes.

An implementation of this architecture is clearly demonstrated in the **ShapeDrawer** example application. However, before getting into the code, it will be helpful to look at the general features required of this architecture:

- The application must be designed so that each `CAknView`-derived Avkon view class owns a Container, and the AppUi then owns each Avkon view.

- The application's AppUi class must be derived from `CAknViewAppUi` instead of just `CAknView`, because the former offers methods to register, activate and deactivate Avkon views.

- All Avkon views must be registered with the View Server.

- The Avkon views have activation/deactivation member functions that can be called directly by the View Server. You must override these functions to provide correct handling of the subordinate Container.

The View Server follows a set of very simple rules. The primary rule is to ensure that *one, and only one, Avkon view is active per application, at any given time*. Upon registration, Avkon views are uniquely identified to the View Server by two UIDs: one to identify the owning application, and one to uniquely identify the view within that application.

This one-active-view-per-application scheme suits typical phone usage—users do not usually rapidly task back and forth between views, so there is no need to hold a number of views open and waste the limited available memory.

Although view switching is not a new paradigm, Series 60 classes encapsulate and simplify this functionality. The process of handling view switches has been made more transparent for the developer than with other Symbian OS platforms, and the Avkon classes have integrated additional features into the paradigm. For example, each Avkon view can automatically have its own menu system by defining one in the `AVKON_VIEW` resource structure. Resource files and structures are covered in more detail in Chapter 5.

For each `CAknView`-base class, the activation/deactivation functions you need to implement are `DoActivateL()` and `DoDeactivate()`, and these are responsible for instantiating and displaying, or deleting the UI controls owned by the Avkon view. Again, this highlights that the View Server does not concern itself with UI controls directly. Note that the View Server will, on its own initiative, call `DeactivateView()` in order to enforce the one-active-view-per-application rule.

How to Use Avkon View-Switching Architecture

This subsection will show how you would go about implementing an Avkon View-Switching Architecture application. Figure 4–1 illustrates the Series 60 classes that a view-switching application must derive from. It is important to note that, when using this architecture, the `CAknViewAppUi` and `CAknView` classes must be used in conjunction with one another. Each Avkon view derives from `CAknView` and must provide an `Id()` function so that it can be identified by the system. It must also implement the `DoActivateL()` and `DoDeactivateL()` function. Additionally, it should implement the

Chapter 4 Application Architecture

HandleForegroundEventL(), HandleCommandL() and HandleStatusPaneSizeChange() functions to handle various events. The class declaration for CShapeDrawerListView shows a typical Avkon view:

```
class CShapeDrawerListView: public CAknView
    {
public: // constructors and destructor
    static CShapeDrawerListView* NewL();
    static CShapeDrawerListView* NewLC();
    ~CShapeDrawerListView();
    CShapeDrawerAppUi& GetAppUi();

private: // from CAknView
    TUid Id() const;
    void HandleCommandL(TInt aCommand);
    CShapeDrawerListView();

    void DoActivateL(
        const TVwsViewId& aPrevViewId,
        TUid aCustomMessageId,
        const TDesC8& aCustomMessage);

    void DoDeactivate();

private: // constructors
    void ConstructL();

private: // data
    CShapeDrawerListViewContainer* iContainer;
    TUid iId;
    };
```

DoActivateL() is called by the View Server when a client requests that your view is activated. Its purpose is to instantiate and display the control that renders the view. As the function prototype above implies, it is possible to pass customized messages to the View being activated. This information could be used, for example, to prepopulate the view with data specified in the message. The passing of these message parameters is entirely optional.

Be aware that DoActivateL() may be called multiple times prior to DoDeactivateL(), and therefore your implementation must not assume that the view's control has not already been created and displayed. Such multiple calls can occur when a client requests reactivation of the view with some additional message parameters. Your DoActivateL() call must be prepared to handle this eventuality.

DoDeactivate() is called when your Avkon view is to be deactivated. This function is responsible for destroying its control. Your view will be deactivated when your application exits, or when another view in the same application is activated. This function must not leave.

Chapter 4 Application Design

`HandleForegroundEventL()` will be called only while your Avkon view is active (in other words, in-between calls to `DoActivateL()` and `DoDeactivate()`). When your view comes to the foreground, it will receive `HandleForegroundEventL(ETrue)`. When your view is removed from the foreground, it will receive `HandleForegroundEventL(EFalse)`. This function will be called only when the foreground state actually changes. Note that it may be called a number of times during your view's active period, as the owning application comes and goes from the foreground. You may want to use this method for setting focus or for controlling screen updates.

`HandleCommandL()` will be called when the view's menu generates a command, and `HandleStatusPaneSizeChange()` will be called when the client rectangle size changes due to a change in the status pane.

This is the typical order of events that an Avkon view will receive over an active period.

Activating a new Avkon view:

1. `DoActivateL()`
2. `HandleForegroundEventL(ETrue)`

Deactivating an Avkon view:

1. `HandleForegroundEventL(EFalse)`
2. `DoDeactivate()`

As previously stated, the pair of view-activation function calls may happen a number of times during a view's active period.

Typically, each Avkon view will require its own menu options. It is easy to give each Avkon view its own unique set by simply associating the view with a resource. Therefore, to let an Avkon view define its own soft keys and/or menu, create an `AVKON_VIEW` resource in your resource (`.rss`) file, then pass the resource ID into the view's `BaseConstructL()` function. Again, further detail on resource files and structures can be found in Chapter 5.

The view resource for **ShapeDrawer** is shown here:

```
RESOURCE AVKON_VIEW r_shapedrawer_graphicview
    {
    menubar = r_shapedrawer_menubar1;
    cba = R_AVKON_SOFTKEYS_OPTIONS_BACK;
    }
```

The menubar `r_shapedrawer_menubar1` is also defined in the resource file and specifies the menu options. Note that **CBA** (or Command Button Area) is an alternative name sometimes used for soft keys—this comes from a Symbian OS

Chapter 4 Application Architecture

convention. `R_AVKON_SOFTKEYS_OPTIONS_BACK` is a standard soft key definition (defined in `avkon.rsg`) and sets the soft keys to **Options** and **Back**. Another definition is `R_AVKON_SOFTKEYS_OPTIONS_EXIT`, which sets the soft keys to **Options** and **Exit**.

The code to associate this resource with its Avkon view would be:

```
void CShapeDrawerGraphicView::ConstructL()
    {
    BaseConstructL(R_SHAPEDRAWER_GRAPHICVIEW);
    }
```

All of the Avkon views in your application would normally be constructed in the AppUi object's `ConstructL()` method. They are registered with the View Server using `AddViewL()`, and finally the initial view is activated by setting it as the default view. The following code from the **ShapeDrawer** application demonstrates the AppUi as the owner of two Avkon views, `CShapeDrawerGraphicView` and `CShapeDrawerListView`, and shows the required initialization process.

```
void CShapeDrawerAppUi::ConstructL()
    {
    BaseConstructL();

    iView1 = CShapeDrawerGraphicView::NewL(iDocument);
    AddViewL(iView1); // transfer ownership

    iView2 = CShapeDrawerListView::NewL(iDocument);
    AddViewL(iView2); // transfer ownership

    SetDefaultViewL(*iView1);
    }
```

As mentioned before, Avkon views have no innate drawing capabilities. Therefore each Avkon view will typically own a Container class that derives from `CCoeControl`. An instance of this will therefore be present in the view:

```
class CShapeDrawerGraphicView: public CAknView
    {
    ...
private: // data

    CShapeDrawerGraphicViewContainer* iContainer; // what this view will display
    ...
    };
```

The currently activated view receives commands via its `HandleCommandL()` method. This is where soft key-generated commands and commands generated from the user interacting with the view's associated menu are handled. For example:

```
void CShapeDrawerListView::HandleCommandL(TInt aCommand)
   {
   switch (aCommand)
       {
       case EAknSoftkeyBack:
           {
           AppUi()->HandleCommandL(EEikCmdExit);
           break;
           }
       case EShapeDrawerRemoveSelectedItem:
           {
           iContainer->RemoveCurrentItemL();
           break;
           }
       default:
           {
           AppUi()->HandleCommandL(aCommand);
           break;
           }
       }
   }
```

Note that commands that are not handled by the Avkon view are passed on to the AppUi. In the **ShapeDrawer** example, view-switching commands are passed to the AppUi:

```
void CShapeDrawerAppUi::HandleCommandL(TInt aCommand)
   {
   switch (aCommand)
       {
       case EShapeDrawerSwitchToListView:
           {
           TUid viewId;
           viewId.iUid = EShapeDrawerListViewId;
           ActivateLocalViewL(viewId);
           break;
           }
       case EShapeDrawerSwitchToGraphicView:
           {
           TUid viewId;
           viewId.iUid = EShapeDrawerGraphicViewId;
           ActivateLocalViewL(viewId);
           break;
           }
           ...
       default:
           User::Panic(KShapeDrawerPanicName,
EShapeDrawerUnknownCommand);
           break;
       }
   }
```

Local view switching, or switching to a view that is owned by your application, is performed by referring to the UID of the Avkon view that you want to switch to.

In order to perform an external view switch, slightly more information is required. Call one of the `CCoeAppUi::ActivateViewL()` functions, providing a `TVWsViewId` containing the target application's UID and the target view UID. The following code excerpt demonstrates this:

```
const TUid KPhoneBookUid = { 0x101f4cce }; //from PbkUID.h
...
const TUid KPhoneBookContactViewUid = { 1 };
ActivateViewL(TVwsViewId(KPhoneBookUid,
KPhoneBookContactsViewUid));
```

As you can see, the mechanism for accessing an external view is similar to that for accessing local views. See the **Using Standard Application Views** section in Chapter 12 for more detailed information on external view switching.

All view-switching applications should publish their application and view UIDs by exporting them to a header file, so that they can be identified by other applications. Of course, if the application holds views that should not be accessible by other applications, then their publication should be withheld.

Choosing the Appropriate Application Architecture

When to Use Avkon View-Switching Architecture

Although view switching operates on a very simple set of rules, it provides a new way of thinking about application design and interapplication interaction. View switching allows applications to make better use of each other, in a way that makes navigation between applications seamless for the user.

Consider how users had to navigate between applications prior to the advent of view switching: Imagine that a user had just received an email from a new colleague. Apart from reading the message, it is plausible that they would want to add this new colleague to their contacts list. Without view switching, it would be necessary for them to manually start up the contacts application, manually navigate to the contact entry view, and then to enter the colleague's details. But with Avkon View-Switching Architecture, it is possible for the email application to provide a direct switch to the contact entry view of the contact application, and also to populate the view with some initial contact details.

NOTE The above scenario suggests using a view switch to access the contacts application in order to enter contact data, but this is not the only way to achieve this type of functionality—Series 60 provides many APIs to manipulate data in standard applications by accessing their application engine components directly. Chapter 12 gives more information on this alternative approach.

With this new architecture, programming this view switch into an application is trivial—you only have to request view activation from the View Server, passing in the application UID, the view UID and a package buffer containing any other relevant information. The View Server will then automatically start up the contacts application (if not already started) and switch directly to the appropriate view. Switching to standard applications is covered in more detail in Chapter 12.

This mechanism allows applications to smoothly interoperate, and in most cases, view switching is the best architecture choice. However, you need to be aware of its limitations. For example, the view-switching scheme has no built-in way to preserve the context of a view switch. That is, it does not provide a standard mechanism for navigating back to the previously active view (think of the "back" button on a Web browser). These limitations mean that if an application executes a remote view switch, the user has no visual clue as to how this view was reached. However, DoActivateL() does receive an identifier for the previously active view, so it is possible to develop your own "back" button functionality.

Another consideration is that you may wish to prevent external applications from being able to switch to certain views within your application. The View Server architecture offers no way to explicitly forbid the external activation of a view. There are workarounds, however. For example, you can make it very difficult to access views by not publishing their UIDs. What is important to understand is that, if it is absolutely necessary for some views to be unreachable by external applications, then the View Server cannot accommodate this requirement, and it will be necessary to use a different UI architecture.

When to Use Traditional Symbian OS Control-Based Architecture

In general, the default architecture for your Series 60 application would probably be the Avkon View-Switching Architecture. However, there are reasons why you might prefer the Traditional Symbian OS Control-Based design:

- Your application requires only one view, and it is unlikely that other applications will want to perform an external switch to it. In this case, you would not get anything out of the Avkon View-Switching scheme; it would just be unnecessary overhead.

- Your application has UI controls whose privacy must be guaranteed. If you are using the Avkon View-Switching Architecture, you can make it very difficult for other applications to switch to a view in your application by not publishing its ID, but there are no guarantees of privacy.

- You are porting an application from a different Symbian OS platform to Series 60. In this case, you may wish to simply stick with the existing UI architecture, especially if it appears unlikely that other applications would want to perform external switches to any views within the application.

Chapter 4   Application Architecture   199

When to Use Dialog-Based Architecture

This approach is very suitable to applications whose UI controls are made up of a straightforward arrangement of traditional OS controls, such as data entry or settings applications. Controls can be defined in resource files, with the dialog handling layout and drawing automatically—this is much easier than implementing custom drawing behavior.

The other consideration is the application's navigation requirements from screen to screen. The application is a candidate for this "dialog-based" approach if *and only if* there are no cyclical navigation paths among the application's views. Figure 4–6, along with the discussion that follows, will clarify this point.

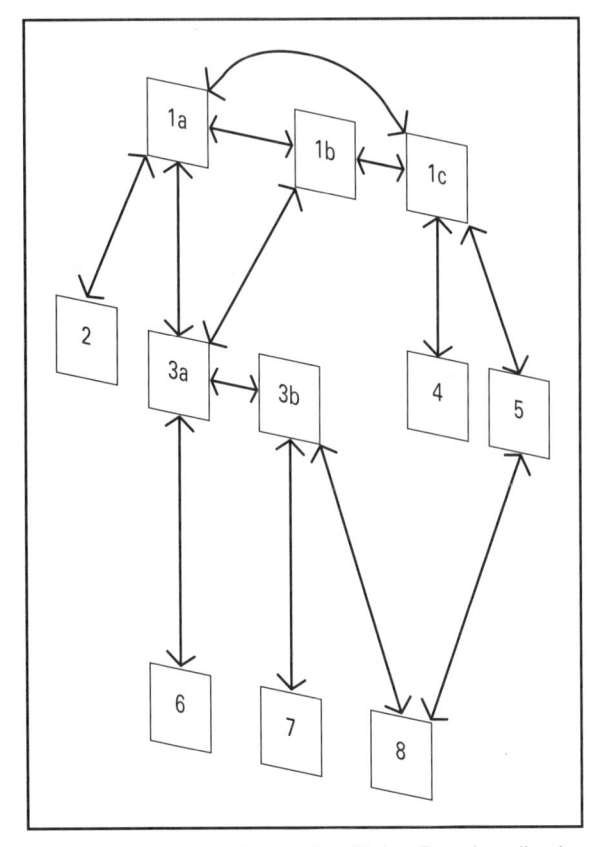

Figure 4–6 Screen navigation in a Dialog-Based application.

Each box in the figure represents a different view in an application, and each number represents a different dialog class. Classes 1 and 3 are multipage dialogs—the lettered variations represent their different pages. The arrows depict the screen navigation possibilities.

The rule is that, in navigating downward in the hierarchy (for example, from 1a to 3a), you must return along the same path when navigating back upward. Therefore, navigating from 1a to 3a to 1b is disallowed. If this were allowed, then the user could navigate in a big circle. Each cycle (1a-3a-1b-1a) would instantiate two new dialog objects (dialog 1 creates an instance of dialog 3, in navigating from 1a to 3a, and dialog 3 creates another instance of dialog 1 in navigating back to 1b), while at the same time still holding objects instantiated in past cycles in memory. So, apart from having no logical benefit, this cyclical navigation is a liability in terms of its undue memory consumption.

If you are considering using dialogs as your main view, then note that their navigation paths are limited, as well as their layout possibilities.

File Handling

In Chapter 3, you were introduced to files, streams and stores, and you learned about using `InternalizeL()`, `ExternalizeL()` and the operators << and >> to save data. In the **Elements** example code, an explicit command to save or retrieve some data was given. However, in a GUI application, if you need to persist application data, the framework can initiate the process automatically.

The document class contains two important functions that can be overwritten: `StoreL()` and `RestoreL()`. The framework calls `RestoreL()` automatically at application startup. This function is then responsible for loading the application's persisted data. The `StoreL()` function is called by the framework when the application exits in order to save application data—you need to include a call to `CEikAppUi::SaveL()` to instruct the framework to make the call. The implementation below is for `CShapeDrawerDocument::StoreL()`:

```
void CShapeDrawerDocument::StoreL(CStreamStore& aStore,
   CStreamDictionary& aStreamDic) const
   {
   // Get the model to save itself to the store
   TStreamId modelStreamId = iModel->StoreL(aStore);

   // Add an entry into the dictionary for the model data
   aStreamDic.AssignL(Application()->AppDllUid(),
modelStreamId);
   }
```

You can see that the first step in the *document's* `StoreL()` function is a call to the *model's* `StoreL()`—where the model will externalize all of its data. The stream ID for the model data is then recorded in the stream dictionary. The **ShapeDrawer** application provides implementations of both the `StoreL()` and `RestoreL()` methods, and a review of this code will clarify their usage.

Series 60 differs from other Symbian OS platforms in that **the default behavior of the document class is not to persist data using a file store**. Therefore, Series 60 does not automatically open a file for document storage at application startup—the implementation of `CAknDocument::OpenFileL()` is empty.

Chapter 4 Splitting the UI and the Engine

However, you can still accomplish this behavior by overriding `OpenFileL()` in your document class and calling the base Symbian OS implementation:

```
CFileStore* CShapeDrawerDocument::OpenFileL(TBool aDoOpen,
    const TDesC& aFilename, RFs& aFs)
    {
    return CEikDocument::OpenFileL(aDoOpen, aFilename, aFs);
    }
```

Note that if document support is not enabled, then `StoreL()` and `RestoreL()` will not be automatically called by the framework.

Also, support for Symbian OS `.ini` (application settings) files is not provided in Series 60 by default. This causes any Series 60 applications that try to open their `.ini` file to fail with a "Not Supported" error. To enable your application to use an `.ini` file, your application class's `OpenIniFileLC()` method must be overridden to call `EikApplication::OpenIniFileLC()`. For example:

```
CDictionaryStore* CMyApplication::OpenIniFileLC(
RFs& aFs) const
    {
    return CEikApplication::OpenIniFileLC(aFs);
    }
```

Please note that the above excerpt is not taken from any of the chapter examples.

> **NOTE** Many standard Series 60 applications use "Shared Data Files" rather than traditional Symbian OS `.ini` files. These consist of user-readable (plain) text rather than the (binary) stream-based file stores (`CDictionaryFileStore`) that Symbian OS `.ini` files use. They are stored under `\system\shareddata\` and named `<UID of application>.ini`—for example, `101FDA63.ini` for an application with a UID of `0x101FDA63`. The fixed location and coherent naming of these files mean that it is possible for other applications to see (and potentially change) the settings of any application that uses them. Note, however, that the Shared Data API used to access such files is not included in the public SDK.

Splitting the UI and the Engine

This section offers a pragmatic approach to application design to ensure that applications' user interfaces are cleanly decoupled from their underlying functionality, thereby ensuring that the non-UI parts of the application are *almost entirely portable*. It explains why you should view the UI and engine as separate components of your application, and it provides details for an example implementation.

Chapter 4 Application Design

The User Interface renders an interpretation of the application's data and allows the user to interact with it. The UI often gives the user the ability to enter, edit or delete the data. The UI, therefore, actually comprises two interfaces—one to handle user input, and another to transfer user requests into the engine.

The engine of an application comprises the data, the algorithms that manage and process that data and routines that persist (store and retrieve) it. Consider the game of chess as an example. If this were to be written as an application, the model would comprise the internal state of the board, the rules of movement for each piece, and algorithms for calculating the game state at each turn (check, checkmate and so on). The UI layer might be complex, but it would have nothing to do with the chess game itself. It would be merely a mechanism for representing the game state and providing a way for users to input their next move.

This application's engine would also encapsulate an AI component so that the computer could calculate its next move. Once the move was made, the UI would be notified so that it could display the game's new state.

All applications can be cleanly divided along these functional lines, and developers should make a clear separation on both the class and component level. From the component perspective, the engine should be written as a stand-alone DLL. Alternatively, it could be written as an `.exe` server if the engine had to be able to service several different applications simultaneously, or if asynchronous handling was required for data-sharing purposes. The application's `.mmp` file would then simply include the library file for the engine.

There are many reasons for doing this. A basic principle of good design is to avoid creating unnecessary dependencies between components. Also, maximizing code reuse so that common functional requirements can be satisfied from shared libraries is very desirable. The engine for your application should be a functional unit that stands apart from whatever GUI layer represents the data to the user and allows them to interact with it. For the purposes of flexibility, it is best to allow the engine to reside in its own DLL and to offer its own API to any UI layer that might want to use it.

This sounds like extra work, and yes, it does require slightly more effort—plus it adds some complexity—but it also affords numerous benefits. Another reason for separating the engine is so that it can be tested without the need to write a UI layer first, and that the engine can be thoroughly tested via test harnesses before the UI enters into the equation. From a project standpoint, this means that the UI layers and engines can be developed in parallel, thereby significantly reducing the minimum required development time.

Software must be maintained. Often this must be performed by developers who were not present during the initial design and development phases. The maintainability of code is directly related to its structure—unstructured code will prove difficult and expensive to maintain. Separation of GUI and engine is an opportunity to cleanly define the boundaries of independent components. This seems like stating the obvious, but applications are most often crafted to

Chapter 4 Splitting the UI and the Engine

tight deadlines—the "get-it-out-the-door" mentality sometimes pays little attention to these longer-term design considerations.

Separation of GUI and engine is certainly not a goal that is specific to mobile computing devices. But it does bear even greater practical relevance for these. You will probably want to ensure that your applications are available on as many devices as possible. In the world of mobile computing, there are a host of Symbian OS devices, but they will differ in the version of Symbian OS, or the UI platform they deploy. However, most of the non-UI services, and the APIs that use these services, are identical. As a result of this transplatform consistency beneath the UI layer, it is possible to write engines that are directly reusable, or that will require only minor tweaks when porting.

The **ShapeDrawer** example application gives practical guidance in the following areas:

- Writing a separate engine with its own API.
- Writing a GUI that invokes this API and alters the state of the engine.
- Demonstrating the built-in class member functions, called by the framework, to help coordinate the UI with the changing state of the engine.

As explained in the chapter introduction, this application allows the user to draw shapes on the screen using one view, and to list the drawn shapes (with their coordinates) in another. The model's job, therefore, is to process user requests to draw squares or circles, and to remember the placement of each shape. In order to achieve this, the model must support:

- A shape class which holds position, size and shape.
- Methods for adding these new shapes.
- Methods that read and write shape data, to and from streams.
- Methods to actually draw the shapes.

This last requirement may seem contrary to the general rationale for separating UI functionality from the engine. However, it is deemed possible to perform this drawing in a generic way, since the actual rendering of the shapes only requires function calls to a `CWindowGc` object, which can be passed into the `Draw()` function by reference. The `CWindowGc` is definitely a UI component, but remember that the UI API classes reside in different tiers: this class resides in a lower-level component that will be consistent across various platforms. Alternatively, the engine could be designed to simply deliver all of the relevant shape data back to the UI, which then renders it. However, the fact that methods called on `CWindowGc` are cross-platform is used to simplify the code.

At application startup, the `RestoreL()` method of the document will be called automatically by the framework. This initiates the reinstatement of the document's model (`CShapeListManager`) from streams stored in the

Chapter 4 Application Design

application's file. In addition to kicking off this process automatically, the framework also provides an automatic update to the AppUI, to notify it that the state of the model has changed (by way of initializing it from a file store). `HandleModelChangeL()` is called in the AppUi as soon as the model has restored itself (completed its `RestoreL()` method). The AppUi then renders the model's data by calling methods on its API. The following code shows the engine's primary API class that the AppUi layer uses to render the data:

```
class CShapeListManager : public CBase
    {
public:
    IMPORT_C static CShapeListManager* NewLC();
    IMPORT_C static CShapeListManager* NewL();
    IMPORT_C ~CShapeListManager();
    IMPORT_C void Clear();
    IMPORT_C void AddShapeL(NShapes::TShape* aShape); //takes ownership
    IMPORT_C void RemoveShapeL(NShapes::TShape* aShape);
    IMPORT_C NShapes::TShape* GetNextShape(); // no ownership xfer
    IMPORT_C TStreamId StoreL(CStreamStore& aStore) const;
    IMPORT_C void ExternalizeL(RWriteStream& aStream) const;
    IMPORT_C void RestoreL(const CStreamStore& aStore, const TStreamId& aStreamId);
    IMPORT_C void InternalizeL(RReadStream& aStream);

protected:
    IMPORT_C CShapeListManager();
    IMPORT_C void ConstructL();

private:
    TInt iNextShapeIndex;
    RPointerArray<NShapes::TShape> iShapeList;
    };
```

The following code shows how `GetNextShape()` is used by the UI layer in order to, in an iterative manner, grab a pointer to each `TShape` object stored in the model, and then to call `TShape::Draw()` in order to represent it:

```
void CShapeDrawerGraphicViewContainer::Draw(const TRect& /*aRect*/) const
    {
    CWindowGc& graphicsContext = SystemGc();

    // Clear the application view
    graphicsContext.Clear();

    // Draw the 'cursor' crosshair.
    // Size is KCrosshairWidth by KCrosshairHeight pixels
    graphicsContext.SetPenSize(TSize(1, 1));
    graphicsContext.SetPenColor(KRgbBlack);
```

Chapter 4   Splitting the UI and the Engine

```
    graphicsContext.DrawLine(
TPoint(iPosition.iX - KCrosshairWidth, iPosition.iY),
TPoint(iPosition.iX + KCrosshairWidth, iPosition.iY));

    graphicsContext.DrawLine(
TPoint(iPosition.iX, iPosition.iY - KCrosshairHeight),
TPoint(iPosition.iX, iPosition.iY + KCrosshairHeight));

    // Draw all the current shapes
    TShape* shape = iDocument->Model()->GetNextShape();
    while (shape)
        {
        shape->Draw(graphicsContext);
        shape = iDocument->Model()->GetNextShape();
        }
    }
```

Here you see the function iterating through each shape contained in the model. As discussed previously, the shape objects perform their own rendering. In a similar fashion, the application's list view also represents the model's data, but it does so textually. Although it calls `TShape::Coordinates()` instead of `Draw()` in order to build a text string that represents each item in the model, its interaction with the model is very similar. The code below shows how the UI can add information to the model (note that the full method is not shown here):

```
TKeyResponse
CShapeDrawerGraphicViewContainer::OfferKeyEventL(const
TKeyEvent& aKeyEvent, TEventCode aType)
    {
    if (aType != EEventKey)
        {
        return EKeyWasNotConsumed;
        }

    // Move left
    if (aKeyEvent.iScanCode == EStdKeyLeftArrow)
        {
        if (iPosition.iX > (Rect().iTl.iX + KCrosshairWidth))
            {
            —iPosition.iX;
            DrawNow();
            }
        return EKeyWasConsumed;
        }

    // Move right
    ...
    // Move up
    ...
    // Move down
    ...
```

```
    // Place a shape
    else if (aKeyEvent.iScanCode == EStdKeyDevice3)
        {
        TShape* newShape = NULL;
        // Update the coordinates in the model to the
        // position at which the event occurred.

        switch (iBrushShapeType)
            {
          case ECircle :
              newShape = new (ELeave) TCircle(iPosition,
KBrushRadius);
              iDocument->Model()->AddShapeL(newShape); // Takes
ownership
              break;
          case ERectangle :
              newShape = new (ELeave) TRectangle(iPosition,
KBrushHeight, KBrushWidth);
              iDocument->Model()->AddShapeL(newShape); // Takes
ownership
              break;
          default :
              User::Panic(KShapeDrawerPanicName,
EShapeDrawerInvalidBrushType);
            }

        DrawNow();
        return EKeyWasConsumed;
        }

    return EKeyWasNotConsumed;
    }
```

Here you see that the view holds a data member that holds the type of shape to draw (iBrushShapeType), which the user can select from a menu. The TCircle and TRectangle classes are both part of the engine's API. Adding a shape to the model is then very simple. Note that the model then takes ownership of the new shapes. It is clear now how the model's API can be used by the UI layer in order to both give input to, and receive necessary output from, the model.

ECom

The recommendation in the **Splitting the UI and the Engine** section to cleanly separate GUI and engine code is founded on good design principles. This section takes the issue of code separation and reusability, which are the pillars of those principles, one step further. It suggests that engines, and other kinds of software components, could be reused with the greatest of ease if there were some kind of framework to allow them to dynamically plug-in, so that they

Chapter 4 ECom

essentially become an extension of the OS. ECom is exactly this—it stands for **Epoc Component Object Model**, and it provides a means of taking new software components that may be widely beneficial and placing them within a delivery framework that is managed by the OS. ECom is a recent development and is available only from Series 60 2.x. Although it is not uniformly available, it is extremely useful and deserves coverage here, in the spirit of maximum code reuse and minimum redundancy.

The ECom framework is a Client/Server architecture that provides a service to instantiate, resolve, and destroy plug-in instances. Essentially it means you can

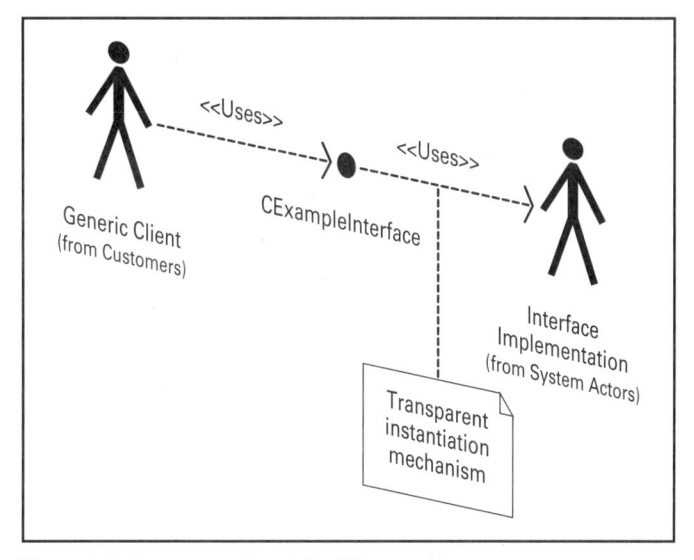

Figure 4-7 Representation of the ECom architecture.

encapsulate common functionality in a DLL that can be accessed by multiple clients through an interface class. The interface is client-side while the server manages the implementation. The framework takes care of the instantiation, as shown in Figure 4-7.

> **NOTE** The ECom framework is a new addition to Series 60 and is available only from Series 60 2.x.

Prior to ECom, the polymorphic DLL was the standard Symbian OS convention for delivering additional functionality at runtime. This convention meant that:

- Modification of the DLL did not enforce recompilation of applications that used it (as opposed to statically linked libraries).

- The loading of the DLL did not have to occur at application startup—it could be loaded as the need arose, thereby preventing applications from needlessly using up memory.

ECom retains all of these runtime advantages. But at the same time, it significantly improves the ease of access to additional functionality. Whereas accessing objects in polymorphic DLLs requires knowledge of the DLL's location, ECom abstracts all such physicalities from developers wishing to use ECom plug-ins.

This section first provides a conceptual overview of the ECom convention, discussing what it does and the advantages it provides. Then it delves into the mechanisms of ECom and exemplifies the work required to design and deploy an ECom plug-in DLL. To illustrate this, the **EComExample** example application has been provided.

ECom Conceptual Overview

Each ECom object has one **interface**, contained within a standard header file. This interface is all that an ECom user requires—the interface definition not only specifies how to interact with the object, but it also provides one or more static methods to instantiate the object.

One key feature of the ECom convention is that an object interface can have more than one implementation. To clarify the benefit of this, consider the example application **EComExample**—the engine offers to draw both circles and rectangles, but each of these classes is derived from CShape. In the ECom paradigm, CShape would be the interface, and CCircle and CRectangle would be the implementations of that shape. Therefore, different functionality can be provided through a common interface. This feature is also useful, for example, to provide different versions of an object through the same interface.

ECom interfaces are typically very similar to those exposed in polymorphic DLLs, in that they are usually **pure abstract** class definitions, and this abstraction is what allows an interface to have more than one implementation.

ECom implementations each have a repository of descriptive information about the object. This includes the object's name, version, description and so on. This information is very useful, because it provides a couple of different ways to get access to a particular implementation—they can be requested either by their unique ID or via a text-based cue. The latter request makes use of an ECom resolver, which searches through all the meta-information for each implementation and then returns the one that best matches the cue. Figure 4–8 broadly demonstrates the ECom convention. The user requests an object via the provided interface definition. It is ECom, however, that actually instantiates the requested object.

ECom Interface

ECom uses Symbian OS Client/Server architecture. The interface resides client-side, whereas the implementations are managed by the server. The following code, taken from **EComExample**, shows an ECom object interface.

```
class CShape : public CBase
   {
   public:
      static CShape* NewL();
      static CShape* NewL(const TDesC8& aMatch);
      virtual ~CShape();

      virtual void Draw(CWindowGc& aGraphicsContext) const = 0;

   private:
      TUid iDestructorIDKey;
   };
```

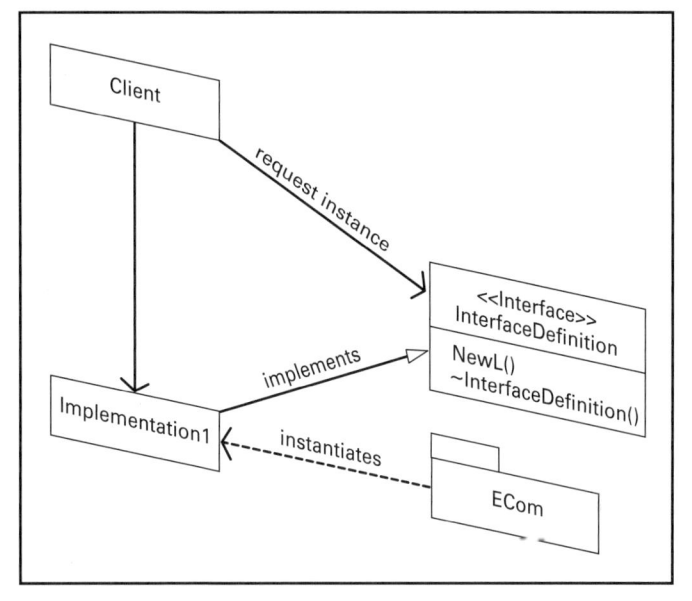

Figure 4–8 Basic ECom architecture.

The ECom-specific aspect of this interface resides in the overloaded `NewL()` functions and the destructor—these use the `REComSession` API, through which actual implementations are requested and delivered. The two overloaded `NewL()` functions demonstrate the two ways in which implementations can be requested: The first `NewL()`, taking no parameters, simply requests the implementation by its known implementation UID (`KCCircleInterfaceUid`), so if this method is invoked, a `CCircle` object will be instantiated:

```
inline CShape* CShape::NewL()
   {
   const TUid KCCircleInterfaceUid = { 0x101FDA61 };

   TAny* interface =
```

Chapter 4 Application Design

```
RecomSession::CreateImplementationL(KCCircleInterfaceUid,
_FOFF(CShape, iDestructorIDKey));

    return reinterpret_cast<CShape*>(interface);
    }
```

The second version, shown below, requests the implementation based upon a free-form text query:

```
inline CShape* CShape::NewL(const TDesC8& aMatch)
    {
    const TUid KCShapeInterfaceUid = { 0x101FDA5D };
    const TUid KCShapeResolverUid = { 0x101FDA5F };

    TEComResolverParams resolverParams;
    resolverParams.SetDataType(aMatch);
    resolverParams.SetWildcardMatch(ETrue);

    TAny* interface =
RecomSession::CreateImplementationL(KCShapeInterfaceUid,
_FOFF(CShape, iDestructorIDKey), resolverParams,
KCShapeResolverUid);

    return reinterpret_cast<CShape*>(interface);
    }
```

To use this version of the `NewL()` function, the client would need to pass through a string that indicates the shape they require. The particular string text necessary for instantiating each type of object is defined in a resource file, `101FDA60.rss`. So if you wanted a `CRectangle` object:

```
    _LIT8(KRectangleText, "Rectangle");
    CShape* shape = CShape::NewL(KRectangleText)
```

Apart from instantiation, the interface interacts with `REComSession` in order to destroy the object. Each object is allocated a unique key, and this key is passed back to ECom so that it can destroy the correct object.

```
inline CShape::~CShape()
    {
    REComSession::DestroyedImplementation(iDestructorIDKey);
    }
```

Therefore, designing an ECom interface requires you to:

- Provide a class definition for the object that will be returned from the ECom Server.

- Provide static `NewL()` function(s) for implementing the object.

Chapter 4 ECom

- Implement the NewL() function, using REComSession to request the object.
- Implement a virtual destructor which will, through REComSession, request to destroy the object.

ECom DLL

The previous subsection described the requirements of an ECom interface. This subsection details the work required to add an ECom plug-in into the framework.

An ECom implementation is simply a collection of one or more implementations for one or more interfaces. Physically, it is made up of a pair of files. The first is the all-important DLL, which contains the actual code. The second is a compiled resource file, which is a map linking user interfaces to their implementations. It also provides descriptive information for each implementation. These two files are associated with one another by their names—the unique UID of the DLL is typically used as the name for both files, as it is unique and relates the files to their implementation.

The resource files are what the ECom Server browses through in order to find the correct implementation for an interface. The following excerpt shows the format of the resource file 101FDA60.rss:

```
RESOURCE REGISTRY_INFO r_theinfo
    {
    // UID for the DLL
    dll_uid = 0x101FDA60;

    // Declare array of interface info
    interfaces =
        {
        INTERFACE_INFO
            {
            // UID of interface that is implemented
            interface_uid = 0x101FDA5D;
            implementations =
                {
                // Info for CCircle
                IMPLEMENTATION_INFO
                    {
                    implementation_uid = 0x101FDA61;
                    version_no         = 1;
                    display_name       = "Circle shape";
                    default_data       = "CIRCLE";
                    opaque_data        = "";
                    },
                // Info for CSquare
                IMPLEMENTATION_INFO
```

```
                {
                implementation_uid = 0x101FDA62;
                version_no         = 1;
                display_name       = "Rectangle shape";
                default_data       = "RECTANGLE";
                opaque_data        = "";
                }
            };
        }
    };
}
```

This demonstrates visibly how the information links together. The implementation DLL UID is defined first. Next, each interface that the implementation DLL contains is defined by the `INTERFACE_INFO` structure. Each of these contains one or more `IMPLEMENTATION_INFO` structures, exposing each implementation UID and accompanying descriptive information. Resource files and structures are covered in more detail in Chapter 5.

The resource file clearly provides the required information to:

- Associate an interface with a particular implementation DLL.
- Associate an interface with its various interface implementations.
- Allow the user to request an implementation either by its unique UID, or by a text description of it.

The final requirement is for ECom to be able to locate a particular implementation within the DLL. In order to achieve this, every ECom DLL implements a table that maps each implementation in the DLL to its memory address, and a single function is exported that gives access to this table:

```
const TImplementationProxy ImplementationTable[] =
    {
        { { 0x101FDA61 },   CCircle::NewL },
        { { 0x101FDA62 },   CRectangle::NewL }
    };

EXPORT_C const TImplementationProxy*
ImplementationGroupProxy(TInt& aTableCount)
    {
    aTableCount = sizeof(ImplementationTable) / sizeof(TimplementationProxy);

    return ImplementationTable;
    }
```

ECom DLLs must possess a UID that identifies them as such. The DLLs, along with the compiled `.rsc` files, must then be placed within the `\system\libs\plugins\` directory in order for them to become registered.

The following example `.mmp` file should clarify any final informational requirements:

```
TARGET ecomexample.dll
TARGETTYPE ECOMIIC

// ECom Dll recognition UID followed by the unique UID for this dll
UID 0x10009D8D 0x101FDA60

SOURCEPATH      .
SOURCE main.cpp
SOURCE proxy.cpp
SOURCE circle.cpp
SOURCE rectangle.cpp

USERINCLUDE     .
SYSTEMINCLUDE           \epoc32\include
SYSTEMINCLUDE           \epoc32\include\ecom

RESOURCE        101FDA60.RSS

LIBRARY euser.lib ECom.lib
```

Internationalization

The software market is becoming increasingly globalized, now spanning multiple languages and cultures. This shift bears directly upon how applications are designed—it requires that software can be easily reengineered for different languages so that products can achieve their full market potential. To this end, this section addresses the issue of Internationalization (I18N)—a proactive development methodology that aims to minimize the effort required to redeploy software in different languages. The process of reworking software for different languages is known as Localization (L10N). Internationalization requires careful consideration of several design issues from the outset. This section introduces these general issues and details the facilities that Series 60 provides to help developers address them.

General Guidelines for Developers

The goal of Internationalization is to make it easy for your application to transcend not languages exactly, but locales. A **locale** is composed of a base language, but it encompasses more than just this. To clarify this concept, consider that the United States (US) and the United Kingdom (UK) use different locales (en_US and en_UK, respectively). Though they share English as their base language, there are still differences in the way that they represent certain information. For example, the US uses the "$" (dollar) currency symbol and the UK uses "£" (pound). Also, date formatting is different for each

Chapter 4 Application Design

(typically `DD/MM/YYYY` for the UK and `MM/DD/YYYY` for the US). Locales, therefore, encompass not just language, but all regional data-representation conventions.

You must take several things into consideration when creating applications that can be localized efficiently. The three basic things to consider are:

- Be sensitive to all data whose representation is locale specific. This includes various conventions such as date and time, monetary and decimal formatting and so on.
- Understand the concept of a locale, and what this means in Series 60.
- Understand how applications use resource files—this is covered in more detail in Chapter 5.
- Understand how to prepare resources for easy translation.

All data that is not globally recognizable (in other words, data whose formatting is locale specific) needs to be considered from the start, and treated differently. Examples of data that will be different according to the locale are dates, times, currency and, of course, all user-visible text. Perhaps less apparent is the fact that sounds, images and filenames may also need to be locale-specific.

The proper way to treat locale-specific data is to use the special classes offered by the **Locale Settings API**. These classes represent data according to the locale that is loaded on the given device. So, these special classes work with the application framework in order to deliver output that is correct for a given locale.

Examples of these classes are:

- `TLocale`
- `TLanguage`
- `TDay`, `TDateName` and `TDayNameAbb`
- `TMonth`, `TMonthName` and `TMonthNameAbb`
- `TDateSuffix`
- `TAmPmName`
- `TCurrencySymbol`

Their names indicate their purpose, but further information is available on each class in the SDK documentation.

Hardware devices come from the factory with a particular default locale loaded on them. Therefore, US phones and UK phones running software that uses these classes will end up displaying data slightly differently. Here is an example of how developers should format a date so that it will be rendered in accordance with the device locale:

Chapter 4 Internationalization

```
void CMyDateField::DateFieldTextL(const TTime aTTime, TDes&
newDateText)
    {
    aTTime.FormatL(newDateText, TShortDateFormatSpec());
    }
```

This function is deceptively simple looking. However, it accomplishes a lot for you behind the scenes. The `FormatL()` function formats the date, using the systemwide locale settings encapsulated by the class `TShortDateFormatSpec`. For the `en_UK` locale, this function would produce a date formatted such as "25/12/2003". As an alternative, you could use `TLongDateFormatSpec`, which would produce the equivalent "25th December 2003".

Developers must keep in mind that all user-visible text and locale-specific data will change, once localized—the obvious consequences are that text may take up more (or less) space in some languages than others, and consequently more (or less) memory. This means that users should avoid fixed-size text buffers. Wherever possible, dynamic buffers should be used instead.

TIP If you have to make decisions about the maximum length of resource strings, then in general you should allow at least 30% extra space for translated text. If the string is just a single word, allow for an expansion of at least 100%.

The following excerpt demonstrates two ways to create dynamically allocated text buffers, with both methods passing ownership out and leaving the created buffer on the Cleanup Stack:

```
HBufC* message = StringLoader::LoadLC(R_DIALOG_TEXT);
HBufC* msg1 = iEikonEnv->AllocReadResourceLC(R_MY_TEXT_01);
```

These functions both dynamically load in a string from a resource, without having to specify the size of the `HBufC` in the code. `CCoeEnv::AllocReadResourceLC()` is the traditional Symbian OS method for achieving this and will be commonly seen in old code. The `StringLoader` class is new to Series 60 but provides a more flexible API for formatting strings as they are read in. Further information on both methods can be found in the SDK documentation.

TIP In your code, do not concatenate strings to produce longer sentences. For example, in Finnish, the phrase "new message received" can be can translated to "uusi viesti saapunut", but if you wanted to say "3 new messages received", the Finnish would be "3 uutta viestiä saapunut". "New" has a different translation depending on the sentence, so just joining words together does not work!

Succeeding in producing an easily localizable application means that developers must also think carefully about the text they display on the screen. Menu items should be kept short and to the point. The longer they are, the greater the chance for significant length variation in other languages, which can pose significant UI problems. Thinking along these lines will greatly increase the localizability of applications.

OS Support for Localization

The most important aspect of internationalization is to understand that an application's resource files are not loaded until runtime. Furthermore, the framework provides a special nomenclature for resource files—it allows multiple resource files, each localized for a particular language, for any given application. The name of each compiled resource file, such as *.r01, *.r02, *.r03 and so on, indicates to the framework the locale that each file represents. For example, English is 01, French is 02 and so on. When the application is loaded, the framework can then, depending on the device's locale, load the appropriate one.

> **NOTE** Installing multiple localized resource files onto a device whose locale setting is unlikely to change is an inefficient use of memory. To remedy this, Series 60's application installation process can optionally prompt users for their preferred language, and then install only that resource file. Refer to the *Multi-Locale Installation* section in Chapter 2 to understand how your application can benefit from this feature.

The typical way to build applications so that they have locale-independent .rss files is to refrain from hard-coding any text strings in the .rss file itself. Instead, the .rss file should include a .loc file. This .loc file then acts as a large switch:

```
#ifdef LANGUAGE_01
#include "Language.101"
#endif
#ifdef LANGUAGE_02
#include "Language.102"
#endif
#ifdef LANGUAGE_03
#include "Language.103"
#endif
#ifdef LANGUAGE_06
#include "Language.106"
#endif
```

Each individual .loc (*.101, *.102 and so on) file then contains all the text strings for each locale. Within the project's .mmp file, you can specify all of the locales for which you want to compile the application. Consider the following .mmp file, which specifies that it should be compiled for six different languages:

```
TARGET    HelloWorldLoc.app
TARGETTYPE  app
UID    0x100039CE 0x101FDA4E
TARGETPATH  \system\apps\HelloWorldLoc
...
LANG 01 02 03
```

The `LANG` instruction tells the resource compiler to compile three different compiled .rsc files. As a result, three compiled files will be produced: (language.r01, language.r02 and so on). Changing the locale of a phone will result in a change in which resource file gets loaded. Note that caption resource files can also be localized and the AIF file written to allow the application icon to be locale dependent.

> **NOTE** Series 60 supports non-Latin-based character sets, such as Chinese. Additionally, bidirectional text support has been added to version 2.x.

Good Application Behavior

It may seem strange to hear the term "behavior" applied to applications, but it is very appropriate. Applications, if designed incorrectly, can potentially waste systemwide resources and therefore hurt the performance of the device as a whole. The application framework classes do not take care of good behavior for you—it is up to you to ingrain this in your applications. This section provides general guidelines to ensure that your application serves its users well, but not to the detriment of the device in general.

Adopt a Skeptical and Critical Approach to Development

You will need a skeptical mindset when designing and writing applications. Always take into consideration how your application would react in undesirable, unexpected or rare circumstances. Cast a critical eye upon all code as you write it, and continually ask yourself questions like: "What would happen to my application if the device ran out of memory?", "What would happen if my application were requested to go into the background?", and "What would happen if the battery were suddenly removed?" Under all such circumstances, you should strive to create the best possible outcome for the user, without affecting other applications, or the OS.

Handle Window Server-Generated Events

There are several different types of Window Server events, such as key events, system events and foreground events, which must be handled appropriately. The AppUi class has several functions that are called by the Window Server to handle each type of event notification—these were previously overviewed in the *Important AppUi Methods* subsection.

What you do within these overridden functions is up to you, but the following discussion details some likely scenarios and suggests ways to handle events responsibly.

It is possible for an application to gain or lose focus at any time. This can happen if the user switches between applications, or if the system brings another application into the foreground. Whenever an application's window group gains or loses focus, the application UI's `HandleForegroundEventL()` method will be called. This method has a single `TBool` parameter, which will be `ETrue` if the application is gaining focus and `EFalse` if it is losing focus.

When an application loses the foreground, it must strive to establish a stable, recoverable state. This means suspending all processing and saving application data—for two reasons. First, processing power must be reserved for the foreground application. It is of primary importance, because it is the application with which the user interacts. If background applications were to continue processing, users might experience an unwanted reduction in performance. Second, background-running applications may be automatically shutdown by the system in low memory conditions. If a forced shutdown were suddenly to occur, and background applications were to continue processing, it might not be possible to establish a steady and recoverable state.

User input must also be handled correctly. When the user carries out an action on the user interface, an event will be sent to the application that owns the window the event occurred in. There are several different types of events, but you will be interested mainly in key events and redraw events. (These are discussed further in Chapter 5.) Good programming behavior recommends that you take into consideration not only the key presses you expect, but also those that you do not.

Always Exit Gracefully

Your application needs to exit gracefully. Exiting can be initiated explicitly by users, or automatically by the system. Under these circumstances, an `EEikCmdExit` instruction will be received by `HandleCommandL()`. This is a last opportunity to save application data and to release system resources that are no longer required. Note that applications must shutdown quickly once they receive this command—there is no time to present options to the user.

It is also possible that the entire system may be shutdown before your application is closed. This could happen, for example, if the user turned off the device, or if the battery was removed accidentally. You must consider the impact of a sudden shutdown on application data. Consider what would happen if the **ShapeDrawer** application were suddenly interrupted. If it was in the background at the time, no data would be lost, because `HandleForegroundEventL()` forces a save of the data at this time. But if the user had added new shapes to the document since application startup and the application had been running in the foreground the entire time, sudden interruption could cause the user's newly entered shapes to be lost. One way around this would be to update the document every time the user inputs a new shape. Although this approach minimizes the amount of data that is likely to be lost, it carries a high performance cost. When designing applications, you need to be aware that performance and data security are two competing aims, and that you need to strike a balance according to your application's requirements.

Check Disk Space before Saving Data

Saving data must be performed responsibly. When the file system becomes too full, it can have a negative affect on system performance. Therefore, all applications must aim to prevent this from happening by checking available disk resource levels before committing data memory. The **ShapeDrawer** application accomplishes this by using the `SysUtil::DiskSpaceBelowCriticalLevelL()` utility function, defined in `sysutil.h`.

This function's return value tells you whether writing *x* bytes to a particular drive will cause disk space to drop below a critical level. Do not ignore this function when it returns `ETrue`—losing your application's data is preferable to potentially corrupting the file system.

Other Hints and Tips

File System—When using files, minimize the number of opening, reading and writing instructions. Write larger data blocks in one go. Persisting one large block of data is much more efficient than multiple writes of smaller blocks of data. As mentioned in the previous subsection, check the free disk space available using the methods provided in `SysUtil`.

System Watchdogs—There are subsystems within Series 60 that may request that your application exit at any time—for example, the OOM (Out-Of-Memory) Watchdog. This closes down background applications under conditions of low memory.

Reserve the `EEikCmdExit` command for system-generated exit requests. When your application is requested to exit by the system, save data, if necessary, and exit the application **without any UI confirmation queries**. The urgency of this message precludes the possibility of presenting options to users before shutting down.

Hard-Coding and Magic Numbers—Avoid these at all costs. Examples of data that are frequently hard-coded are directory paths, key codes, calculation variables and so on.

You should use the `AknUtils` API `CompleteWithAppPath()` (as described in the SDK documentation) to access data files within the user data area, without assuming that your application is installed on any particular drive. Also, do not assume that `z:` is the system drive—instead, use `BaflUtils::GetSystemDrive()`.

Use header files to share constants, and resource files to provide all user-visible text, as explained in the **Internationalization** section. Also, always query the system for values whenever possible—for example, to obtain screen sizes—rather than hard-coding such values.

Timers—Do not use continuous, fast-ticking timers. All continuous timers ticking faster than once every five seconds are problematic—they can continually interrupt the device from its low-power mode and shorten battery life. Remember to stop all timers and pause the application when it goes into the background.

Active Objects and Responsiveness—If an application hangs, or otherwise does not respond for more than ten seconds, then the View Server will terminate it—so ensure you do not have long-running active objects.

Summary

In this chapter the architecture of a Series 60 GUI application has been discussed. Series 60 offers three design choices with regard to UI development—namely, Avkon View-Switching Architecture, Traditional Symbian OS Control-Based Architecture and Dialog-Based Architecture, and each of these has been considered. You should now understand the benefits and drawbacks of each approach and be able to create an application using your chosen option, deriving from the appropriate framework classes.

So that you get into good habits as soon as you start writing Series 60 applications, some important development principles have been explained. Separating the UI from the engine allows you to develop both components independently. It also means that porting your application to multiple Symbian OS platforms should be simple. The ECom framework was introduced as a method of implementing DLL plug-ins, and it should be considered if you are writing a DLL for a Series 60 2.x device. Although this would be overkill for most UI applications, it does provide a means of dynamic expandability, where new implementations of interfaces can be added (as a separate library) at a later date, without the need for changes to client code.

Chapter 4 Summary

Making your application available to an international market may not be an immediate consideration for you, but if you follow the Internationalization guidelines outlined in this chapter, you will be able to easily add support for other languages.

You are now in a position to create a simple GUI application. When you tackle more complex projects, you will be able to draw upon the ideas in this chapter to produce efficient, well-designed applications. The next chapter supplements the lessons learned here by showing you how to develop UI controls. These two chapters, combined, provide enough information for you to begin developing real applications.

chapter 5

Application UI Components

Constructing a basic user interface

Chapter 5 Application UI Components

The previous chapter focused on the basic architecture and functional operations common to all applications. This chapter builds on that to cover the fundamentals of user interface (UI) development.

The topics covered in this chapter are as follows:

- **Controls**—Controls are rectangular areas of the screen that can display information and/or receive user input. This chapter shows you how to write both simple and compound controls.

- **Skins**—Skins allow the user to dynamically apply a new look and feel to the UI of a Series 60 2.x application. This section shows the APIs used to ensure that your applications support skins where appropriate.

- **Event Handling**—Series 60 is a graphical, event-driven environment. Applications receive notification of events from various system servers and update themselves accordingly. This section shows the APIs used to handle different kinds of events.

- **Resource Files**—Resource files allow UI elements, such as controls and user-visible strings, to be defined separately from source code and easily localized. This section covers the syntax and structure of Series 60 resource files, plus the APIs required to access them in C++ code.

- **Menus**—Menus are standard UI elements that allow a user to select and execute commands within an application. This section covers the use of menus, submenus, and context-sensitive menus, showing how they can be constructed from resources and modified dynamically.

- **Panes**—The Series 60 screen is made up of a number of areas known as panes, each with its own specific purpose. Many of these can be modified dynamically by an application. This section details the layout and purpose of each pane and describes how to programmatically control their behavior.

The following example applications are used in this chapter to illustrate the features introduced:

- **Controls**—A simple application that defines both a simple and a compound control.

- **SimpleMenu**—An application with a simple menu, illustrating basic menu resources.

- **ContextMenu**—An application with a context-sensitive menu.
- **StatusPane**—Shows you how to dynamically alter the contents of the status pane in your application.
- **TitlePane**—Demonstrates the APIs for dynamically changing the title pane contents.
- **NavigationPane**—Demonstrates the APIs for dynamically changing the contents of the navigation pane.

By the end of this chapter, you will understand the features and APIs required to construct a basic application user interface. However, Series 60 provides you with a large number of standard controls that can be used to quickly develop a more sophisticated UI, with a consistent Series 60 look and feel. These standard controls are covered in detail in Chapters 6, 7 and 8.

Controls

Controls and Windows

Controls are the basic elements for user interaction in the user interface. Most of the user interface elements you see are controls—this includes buttons, menus, views, and most other UI elements. Every control is derived from the abstract base class `CCoeControl` (defined in the system header file `coecntrl.h`), which gives it its fundamental features and capabilities. When it is visible, a control represents a rectangular area of the screen which can accept user input and/or display output. Controls may be nested within other controls to an arbitrary level (although practical considerations will usually limit the extent to which this is sensible).

A **window** is a system resource (owned by the Window Server) representing a rectangular area of the screen that contains one or more controls. It provides the mechanism for displaying its controls on screen. Be aware, though, that application UI code does not draw directly to a window—it only draws to controls. A number of overlapping windows can be displayed simultaneously—each window has a z-order attribute associated with it to represent its ordinal position (that is, the position between nearest the viewer and furthest away from the viewer). There is a family of classes representing windows, derived from `RWindowTreeNode`. A control's window can be obtained by calling the `CCoeControl::Window()` method.

Simple and Compound Controls

Controls are said to be either simple or compound. A simple control is one which does not contain any other controls. In other words, it is solely responsible for drawing to the rectangle which it represents.

Chapter 5 Application UI Components

A compound control contains one or more other controls. Whereas simple controls take sole responsibility for drawing to an area of the screen, complex controls delegate this responsibility, in part or totally, to one or more subcontrols. The controls contained by a compound control may in turn be either simple or compound.

NOTE In C++ terms, a compound control object owns other control objects.

It is a compound control's responsibility to manage the controls it contains and provide access to them. The controls contained by a compound control are known as its component controls.

When you are laying out the screen, compound controls provide more flexibility than simple controls. You can divide the screen into separate component controls and then redraw each control separately when it needs to be updated. This approach has the advantage that only the areas of the screen that have changed need to be redrawn, which can improve efficiency and reduce screen flicker. For these reasons, it is a good idea to place areas of the screen that need to be updated frequently into their own controls, keeping them separate from areas of the screen that change infrequently, or not at all.

In order to illustrate this, imagine an application that simultaneously displays the local time at various locations around the world—for example, New York City, London, and Tokyo, as shown in Figure 5–1. Clearly, the area of the screen displaying the times will need to change once a minute. However, the labels containing the city names will not normally need to be updated.

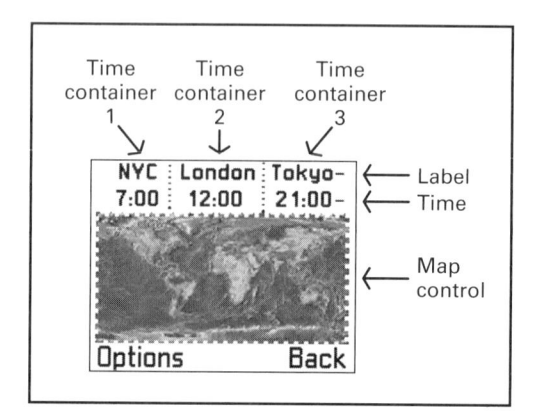

Figure 5–1 A fictitious application that uses compound controls.

Imagine that the application also displays a map of the world, showing roughly the areas currently in daytime or nighttime. As time progressed, the application would need to update the highlighted and darkened areas of the map, but with

Chapter 5 Controls

much less frequency than the times. To summarize, the application therefore has areas on screen that:

- Do not need updating (the location labels).
- Need updating frequently (the actual time).
- Need updating infrequently (the background map).

Figure 5–2 shows the hierarchy of controls, contained within a single compound control, which might be used to address this requirement. The system is envisaged with two levels of compound control. First, the screen is split into the map and three "time-container" controls. Each of these time-container controls is then divided into a label (for the city name) and a control to display the time.

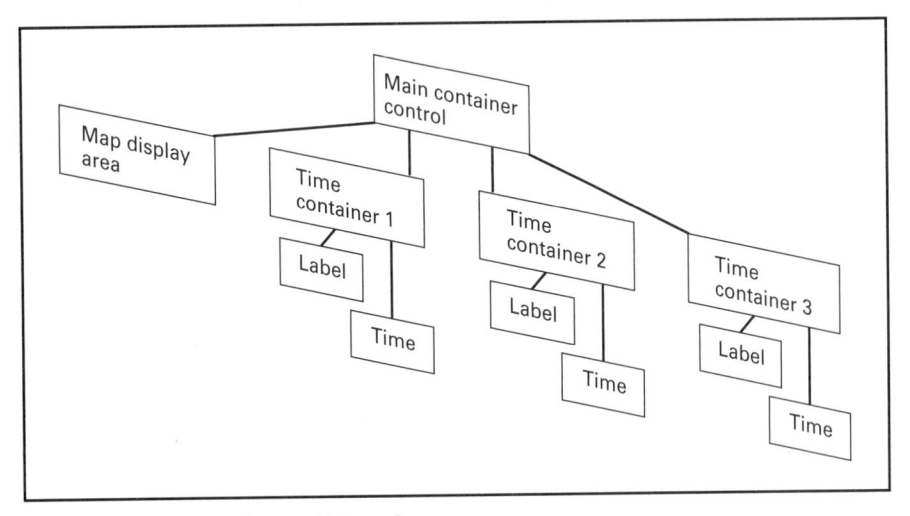

Figure 5–2 A compound control hierarchy.

Keeping the time control separate from its associated label will allow it to be redrawn at regular intervals, without the label needing be redrawn at the same time. The "time-container" compound control that owns them also provides a number of benefits:

- The behavior of the two controls is now encapsulated into a single, higher-level class, making it easier for developers to reuse.
- The compound control will determine the relative layout of the two controls, to ensure that all time controls have a consistent look and feel.
- The compound control also hides the specific drawing code for the two smaller controls. Once this drawing has been optimized (for example, ensuring that the time control is redrawn frequently but the label is not), developers are provided with a single, high-level compound control which they just need to add to their UI, and which will handle all of its own specific drawing requirements.

Series 60 provides a huge number of reusable compound controls which have exactly these benefits: they hide specific and often complicated layout and drawing code from the developer, to produce a simple, encapsulated control that fulfils a particular purpose.

> **NOTE** A great feature of the control framework is that a developer does not need to know whether a control is simple or compound—that is an internal implementation detail which is hidden behind the API. This means that any control can be reused in exactly the same way, regardless of how complex it may be.

Window Ownership

Controls draw onto windows, and so every control must have an associated window. Rather than give each control its own window, however, you can share a window between controls. Controls can be either "window-owning" (having their own window) or "non-window-owning" (sharing a window-owning control's window).

Non-Window-Owning Controls

Because a window is an expensive resource, it is not a good idea for every control to create its own window. You can reduce the number of windows needed by using non-window-owning controls. This reduces the number of resources used, the amount of Client/Server traffic, and the number of process switches. Text editing controls and labels are examples of controls that are usually non-window-owning.

> **NOTE** Non-window-owning controls are also known as lodger controls, because they "lodge" in another control's window.

There are, however, some restrictions on the use of non-window-owning controls, as outlined here:

- They must be positioned so that they do not overlap each other.
- They must be contained within the extent of their associated window.
- They must have a container.

In order to construct a non-window-owning control, you need to call the `SetContainerWindowL()` method, passing the control whose associated window you want to share. For example:

```
CMyNonWindowOwningControl::ConstructL(
    const CCoeControl& aControl)
    {
    SetContainerWindowL(aControl);
    // Do other operations needed to construct the control
    }
```

Window-Owning Controls

Unlike a non-window-owning control, a window-owning control cannot share another control's window. Instead, it has to create its own window when it is constructed. Window-owning controls have the same size and position as the window they create. Window-owning controls can also overlap each other and can be moved anywhere within the boundaries of their parent window.

It is also possible for a window-owning control to be a component of a compound control. In this case, what limits the movements of the control will be the window-owning control's window, not the compound control's window. Dialogs and menus are examples of controls that are usually window-owning.

In order to construct a window-owning control, you need to call the `CreateWindowL()` method—this is usually performed in the control's `ConstructL()` method. For example:

```
CMyWindowOwningControl::ConstructL()
    {
    CreateWindowL();
    // Do other operations needed to construct the control
    }
```

`CreateWindowL()` will most often be called by the constructor of either a simple control, or of the top-level parent of a complex control. In the case of a simple control, this function provides it with a window over which the control will have exclusive drawing responsibilities. In the case of a complex control, the top-level parent will call this function to provide a window for which it will allocate areas of drawing responsibility to its constituents.

In general it is recommended that controls should be non-window-owning wherever possible, in order to conserve system resources. A control should be window-owning only when it requires the properties of a window, such as being able to overlap other controls.

Creating a Simple Control

This subsection and the next describe how to implement a basic simple and compound control, with reference to the example application **Controls**. This application defines two controls: a simple control which just draws a red ellipse on the screen, and a compound control which contains two of these simple controls.

Creating a simple control involves the following steps:

- Create a class derived from `CCoeControl`.
- Implement construction code (typically in a `ConstructL()` function).
- Override the virtual function `Draw()` to provide the drawing code for your control.
- Override the virtual function `SizeChanged()` to layout the control again if its size changes.

All of these steps can be seen in the `CSimpleControl` class in **Controls**. The relevant sections of code from this class are shown here.

Creating a simple control class is quite straightforward. Here is the complete declaration of `CSimpleControl`—note that apart from construction and destruction the only functions implemented are the two virtual functions from `CCoeControl` mentioned above.

```
class CSimpleControl : public CCoeControl
    {
public: // constructors and destructor

    static CSimpleControl* NewL(const TRect& aRect,
        const CCoeControl* aParent);
    static CSimpleControl* NewLC(const TRect& aRect,
        const CCoeControl* aParent);
    ~CSimpleControl();

private: // constructor

    void ConstructL(const TRect& aRect, const CCoeControl* aParent);

private: // from CoeControl
    void Draw(const TRect& aRect) const;
    void SizeChanged();

private: // data
    };
```

Every control should have a `ConstructL()` method to carry out its second-phase construction. For most controls this will carry out at least the following basic steps:

- Set the control's window. This can either be a new window (in which case the control becomes window-owning) or an already existing window (in which case the control is non-window-owning).
- Set the control's client rectangle—in other words, the area of screen that this control is responsible for drawing to.

Chapter 5   Controls

- Activate the control. This informs the UI framework that the control is ready to be drawn.

You can see all three of these steps in the `ConstructL()` of `CSimpleControl`:

```
void CSimpleControl::ConstructL(const TRect& aRect, const
CCoeControl* aParent)
    {
    if (aParent == NULL)
        {
        // No parent control, so create as window-owning
        CreateWindowL();
        }
    else
        {
        // Part of a compound control, so just share
        // the parent's window
        SetContainerWindowL(*aParent);
        }

    SetRect(aRect);
    ActivateL();
    }
```

You will notice that `CSimpleControl` can behave either as a window-owning or a non-window-owning control. These lines in the `ConstructL()` are the only ones required to make this distinction—the rest of the code for the control will be the same. It is good practice to write controls in this way, unless there is a compelling reason for them to be window-owning or not. (A dialog control, for example, will always be window-owning.) Writing controls generally, where possible, allows them to be easily used in different circumstances.

After construction, all that remains is to implement the two virtual functions `Draw()` and `SizeChanged()` from `CCoeControl`. The first of these is shown here:

```
void CSimpleControl::Draw(const TRect& /*aRect*/) const
    {
    CWindowGc& gc = SystemGc();
    gc.Clear(Rect());

    // Set GC pen & brush styles
    gc.SetPenStyle(CGraphicsContext::ESolidPen);
    gc.SetPenColor(KRgbBlack);
    gc.SetBrushStyle(CGraphicsContext::ESolidBrush);
    gc.SetBrushColor(KRgbRed);

    gc.DrawEllipse(Rect());
    }
```

The parameter to this function, `aRect`, specifies the area of the control that requires redrawing. This may be the entire control or just be a region of it (for example, if another window appeared over part of the control and has been dismissed). A more complex control may take care just to redraw this part of the control. However, in order to keep the **Controls** example simple, it is ignored here, and the whole control is redrawn each time.

Drawing is achieved by obtaining a graphics context and using Series 60 graphics APIs to draw to it. This example draws an ellipse, with black outline and red fill, to occupy the full height and width of the control's client rectangle, as shown in Figure 5-3.

Finally, you just need to implement the `SizeChanged()` function to perform any steps required when your control changes size. In this case you will notice that the `Draw()` method above is entirely dynamic—it gets the control's current size using the `Rect()` method, and all drawing is relative to that. Therefore, no specific handling is required for changes to the control's size, and `SizeChanged()` is empty:

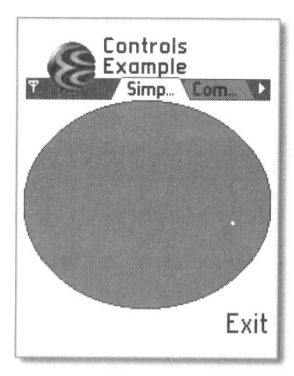

Figure 5-3 A simple control.

```
void CSimpleControl::SizeChanged()
    {
    }
```

You will see in the next subsection how `SizeChanged()` is implemented for a compound control. Further details of the Series 60 graphics APIs can be found in Chapter 11.

Creating a Compound Control

Implementing a compound control starts with the same steps as for a simple control:

- Create a class derived from `CCoeControl`.
- Implement construction code (typically in a `ConstructL()` function).
- Override the virtual function `Draw()`.
- Override the virtual function `SizeChanged()`.

In addition, you will need to do the following:

- Add component controls to the control, usually as member data of the compound control class.
- Override the virtual functions `CountComponentControls()` and `ComponentControl()` to ensure the component controls are drawn.

Chapter 5   Controls

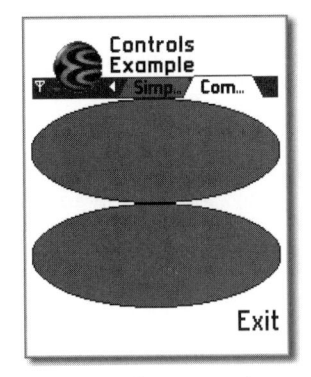

This subsection covers these steps with reference to the `CCompoundControl` class in the **Controls** application. This is a compound control which contains two instances of `CSimpleControl` and draws them in the top and bottom halves of the screen, as shown in Figure 5–4.

As before, the first step is to create a control class derived from `CCoeControl`. Notice that this time the control owns other controls as member data:

Figure 5–4 A compound control.

```
class CCompoundControl : public CCoeControl
    {
public: // constructors and destructor
    static CCompoundControl* NewL(const TRect& aRect, const CCoeControl* aParent);

    static CCompoundControl* NewLC(const TRect& aRect, const CCoeControl* aParent);

    ~CCompoundControl();

private: // constructor
    void ConstructL(const TRect& aRect, const CCoeControl* aParent);

private: // from CoeControl
    TInt CountComponentControls() const;
    CCoeControl* ComponentControl(TInt aIndex) const;
    void Draw(const TRect& aRect) const;
    void SizeChanged();

private:
    void CalculateRects();

private: // data
    CSimpleControl* iTop;
    CSimpleControl* iBottom;

    TRect iTopRect;
    TRect iBottomRect;

private: // member enum
    enum TComponentControls
        {
        ETop = 0,
        EBottom,
        ENumberOfControls
        };
    };
```

Construction of `CCompoundControl` looks very similar to that of `CSimpleControl`, with the addition of some extra code to construct the components controls:

```
void CCompoundControl::ConstructL(const TRect& aRect, const
CCoeControl* aParent)
    {
    if (aParent == NULL)
        {
        CreateWindowL();
        }
    else
        {
        SetContainerWindowL(*aParent);
        }

    // Size and construct the two component controls
    CalculateRects();
    iTop = CSimpleControl::NewL(iTopRect, this);
    iBottom = CSimpleControl::NewL(iBottomRect, this);

    SetRect(aRect);
    ActivateL();
    }
```

`CalculateRects()` is a private utility function provided to calculate the client rectangles for the two component controls. It gets the current client rectangle for this control and splits it into two halves, which are stored in the member variables `iTopRect` and `iBottomRect`:

```
void CCompoundControl::CalculateRects()
    {
    TRect outerRect = Rect();

    // Calculate dimensions of inner rectangles
    const TInt innerRectWidth = outerRect.Width();
    const TInt innerRectHeight = outerRect.Height() / 2;

    // Set rectangle for top control
    iTopRect.SetRect(outerRect.iTl,
         TSize(innerRectWidth, innerRectHeight));

    // Set bottom rectangle. Easiest way is to copy the top
    // rectangle, then move it down by innerRectHeight:
    iBottomRect = iTopRect;
    iBottomRect.Move(0, innerRectHeight);
    }
```

Next you need to think about how your compound control will be drawn. For a compound control this is the joint responsibility of three functions: `Draw()`, `CountComponentControls()` and `ComponentControl()`. It helps at this stage to understand the order of events that occur inside the UI framework when a control is drawn:

Chapter 5 Controls

- The compound control's `Draw()` method is called.
- The framework then calls `CountComponentControls()` on the compound control to determine the number of component controls it has.

The framework iterates through each of the component controls by calling `ComponentControl()` to obtain a pointer to each. For each control it repeats these three steps, eventually stopping when it reaches a control with no component controls. In this way it is able to draw a compound control of arbitrary complexity.

NOTE This process is the same for all controls, both simple and compound. The difference is that simple controls do not override `CountComponentControls()` and `ComponentControl()`. Instead, they inherit the default implementations from `CCoeControl`, and these return 0 and `NULL`, respectively, so that the framework knows there are no component controls to draw.

For a compound control, the `Draw()` method is often trivial, as usually the real drawing is delegated to component controls. However, it is good practice for a compound control to at least fill its client rectangle with a default color before the component controls are drawn. This ensures that any gaps between the component controls will not be left transparent. This is what `CCompoundControl::Draw()` does:

```
void CCompoundControl::Draw(const TRect& aRect) const
    {
    CWindowGc& gc = SystemGc();
    gc.Clear(aRect);
    }
```

TIP If your container control does no drawing at all, you do not need to override the `Draw()` method at all. However, if you do not have a `Draw()` method, you should call the `SetBlank()` method (inherited from `CCoeControl`) during the compound control's construction to ensure its background is filled in. Otherwise the contents of the screen behind the control will not be overdrawn.

Now you need to implement `CountComponentControls()` and `ComponentControl()` appropriately, otherwise your component controls will not be drawn. This is often done with the help of an `enum` which enumerates the component controls.

In the **Controls** example this is the `TComponentControls` enumeration, defined as a private member of the `CCompoundControl` class. `CountComponentControls()` is then implemented as follows:

```
TInt CCompoundControl::CountComponentControls() const
    {
    return ENumberOfControls;
    }
```

And `ComponentControl()` is implemented as:

```
CCoeControl* CCompoundControl::ComponentControl(TInt aIndex)
const
    {
    switch (aIndex)
        {
        case ETop:
            return iTop;
        case EBottom:
            return iBottom;
        default:
            return NULL;
        }
    }
```

The final element that needs adding to make the compound control work properly is the `SizeChangedL()` function. Unlike `CSimpleControl`, `CCompoundControl` requires a non-trivial implementation to ensure its component controls are correctly resized. To achieve this, it re-uses the `CalculateRects()` function, as shown here:

```
void CCompoundControl::SizeChanged()
    {
    CalculateRects();
    iTop->SetRect(iTopRect);
    iBottom->SetRect(iBottomRect);
    }
```

This is all that needs to be done. Now when the framework draws the component controls, they will have their client rectangles set correctly and will be drawn appropriately.

SEE ALSO
The **Event Handling** section of this chapter contains further details on how a compound control should distribute events to its component controls. For the sake of simplicity, the examples in this section have not handled input events at all.

Establishing Relationships between Controls

The unique screen layout of Series 60 has an impact on the API design of its UI controls. Consider an application that provides a scrolling listbox of items. In Series 60, such scrollable controls do not own their scrollbar. Instead, the scrollbar appears as part of the dedicated menu pane, as in Figure 5–5 (the inverted triangle is the scroll indicator).

This requires the application's control to be able to reference the menu pane, in order to update the scrollbar—as the user navigates the list, the scrollbar indicator must change to denote relative list position. As you have seen, parent-child relationships already exist between controls due to their implementation, but this doesn't provide a means of establishing communication between controls that do not possess the conventional hierarchical relationship. To satisfy this scenario, and in anticipation of other situations in which unrelated UI controls may need to interact, the `MObjectProvider` interface has been implemented.

This mechanism provides a relationship between all registered controls and allows a control to request access to others via its parent—the object request is continually passed upward (from child to parent, in an iterative fashion, with parent controls querying their child controls) until the request can be satisfied (or not, as the case may be).

The implementation of this interface for standard controls is provided by the system—the only thing you need to do to take advantage of it is to call `SetMopParent()` for your control to establish the child-parent relationship necessary for the scheme to work.

Skins

Skins provide a way of changing the appearance of the UI at runtime. They allow you to customize the default background for applications and modify colors and icons. Skins are a new feature in Series 60 2.x. Some example skins are shown in Figure 5–5.

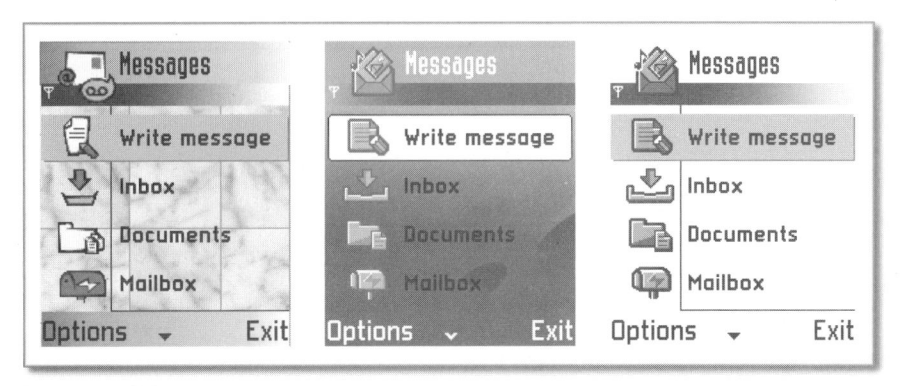

Figure 5–5 Skins.

Not all controls need to provide the same level of support for skins. Series 60 UI controls (sometimes known as Avkon controls) offer different levels of support for skins, depending on their needs. These controls can be compulsory skin-providing, optionally skin-providing, skin-observing, or non-skin-aware.

Compulsory Skin-Providing Controls

Compulsory skin-providing controls always display skins and dynamically comply with skin changes. You do not have to do anything special to get compulsory skin-providing controls to display skins—this is their default behavior. The control and status panes are examples of this type of control.

Optionally Skin-Providing Controls

The default behavior of optionally skin-providing controls is not to display skins, but they can be enabled to do so if required. Lists and grids are examples of this type of control. Optionally skin-providing controls can be skin-enabled in one of two ways: either each control is enabled individually, or all controls within an application can be enabled simultaneously.

The typical case is to enable skins for all controls together, to give your application a consistent look and feel. This is performed in the AppUI's second-phase constructor by calling the `BaseConstructL()` method and passing `EAknEnableSkin` as its argument:

```
void CMyAppUi::ConstructL()
    {
    // Enables skins in all the optionally
    // skin-providing controls in this app
    BaseConstructL(EAknEnableSkin);
    }
```

This will ensure that all optionally skin-providing controls will have skins enabled. If you do not want to enable skins, then simply leave out the `EAknEnableSkin` argument.

To enable skins for optionally skin-providing controls individually, you should call the `SetSkinEnabledL()` method for the control in question:

```
    // Enables skin in a single optionally
    // skin-providing control
    iSkinProvidingControl->SetSkinEnabledL(ETrue);
```

Skin-Observing Controls

Skin-observing controls do not provide any information themselves as to what skin should be used to draw them, but rely instead on their container control to provide this information. If this information is available, then they will use it to draw themselves. In other words, they will use the same skin as used by their parent control.

Non-Skin-Aware Controls

Non-skin-aware controls provide no support for skins, but it is still possible to use them in layouts that use skins. For example, controls such as labels can be instructed not to draw their backgrounds, and so they can be drawn over a skinned background without disrupting it.

Defining Skin-Aware Controls

If you define any controls in your application that need to be made skin-aware, it is up to you to write them so that they explicitly support skins. This may mean changing your control's drawing code so that it uses the methods provided by the skins utility classes `AknsUtils` and `AknsDrawUtils`. If a control clears its background, it must do so using methods provided by the `AknsDrawUtils` class. When setting up your controls, you must also make sure that you have correctly set up the `MObjectProvider` chain for all skin-aware controls.

Controls are notified of skin changes via their `HandleResourceChange()` method—this is called by the application framework with a value of `KAknsMessageSkinChange` when the skin is changed. If you are using the skin utility classes, then you will have to link against `aknskins.lib`.

Creating skin-aware controls or actual skins is beyond the scope of this book. However, further information on implementing skin-aware controls is available in the *Avkon Skins User's Guide* in the Series 60 2.x SDK Documentation and **Series 60 Theme Studio** from Forum Nokia (**http://www.forum.nokia.com**) provides a means to create new skin (or "theme") designs.

Event Handling

As the user interacts with the user interface, events will be sent to the application that owns the window in which the event occurred. There are several different types of events, but the main ones you will be interested in are key events and redraw events. Key events occur when the user presses, holds or releases a key. Redraw events occur when a control (or part of a control) is revealed on screen and needs to redraw itself.

Key Events

A key event is a system event that is generated whenever a key on the device keypad is pressed, held down or released. Key events are routed to applications by the Window Server by means of the *control stack*, a structure which maintains information on the order in which key events should be passed to applications. A control is informed of a key event by a call to its `OfferKeyEventL()` method—this is a virtual function defined in `CCoeControl` which a control should override to provide its own key-handling behavior. This process is described in more detail below.

Focus

As far as the Window Server is concerned, only applications, and not windows, have keyboard focus. This means that only the application that currently has focus will receive key events. The application itself is then responsible for distributing these key events to the appropriate control. In order to help with this, the control framework provides the control stack, which supports the distribution of key presses to different controls within an application.

An application can gain or lose focus at any time. This can happen, for example, if the user switches between applications, or if the system brings another application into the foreground. Whenever an application gains or loses focus, the AppUI's `HandleForegroundEventL()` method will be called, with a `TBool` parameter indicating whether focus was lost or gained. You do not normally have to redraw controls when your application loses focus, because applications that do not have focus are not visible on the screen. However, it may be useful to perform other activities, such as pausing a game when focus is lost and starting it again when focus is gained. The following code shows how you would achieve this:

```
void CExampleAppUi::HandleForegroundEventL(TBool aForeground)
    {
    if (aForeground)
        {
        // focus gained, perform appropriate actions here
        }
    else
        {
        // focus lost, perform appropriate actions here
        }

    CAknAppUi::HandleForegroundEventL(aForeground);
    }
```

Compound controls should handle focus changes within their component controls. It is up to the compound control to call the component control's `FocusChanged()` method when it changes the component control's focus. It is up to you to implement this in your own compound controls. You give focus to a control by calling its `SetFocus()` method.

TIP A control's current focus state can be obtained by calling the `IsFocused()` method. If you change focus from one control to another, you should remember to call `SetFocus(EFalse)` on the control that is losing focus as well as calling `SetFocus(ETrue)` on the control that is gaining focus.

Chapter 5 Event Handling

Controls may indicate visually whether they are focused—for example, by adding a border. The system-provided controls will highlight themselves in an appropriate way when their focus changes—all you need to do is call the `SetFocus()` method on them. If you write your own controls, it is up to you to make sure that the control redraws itself when focused if necessary.

The Control Stack

The control framework provides a control stack used for routing key events. Each control has a priority that determines its position on the stack. A control's position on the stack determines the order in which it will be offered key events.

The control with the highest priority is placed at the top of the stack, with controls arranged in decreasing order of priority. The relative placement of controls with equal priorities is determined by the order in which they were added, with the most recently added control highest on the stack. Events are offered to controls from the top of the stack downward until they are consumed (that is, handled by a control).

In order for a control to receive key events, it must explicitly be added to the control stack—by default, controls are not on the stack. A control is added to the stack using the function `CAknAppUi::AddToStackL()`. In the case of compound controls, only the compound control should be added to the stack— its component controls should not. Instead, the compound control should be responsible for passing events on to its component controls as required.

Offering Key Events

Key events are passed to a control by a call to its `OfferKeyEventL()` method. Again, this is a virtual method in `CCoeControl` which you need to override in your own control to provide appropriate behavior.

A control can decide whether or not it wants to process a key event. If it does process the event, it should return the value `EKeyWasConsumed`; otherwise it should return `EKeyWasNotConsumed`. The event will be passed to each control on the stack in turn until one of the controls returns `EKeyWasConsumed` or there are no more controls left.

The `OfferKeyEventL()` method has two arguments: a key event (`const TKeyEvent& aKeyEvent`) and a key type (`TEventCode aType`). The key event is a structure that indicates, among other things, which key has been pressed. The key type is an enumeration that indicates the nature of the key event. There are three different types of key events: `EEventKeyDown`, `EEventKey` and `EEventKeyUp`. An `EEventKeyDown` event is received when the key is first pressed down, followed by one or more `EEventKey` events as it is held down, then an `EEventKeyUp` event when it is released.

The following piece of code shows how a control would handle key up events from the **Up** and **Down** arrow keys and discard everything else:

Chapter 5 Application UI Components

```
TKeyResponse CMyControl::OfferKeyEventL(const TKeyEvent&
aKeyEvent, TEventCode aType)
    {
    if (aType == EEventKeyUp)
        {
        switch (aKeyEvent.iCode) // check the key code
            {
            case EKeyUpArrow: // Up arrow key
                // handle up arrow key press here
                return EKeyWasConsumed;

            case EKeyDownArrow: // Down arrow key
                // handle down arrow key press here
                return EKeyWasConsumed;

            default:
                return EKeyWasNotConsumed;
            }
        }
    return EKeyWasNotConsumed;
    }
```

SEE ALSO A list of all the key code enumerations can be found in the system header file e32keys.h. Uikon.hrh redefines some of these to have Series 60-specific names—for example, EKeyDevice3 is redefined as EKeyOk.

As mentioned above, a compound control is responsible for offering key events to its component controls. An example of how this would be implemented is shown below:

```
TKeyResponse CMyCompoundControl::OfferKeyEventL(
    const TKeyEvent& aKeyEvent, TEventCode aType)
    {
    TKeyResponse response(EKeyWasNotConsumed);

    if (iComponentControl1 && iComponentControl1->IsFocused())
        {
        response = iComponentControl1
->OfferKeyEventL(aKeyEvent, aType);
        }
    else if (iComponentControl2 && iComponentControl2
->IsFocused())
        {
        response = iComponentControl2
->OfferKeyEventL(aKeyEvent, aType);
        }

    return response;
    }
```

Notice that if there is more than one component control that can handle the event, then the event should be sent to the control that currently has focus.

Redraw Events

Redraw events tell a control that it needs to update itself on the screen. They are generated when a control is first displayed, whenever part or all of it is revealed after being covered by another window, or whenever the application's data changes and the view needs to be refreshed appropriately. A redraw event can occur in one of two different ways:

- **System-Initiated Redraws**—The window server sends the redraw event because a region of the screen has become invalid. This can happen, for example, when a window that was obscuring the control has disappeared.

- **Application-Initiated Redraws**—The application itself sends the redraw event because, for example, some application data has changed and the control needs to be updated to reflect this.

System-initiated redraws result in a call to the `Draw()` method of every control that intersects the invalid region. In this case, all you have to worry about is implementing the `Draw()` method of your controls so that they correctly redraw themselves.

Generating an application-initiated redraw is performed by calling either `DrawDeferred()` or `DrawNow()` on the appropriate control. The reason for having two draw functions is that drawing involves sending an instruction to the Window Server, and transactions with the Window Server may be buffered to avoid the inefficiency of too many Client/Server calls. `DrawDeferred()` adds a draw request to the current Window Server buffer, to be sent to the Window Server once the buffer is full. In practice this is often instantaneous, since the nature of Series 60 as a highly graphical environment means that the buffer tends to fill quickly. You should certainly use `DrawDeferred()` for any non-time-critical updates. `DrawNow()` does the same as `DrawDeferred()`, but then forces a flush of the command buffer. You should generally use this only if it is critical that your update happens immediately (for example, in a precise clock control). Overuse of `DrawNow()` is inefficient and should be avoided.

It is important to note that if you want to redraw a control explicitly, you must call `DrawDeferred()`, or `DrawNow()` on it—you must not call its `Draw()` method directly. `DrawDeferred()` and `DrawNow()` ensure that the Window Server is informed of the control's need to redraw itself, so that the drawing you perform in `Draw()` will appear on screen. They will also cause a compound control's component controls to be redrawn correctly. Calling `Draw()` directly does neither of these, and often it will not produce the result you are expecting.

NOTE Unlike key events, compound controls do not need to distribute redraw events to their component controls. As you saw earlier in this chapter, the control environment does this using the `CountComponentControls()` and `ComponentControl()` methods of the compound control.

Observers

A final point about event handling is that it is possible for a control to send events to another object. This can be useful when dealing with compound controls—for example, if you need a component control to report events to the compound control which contains it.

Events are passed out of controls by means of a control observer interface. A control observer does not need to be a control itself—any class can be set up as a control observer by deriving from MCoeControlObserver and setting it as the observer for the control (or controls) it needs to observe. A control's observer is set by calling the control's SetObserver() method, passing a pointer to the observer. An observer can be set to observe multiple controls, but each control can have at most one observer.

A control sends an event to its observer using its ReportEventL() method. This will result in a call to the observer's HandleControlEventL() method (inherited from MCoeControlObserver). The observer can then process the event in any way it desires. For example:

```
void CMyObserver::HandleControlEventL(CCoeControl* aControl,
TCoeEvent aEventType)
    {
    if (aControl == iMyControl1 &&
    aEventType == EEventStateChanged)
        {
        // handle event here
        }
    }
```

Notice that HandleControlEventL() receives two arguments. The first is a pointer to the control from which the event originated. This is useful if an observer is observing multiple controls. (If you are observing only a single control, then you can ignore this parameter.) The second parameter is the event type.

Dangling Pointers

When using control observers, you should be careful to avoid dangling pointers. If the observer owns a pointer to the control, you need to avoid a situation where the control is deleted but the observer still has a dangling pointer to it. This can be avoided by having the control send an appropriate event to the observer just before it is deleted.

Likewise, in an observer's destructor you should ensure that it deregisters itself with any controls it observes, so that they do not try to send it events after it has been deleted. This is performed by calling SetObserver(NULL) on each control.

Resource Files

A resource file is a text file with an .rss file extension, used to specify user-visible elements of your application separately from the source code. Resource files are used to specify layout of UI elements such as menus, dialogs and lists, as well as any user-visible text used by the application. Every Series 60 application has at least one resource file associated with it.

A resource file is compiled into a binary file and accompanying header (.rsg) file by the resource compiler, as part of the standard abld build process. This binary file is then opened by the application framework upon application startup, and individual resources are loaded into the C++ code as needed, by means of the resource identifiers created in the .rsg file.

Keeping resource information separate from source code offers a major advantage to developers, namely that the UI appearance of the application can be substantially modified and rebuilt without the need to change the source code or recompile the application. This in turn makes applications much easier to localize (that is, convert to different languages and regional settings), as for a well written application only the resource file needs to be recompiled.

Resource File Syntax

The syntax of resource files is similar to that of C++, but not identical. Because the C preprocessor is used as the initial step in processing resource files, you can use both C and C++ style comments in resource files, as well as preprocessor statements such as #include, #define, #if, #else and #endif. All white space appearing in the resource file is ignored, unless it is within a string. The following first few lines of a resource file show preprocessor statements in use:

```
/**
 *
 * @brief Example resource file
 *
 * @copyright EMCC Software Ltd
 * @version 1.0
 */

//   RESOURCE IDENTIFIER
NAME   RSF1 // 4 letter ID

//   INCLUDES

#include <avkon.rh>
#include <avkon.rsg>
#include <eikon.rh>
#include <eikon.rsg>
...
```

Chapter 5 Application UI Components

Additionally, resource files can also contain `enum` definitions. This means that a header file containing only `enum` definitions can be shared between resource files and C++ source files, and this is a common way of defining values shared between both types of file. Such a file is conventionally given the file extension `.hrh`. The following is an example of a simple `.hrh` file, taken from the **SimpleMenu** example:

```
/**
*
* @brief Constants file for SimpleMenu application
*
* Copyright (c) EMCC Software Ltd 2003
* @version 1.0
*/

#ifndef SIMPLEMENU_HRH
#define SIMPLEMENU_HRH

// Command ids
enum
    {
    ESimpleMenuCmdNewGame = 0x6000,
    ESimpleMenuCmdViaIR,
    ESimpleMenuCmdViaBluetooth,
    ESimpleMenuCmdViaSMS
    };

#endif// SIMPLEMENU_HRH

// End of File
```

> **NOTE** `.hrh` files must contain only enumerations and preprocessor statements. Any other C++ syntax—such as class declarations, constant integer values, and so on—will cause the resource compiler to fail.

Any strings appearing in the resource file need to be enclosed in double quotes. For example:

```
RESOURCE TBUF r_simple_string
    {
    buf = "Hello world";
    }

RESOURCE TBUF r_quoted_string
    {
    buf = "includes a \"phrase in quotation marks\"";
    }
```

```
RESOURCE TBUF r_copyright
    {
    // Create the string "© 2003"
    buf = <0x00A9>" 2003";
    }
```

Double-quote and backslash characters should be preceded by a backslash escape character: \" and \\, respectively. Unicode characters can be included in a string by specifying the Unicode value between < and > symbols. Characters specified this way should not be enclosed within double quotes. The `r_copyright` resource in the example above shows you how to achieve this.

Note that it is bad practice to actually specify text strings in a resource file—any user-visible text should instead be supplied in a localization file, as described in the **Internationalization** section of Chapter 4. The examples above are for illustration only and are not taken from a real example application.

Resource File Structure

This subsection covers the structure of Series 60 resource files. This structure can be divided broadly into two parts:

- A **header part** containing include statements and some standard information about the resource file which is used by the resource compiler and application framework.

- A **body part** defining a number of resources. Each resource declaration starts with the keyword `RESOURCE`.

The following describes each of these parts in detail:

Header Part

A resource file's header part consists of the following elements:

- **Name**—This is defined in a `NAME` statement, which must be the first meaningful line in your resource file (except, that is, for comments and white space). It specifies a four-letter name used to differentiate resource files in cases where an application uses more than one. The name therefore needs to be unique within an application, but need not be globally unique. (In other words, two resource files used by the same application must have different `NAME` values, but two resource files used by different applications can have the same `NAME` value.) You should also make sure that the name used is different from the names used in the system resource files, which are available for use by all applications.

- **Include statements**—As in C and C++, the `#include` statement imports other files, allowing a resource file to use symbols and structures defined elsewhere.

- **Signature**—The RSS_SIGNATURE is required by the application framework, and failure to include it in your resource file can cause a panic at runtime when the binary resource file is loaded. However, its contents are actually ignored, so you should just leave this resource empty, as shown below.

- **A document name buffer**—This is a TBUF resource which specifies the name of the default document for an application. Most Series 60 applications do not use documents, in which case the value is unimportant, but the resource must still be included. (Otherwise the application architecture will panic when loading the resource file.) Remember that you do not need to specify a file extension, as Series 60 native documents do not use extensions.

- **Application information resource**—The EIK_APP_INFO resource specifies various standard controls for an application, such as the status pane. It is common to create a resource specifying the new status pane contents and then refer to it using the status_pane field of the EIK_APP_INFO resource.

All of these parts can be seen in the following resource file example, taken from **Controls**:

```
/**
*
* @brief Resource file for Controls application
*
* Copyright (c) EMCC Software Ltd
* @version 1.0
*/

//    RESOURCE IDENTIFIER
NAME   CONT // 4 letter ID

//    INCLUDES
#include <avkon.rh>
#include <avkon.rsg>
#include <eikon.rh>
#include <eikon.rsg>
#include "Controls.hrh"
#include "Controls.loc"

// RESOURCE DEFINITIONS
RESOURCE RSS_SIGNATURE { }

RESOURCE TBUF { buf = "Controls" } // Default document name

RESOURCE EIK_APP_INFO
    {
    status_pane = r_controls_status_pane;
    }
```

Body Part

The body of the resource file defines the resources that the application will use. Each resource is defined using the RESOURCE statement, as follows:

```
RESOURCE STRUCTNAME resource-name
    {
    resource-initializer-list
    }
```

You should replace STRUCTNAME with the type of the resource structure (STRUCT) you require. These are typically defined in .rh files, which you will need to #include in your .rss file. The system resource types are defined in avkon.rh, uikon.rh and eikon.rh.

> **NOTE** The resources in the header part (RSS_SIGNATURE, TBUF and EIK_APP_INFO, as seen above) are anonymous and should leave the *resource-name* blank.

For named resources, you should also replace *resource-name* with a unique name for your resource. The resource name you choose must be in lower case. By convention names begin with r_, so, for example, to define a buffer called r_buffer_1, you would create the following resource:

```
RESOURCE TBUF r_buffer_1
    {
    buf = "text here";
    }
```

When the resource file is compiled, a constant representing the resource will be added to an .rsg header file generated by the resource compiler. The name given to this constant is the same as the name given to the resource in the resource file, but in uppercase. So, for the example resource defined above, a constant named R_BUFFER_1 will be created. Note again that user-visible text would not normally be inserted into a resource file.

The .rsg file is generated in \epoc32\include, with the same base name as your application's .app file. This is included (as a system include) in your C++ source, and the values defined in it are used to access individual resources. These values are known as resource IDs.

The following code shows how the buffer defined above might be read in from resource:

```
#include <example.rsg>
...

HBufC* buf = iCoeEnv->AllocReadResourceLC(R_BUFFER_1);
```

> **TIP**
> Unfortunately there is no special type defined for resource IDs—they are just handled as simple `TInt` values. By convention, however, functions that expect a resource ID as a parameter often name the parameter `aResourceId`. The context will also usually make it clear that a resource ID is expected.

The *resource-initializer-list* defines a set of values for the fields of the resource. Each initializer in this list has to be terminated with a semicolon. Each type of resource has different fields that you can include in the initializer list.

> **SEE ALSO**
> Specific details of the fields used by each control will be covered along with each individual control in the following chapters.

There are three different ways of initializing a field, depending on the type of the field you are initializing: simple initializers, array initializers and struct initializers. All three are shown in this example:

```
RESOURCE EXAMPLESTRUCT r_my_example_struct
    {
    simple = EEikCtLabel;           // simple initializer
    array = {1, 2, 3};              // array initializer
    structmember = OTHERSTRUCT      // struct initializer
        {
        simple1 = "hello";
        simple2 = "goodbye";
        };
    }
```

- Simple initializers assign a single value or string to a field.

- Array initializers assign one or more values to an array field, formatted as a list of comma-separated elements inside braces.

- Struct initializers assign one or more values to a `STRUCT` field. A `STRUCT` field is initialized by giving it the name of the required `STRUCT` and then specifying each field of that `STRUCT`. You need to be careful when doing this, as the resource compiler does not carry out type checking. This means that if you use the wrong type of `STRUCT` to initialize a `STRUCT` field, your program will compile, but it may panic at runtime. It is up to you to make sure you are using the correct `STRUCT`.

String Resources

As you have seen, strings can be included in a resource file using the `TBUF` resource. All user-visible, locale-specific text must be specified in resources and not in source code, in order to facilitate translation to other languages.

Chapter 5 Resource Files

Additionally, it is usual to define string literals in a `.loc` file, or a language-specific `.lxx` file, rather than in the `.rss` file. This makes translation even simpler, as translators only need to see a file containing text strings, without all the resource definitions in the `.rss` file.

A `.lxx` file takes this principle one step further and allows for multiple files to be provided, each defining the same text strings in a different language. The *xx* should be replaced with a two-digit locale code specified by the `TLanguage` enum in `e32std.h`. The `.lxx` files are then included by the `.loc` file according to the current build locale set in the project `.mmp` file. Examples of each file discussed here are shown below.

First, the `.rss` file includes the `.loc` file for all of its string definitions:

```
#include "MyApp.loc"
...

RESOURCE TBUF r_hello
   {
   buf = STR_HELLO; // note: no string literals in the .rss!
   }
```

The `.loc` file in turn includes multiple `.lxx` files, depending on the current locale:

```
#ifdef LANGUAGE_01        // 01 = (British) English
#include "MyApp.101"
#endif

#ifdef LANGUAGE_02        // 02 = French
#include "MyApp.102"
#endif

...
```

And finally, the `.101` and `.102` files define the strings in the respective languages. The `.101` (UK English) file would look something like this:

```
#define STR_HELLO "Hello world!"
...
```

To ensure that the correct strings are used when the resources are compiled, you should include one or more `LANG` statements in your `.mmp` file. The following lines in an `.mmp` file will ensure that resources are built for both English and French, resulting in two binary resource files being built: `.r01` and `.r02`:

```
LANG 01
LANG 02
```

Chapter 4 contains more information on localization.

Punctuation

You may have noticed that some lines in the resource file end in semicolons and some do not. Use the following rules when deciding whether or not to punctuate:

- A semicolon is needed after any assignment.
- Elements in a list are separated by commas.
- There should be no semicolon after a resource definition, or after the last element in a list.

For example:

```
RESOURCE AVKON_VIEW r_myapp_view
   {
   menubar = r_myapp_menubar; // assignment: semicolon needed
   cba = r_myapp_cba;         // assignment: semicolon needed
   }                          // end of resource definition:
                              // no
                              //    semicolon needed
...

RESOURCE TAB_GROUP r_myapp_tabgroup
   {
   tab_width = EAknTabWidthWithTwoTabs;
   active = 0;
   tabs = {                   // start of list
      TAB                     // first TAB STRUCT in the list
         {
         id = ENavigationPaneTab1;
         txt = TAB1_TEXT;
         },                   // comma between elements in a list
      TAB                     // second TAB STRUCT in the list
         {
         id = ENavigationPaneTab2;
         txt = TAB2_TEXT;
         }                    // end of list: no semicolon needed
      };                      // end of assigning list to "tabs":
                              // semicolon needed
   }                          // end of resource definition:
                              // no semicolon needed
```

The **Menus** and **Standard Panes** sections provide more information on these resources.

Creating Resource Structures

While the `RESOURCE` statement creates an instance of a specific resource, the `STRUCT` statement defines a type of resource. This allows you to create your own resources for your own controls. You should store any `STRUCT` definitions you create in a file with an .rh extension.

You can think of the `STRUCT` statement as being the resource file's version of the C++ `struct` or `class` statement, used to define a structure or class, and the `RESOURCE` statement as being analogous to creating a new object of the specified class.

The `STRUCT` statement is followed by the `STRUCT`'s name and then by a list of fields defined for this structure:

```
STRUCT MYCONTROL
   {
   BUF buf;           // non-zero-terminated text string
   LONG value = 0;    // a 32-bit integer with a default value
   }
```

Note that a `STRUCT`'s name must be all uppercase, with no spaces, and start with an alphabetic character. Subsequent characters can be letters, numbers or underscores. The name must also not start with any of: `GLOBAL`, `STRUCT`, `LEN`, `RESOURCE` or any of the `STRUCT` field types listed in Table 5–1. The `STRUCT`'s fields are listed within braces.

Each field in a `STRUCT` is defined with a field type, followed by a field name, an optional initializer and a semicolon. The type must be all uppercase, and the field name all lowercase. If a default value is provided, then this field can be omitted in a `RESOURCE` definition using this `STRUCT`, in which case the default value will be used.

As well as these simple fields, a field can also be defined as an array of values of the same type. To define a field to be an array, add a pair of square brackets after the field name, as follows:

```
STRUCT MENU_PANE
   {
   STRUCT items[]; // MENU_ITEMs
   LLINK extension = 0;
   }
```

This shows the `MENU_PANE` `STRUCT` defined in \epoc32\include\uikon.rh. The **Menus** section of this chapter will explain the `MENU_PANE` and `MENU_ITEM` `STRUCT`s in more detail; but for now, the `items` field provides an adequate example of an array field.

The `items` field allows you to have any number of `MENU_ITEM` `STRUCT`s in this resource, each separated by a comma. Each element in this array should be a `STRUCT` of type `MENU_ITEM`s. Further information on resource files can be found in the **Resource File Format** section of the SDK documentation.

Table 5–1 Resource File STRUCT Field Types

Field Type	Description
BYTE	A single byte. This can be interpreted as either a signed or unsigned integer.
WORD	Two bytes. This can be interpreted as either a signed or unsigned integer.
LONG	Four bytes. This can be interpreted as either a signed or unsigned integer.
DOUBLE	Eight bytes. This represents a double-precision floating-point number.
TEXT	A NULL-terminated string. This is deprecated: use LTEXT instead.
LTEXT	A Unicode string. This has a leading byte holding the length of the string, and no terminating NULL.
BUF	A Unicode string. This has no leading byte and no terminating NULL.
BUF8	An 8-bit character string. This has no leading byte and no terminating NULL. Use this to put 8-bit data into a resource.
BUF<n>	A Unicode string with a maximum length of n. This has no leading byte and no terminating NULL.
LINK	A 16-bit ID of another resource. This is like having a reference to the specified resource.
LLINK	A 32-bit ID of another resource.
SRLINK	The 32-bit ID of the resource in which this field is defined. This allows you to create a STRUCT with a reference to itself (SRLINK stands for Self-Referencing LINK). The value for a field of this type is assigned automatically by the resource compiler—you cannot supply an initializer yourself.
STRUCT	A STRUCT. This creates a field that is itself a STRUCT, allowing you to embed STRUCTs within STRUCTs.

Menus

A menu is a window that presents a list of commands to the user. A menu is arranged into lines, known as menu items, each of which contains a text label. When a user selects a menu item, the appropriate command is invoked. Figure 5–6 shows an example menu from a game application.

Chapter 5 Menus

Series 60 applications usually group commands together on the **Options** menu, which appears when the **Options** soft key is pressed. Applications can also display context-sensitive menus, activated when a particular key—for example, the **Selection** key—is pressed. You would use this type of menu when there is no single intuitive function that could occur when a particular key is pressed. For example, you might be writing an application that displays a list of text items—you must decide what the user would expect to happen when they press a particular key on a focused item. In some applications, it would be obvious that the item should open, in which case there would be no need for a context-sensitive menu. In other applications, it might be appropriate to either open or delete the item—in this case, you could use a context-sensitive menu to give the user both options.

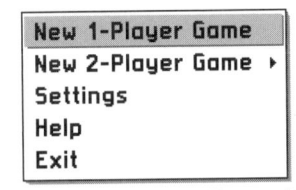

| New 1-Player Game |
| New 2-Player Game ▸ |
| Settings |
| Help |
| Exit |

Select   Cancel

Figure 5–6 Example menu.

You can add menus to the main application screen, its views, dialogs, and some of its other controls. They appear in a pop-up window just above the left soft key. As you can see in Figure 5–6, once the menu is displayed, the soft key labels change, so that the user can select a menu item with the left (**Select**) soft key and close the menu with the right (**Cancel**) soft key. Pressing the **Selection** key also opens the focused menu item.

Submenus

A menu item can contain a submenu. A submenu groups a further set of commands, which are relevant to a menu item. For example, a two-player card game might have an additional menu item to play a two-player game, which has a submenu offering to play via IR, or via SMS, as shown in Figure 5–7.

New 1-Player Game	
New 2-Player	Via IR
Settings	Via SMS
Help
Exit

Select   Cancel

Figure 5–7 A submenu.

You can open the submenu by focusing the item and pressing the left (**Select**) soft key, **Selection** key, or right direction key. You close it by pressing the right (**Cancel**) soft key (which closes the entire menu) or the left direction key (which closes just the submenu).

> **NOTE** It is only possible to have one level of submenu—menu items in submenus may not themselves have submenus.

To make a menu as usable as possible, you should try to keep the number of items on a submenu small. You should also ensure that a submenu never contains an item that is available at the previous level.

Menu Basics

This subsection looks at the **SimpleMenu** example application, which makes basic use of menus. The application illustrates how to:

- Define a menu using resources.
- Define a submenu using resources.
- Use a menu in a control.
- Handle menu commands.
- Handle submenu commands.

The **SimpleMenu** example displays a blank screen and has **Options** and **Back** soft keys. The **Options** soft key displays the **Options** menu, and this has two menu items: **New Game** and **Exit**, as shown in Figure 5–8.

Exit closes the application. (Note that you can also close the application by selecting the right soft key, which is set to **Back** on the main screen, but is also often set to **Exit** in many other applications.) **New Game** has a submenu with items allowing you to choose a connection method for playing a multiplayer game, as shown in Figure 5–9.

Figure 5–8 SimpleMenu Options menu.

Figure 5–9 SimpleMenu New Game submenu.

Defining a Menu Using Resources

You define the menu in a MENU_BAR resource, as shown in SimpleMenu.rss:

```
RESOURCE MENU_BAR r_simplemenu_menu_bar
   {
   titles =
      {
      MENU_TITLE
         {
         txt = ""; // the text is not used in Series 60
         menu_pane = r_simplemenu_menu_pane;
         }
      };
   }
```

A `MENU_BAR` usually contains a single `MENU_TITLE`, contained within the `titles` array. This specifies the title of the menu, and a menu pane containing its menu items. Series 60 does not use the menu title's `txt` field, so you can set it to blank. It is sometimes helpful to provide a label, however, as a reminder of the menu function, and to increase portability to other Symbian OS platforms which require it.

The `MENU_PANE` resource referenced from the `MENU_BAR` resource contains the menu items, each represented by a `MENU_ITEM` resource:

```
RESOURCE MENU_PANE r_simplemenu_menu_pane
    {
    items =
        {
        MENU_ITEM
            {
            command = ESimpleMenuCmdNewGame;
            txt = NEW_GAME_TEXT;
            cascade = r_simplemenu_game_submenu_menu_pane;
            },
        MENU_ITEM
            {
            command = EAknCmdExit;
            txt = EXIT_TEXT;
            }
        };
    }
```

In this case, there are two menu items: one to play a new game and one to exit the application. As a minimum, each `MENU_ITEM` has the following fields: a text field for the menu item caption (`txt`) and the ID of a command that will be invoked when the menu item is selected (`command`). A command ID is an enumerated value which is unique within the application. Command IDs are defined in a separate header file with an .hrh extension. System .hrh files—for example, `uikon.hrh` and `avkon.hrh`—provide common command IDs. .hrh files can be included in both resource files and C++ source files, providing a link between the menu definitions in resources and the command-handler functions in the C++ code.

Typically, command names are of the format `E<AppName>Cmd<CommandName>`—for example, `EAknCmdHelp` is the Help command for Series 60 applications. If you want to use your own application-specific commands, you need to define an enumeration in your own .hrh file. In the **SimpleMenu** example, an application-specific command, `ESimpleMenuCmdNewGame`, is defined in `SimpleMenu.hrh`:

```
enum
    {
    ESimpleMenuCmdNewGame = 0x6000,
    ...
    }
```

In addition, the `cascade` item can optionally be used to specify a submenu that will be displayed when this item is selected. This should just give the name of another `MENU_PANE` resource, which is defined in exactly the same way as this one (with the one exception that it cannot itself reference any submenus).

> **NOTE** You should start the enumeration from `0x6000` to prevent clashes with system commands.

The `MENU_TITLE` and `MENU_PANE` resource `STRUCT`s have a number of fields which Series 60 ignores, such as `bmpfile`, `bmpid` and `bmpmask`. These are ignored because Series 60 menus cannot contain bitmaps. The `MENU_TITLE`, `MENU_PANE` and `MENU_ITEM` `STRUCT`s also specify an `extension` field, which you should not change. The `MENU_ITEM` has an `extratext` field, which Series 60 also ignores.

Using a Menu in a Control

You need to add the menu bar to the screen on which you want it to be available. In the **SimpleMenu** example, there is only one screen, so the options menu is going to belong to the AppUI. You refer to the menu in the `menubar` field of the appropriate resource and ensure that you have appropriate soft keys to access the menu. In the **SimpleMenu** example, the `menubar` and `cba` in `EIK_APP_INFO` are set to achieve this:

```
RESOURCE EIK_APP_INFO
    {
    menubar = r_simplemenu_menu_bar;
    cba = R_AVKON_SOFTKEYS_OPTIONS_BACK;
    }
```

> **TIP** UI components, which can have a menu, have a `menubar` field in their resource `STRUCT`—for example, the view (`R_AVKON_VIEW`) and dialog (`R_AVKON_DIALOG`) components. Specifying a `menubar` for these components allows you to make menus specific to individual screens in an application.

Handling Menu Commands

In order for the menu to be useful, you need to handle its commands. Where you handle commands depends on the type of object containing the menu. In a `CAknAppUi`- or `CAknView`-derived class, you use the `HandleCommandL()` method. In `CAknDialog`-derived classes, you use the `ProcessCommandL()` method.

> **NOTE** Objects that handle commands inherit from the `MEikCommandObserver` mixin. This defines a pure virtual method `ProcessCommandL()`. When you select a menu item, the framework calls `ProcessCommandL()`. `CAknAppUi` and `CAknView` provide a default implementation of `ProcessCommandL()`, which calls `HandleCommandL()`.

As the **SimpleMenu** example uses the menu in the AppUi, the `HandleCommandL()` method is overridden in the `SimpleMenuAppUi` class:

```
void CSimpleMenuAppUi::HandleCommandL(TInt aCommand)
    {
    switch (aCommand)
        {
        case ESimpleMenuCmdViaIR:
            iAppContainer->PlayNewGameVia(aCommand);
            break;
        case ESimpleMenuCmdViaBluetooth:
            iAppContainer->PlayNewGameVia(aCommand);
            break;
        case ESimpleMenuCmdViaSMS:
            iAppContainer->PlayNewGameVia(aCommand);
            break;
        case EAknSoftkeyBack:
        case EEikCmdExit:
            Exit();
            break;
        default:
            break;
        }
    }
```

This code shows a standard way of handling commands—a switch statement handles each command ID on a case-by-case basis. In the **SimpleMenu** example application, handling must be provided for the commands in the submenu and the exit command. The new game command does not involve extra handling code—the framework takes care of this, popping up the submenu when you select the **New Game** item.

You should handle the exit command by calling the `Exit()` method, which closes the application. You should handle the **Back** soft key in the same way. You may have noticed that the **Exit** command handled above (`EEikCmdExit`) is not the command associated with the **Exit** menu item (`EAknCmdExit`). There are two predefined commands for exiting an application: `EAknCmdExit` (in `avkon.hrh`), and `EEikCmdExit` (in `uikon.hrh`).

When defining exit menu items in the application resource file, you should use the `EAknCmdExit` version. For example:

```
MENU_ITEM
    {
    command = EAknCmdExit;
    txt = EXIT_TEXT;
    }
```

However, when handling the command in `ProcessCommandL()`, or `HandleCommandL()`, you should always handle `EEikCmdExit`. For example:

```
void CSimpleMenuAppUi::HandleCommandL(TInt aCommand)
    {
    switch (aCommand)
        {
        case EAknSoftkeyBack:
        case EEikCmdExit:
            {
            Exit();
            break;
            }
    ...
```

This is because, when `EAknCmdExit` activates, the application framework needs to close the current application and any related applications. The term "related applications" refers to situations where another application either embeds the current application or contains embedded applications. All closeable applications are required to handle the `EEikCmdExit` command by calling `CEikAppUi::Exit()`, which closes the application. To ensure that all related applications are closed, the framework sends an `EEikCmdExit` to each of the related applications, and expects them to handle the command by calling the `Exit()` method.

Note that the `EEikCmdExit` command is reserved for system use (such as Out-Of-Memory watchdogs), and so must call `Exit()` without querying the user. This is covered in more detail in the **Good Application Behavior** section of Chapter 4.

The **Back** soft key will always activate the `EAknSoftkeyBack` command. The behavior invoked by the **Back** soft key depends on the screen currently being displayed. Screens that are not at the top level should handle the back command by returning to the previous screen.

The top-level screen should handle the back command by exiting the application in the same way as the exit command does. You can see this in the `HandleCommandL()` code above, which handles both the `EAknSoftKeyBack` and `EEikCmdExit` commands in the same way.

> **NOTE** When you configure the soft keys such that the right key is **Exit** rather than **Back** (`R_AVKON_SOFTKEYS_EXIT`), the command that should be specified and handled is `EAknSoftkeyExit`.

Dynamic Menus

Menus are dynamic—the number of items displayed in them can vary throughout their lifetime. If the number of menu items in a menu is too great to fit on one screen, the framework automatically provides scroll adjusters so that you can scroll through them.

You can control the number of menu items in a menu by hiding and revealing them programmatically. This is useful in order to prevent users from accessing commands at inappropriate times (you would want to hide a "Check Spelling" menu option in a word processor application, for example, if the user had not yet opened a document).

NOTE It is not always desirable to hide menu items. This can be confusing to users, as they may be left wondering where a menu item has gone, or may find it difficult to recall where to locate particular menu items. If you feel that this may be an issue, consider leaving all menu items visible and displaying an informational message when the menu item is selected instead. See the **Notes** section in Chapter 6 for details of displaying informational messages.

All components that can have menus inherit from the `MEikMenuObserver` mixin class. This class defines the virtual method `DynInitMenuPaneL()`, which you can override in order to dynamically change the menu items. For example:

```
void CMyAppUi::DynInitMenuPaneL(
    TInt aResourceId, CEikMenuPane* aMenuPane)
    {
    if (aResourceId == R_MY_NEW_GAME_SUBMENU_MENU_PANE)
        {
        aMenuPane->SetItemDimmed(
            EMyCmdViaBluetooth, DimBluetooth());
        }
    }
```

The `DynInitMenuPaneL()` method has two parameters, which identify the currently active menu pane. You should check that the `aResourceId` argument specifies the menu pane you wish to alter. If it does, then use the `aMenuPane` argument to alter it. You can set the visibility of items using the menu pane's `SetDimmed()` method, passing it the required menu item's command ID, along with `ETrue` to make it invisible, or `EFalse` to make it visible again.

NOTE The SDK documentation for `CEikMenuPane::SetItemDimmed()` suggests that this method will dim (gray out) the menu item rather than hide it. As you have seen in this step, this is not the case in Series 60, but it is in other Symbian OS platforms, such as Series 80.

Context-Sensitive Menus

Context-sensitive menus are secondary menus that are typically launched by pressing the **Selection** key (as opposed to the primary **Options** menu accessed via the left soft key). These menus are sensitive to the currently displayed view, and furthermore can be sensitive to the internal state of that view. For example, a view that constitutes a list box may invoke different menus, depending on the currently selected list item.

This subsection looks at the example application **ContextMenu**, which shows you how to define a context-sensitive menu in a resource, and how to display a context-sensitive menu. The example displays a list of saved games, as shown in Figure 5–10.

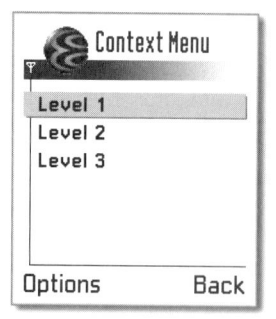

Figure 5–10 ContextMenu screen.

The **Options** menu has only one item, which allows you to exit the application. However, when you press the **Selection** key, a context-sensitive menu is displayed for the focused item, allowing you to play or delete the saved game, as shown in Figure 5–11.

Actual game-playing code has been omitted; however, an empty method called `PlaySelectedGame()` is provided to indicate where such code would reside. Lists are explained in detail in Chapter 7.

Figure 5–11 ContextMenu context-sensitive menu.

> **NOTE** **OK Options Menus** is another name used for context-sensitive menus invoked by the **Selection** key, so called because the **Selection** key is labeled **OK** on the Series 60 emulator.

Defining a Context-Sensitive Menu Using Resources

You define a context-sensitive menu in the same way you define any other application menu—that is, by defining a `MENU_BAR` and associated resources. This has already been covered in "Menu Basics." In the **ContextMenu** example, a menu is defined with two items: one to **Play**, and one to **Delete** a selected game.

Displaying a Context-Sensitive Menu

A context-sensitive menu displays when you press a particular key, typically the **Selection** key. Therefore, you should override the `OfferKeyEventL()` method to handle the key press and display the menu at that point. In the **ContextMenu** example, the container class overrides this method:

Chapter 5 Menus

263

```
TKeyResponse CContextMenuContainer::OfferKeyEventL(
    const TKeyEvent& aKeyEvent, TEventCode aType)
    {
    TBool selectKeyPressed = (aType == EEventKey) &&
(aKeyEvent.iCode == EKeyOK);

    TBool savedGameListNotEmpty = (iSavedGamesListBox) &&
(iSavedGamesListBox->Model()->NumberOfItems() > 0);

    if (selectKeyPressed && savedGameListNotEmpty)
        {
        CEikMenuBar* parentMenuBar = iEikonEnv->AppUiFactory()
->MenuBar();
        parentMenuBar->SetMenuTitleResourceId(
R_CONTEXTMENU_SAVED_GAMES_MENU_BAR);

        if (parentMenuBar)
            {
            parentMenuBar->TryDisplayMenuBarL();
            }
        parentMenuBar->SetMenuTitleResourceId(
R_CONTEXTMENU_MENU_BAR);
        return EKeyWasConsumed;
        }
    else
        return iSavedGamesListBox->OfferKeyEventL(aKeyEvent,
aType);
    }
```

SEE ALSO   The **Selection** key could also have been handled by registering the container as an observer of the list, and handling its events. This is covered in more detail in the Chapter 7.

In order to display the context-sensitive menu, you need to:

- **Check that the key that should activate the menu has been pressed.** You need to check the `TKeyEvent` and `TEventCode` arguments supplied. The **ContextMenu** example checks that they are `EEventKey` and `EKeyOk`, respectively. Event codes are defined in the `TKeyCode` enum in `E32Keys.h`. The event code `EKeyOK` is defined in `uikon.hrh` and is simply a `#define` of `EKeyDevice3`.
- **Get a handle on the menu bar.** If the key is appropriate, then you can get a handle to the menu bar of the container's parent. In this case, the menu bar for the application is accessed by using the `AppUiFactory()` method of `iEikonEnv` to get a handle on the AppUi, and then calling the `MenuBar()` method to get the menu bar. Other parents of containers have similar methods for accessing the menu bar; for example, `CAknView` has a `MenuBar()` method. Please refer to the SDK documentation for further details.

- **Change the menu bar's resource to use the context-sensitive menu.** Once you have a handle on the menu bar, you can change its resource using the `SetMenuTitleResourceId()` method. You pass in the `MENU_BAR` resource you defined for the context-sensitive menu in a previous step.

- **Display the context-sensitive menu** by calling `CEikMenuBar::TryDisplayMenuBarL()`.

- **Set the menu bar back to its original form.** When you changed the menu bar's resource using `SetMenuTitleResourceId()`, the change was permanent. This means that the left soft key will always display that menu until you reset it. The last thing you must do, therefore, is to set it back to the original menu bar by calling the method again, this time passing in the original menu bar value.

> **NOTE** The **ContextMenu** example code also verifies that the list of games is not empty to ensure that the context-sensitive menu displays only if an item on the list is focused. This code is necessary only for context-sensitive menus displayed for lists. See Chapter 7 for further details of list methods and usage.

Panes

This section looks at the standard panes that make up a Series 60 application window—namely, the status, main, and soft key panes, as shown in Figure 5–12. It will concentrate on the status pane, as this is of most interest to developers.

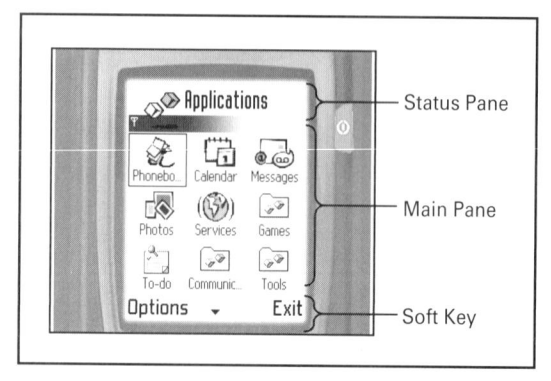

Figure 5–12 The standard panes.

Status Pane

The status pane occupies an area at the top of the screen and displays information regarding the status of the current application, and of the device itself. For example, it might show which tab of a tabbed view is currently selected or how much battery power is remaining.

Chapter 5 Panes

TIP The status pane can be hidden if full-screen mode is required—for example, in a game application. For a true full-screen application, the soft key pane would also need to be hidden.

The status pane itself is divided into the following subpanes:

- Title Pane
- Context Pane
- Navigation Pane
- Signal Pane
- Battery Pane
- Universal Indicator Pane (also known as the Small Indicator Pane)

These parts are all shown in Figure 5–13.

Figure 5–13 The status pane subpanes.

TIP The status pane has a different layout in idle mode. However, this does not affect the way you write code for populating the status pane.

You can change the contents of the following panes in your application:

- **Title Pane**—This shows the application name by default. You can change it to either display your own text or display a bitmap image representing the title instead.

- **Context Pane**—Initially, the context pane shows the default application icon specified in the AIF, but you can change this to display a different icon if you wish.

- **Navigation Pane**—The main uses of the navigation pane are to display information about the current state of the application, and to indicate which view is currently being displayed. It is also used to help the user navigate within the application. The navigation pane is blank by default, but it can be decorated with tabs, a label, an image, an indicator, or a custom control. The type of control you use depends on the requirements of your application. For example, you could use tabs to represent different views within an application, use a label to represent a view of a calendar month, use a volume control indicator to control recording levels in a video application, and so on. You can add horizontal scroll adjusters programmatically—for example, to allow horizontal scrolling through calendar months, or they may be added automatically by the framework—for example, to control the volume indicator or to switch between application views when there are too many tabs to display at once.

The remaining panes are available only to system applications.

Status Pane Basics

This subsection looks at the **StatusPane** application, which shows how to change the visibility of the status pane and handle a change in the status pane size.

The **StatusPane** example simply displays "Hello" on the screen. The **Options** menu has an item, **Toggle Visibility**, which you can use to switch the visibility of the status pane on or off, as shown in Figures 5–14 and 5–15.

Figure 5–14 StatusPane—visibility on.

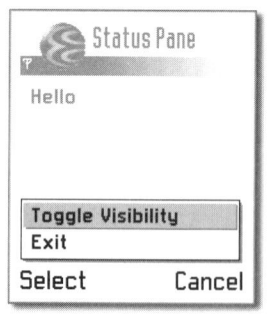

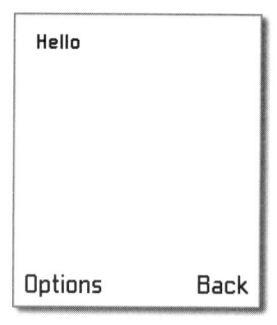

Figure 5–15 StatusPane—visibility off.

Changing the Visibility of the Status Pane

To change the visibility of the status pane, you need to get a handle on it and then call its SwitchLayoutL() method. In the **StatusPane** example, this occurs when handling the **Toggle Visibility** item on the **Options** menu, in CStatusPaneAppUi::HandleCommandL():

```
void CStatusPaneAppUi::HandleCommandL(TInt aCommand)
    {
    switch (aCommand)
        {
        case EStatusPaneToggleVisibility:
            {
            CEikStatusPane* statusPane = StatusPane();
```

Chapter 5 Panes

```
            if (statusPane->CurrentLayoutResId() !=
R_AVKON_STATUS_PANE_LAYOUT_EMPTY)

              {
              statusPane
->SwitchLayoutL(R_AVKON_STATUS_PANE_LAYOUT_EMPTY);
              }
            else
              {
              statusPane
>SwitchLayoutL(R_AVKON_STATUS_PANE_LAYOUT_USUAL);
              }
            break;
            }
          case EEikCmdExit:
            {
            Exit();
            break;
            }
          default:
            break;
          }
    }
```

In the **StatusPane** example, the **Toggle Visibility** menu item is defined using resources, with its command ID set to `EStatusPaneToggleVisibility`. On handling this command, you need to obtain a reference to the status pane by calling `CAknAppUi::StatusPane()`. This delivers a pointer to a `CEikStatusPane`, which can then be manipulated.

> **TIP** Although `CAknAppUi::StatusPane()` returns a pointer to a `CEikStatusPane` object, the `CAknAppUi` object maintains ownership. Therefore, it would be inappropriate to place the object on the cleanup stack.

To make the status pane invisible in the **StatusPane** example, two methods are used:

- `CEikStatusPane::CurrentLayoutResId()` — This returns the resource ID for the current layout.

- `CEikStatusPane::SwitchLayoutL()` — This changes the status pane layout to the resource ID specified.

The resource ID `R_AVKON_STATUS_PANE_LAYOUT_EMPTY` specifies the status pane as "hidden", and the resource ID `R_AVKON_STATUS_PANE_LAYOUT_USUAL` specifies the status pane as shown with the standard layout. Note that the `CEikStatusPane` methods `IsVisible()` and `MakeVisible()` should not be used to test and set the visibility of the pane, as they will also hide the in-call icon.

Handling a Change in the Status Pane Size

By following the previous step, you will have all the code needed to switch the status pane's visibility on and off. However, if this code was executed, when the status pane was made invisible you would still be able to see the status pane from the previous screen. This is because the container would not have been resized—that is, its `SizeChanged()` method would not have been called.

To ensure that this happens, you need to override the `CAknView::HandleStatusPaneSizeChange()` method in the view class. In the **StatusPane** example, this occurs in the `CStatusPaneView1` class:

```
void CStatusPaneView1::HandleStatusPaneSizeChange()
    {
    iContainer->SetRect(ClientRect());
    }
```

The framework calls this method whenever the size of the status pane changes (that is, by changing its visibility or layout). When the status pane has been made invisible, the area available to the container will have increased.

You need to reset the rectangle for the container with the new client rectangle, forcing the container to redraw itself. You achieve this by calling the container's `SetRect()` method, passing in the client rectangle, which you can obtain using `CAknView::ClientRect()`. This will ensure that the container makes use of the area previously occupied by the status pane.

In the **StatusPane** example, the status pane is changed dynamically. It would also have been possible to set the status pane statically in the resource file by defining a `STATUS_PANE_APP_MODEL` and referring to it in the `EIK_APP_INFO`. This would set the status pane on application launch to the fields set in this resource. See the SDK documentation for further details of the fields that can be set in a `STATUS_PANE_APP_MODEL` resource.

> **NOTE** The client rectangle is the area of the screen remaining after the status and soft key panes have been taken into account.

Title Pane

By default, the title pane will show the application's name; however, you can change this to display a title of your choosing, or even an image. If your choice of title is too long to fit on one line, it will be displayed on two lines instead. If it is too long for two lines, then it will be truncated. You can also set the title to an empty string, in which case no title will be displayed.

Title Pane Basics

This subsection looks at the **TitlePane** example, which shows how to change the title pane text and display an image in the title pane. The title pane in the example initially displays the text "Title Pane". The **Options** menu has an item, **Title Text,** which you use to change the text displayed in the title pane, as shown in Figure 5–16.

Figure 5–17 shows the application with new title pane text.

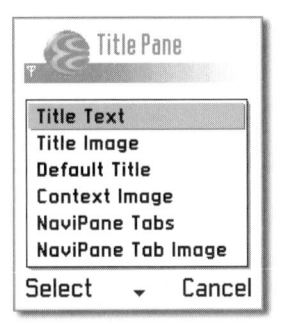

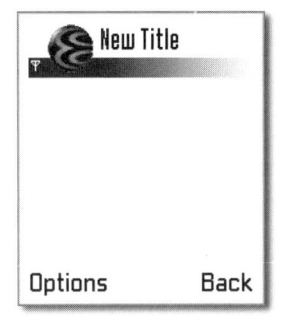

Figure 5–16 TitlePane menu.

Figure 5–17 Status pane with a new title.

Changing the Title Pane Text

The **TitlePane** example demonstrates how to change the text displayed in the title pane. It accomplishes this when handling the **Title Text** item on the **Options** menu, in `CTitlePaneAppUi::HandleCommandL()`:

```
void CTitlePaneAppUi::HandleCommandL(TInt aCommand)
    {
    switch (aCommand)
        {
        // change the title pane text
        case ETitlePaneSetTitleText:
            {
            TUid titlePaneUid;
            titlePaneUid.iUid = EEikStatusPaneUidTitle;

            CEikStatusPane* statusPane = StatusPane();

            CEikStatusPaneBase::TPaneCapabilities subPane = 
statusPane->PaneCapabilities(titlePaneUid);

            // if we can access the title pane
            if (subPane.IsPresent() && subPane.IsAppOwned())
                {
                CAknTitlePane* titlePane = (CAknTitlePane*)
statusPane->ControlL(titlePaneUid);

                // read the title text from the resource file
                HBufC* titleText = 
StringLoader::LoadLC(R_TITLE_TEXT);

                // set the title pane's text
                titlePane->SetTextL(*titleText);
```

```
            CleanupStack::PopAndDestroy(titleText);
            }

        break;
        }
    ...
```

As shown previously, the first required task is to get a pointer to the status pane. Afterward, the existence and readiness of the title pane must be verified. This can be accomplished by calling the pane's `PaneCapabilities()` method, and passing it the title pane's ID, which provides a `TPaneCapabilities` object. Then the `TPaneCapabilities::IsPresent()` method can be used to find out if the title pane is present, and `TPaneCapabilities::IsAppOwned()` can be used to find out if the title pane can be modified by the application. You can obtain information about any of the status pane's subpanes by calling the status pane's `PaneCapabilities()` method, passing it the appropriate ID.

Once the title panes is known to be present and ready for modification, it can then be accessed by calling the status pane's `ControlL()` method, passing it the title pane's ID. As with the status pane, although a pointer to the `CAknTitlePane` is returned, there is no transfer of ownership. Therefore, there is no need to use the cleanup stack.

Now that a pointer to the title pane has been obtained, it can be used to set the text it displays. Before doing that, however, it is first necessary to obtain a pointer to the desired text. This pointer is then passed into title pane's `SetText()` method. The `SetText()` method takes ownership of its argument, so you don't have to worry about deleting the pointer yourself. Alternatively, the `SetTextL()` method could have been used, which makes a copy of its argument instead of taking ownership. It is then up to the application to delete the text when appropriate.

> **NOTE** To reset the title pane back to its default value (that is, to the application's name), call the title pane's `SetTextToDefaultL()` method.

Displaying an Image in the Title Pane

The process for getting the title pane to display an image is similar to that used to modify the text it displays. This occurs when handling the **Title Image** item on the **Options** menu, in `CTitlePaneAppUi::HandleCommandL()`:

```
void CTitlePaneAppUi::HandleCommandL(TInt aCommand)
    {
    switch (aCommand)
        {
        ...
        // display a bitmap in the title pane
        case ETitlePaneSetTitleImage:
            {
```

Chapter 5 Panes

```
            TUid titlePaneUid;
            titlePaneUid.iUid = EEikStatusPaneUidTitle;

            CEikStatusPane* statusPane = StatusPane();

            CEikStatusPaneBase::TPaneCapabilities subPane =
statusPane->PaneCapabilities(titlePaneUid);

            // if we can access the title pane
            if (subPane.IsPresent() && subPane.IsAppOwned())
              {
              CAknTitlePane* titlePane =
(CAknTitlePane*)statusPane->ControlL(titlePaneUid);

              CFbsBitmap* bitmap = iEikonEnv
->CreateBitmapL(KTitleBitMapFile, EMbmTitlepaneTitle);

              // set the title pane's image
              titlePane->SetPicture(bitmap);
              }
            break;
            }
...
```

As before, the first step is to obtain a pointer to the status pane and check that the title pane is present and that it can be modified—only then is a pointer to the title pane itself acquired. The next step is to load a suitable image and tell the title pane to display it. This is accomplished by first creating a bitmap and then passing the bitmap to the title pane's `SetPicture()` method. The `SetPicture()` method will take ownership of the bitmap passed to it.

Changing the Title Pane Text Using Resources

As well as setting the title pane's contents dynamically, you can also set them in the resource file. You achieve this by specifying a TITLE_PANE resource, such as:

```
// Title Pane containing text
RESOURCE TITLE_PANE r_my_title_pane
    {
    txt = TITLE_TEXT;
    }
```

The `txt` field allows you to specify the text to be displayed in the title pane. In the **TitlePane** example, TITLE_TEXT is `#defined` in the localization file:

```
...
#define TITLE_TEXT "New Title"
...
```

Localization is covered in further detail in Chapter 4.

Chapter 5 Application UI Components

Your `TITLE_PANE` resource should then be referenced from the status pane resource, which in turn needs to be referenced from the `EIK_APP_INFO` resource:

```
RESOURCE EIK_APP_INFO
   {
   status_pane = r_my_status_pane;
   }

RESOURCE STATUS_PANE_APP_MODEL r_my_status_pane
   {
   panes =
      {
      SPANE_PANE
         {
         id = EEikStatusPaneUidTitle;
         type = EAknCtTitlePane;
         resource = r_my_title_pane;
         }
      };
   }
```

Displaying an Image in the Title Pane Using Resources

You can define an image to be displayed in the title pane by specifying the `bmpfile`, `bmpid` and `bmpmask` fields in the `TITLE_PANE` structure. The `bmpfile` field identifies a `.mbm` file from which to obtain the image. The `bmpid` field specifies the index of the bitmap to use from the multibitmap file, and the `bmpmask` field specifies the index of the mask to use, if needed:

```
//
// Title Pane containing image
//
RESOURCE TITLE_PANE r_my_title_pane
   {
   bmpid = EMbmMyBitmapIndex;
   bmpmaskid = EMbmMyBitmapMaskIndex;
   bmpfile =
"\\system\\apps\\myapplication\\myapplication.mbm";
   }
```

The SDK documentation contains further details of all the API methods available in the `CAknTitlePane` class. Chapter 11 contains more information on `.mbm` files.

Context Pane

The context pane, by default, will display the current application's icon, as specified in the AIF file. If this doesn't suit your needs, then the icon displayed can easily be changed. The process for doing this is similar to that used to change the title pane's contents.

Chapter 5 Panes

Context Pane Basics

This subsection continues looking at the **TitlePane** example, showing how to change the image displayed in the context pane. The context pane in the example initially displays the default image. The **Options** menu has an item, **Context Image**, which you use to change the image displayed in the context pane. Figure 5–18 shows the title pane with a new context image.

Figure 5–18 Context pane displaying a new image.

The code that handles the image change resides in `CTitlePaneAppui::HandleCommandL()`. This function is called when the user selects the **Context Image** item on the **Options** menu:

```
void CTitlePaneAppUi::HandleCommandL(TInt aCommand)
    {
    switch (aCommand)
        {
        ...
        // display a bitmap in the context pane
        case ETitlePaneSetContextImage:
            {
            TUid contextPaneUid;
            contextPaneUid.iUid = EEikStatusPaneUidContext;

            CEikStatusPane* statusPane = StatusPane();

            CEikStatusPaneBase::TPaneCapabilities subPane =
    statusPane->PaneCapabilities(contextPaneUid);

            // if we can access the context pane
            if (subPane.IsPresent() && subPane.IsAppOwned())
                {
                CEikStatusPane* statusPane = StatusPane();

                CAknContextPane* contextPane = (CAknContextPane*)
    statusPane->ControlL(contextPaneUid);

                CFbsBitmap* bitmap = iEikonEnv
    ->CreateBitmapL(KTitleBitMapFile, EMbmTitlepaneContext);

                // set the context pane's image
                contextPane->SetPicture(bitmap);
                }
            break;
            }
        ...
```

First, you obtain a pointer to the status pane and use this to check that the context pane exists and that applications can interact with it. Then you obtain a pointer to the context pane, and use this to change the icon displayed by calling the `SetPicture()` method, passing it a bitmap.

Displaying an Image in the Context Pane Using Resources

Alternatively, you can also set the context icon in the resource file by specifying a `CONTEXT_PANE` resource. This works in the same way as the `TITLE_PANE` resource. The `bmpfile` field identifies an `.mbm` file from which to obtain the image. The `bmpid` field specifies the index of the bitmap to use from the `.mbm` file, and the `bmpmask` field specifies the index of the mask to use, if needed:

```
RESOURCE CONTEXT_PANE r_my_context_pane
    {
    bmpid = EMbmMyBitmapIndex;
    bmpmaskid = EMbmMyBitmapMaskIndex;
    bmpfile =
"\\system\\apps\\myapplication\\myapplication.mbm";
    }
```

The SDK documentation contains further details of all the API methods available in the `CAknContextPane` class.

Navigation Pane

The navigation pane's main purpose is to display information to the user about the application's current view and state, as well as to help the user navigate around the application. The navigation pane is empty by default, but you can use it to display tabs, a label, an image, indicators, or custom controls. As the following code from the **TitlePane** example demonstrates, in order to decorate the navigation pane with controls, you need to use either a `CAknNavigationControlContainer` object in your code, or a `NAVI_DECORATOR` resource in your resource file.

Tabs

You can use the navigation pane to display tabs, as shown in Figure 5–19. This provides the user with information about the views available in your application, with the currently highlighted tab representing the current view. Each tab can display text, an image, or both.

Figure 5–19 Navigation pane displaying tabs.

This subsection continues looking at the **TitlePane** example, showing how to display tabs in the navigation pane and how to switch between tabs when the user presses the **Left** or **Right** direction keys. Initially, the navigation pane in the **TitlePane** example is empty. The **Options** menu has an item, **NaviPane Tabs**, which you use to display tabs in the navigation pane.

Displaying Tabs in the Navigation Pane

This occurs when handling the **NaviPane Tabs** item on the **Options** menu, in `CTitlePaneAppUi::HandleCommandL()`:

Chapter 5 Panes

```
void CTitlePaneAppUi::HandleCommandL(TInt aCommand)
   {
   switch (aCommand)
      {
      ...
      // display two tabs in the navigation pane
      case ETitlePaneSetNaviPane:
         {
         TUid naviPaneUid;
         naviPaneUid.iUid = EEikStatusPaneUidNavi;

         CEikStatusPane* statusPane = StatusPane();

         CEikStatusPaneBase::TPaneCapabilities subPane =
statusPane->PaneCapabilities(naviPaneUid);

         // if we can access the navigation pane
         if (subPane.IsPresent() && subPane.IsAppOwned())
            {
            CAknNavigationControlContainer* naviPane =
CAknNavigationControlContainer*) statusPane
->ControlL(naviPaneUid);

            delete iNaviDecorator;
            iNaviDecorator = NULL;

            // ownership is transferred to us here
            iNaviDecorator = naviPane->CreateTabGroupL();

            // ownership is not transferred here
            CAknTabGroup* tabGroup = (CAknTabGroup*)
iNaviDecorator->DecoratedControl();

            // Display two tabs of normal length on the
navigation pane at a time
            tabGroup->SetTabFixedWidthL(KTabWidthWithTwoTabs);

            TInt tabId = 0;

            // load the text to be displayed in the tabs
            HBufC* tab1Text =
StringLoader::LoadLC(R_TAB1_TEXT);
            tabGroup->AddTabL(tabId++, *tab1Text);
            CleanupStack::PopAndDestroy(tab1Text);

            HBufC* tab2Text =
StringLoader::LoadLC(R_TAB2_TEXT);
            tabGroup->AddTabL(tabId++, *tab2Text);
            CleanupStack::PopAndDestroy(tab2Text);

            // highlight the first tab
            tabGroup->SetActiveTabByIndex(0);
```

```
            naviPane->PushL(*iNaviDecorator);
            }
        break;
        }
...
```

First a pointer to the navigation pane is obtained by calling the status pane's `ControlL()` method, passing the navigation pane's ID as an argument. After checking that the navigation pane exists and that applications can interact with it, the navigation pane's `CreateTabGroupL()` method is then used to create a new navigation decorator that contains a tab group. You can then use this tab group to add tabs to the navigation control.

In the **TitlePane** example, two tabs are added to the tab group, and the first tab is set to be active using the `SetActiveTabByIndex()` method. This highlights the tab with the specified index, which in this case is the first one. Finally, the tab group is pushed onto the navigation pane's object stack. You must do this for the tab group to actually appear on the navigation pane.

It is also possible to display a bitmap in a tab. To achieve this, use the appropriate overload of the `AddTabL()` method:

```
...
// load the bitmap to be displayed in the tabs
CFbsBitmap* bitmap = iEikonEnv
->CreateBitmapL(KTitleBitMapFile, EMbmTitlepaneTab);

// display a bitmap in the tab
tabGroup->AddTabL(tabId++, bitmap);
...
```

Here, the `bitmap` variable is a pointer to a bitmap of type `CFbsBitmap*`.

You can control the number and size of the tabs displayed on the navigation pane by using the `CAknTabGroup::SetTabFixedWidthL()` method. This allows you to have anywhere between one and four tabs displayed on the navigation pane, and these can be either long or short tabs. The type of tabs you use will depend on the information you want to display in them. Long tabs have more space, making it easier to display text. When using long tabs, however, the currently highlighted tab will obscure most of the area available for the remaining tabs. This means that the user will not be able to see the text or images contained in the other tabs. Short tabs, on the other hand, have a much more limited amount of space available, but they will not obscure the other tabs, giving the user a better view of the available options.

Although you are limited to displaying a maximum of four tabs on the navigation pane at any point in time, you can actually have more tabs than this. Scroll indicators will automatically be displayed if there are more tabs available than are displayed on the navigation pane. The user will then be able to scroll to any of the available tabs. It is generally recommended, though, that you keep the total number of tabs low—six tabs is the recommended maximum.

Chapter 5 Panes

NOTE In Series 60 1.x, tabs always appear in a left-to-right order. The leftmost tab is always the first tab, and the rightmost is always the last tab. Series 60 2.x provides support for bidirectional text (in other words, support for languages that read right-to-left), which means that the order of tabs may be reversed, depending on the locale.

When the tabs are reversed (mirrored), they are traversed from right-to-left. In order to support bidirectional text, you should now pass key events on to the tab group, and then let the tab group notify your application if the view changes, rather than handle tab changes yourself. The old key-handling logic will still work in Series 60 2.x if used, but tabs won't be mirrored correctly if a mirrored language is used.

It is up to you to implement the code allowing the user to switch between tabs. For example:

```
TKeyResponse CTitlePaneAppUi::HandleKeyEventL(
    const TKeyEvent& aKeyEvent, TEventCode aType)
    {
    if (iTabDecorator == NULL)
        {
        return EKeyWasNotConsumed;
        }

    CAknTabGroup* tabGroup = (CAknTabGroup*)iTabDecorator
->DecoratedControl();

    if (tabGroup == NULL)
        {
        return EKeyWasNotConsumed;
        }

    return tabGroup->OfferKeyEventL(aKeyEvent, aType);
    }
```

This function checks whether the navigation decorator exists, then obtains the tab group from it. If either of these operations fails, it means that a tab group is not available, and that the control event cannot be processed by this function. In this case, the function then returns `EKeyWasNotConsumed`. Otherwise, the key press is passed on to the tab group, and it will handle switching tabs.

NOTE The application UI's `HandleKeyEventL()` method will be called if none of the controls on the Control Stack consume the event. See the `CAknTabGroup` help page in the SDK documentation for a full list of methods available.

Chapter 5 Application UI Components

The **NavigationPane** example illustrates how to display tabs, a label, or an image in the navigation pane using resources, as shown in Figure 5–20.

The navigation pane in the **NavigationPane** example initially displays two tabs that are defined using resources. The **Options** menu also has a **Set NaviPane Tabs** item, which you can use to display the tabs in the navigation pane.

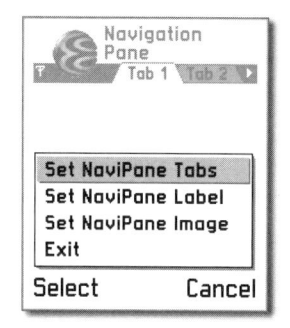

Figure 5–20 Navigation pane display options in NavigationPane.

Displaying Tabs in the Navigation Pane Using Resources

The **NavigationPane** example shows you how to display tabs in the navigation pane using a resource file, as shown in the following code snippets taken from `NavigationPane.rss`. First, the `status_pane` field is defined in the application resource:

```
RESOURCE EIK_APP_INFO
    {
    status_pane = r_navigationpane_status_pane;
    }
```

Then the status pane itself is defined:

```
RESOURCE STATUS_PANE_APP_MODEL r_navigationpane_status_pane
    {
    panes =
        {
        SPANE_PANE
            {
            id = EEikStatusPaneUidNavi;
            type = EAknCtNaviPane;
            resource = r_navigationpane_navi_tabgroup;
            }
        };
    }
```

Finally the navigation pane resource is defined:

```
RESOURCE NAVI_DECORATOR r_navigationpane_navi_tabgroup
    {
    type = ENaviDecoratorControlTabGroup;
    control = TAB_GROUP
        {
        // display two tabs
        tab_width = EAknTabWidthWithTwoTabs;
        active = 0;
        tabs = {
            TAB
```

```
            {
            id = ENavigationPaneTab1;
            txt = TAB1_TEXT;
            },
        TAB
            {
            id = ENavigationPaneTab2;
            txt = TAB2_TEXT;
            }
        };
    };
}
```

You can define tabs that will appear upon application startup in the resource file with a `NAVI_DECORATOR` resource. The `tab_width` field allows you to specify the number and width of the tabs that will appear. The `active` field specifies the index of the tab that will be initially highlighted. The `tabs` field then specifies all the tabs, their IDs, and contents. You can specify text content using the `txt` field. If you want to display images as well as, or instead of, text, then you need to use the `bmpfile`, `bmpid` and `bmpmask` fields. These work in the same way as in the `TITLE_PANE` resource described previously. The `bmpfile` field identifies an `.mbm` file from which to obtain the image. The `bmpid` field specifies the index of the bitmap to use from the `.mbm` file, and the `bmpmask` field specifies the index of the mask to use, if needed.

As well as dynamically creating the tabs in your application and defining the tabs in the resource file, you can also create tabs at runtime using a resource specified in the resource file. To achieve this, you first need to specify a `TAB_GROUP` resource in your resource file:

```
RESOURCE TAB_GROUP r_navigationpane_tabgroup
    {
    tab_width = EAknTabWidthWithTwoTabs;   // display two tabs
    active = 0;

    tabs =
        {
        TAB
            {
            id = ENavigationPaneTab1;
            txt = TAB1_TEXT;
            },
        TAB
            {
            id = ENavigationPaneTab2;
            txt = TAB2_TEXT;
            }
        };
    }
```

Chapter 5 Application UI Components

You can then read this resource from your code by using a `TResourceReader` object:

```
void CNavigationPaneAppUi::HandleCommandL(TInt aCommand)
    {
    switch (aCommand)
        {
        // display a tab group in the navigation pane from a resource
        case ENavigationPaneSetNaviPane:
            {
            TUid naviPaneUid;
            naviPaneUid.iUid = EEikStatusPaneUidNavi;

            CEikStatusPane* statusPane = StatusPane();

            CEikStatusPaneBase::TPaneCapabilities subPane =
            statusPane->PaneCapabilities(naviPaneUid);

            // if we can access the navigation pane
            if (subPane.IsPresent() && subPane.IsAppOwned())
                {
                CAknNavigationControlContainer* naviPane =
                (CAknNavigationControlContainer*) statusPane
                ->ControlL(naviPaneUid);

                // read the tab group resource
                TResourceReader reader;
                iCoeEnv->CreateResourceReaderLC(reader,
                R_NAVIGATIONPANE_TABGROUP);

                if (iNaviDecorator)
                  {
                  delete iNaviDecorator;
                  iNaviDecorator = NULL;
                  }

                // set the navigation pane tab group
                iNaviDecorator = naviPane
                ->CreateTabGroupL(reader);
                CleanupStack::PopAndDestroy(); // pushed by
                CreateResourceReaderLC

                naviPane->PushL(*iNaviDecorator);
                }
            break;
            }
    ...
```

Here, the `TAB_GROUP` resource specified previously is used to create the tabs that will appear on the navigation pane. This avoids having to create each tab dynamically.

Chapter 5 Panes

After creating a `TResourceReader` object, the control environment's `CreateResourceReaderLC()` method is called. This method will read the specified resource from the resource file, set its resource reader, and then push the reader onto the cleanup stack. This reader is then used to create a tab group by calling the navigation pane's `CreateTabGroupL()` method, which creates a new navigation decorator containing the tab group. Then the reader that was pushed onto the cleanup stack is removed and destroyed, before pushing the newly created navigation decorator onto the navigation pane's object stack.

Labels

You can use the navigation pane to display a label. This is usually employed as an alternative to using tabs and views. One example of this is a calendar application, where you can display dates on a month-by-month basis, and have a label telling the user which month is currently displayed. The label can contain as much text as you wish, but if the label is too long to fit in the navigation pane, then it will be truncated.

Figure 5-21 Navigation pane displaying a label.

This subsection continues looking at the **NavigationPane** example, showing how to display a label in the navigation pane using resources, as shown in Figure 5-21.

The **Options** menu has an item, **Set NaviPane Label**, which you use to display a label in the navigation pane.

Displaying a Label in the Navigation Pane Using Resources

You can specify a label for the navigation pane in the resource file using the `NAVI_LABEL` resource. This resource has a single field, `txt`, which specifies the text to appear in the navigation pane, as shown in `NavigationPane.rss`:

```
RESOURCE NAVI_LABEL r_navigationpane_navi_text
    {
    txt = LABEL_TEXT;
    }
```

This resource can then be read in your application and used to set the navigation pane's label:

```
void CNavigationPaneAppUi::HandleCommandL(TInt aCommand)
    {
    switch (aCommand)
        {
        ...
        // display a label in the navigation pane from a resource
        case ENavigationPaneSetNaviPaneLabel:
            {
            TUid naviPaneUid;
```

Chapter 5 Application UI Components

```
            naviPaneUid.iUid = EEikStatusPaneUidNavi;
            CEikStatusPane* statusPane = StatusPane();

            CEikStatusPaneBase::TPaneCapabilities subPane =
   statusPane->PaneCapabilities(naviPaneUid);

            // if we can access the navigation pane
            if (subPane.IsPresent() && subPane.IsAppOwned())
              {
              CAknNavigationControlContainer* naviPane =
   (CAknNavigationControlContainer *) statusPane
   ->ControlL(naviPaneUid);

              // read the navigation pane text resource
              TResourceReader reader;
              iCoeEnv->CreateResourceReaderLC(reader,
   R_NAVIGATIONPANE_NAVI_TEXT);

              if (iNaviDecorator)
                {
                delete iNaviDecorator;
                iNaviDecorator = NULL;
                }

              // set the navigation pane label
              iNaviDecorator = naviPane
   ->CreateNavigationLabelL(reader);
              CleanupStack::PopAndDestroy(); // pushed by
   CreateResourceReaderLC

              naviPane->PushL(*iNaviDecorator);
              }
            break;
            }
         ...
```

This code is very similar to that used to create a tab group from a resource. The only differences are that when the `CreateResourceReaderLC()` method is called, the name of the `NAVI_LABEL` resource is used, and instead of calling the navigation pane's `CreateTabGroupL()` method, the `CreateNavigationLabelL()` method is called instead.

> **TIP** If, instead of creating the navigation label from a resource file, you need to create the text for the label dynamically, you can use the overloaded version of `CreateNavigationLabelL()` that takes a `const TDesC&` argument.

Images

You can also use the navigation pane to display an image.

Chapter 5 Panes

Figure 5-22 Navigation pane displaying an image.

This subsection continues looking at the **NavigationPane** example, showing how to display an image in the navigation pane using resources, as shown in Figure 5-22.

The **Options** menu has an item, **Set NaviPane Image**, which you use to display an image in the navigation pane.

Displaying an Image in the Navigation Pane Using Resources

You can specify an image for the navigation pane in the resource file using the `NAVI_IMAGE` resource. This resource has three fields: `bmpfile` specifying a `.mbm` file to use, `bmpid` specifying the index of the bitmap to use within the `.mbm` file, and `bmpmask` optionally specifying the index of the mask to use within the `.mbm` file. The following code snippet is taken from `NavigationPane.rss`:

```
...
RESOURCE NAVI_IMAGE r_navigationpane_navi_image
    {
    bmpfile = "z:\\system\data\avkon.mbm";
    bmpid = EMbmAvkonQgn_stat_keyguard;
    }
...
```

This code uses one of the standard images provided by Series 60. The `bmpfile` field specifies the `.mbm` file's location, and the specific bitmap to use from the file is specified by the `bmpid` field. Note that the `bmpfile` here is hard-coded to the `z:` drive, as it is a system file—if you were using your own bitmap file, then the drive letter would be omitted, and the file would be loaded from the drive that the application is installed on. (Ideally you would use `BaflUtils::GetSystemDrive()` rather than hardcoding `z:`, but that would require a custom resource reader and unnecessarily complicate the example!)

NOTE The full list of images provided by Series 60 can be found in the system header file `\epoc32\include\avkon.mbg`, or can be viewed using the **mbmviewer** described Chapter 2.

The process for dynamically displaying an image in the navigation pane is very similar to that used for displaying a label—the only difference is that you call the `CreateNavigationImageL()` method, passing it the bitmap to use instead of the `CreateNavigationLabelL()` method:

```
void CNavigationPaneAppUi::HandleCommandL(TInt aCommand)
    {
    switch (aCommand)
        {
        ...
        // display an image in the navigation pane from a resource
```

Chapter 5 Application UI Components

```
         case ENavigationPaneSetNaviPaneImage:
            {
            TUid naviPaneUid;
            naviPaneUid.iUid = EEikStatusPaneUidNavi;

            CEikStatusPane* statusPane = StatusPane();

            CEikStatusPaneBase::TPaneCapabilities subPane =
  statusPane->PaneCapabilities(naviPaneUid);

            // if we can access the navigation pane
            if (subPane.IsPresent() && subPane.IsAppOwned())
               {
               CAknNavigationControlContainer* naviPane =
  (CAknNavigationControlContainer*) statusPane
  ->ControlL(naviPaneUid);

               // read the navigation pane image resource
               TResourceReader reader;

               iCoeEnv->CreateResourceReaderLC(reader,
  R_NAVIGATIONPANE_NAVI_IMAGE);

               if (iNaviDecorator)
                  {
                  delete iNaviDecorator;
                  iNaviDecorator = NULL;
                  }

               // set the navigation pane image
               iNaviDecorator = naviPane
  ->CreateNavigationImageL(reader);
               CleanupStack::PopAndDestroy(); // pushed by
  CreateResourceReaderLC

               naviPane->PushL(*iNaviDecorator);
               }
            break;
            }
         ...
```

Indicators

The navigation pane can also contain indicators. These are used to inform the user that they can scroll left and right in tabbed windows, when there are more tabs than will fit on the screen; or in a volume control to change the current volume.

Figure 5-23 Navigation pane displaying a volume indicator.

This subsection continues looking at the **TitlePane** example, showing you how to display a volume indicator in the navigation pane, as shown in Figure 5-23.

Initially, the navigation pane in the **TitlePane** example is empty. The **Options** menu has an item, **NaviPane Indicator**, which you use to a volume indicator in the navigation pane. The following code demonstrates the addition of the volume control:

```
void CTitlePaneAppUi::HandleCommandL(TInt aCommand)
    {
    switch (aCommand)
        {
        ...
        // display a volume control in the navigation pane
        case ETitlePaneSetNaviPaneIndicator:
            {
            TUid naviPaneUid;
            naviPaneUid.iUid = EEikStatusPaneUidNavi;

            CEikStatusPane* statusPane = StatusPane();

            CEikStatusPaneBase::TPaneCapabilities subPane =
    statusPane->PaneCapabilities(naviPaneUid);

            // if we can access the navigation pane
            if (subPane.IsPresent() && subPane.IsAppOwned())
                {
                CAknNavigationControlContainer* naviPane =
    (CAknNavigationControlContainer*)statusPane
    ->ControlL(naviPaneUid);

                delete iNaviDecorator;
                iNaviDecorator = NULL;

                // create a volume indicator on the navigation
    pane
                iNaviDecorator = naviPane
    ->CreateVolumeIndicatorL(R_AVKON_NAVI_PANE_VOLUME_INDICATOR);

                naviPane->PushL(*iNaviDecorator);
                }

            CAknVolumeControl* volumeControl =
    (CAknVolumeControl*) iNaviDecorator->DecoratedControl();

            // Get the current volume level
            TInt curVolume = volumeControl->Value();

            // Increase the volume level by one
            volumeControl->SetValue(++curVolume);
            break;
            }
        ...
```

Again, this code is very similar to previous examples. After obtaining a pointer to the status pane, the `CreateVolumeIndicatorL()` method is called, in order to create the volume control. This method takes an argument that specifies what type of volume indicator you want to create. In this case, the `R_AVKON_NAVI_PANE_VOLUME_INDICATOR` type is used. You can also use the `R_AVKON_NAVI_PANE_RECORDER_VOLUME_INDICATOR`, which will display a microphone icon next to the volume bar, and `R_AVKON_NAVI_PANE_EARPIECE_VOLUME_INDICATOR`, which will display a headphones icon next to the volume bar. Then the newly created decorator is pushed onto the navigation pane's object stack.

Finally, the current volume level is increased by one. This is accomplished by calling the volume control's `DecoratedControl()` method and casting the result to a `CAknVolumeControl*` to obtain a pointer to the control itself. Then the volume control's `Value()` method is called to get the current volume. This will return a `TInt` representing the current setting. To increase the volume by one, the `SetValue()` method is called, passing it the current volume increased by one.

Main Pane

The main pane is the main area of the screen where an application is displayed. This area can be freely used by applications to display their data, and the layout of this area is up to the application designer. The Series 60 UI Style Guide in the SDK documentation provides guidelines on the layout of the main pane.

Soft Key Pane

The soft key pane, also known as the control pane, displays the labels that are associated with the soft keys. These tell the user what pressing a given soft key will do. The soft key pane also displays the scrolling indicator arrows used when there is a control displayed that can be scrolled. The **Menus** section of this chapter contains more information on how to define the soft keys.

Summary

This chapter has provided an overview of many different aspects of UI development. It started out by explaining how controls work. It showed how UI components fit in with the larger application framework, and how controls work in conjunction with system windows.

You have learned how to create new simple and compound controls by deriving from `CCoeControl`, and also what functions must be overridden in order to provide the required standard behavior. You have learned how controls should respond to system events, and also how your Series 60 2.x applications can be made "skin-aware".

Chapter 5 Summary

The section on resource files showed you how they could be used in order to facilitate UI development, and subsequent sections highlighted the alternative approaches of creating UI components both dynamically in C++ code and statically in resource files.

In addition, other UI elements such as menus and the various Series 60 panes were covered. These topics, and the many example applications provided, should help to round out your understanding of the different ways that you can provide UI-based information to the user, and also allow user interaction, in your own Series 60 applications.

The next three chapters cover some specific common UI components (dialogs, lists, and editors) in detail. The information gained in this chapter will have provided you with the necessary background to get the maximum benefit out of these upcoming topics.

chapter 6

Dialogs

Dialogs are used extensively by system and application user interfaces for simple notification through to highly sophisticated data presentation and capture

Dialogs provide a wide variety of ways to interact with a user. They can be used to notify, obtain a response, present fixed information, or to allow the user to enter data. Series 60 provides a comprehensive set of dialog classes and base classes that support the typical dialog functionality required by most applications. You can use these classes to create your own custom dialogs and to develop forms, notes, queries, and list dialogs.

This chapter looks at how you can create dialogs in your applications. It covers:

- **Standard Dialogs**—A simple dialog can be constructed by defining its layout and then writing a dialog class to handle the data. Data from the dialog can be validated and saved. More features can be added to a dialog, such as defining a menu for it or adding a custom control. Multipage dialogs can also be created.

- **Forms**—A form can be used if you have a collection of related data that you want the user to edit.

- **Notes**—Notes provide a convenient way to convey information to the user. Wrapped Notes offer a very simple way of communicating with the user, providing a standard format for common types of notes, such as a confirmation note or an information note. Custom notes can also be constructed.

- **Queries**—Queries are a specialized type of dialog to be used when you wish to ask the user a question. In the simplest case, this is a confirmation query asking the user to say yes or no to a question. However, they can be more complex; for example, list queries can be constructed for the user to select items from.

- **List Dialogs**—Two types of list dialogs are available in Series 60: selection lists—which allow one item to be selected, and markable lists—which allow multiple selections.

Each section in this chapter is supported by code snippets from example applications. Details of how to build and run the applications can be found in the Preface. All the examples in this chapter are based around the idea of game playing. For example, one of the notes displays how many games have been loaded; a dialog asks for a player name. However, the applications are intended only to demonstrate the ideas discussed in this chapter—no game-playing functionality has been implemented. Each application will be introduced at the appropriate point in the text.

Common Dialog Characteristics

All dialogs share some basic properties. They are window-owning controls; virtually all dialog classes are ultimately derived from `CCoeControl`, as described in Chapter 5. A dialog framework manages many aspects of their behavior, including layout, drawing and the management of the user interaction with their component controls. Typically, most dialogs of any complexity are fully defined in a resource file and, after dynamic instantiation; their construction is completed by having the dialog framework load the definition from the resource file. The layout and positioning of all the elements of the dialog is usually automatic—the developer can influence the process, as described later. Full dynamic construction is possible for very simple dialogs.

Dialogs can be **modal** or **modeless**, as well as **waiting** or **nonwaiting**. They have a number of lines arranged vertically, each containing one or more controls.

A modal dialog prevents you from interacting with other parts of the application's UI until you dismiss it, while a modeless dialog does allow you to interact with other parts of the application's UI while it is active.

A nonwaiting dialog allows the application to continue processing in the background, whereas a waiting dialog prevents an application from doing any further processing until the dialog is dismissed.

Series 60 dialogs are modal and nonwaiting by default. Modeless dialogs are rarely used in Series 60, as any dialog shown will generally have input focus until it is dismissed. Dialogs used as the main view in a Dialog-Based Application Architecture are the main exception—see Chapter 4 for further details.

Standard Dialogs

The base class for most Series 60 dialogs is `CAknDialog`. Series 60 does provide specialized dialog classes that can be used to construct forms, queries, and the like, but you can also create your own dialog classes based on `CAknDialog`. As you progress through this chapter, you will learn more about the specialized dialogs. It is important to note that, where appropriate, these dialog classes should be implemented in preference to writing your own dialog from scratch. There are, however, times when you will want to use a basic `CAknDialog` dialog—for example, as your main application window for a dialog-based application—so the mechanics of how to do so will be explained next.

In addition to creating a simple dialog based on `CAknDialog`, this section will also include information on how to create multipage dialogs, how to associate menus with your dialogs and now to insert a custom control into a dialog. The example application projects used in this section are **SimpleDlg** and **CustomCtrlDlg**.

Creating a Simple Dialog

The application **SimpleDlg** was introduced in Chapter 4 to explain the Dialog Architecture. It is used here to demonstrate the steps needed to create a standard dialog.

The example application is very simple—its main application window is a dialog with a label showing the name of a game, as shown in Figure 6–1. It has an **Options** menu, with **New Game** and **Exit** as the menu choices. Selecting **New Game** from the menu causes a small dialog to appear which requests a player name, as shown in Figure 6–2. Selecting **Ok,** to dismiss the player name dialog, would theoretically start the game, but, as this is just a trivial example application, the associated game has not been written!

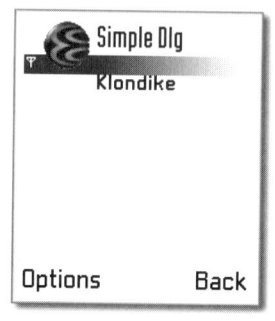

Figure 6–1 SimpleDlg main screen.

Figure 6–2 SimpleDlg player name dialog.

Defining the Resource

The key steps to create a simple dialog are to define a `DIALOG` resource and then create a dialog class derived from the base class `CAknDialog` that can execute the dialog. The resource specifies the layout of the dialog—it is where you define the number of lines your dialog should have, the controls to be used, whether the dialog should be modal or not, the soft keys to be used, and the like. In **SimpleDlg**, the resource for the player name dialog is `r_simpledlg_dialog`, which you will find in the resource file `SimpleDlg.rss`:

```
RESOURCE DIALOG r_simpledlg_player_name_dialog
    {
    flags = EEikDialogFlagNoDrag | EEikDialogFlagCbaButtons |
EEikDialogFlagWait;
    buttons = R_AVKON_SOFTKEYS_OK_CANCEL;

    items =
      {
      DLG_LINE
        {
        id = EConfirmationNoteDlgCIdPlayerName;
```

Chapter 6 Standard Dialogs

```
                type = EEikCtLabel;
                control = LABEL
                    {
                    };
                },
            DLG_LINE
                {
                id = EConfirmationNoteDlgCIdPlayerNameEditor;
                type = EEikCtEdwin;
                control = EDWIN
                    {
                    maxlength = KMaxPlayerNameLength;
                    };
                }
            };
        }
```

There are three components defined for this particular dialog (you will see other options later in the chapter): `flags`, `buttons`, and `items`. The `flags` describe the properties of the dialog. In the **SimpleDlg** example they indicate that the dialog:

- `EEikDialogFlagNoDrag`—cannot be dragged
- `EEikDialogFlagCbaButtons`—uses soft keys
- `EEikDialogFlagWait`—is waiting

TIP It is good practice to define a standard dialog as waiting. You should only make it nonwaiting if it is justified—for example, in a main application window.

The `buttons` component specifies the soft keys to be used. As you can see from Figure 6–2, R_AVKON_SOFTKEYS_OK_CANCEL, labels the left soft key as **Ok** and the right soft key as **Cancel**.

The actual detail of what is contained in your dialog is defined by the `items` component. For each control you want in your dialog—each label, edit box, and so on—you define a `DLG_LINE`. In the **SimpleDlg** example, a label and an `EDWIN` control are displayed, so two `DLG_LINE`s are necessary. As a minimum, you should specify the following fields in each `DLG_LINE`:

- `id`—Use this to reference the dialog line within your application. You must enumerate in the application `.hrh` file.
- `type`—The type of control it contains (defined in an `.hrh` file, for example, `avkon.hrh`, or `uikon.hrh`).
- `control`—The control used in this dialog line.

WARNING Unlike other Symbian OS platforms, Series 60 insists that each dialog line has a unique ID. If you do not specify one, or specify duplicates, the application will panic when you attempt to execute the dialog.

Optionally an `itemflags` field can be specified in the `DLG_LINE` resource. This field determines the behavior of the line. For example, a pop-up field text control could have `itemflags`, indicating that it should open a pop-up window when the **Selection** key is pressed—for example, `EEikDlgItemTakesEnterKey | EEikDlgItemOfferAllHotKeys`.

TIP You can find values for `itemflags` in `uikon.hrh`. Apart from the flags already seen, probably the most useful is the `EEikDlgItemSeparatorBefore` flag, which inserts a horizontal "separator" line before the dialog line.

Writing a Dialog Class

You need to write a class that can construct and execute the dialog, initialize the data in the controls, handle the data received from them, and determine how the dialog is dismissed. Other functionality can be added to the dialog class, such as methods for validating and saving the dialog data. It is also possible within this class to initialize the dialog with values at runtime, rather than statically defining them in resource. This allows you to use the same resource for different uses of the dialog. For example, in **SimpleDlg**, the text of the dialog's label is set dynamically.

It is common to provide a static method to execute a dialog, which wraps up the first-phase constructor and `ExecuteLD()`—the function that loads, displays, and destroys the dialog. By convention, this method is called `RunDlgLD()`, and it passes on the `TBool` that is returned from `ExecuteLD()`.

The dialog class illustrated by the **SimpleDlg** example is called `CSimpleDlgPlayerNameDialog`, and it inherits from `CAknDialog`. It provides a static `RunDlgLD()` method as a means of constructing the dialog:

```
TBool CSimpleDlgPlayerNameDialog::RunDlgLD(TDes& aPlayerName)
    {
    CSimpleDlgPlayerNameDialog* playerNameDialog = new (ELeave) CSimpleDlgPlayerNameDialog(aPlayerName);

    return playerNameDialog->ExecuteLD(
R_SIMPLEDLG_PLAYER_NAME_DIALOG);
    }
```

Chapter 6 Standard Dialogs

As the dialog is modal and waiting, you do not need to make it member data, as you will not be responsible for its deletion or for ensuring that it receives key events. It can therefore be declared locally, and it will delete itself as the last step of `ExecuteLD()`.

It may seem strange that the dialog is not placed onto the Cleanup Stack following its construction, especially given that it is held by a local stack-based pointer to an object on the heap and that `ExecuteLD()` is a leaving function. However, `ExecuteLD()` takes ownership of the dialog, wrapping up two further method calls: `PrepareLC()` and `RunLD()`. `PrepareLC()` puts a pointer to the dialog onto the Cleanup Stack and then completes the construction of the dialog. `RunLD()` displays the dialog and pops it off the Cleanup Stack. However, if you need to call any (potentially) leaving code before calling `ExecuteLD()`, you should put the dialog onto the Cleanup Stack and pop it off before the call to `ExecuteLD()`.

TIP If your dialog is nonwaiting, then it will return immediately from `ExecuteLD()` without being deleted. Nonwaiting dialogs do not delete themselves at all, so it is quite safe to delete them in the destructor. Note that many Series 60 notes and dialogs use self-pointers to delete themselves—see the subsection on **Progress Notes** later in this chapter.

Saving and Validating Dialog Data

For a dialog to be able to update application data, it will need a reference to the data. In the **SimpleDlg** application, `RunDlgLD()` takes a reference to a descriptor, which the dialog then modifies using the value in its editor control. The reference is stored as member data by the dialog during construction. A common place to validate and update the data is in the `OkToExitL()` method. This function will be called by the framework when any soft key other than **Cancel** is pressed. (You can force the framework to call `OkToExitL()` when **Cancel** has been pressed by setting the dialog's `flags` to include `EEikDialogFlagNotifyEsc`.) `OkToExitL()` must return `ETrue` if the dialog can be allowed to exit, and return `EFalse` if not.

The implementation of `CSimpleDlgPlayerNameDialog::OkToExitL()` is shown:

```
TBool CSimpleDlgPlayerNameDialog::OkToExitL(TInt aButtonId)
    {
    if (aButtonId == EAknSoftkeyOk)
        {
        CEikEdwin* editor = static_cast<CEikEdwin*>(
ControlOrNull(EConfirmationNoteDlgCIdPlayerNameEditor));
```

```
            if (editor)
                {
                editor->GetText(iPlayerName);
                }
            }
        return ETrue;
        }
```

Remember that each line in a dialog has an ID and a control. A handle on a particular control can be obtained using `CEikDialog::ControlOrNull()`, passing in the ID for the dialog line. In the **SimpleDlg** example, `EConfirmationNoteDlgCIdPlayerNameEditor` is passed in to the function to get the editor control. If the ID is valid, it returns a pointer to a `CCoeControl`, and this must be cast to the appropriate type—in this case, to a `CEikEdwin*`.

> **NOTE** You do not take ownership of the control despite the fact that the method returns a pointer to it.

`CEikDialog::ControlOrNull()` returns `NULL` if the supplied ID is invalid. This technique is useful, for example, if a control is available only in a particular locale. If however, an invalid ID is considered to be a programming error in your application, you should use the alternative `CEikDialog::Control()` to get a handle on the control—this panics if the ID is invalid.

Once a handle on the control is available, its data can be obtained. In the **SimpleDlg** example, the text entered by the user for the player name is retrieved using `CEikEdwin::GetText()`. It is at this point that any validation should be performed on the data entered. In the **SimpleDlg** example, no validation occurs, and `iPlayerName` is set to the value in the editor.

Once the contents of the dialog are considered valid, `ETrue` should be returned to allow the dialog to dismiss, otherwise `EFalse` should be returned.

Initializing a Standard Dialog Dynamically

It is a common requirement to set items in a dialog dynamically—for example, based on data passed into the dialog. This is performed in the `PreLayoutDynInitL()` method. The dialog framework calls this method prior to execution of the dialog. In the example player name dialog, this is where the text for the label is set:

```
void CSimpleDlgPlayerNameDialog::PreLayoutDynInitL()
    {
    CEikLabel* label = static_cast<CEikLabel*>(
ControlOrNull(EConfirmationNoteDlgCIdPlayerName));

    if (label)
        {
```

```
        HBufC* labelText =
StringLoader::LoadLC(R_ENTER_NAME_TEXT);
        label->SetTextL(*labelText);
        CleanupStack::PopAndDestroy(labelText);
        }
    }
```

A handle on the appropriate control is obtained using the `ControlOrNull()` method. Its value is then set using an appropriate method. In the **SimpleDlg** example the data is set via a call to `CEikEdwin::SetTextL()`, passing in a value read from the application resource file.

> **TIP** If necessary, you can override another method, `PostLayoutDynInitL()`, in order to change the layout and sizes of a dialog's controls, or if you wish to start a timer just before the dialog is displayed.

Constructing and Executing a Dialog

A `RunDlgLD()` method encapsulates construction and execution of a dialog. Writing one allows you to construct and execute the dialog, and also determine how it is closed via its return value, in a single step. Alternatively, the functionality encapsulated in that method can be separated out by making a call to the first-phase constructor followed by a call to `ExecuteLD()`.

In **SimpleDlg**, `CSimpleDlgAppUi::RunDlgLD()` is called from the AppUi, when a new game command (`ESimpleDlgCmdNewGame`) is received.

```
void CSimpleDlgAppUi::HandleCommandL(TInt aCommand)
    {
    switch (aCommand)
        {
        case ESimpleDlgCmdNewGame:
            {
            if
(CSimpleDlgPlayerNameDialog::RunDlgLD(iPlayerName))
                {
                StartNewGameL();
                }
            break;
            }
        ...
```

As the player name dialog is waiting and modal, `ExecuteLD()`, and therefore `RunDlgLD()`, will delete the dialog on return—in other words, when the dialog has been dismissed. The dialog can be instantiated and held as a local variable.

Determining how a dialog was closed is done by checking the return value of `ExecuteLD()`, or `RunDlgLD()` if this has been implemented. A return value of `EFalse` indicates that **Cancel**, **Back** or **No** was pressed to dismiss the dialog; `ETrue` is returned otherwise—for example, by the user selecting **Ok** or **Accept**.

> **NOTE**
>
> Series 60 has a "dialog shutter," which it uses to shut all open dialogs when it terminates an application. All dialogs must respond to an escape key by dismissing themselves—the dialog shutter sends up to 50 escape key events (`EEikBidCancel` or `EKeyEscape`) to the application to ensure that all dialogs specifying the `EEikDialogNotifyEsc` flag are dismissed.

The **SimpleDlg** example shows how to create a simple dialog in an application. In practice, the particular dialog constructed here would be better implemented using a data query (see the **Queries** section later in this chapter), but it serves to illustrate the basic concepts involved in using dialogs. In the remainder of this chapter, you will learn about the specialized dialog classes provided by Series 60 to meet common needs. Where possible, you should use these classes in preference to creating your dialog from scratch.

Multipage Dialogs

A dialog can be split into logical sections by using a **multipage dialog**. This allows the presentation of a dialog in an organized manner, with labeled tabs displayed in the navigation pane to clearly indicate the purpose of each page to the user. An example is shown in Figure 6-3. The navigation pane is explained further in Chapter 5.

Figure 6-3 A multipage dialog.

The framework provides the tabs and handles them automatically—you just need to specify a pages field instead of an items field in your `DIALOG` resource. You also need to make sure that the dialog fills the entire application client rectangle by specifying the `EEikDialog-FlagFillAppClientRect` flag. Otherwise, the tabs will appear but will be dimmed. An example resource definition for a multipage dialog is shown below:

```
RESOURCE DIALOG r_myapp_player_dialog
    {
    flags = EEikDialogFlagNoDrag | EEikDialogFlagCbaButtons |
EEikDialogFlagFillAppClientRect | EEikDialogFlagWait;
    buttons = R_AVKON_SOFTKEYS_OPTIONS_BACK;
    pages = my_pages;
    }
```

The `pages` field of the resource is set to equal an `ARRAY` resource, `my_pages`. The array resource should also be defined in the resource file:

```
RESOURCE ARRAY my_pages
    {
    items =
      {
      PAGE
```

```
            {
            text = "Name";
            lines = my_name_lines;
            },
        PAGE
            {
            text = "Age";
            lines = my_age_lines;
            }
        };
    }
```

The ARRAY has two PAGE resources, each page containing:

- text, which will appear on the tab.
- lines, which refers to an ARRAY of DLG_LINE resources.

```
RESOURCE ARRAY my_name_lines
    {
    items =
        {
        DLG_LINE
            {
            id = EMyAppDlgCIdNameLabel;
            type = EEikCtLabel;
            control = LABEL
                {
                txt = "Enter your name:";
                };
            },
        DLG_LINE
            {
            id = EMyAppDlgCIdNameEditor;
            type = EEikCtEdwin;
            control = EDWIN
                {
                avkon_flags = EAknEditorFlagNoEditIndicators;
                maxlength = KMaxPlayerNameLength;
                };
            }
        };
    }
...
```

Within the array, the dialog lines are almost the same as they would be in a single-page dialog. The exception is that editors should have turned off their edit indicator by setting their avkon_flags to EAknEditorFlagNoEditIndicators. If the edit indicators are not switched off, they will appear on top of the tab pane, thus obscuring it. The reason for this is that they have the highest priority on the navigation pane control stack. Chapters 5 and 8 provide further details on the navigation pane editor indicators.

NOTE For clarity, the example code used here does not take account of the need for localization, so the text for various fields is supplied literally. For commercial applications, you should supply all such user-visible text in separate localized files, as described in Chapters 2 and 4.

Defining a Menu for Your Dialog

When the player name dialog is displayed in the **SimpleDlg** application, the soft keys are labeled **Ok** and **Cancel**, as shown in Figure 6–2. A more complex dialog may need to have a menu associated with it, the menu being activated by an **Options** soft key.

To do this, define the menu in a `MENU_BAR` resource, ensuring that it has an **Exit** option. Set the `DIALOG` resource buttons to an appropriate value—this means one containing an **Options** soft key, such as `R_AVKON_SOFTKEYS_OPTIONS_BACK`. The resource is then passed through to the dialog in its `ConstructL()` method. To dynamically configure the menu you have to override `DynInitMenuPaneL()`.

Note that if it is necessary to override the `OkToExitL()` function, then the overridden version of this function is responsible for displaying the menu. To handle custom menu commands, implement the `ProcessCommandL()` function. Further details on menus are available in Chapter 5.

Custom Controls in Dialogs

This subsection looks at an example, **CustomCtrlDlg**, which shows how a custom control can be added to a dialog.

The **CustomCtrlDlg** example displays a dialog in its main application window, which has a label showing the name of a game, as shown in Figure 6–4. When **New Game** is selected from the **Options** menu, a dialog appears showing a custom control. The custom control in this case simply draws a "star" on the screen, as shown in Figure 6–5. The new control is defined in the class `CCustomCtrlDlgCustomControl` and implements the `Draw()` function to draw the star shape. In a real application, this could be a control written elsewhere, which could be integrated into a dialog-based application.

The main architecture of this application should be familiar by now, as it is identical to that of **SimpleDlg**, so this subsection will just concentrate on the construction of the control. The first step, as before, is to define a layout in a `DIALOG` resource. A custom control is added to a `DLG_LINE` by defining an `id` and `type`. With a custom control, there is no need to define the `control=` element, as you would normally. The control ID should be defined in the `.hrh` file, and, as this is not a standard Series 60 control, you will need to define a `type` for your control in the `.hrh` file. An `enum` is created for each of the custom controls. To avoid clashing with standard controls, the ID numbering starts at 1000.

Figure 6–4 CustomCtrlDlg main dialog.

Figure 6–5 CustomCtrlDlg custom control dialog.

```
RESOURCE DIALOG r_customctrldlg_custom_control_dialog
    {
    flags = EEikDialogFlagNoDrag | EEikDialogFlagCbaButtons | EEikDialogFlagWait;
    buttons = R_AVKON_SOFTKEYS_OK_CANCEL;

    items =
        {
        DLG_LINE
            {
            id = ECustomCtrlDlgDlgCIdCustomControl;
            type = ECustomCtrlDlgCtCustomControl;
            }
        };
    }
```

Creating a Custom Control in a Dialog Class

In order to display a custom control, a dialog class is created, derived from `CAknDialog`, that provides an overridden version of `CreateCustomControlL()`. In **CustomCtrlDlg** it is `CCustomCtrlDlgCustomControlDialog` that provides the implementation:

```
SEikControlInfo
CCustomCtrlDlgCustomControlDialog::CreateCustomControlL(
    TInt aControlType)
    {
    SEikControlInfo controlInfo;
    controlInfo.iControl = NULL;
    controlInfo.iTrailerTextId = 0;
    controlInfo.iFlags = 0;

        switch (aControlType)
        {
        case ECustomCtrlDlgCtCustomControl:
            controlInfo.iControl = new (ELeave)
CCustomCtrlDlgCustomControl();
```

```
              break;
        default:
              break;
        }
    return controlInfo;
}
```

The dialog uses a control factory to create the controls it contains. The dialog framework will call `CreateCustomControlL()` if it finds a type in the `DLG_LINE` resource that it does not recognize.

> **NOTE** `CreateCustomControlL()` panics by default, so you must override it if you define a `DIALOG` with a custom control.

As the method name suggests, `CreateCustomControlL()` constructs the custom control on behalf of the control factory. It has a parameter, `aControlType`, which indicates the control `type`. By convention, the value of this is used in a `switch` statement to determine which type of control to construct. In **CustomCtrlDlg**, the control `CCustomCtrlDlgCustomControl` is constructed when the argument is `ECustomCtrlDlgCtCustomControl`.

Interestingly, the method returns a `struct`, as opposed to an object, of type `SEikControlInfo`. The `iControl` field of this `struct` must be populated with your custom control object and then it can be returned.

To see the custom control displayed in the dialog, run the **CustomCtrlDlg** application and select **NewGame** from the **Options** menu.

The first part of this chapter has concentrated on creating standard dialogs by defining a layout in a `DIALOG` resource and implementing a dialog class derived from `CAknDialog`. In the remainder of the chapter the specialized dialog classes will be introduced, and example applications containing forms, notes, queries, and list dialogs will be analyzed.

Forms

Forms provide a way to allow the user to quickly and easily enter or edit many items of data in one process. Forms can have a number of fields, displayed in a similar way to a list, which the user can scroll up and down. Unlike list items, form lines are editable when they are in focus.

Forms have either a view mode and an edit mode, or just an edit mode. Use of a view mode implies that there is some existing information to display in the form, which the user can then edit. For example, in a game application a form could be used in view mode to display information about existing details for an opponent, and then in edit mode to change the details.

Chapter 6 Forms

If a form has a view mode, then that will be its default mode. In view mode, the focus appears as a solid block, as shown in Figure 6–6. Switching to the edit mode is achieved by selection of **Edit** from the **Options** menu.

Generally, a form will be used with just an edit mode to create a new piece of information—for example, to create new details for an opponent in a game. As there are no previous details for the opponent, there would be little point in having a view mode. In edit mode, the highlight appears as an outline to the field, with a flashing cursor for text editors, as shown in Figure 6–7.

Figure 6–6 Form in view mode.

Figure 6–7 Form in edit mode.

Form Lines

Each line in a form dialog must have a label and a control for editing the value of the item. The label can contain text, an icon, or both. The most common controls used are sliders, editors and pop-up fields, as shown in Figure 6–8, Figure 6–9, and Figure 6–10. As with other dialogs, custom controls can be used in forms.

The label may occupy the same line as the control, as shown in Figure 6–9, or a separate one, as shown in Figure 6–11. This can also vary between modes, such that the view mode displays the label on the same line, and the edit mode displays it on a separate line.

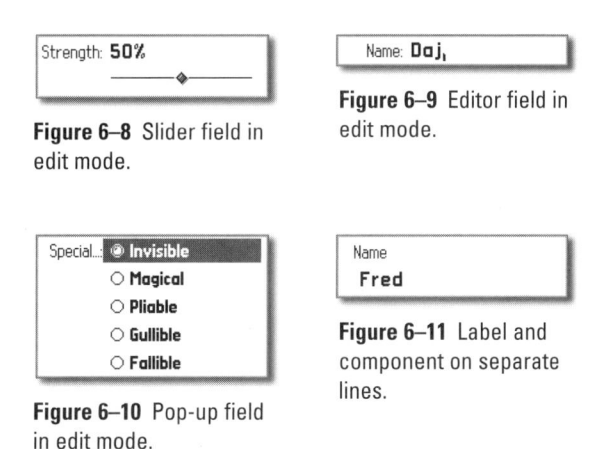

Figure 6–8 Slider field in edit mode.

Figure 6–9 Editor field in edit mode.

Figure 6–10 Pop-up field in edit mode.

Figure 6–11 Label and component on separate lines.

Chapter 6   Dialogs

TIP — It is possible to make a form hide lines whose values are not currently set—obviously, this makes sense only in view mode. It is done using the flags field in the resource definition—use `EEikFormHideEmptyFields` to make empty data fields invisible.

Form Soft Keys

The behavior of the soft keys and the contents of the **Options** menu depend on the mode of the form. In *view* mode, the **Back** key dismisses the form and returns to the previous screen. The **Options** menu, by default, contains an item to switch the form to *edit* mode (**Edit**). In *edit* mode, if no line values have changed, then the **Back** key returns the form to view mode—if a view mode has been defined for the form. If the form has no view mode, then the **Back** key dismisses the form and returns to the application's previous screen.

If one or more line values have changed, the **Back** key displays a query asking the user if they wish to save their changes, as shown in Figure 6–12. Selecting **Yes** saves the form data, selecting **No** discards it. The form then returns to the view mode, if there is one, or to the previous screen if not.

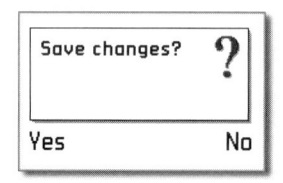

Figure 6–12 Save changes query.

In edit mode, the **Options** menu contains a number of items by default, as shown in Figure 6–13.

The basic meaning of all these items is self-explanatory, but for clarity a more detailed description is given here:

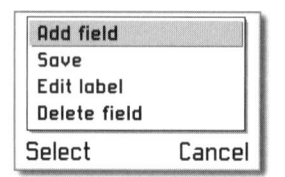

Figure 6–13 Options menu for a form in edit mode.

- **Add Field**—Adds a line to the form and redraws the display. The application developer determines the exact behavior of this item.

- **Save**—Saves the form's data and returns to the view mode if there is one, or to the previous screen if not.

- **Edit Label**—Displays a pop-up window, which allows you to change the label for the current item, as shown in Figure 6–14.

- **Delete Field**—Displays a query to check with the user whether or not to delete the field, as shown in Figure 6–15. If the answer is "Yes," the field is deleted and the form redrawn; if "No," the query is dismissed.

Chapter 6 Forms   305

Figure 6–14 Form label editor.

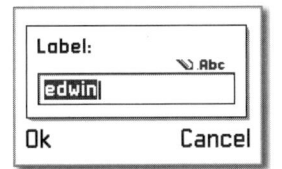

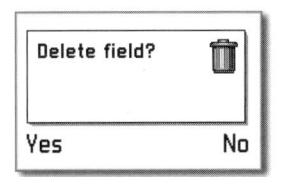

Figure 6–15 Form delete field query.

> **TIP** You can suppress or replace the default menu items, as you can with all other menu items. See the **Menus** section in Chapter 5 for details of how to do this. The enumeration for the items is in `avkon.hrh` under the heading "`FORM default menu constants`".

Editing Forms

The example application **OpponentForm** has a form that allows the user to enter and edit details for opponents in a multiplayer game. It displays brief details of the opponents in a list, as shown in Figure 6–16.

Selecting **New** from the **Options** menu allows the user to add a new opponent, by filling in a form. Selecting **Open** allows editing of an existing opponent's details, as shown in Figure 6–17.

Figure 6–16 List of opponents.

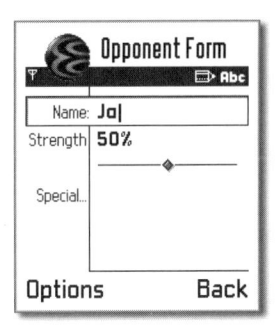

Figure 6–17 Form edit screen.

With the exception of **Save**, all the default items have been removed from the **Options** menu. Table 6–1 details the methods that are called when invoking the other default commands. These methods can be overridden in your code if required.

Table 6–1 Default Options Menu Items for Forms in Edit Mode

Item	Enumeration	Method Called on Invocation
Add Field	EAknFormCmdAdd	AddItemL
Edit Label	EAknFormCmdLabel	EditCommandLabelL
Delete Field	EAknFormCmdDelete	DeleteCurrentItemL

Creating a Form in an Application

The **OpponentForm** example demonstrates how to add a form to an application. This subsection discusses that application and shows how to define a form in resource, create a derived form class, and initialize form values. Additionally, it demonstrates how to save changes to a form, cancel changes, and finally execute a form.

Since forms are a specific type of dialog, the general process for creating them should be familiar. First, the layout of the form is specified in a resource file, and then a class derived from `CAknForm` should be created to initialize it and handle the form's data. The form is then executed using `CAknDialog::ExecuteLD()`.

In the **OpponentForm** example, the form uses a slider to set the strength of an opponent, an editor to set the name, and a pop-up field to select a special power. The form data is validated on saving to ensure that both a user name and power have been set for the player.

Defining a Form in a Resource

Forms are contained within a `DIALOG` resource, specifying the `form` components.

```
RESOURCE DIALOG r_opponentform_form_dialog
    {
    flags = EEikDialogFlagNoDrag | EEikDialogFlagFillAppClientRect | EEikDialogFlagNoTitleBar | EEikDialogFlagWait | EEikDialogFlagCbaButtons;

    buttons = R_AVKON_SOFTKEYS_OPTIONS_BACK;
    form = r_opponentform_form;
    }
```

Set the `flags` to indicate the following about the dialog:

- `EEikDialogFlagWait`—the dialog is waiting
- `EEikDialogFlagNoDrag`—cannot be moved

Chapter 6 Forms

- `EEikDialogFlagNoTitleBar`—has no title
- `EEikDialogFlagFillAppClientRect`—fills the whole screen
- `EEikDialogFlagCbaButtons`—uses soft keys

As usual, the `buttons` field determines the soft keys—in this case **Options** and **Back**. The `form` field refers to a FORM resource.

```
RESOURCE FORM r_opponentform_form
   {
   flags = EEikFormEditModeOnly;

   items =
      {

      DLG_LINE
         {
         type = EEikCtEdwin;
         prompt = NAME_TEXT;
         id = EOpponentFormDlgCIdEdwin;
         control = EDWIN
            {
            width = KMaxNameLength;
            maxlength = KMaxNameLength;
            };
         },
      DLG_LINE
         {
         type = EAknCtSlider;
         prompt = STRENGTH_TEXT;
         id = EOpponentFormDlgCIdSlider;
         control = SLIDER
            {
            layout = EAknFormSliderLayout1;
            minvalue = 0;
            maxvalue = 100;
            step = 5;
            minlabel = MIN_TEXT;
            maxlabel = MAX_TEXT;
            valuetype = EAknSliderValuePercentage;
            };
         },
      DLG_LINE
         {
         type = EAknCtPopupFieldText;
         prompt = POWER_TEXT;
         id = EOpponentFormDlgCIdPopup;
         itemflags = EEikDlgItemTakesEnterKey |
EEikDlgItemOfferAllHotKeys;
         control = POPUP_FIELD_TEXT
            {
            popupfield = POPUP_FIELD
               {
```

Chapter 6 Dialogs

```
                    width = KMaxPowerLength;
                };
            textarray =
r_opponentform_power_popup_field_textarray;
            active = 0;
            };
        }
    };
}
```

As a minimum, the FORM resource should define the flags and items components. In the **OpponentForm** example, the flags used are:

- EEikFormEditModeOnly—indicates that the form should have an edit mode only.
- EEikFormHideEmptyFields—hides empty fields.
- EEikFormShowBitmaps—displays bitmaps in labels.
- EEikFormUseDoubleSpacedFormat—splits labels and controls onto two lines.

The items field refers to an array of dialog lines (DLG_LINE), which are the fields in the form. Each dialog line should contain a minimum of type, prompt, id and control. In the **OpponentForm** example, the third dialog line contains a pop-up field text control—this has the optional itemflags indicating that it should open the pop-up window when the **Selection** key is pressed, as shown in Figure 6–18.

Figure 6–18 Pop-up field control.

Creating a Form-Derived class

In order to use the form dialog defined in the resource file, it is necessary to derive from CAknForm and override some of its methods. In **OpponentForm**, the class COpponentFormForm illustrates this process.

```
class COpponentFormForm : public CAknForm
    {
public: // Constructor
    static COpponentFormForm* NewL(TOpponentFormOpponent&
aOpponent);
    ...
private: // Constructor
    COpponentFormForm(TOpponentFormOpponent& aOpponent) :
iOpponent(aOpponent){};
    ...
```

Chapter 6 Forms

On construction, it is usual to provide a reference to the data the form is to modify. In the **OpponentForm** example, a reference to a T-class, `TOpponentFormOpponent`, is passed through. This class has member data for the name, strength (defaulting to 50), and power of an opponent. A public method, `IsValid()`, determines whether the object contains valid data.

As with all dialogs, the form values can be initialized in `PreLayoutDynInitL()`, using `ControlOrNull()` to obtain a handle on each control as required. In `COpponentFormForm`, `PreLayoutDynInitL()` calls a private convenience method, `LoadFormValuesFromDataL()`, to perform the initialization. This is convenient, as it means `LoadFormValuesFromDataL()` can also be used to discard any changes if the user cancels a save.

When changes to the form data are to be saved—for example, if **Save** has been selected from the **Options** menu—the framework calls `SaveFormDataL()`. Therefore, this function must be overridden in the derived form class to ensure that the values in the controls are saved back into the form's member data.

```
TBool COpponentFormForm::SaveFormDataL()
    {
    CAknSlider* slider = static_cast<CAknSlider*>(
ControlOrNull(EOpponentFormDlgCIdSlider));

    if (slider)
        {
        iOpponent.SetStrength(slider->Value());
        }
    ...
    return ETrue;
    }
```

As usual, the `ControlOrNull()` method provides access to the form's controls. The data in each control can then be obtained. If all saves have been successful—for example, if valid values have been entered—then `SaveFormDataL()` should return `ETrue`, otherwise it should return `EFalse`.

If the user chooses to cancel the changes made to the form—for example, when selecting **No** in the save query—the framework calls `DoNotSaveFormDataL()` to discard any changes. Override this function in the derived form class to ensure that the values of the form's controls revert to those in the form data. This is done in the same way as when initializing the form values. In `COpponentFormForm` it is achieved by simply calling the private convenience method, `LoadFormValuesFromDataL()`, which resets the values.

Executing a Form

To construct and execute a form, use the standard dialog methods—for example, `ExecuteLD()`. In **OpponentForm** this is performed in `COpponentFormContainer::NewOpponentL()`, shown below, and also in

Chapter 6 Dialogs

`COpponentFormContainer::OpenOpponentL()`. The AppUi calls these functions in response to the **New** and **Open** commands, respectively.

```
void COpponentFormContainer::NewOpponentL()
    {
    TOpponentFormOpponent opponent;
    COpponentFormForm* form =
COpponentFormForm::NewL(opponent);
    form->ExecuteLD(R_OPPONENTFORM_FORM_DIALOG);
    …
```

Notes

Series 60 uses notes to convey information to users about a current situation. In general, a note is displayed on screen for a couple of seconds, but it can be dismissed earlier by the user.

A number of predefined notes exist, which give the functionality and appearance required by most applications. These predefined notes range from confirming actions to providing warnings to a user and are outlined in Table 6–2.

Wrapped notes and global notes offer an easy way to display simple notification messages. (However, global notes should be used sparingly as they take system focus even when the application that called them is in the background.) Wait and Progress notes are more complex and require a little more work.

Example implementations of all these note types are featured in later sections of this chapter. The applications **ConfirmationNote**, **WaitNote**, and **ProgressNote** are used as the basis of the discussion that follows.

Bear in mind that notes should be used sparingly to achieve maximum effect in an application. This is particularly true of the more serious error notes and warning notes—an information note should be used in preference to an error note unless something serious has happened. Please refer to the Series 60 UI style guide, provided with the SDK documentation, for further details.

Wrapped Notes

To standardize the notifications displayed by applications and to make life easier for developers, Series 60 offers **wrapped notes**. These are standard notes where you just need to supply the text you want to display. It is not necessary to specify a resource—a wrapped note will already have a resource associated with it, and this will define the layout, particular image and tone sound for it to use. There are four different types of wrapped notes: Confirmation Note, Information Note, Warning Note and Error Note. Table 6–2 shows an example note of each type and describes the tone associated with each.

Table 6–2 Predefined Notes

Type of Note	Class Name	Description	Example
Confirmation	`CAknConfirmationNote`	Informs that an action was successful (for example, a game saved successfully). A short, quiet tone may play when the note displays.	Confirmation Note
Information	`CAknInformationNote`	Informs that an error occurred which can be rectified, usually in response to user input (for example, entering a duplicate player name). A medium tone may play when the note displays.	Information Note
Warning	`CAknWarningNote`	Informs that a situation occurred that may require user intervention (for example, reaching the maximum number of saved games for an application). A long, loud tone plays when the note displays.	Warning Note
Error	`CAknErrorNote`	Informs that a serious situation occurred, which could cause considerable problems if the user does not intervene (for example, entering an incorrect PIN, causing a lock out). A long, distinctive tone plays when the note displays.	Error Note
Wait	`CAknWaitDialog` / `CAknWaitNoteWrapper`	Informs a user that an event is occurring which may take some time, but gives no indication of the progress or length of the event. An animation shows while the event is occurring. Pressing the right soft key cancels the note and the processing. Otherwise, the wait note closes when the processing completes.	Saving Game

Table 6–2 continued

Type of Note	Class Name	Description	Example
Progress	CAknProgressDialog	Informs a user that an event is occurring which may take some time. The progress or length of the event is indicated by means of a progress bar, which increments at regular intervals. Pressing the right soft key cancels the note and the processing. Otherwise, the progress note closes when the processing completes.	Saving Game
Global	CAknGlobalNote	Displays even if the application launching the note is not currently displayed. You can use such notes to display information, warnings, confirmations, errors, information about the current battery state, and so on. A permanent note does not timeout or disappear when the user presses a key, and must be cancelled in code.	Battery low
Custom	CAknNoteDialog	You can customize notes if required (for example, to display a different icon). Text can be either static (for example, "Saving Game") or variable (for example, "Delete saved game 3," where the number of games is determined at runtime). Note that the variable part may only be an integer. CAknNoteDialog does not derive from CAknDialog and so has no menu capabilities.	Delete saved game 3 Yes No

Chapter 6 Notes

The application **ConfirmationNote** shows how to use a wrapped note. This particular example displays a confirmation note, but all wrapped notes are executed similarly. When **Save Game** is selected from the **Options** menu, the current game "saves," and, if successful, a note displays to confirm this, as shown in Figure 6–19. The code for "saving the game" is not actually implemented—an empty method, `SaveGame()` is used to indicate where the saving would be performed.

Figure 6–19 Game saved confirmation note.

A wrapped note is initially created using the first-phase constructor. For a confirmation note, the class of interest is `CAknConfirmationNote`. Construction is then completed using `ExecuteLD()`, passing in a descriptor for the note to display. This function completes the second phase of construction, displays the note and destroys it on exit. In **ConfirmationNote**, the note is executed in the `CConfirmationNoteAppUi::HandleCommandL()` function:

```
HBufC* noteText;
noteText = StringLoader::LoadLC(R_SAVED_GAME_NOTE_TEXT);
CAknConfirmationNote* note = new (ELeave) CAknConfirmationNote();
note->ExecuteLD(*noteText);
CleanupStack::PopAndDestroy(noteText);
```

TIP The construction and execution of all standard wrapped notes follows this same pattern. The only thing that differs is the name of the class itself. This name can be `CAknConfirmationNote`, `CAknInformationNote`, `CAknWarningNote` or `CAknErrorNote`.

Customized Notes

If you need to deviate from the standard wrapped notes—for example, to provide a custom bitmap or a different tone—you can define your own resource for the note and execute this instead. In the **ConfirmationNote** example, when the application is started, a customized confirmation note is displayed to indicate how many saved games have been loaded, as shown in Figure 6–20.

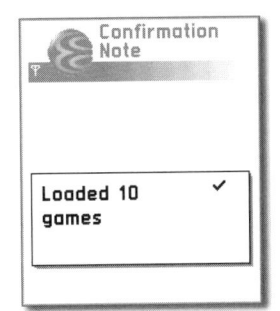

Figure 6–20 Customized confirmation note.

Although this note appears similar to the wrapped note seen earlier, it has been customized to display a different bitmap icon—as defined by the following resource:

Chapter 6 Dialogs

```
RESOURCE DIALOG r_confirmationnote_loaded_games_note
    {
    flags = EAknConfirmationNoteFlags;

    items =
        {
        DLG_LINE
            {
            type = EAknCtNote;
            id = EConfirmationNote;
            control = AVKON_NOTE
                {
                layout = EGeneralLayout;
                singular_label = LOADED_GAMES_SINGULAR_TEXT;
                plural_label = LOADED_GAMES_PLURAL_TEXT;
                imagefile = "z:\\system\data\avkon.mbm";
                imageid = EMbmAvkonQgn_indi_marked_add;
                imagemask = EMbmAvkonQgn_indi_marked_add_mask;
                };
            }
        };
    }
```

The resource uses standard `flags`, which determine to some extent the type of note that displays. The `EAknConfirmationNoteFlags` flag is used to indicate that it is a confirmation note.

TIP By default, a confirmation note has no soft key labels, but it cancels if the user presses either key. To explicitly label the right soft key as **Cancel** you can specify the soft keys for the note to be `R_AVKON_SOFTKEYS_CANCEL` in the `buttons` field of the `DIALOG` resource.

The dialog should have a single line, which as usual has a `type` (`EAknCtNote`), an `id`, and a `control` (of type `AVKON_NOTE`).

For the `AVKON_NOTE` control, the following need to be defined:

- `layout`—This determines the note's appearance. The general layout has been specified for the **ConfirmationNote** example, but other options can be found in `avkon.hrh`.

- `singular_label` and `plural_label`—These set different text to be displayed for singular and plural labels. In the **ConfirmationNote** example, the labels are set to "`Loaded %d game`" and "`Loaded %d games`", respectively. The value for the "`%d`" is set when the note is constructed. Note that using the `StringLoader` class and the `SetTextL()` method can produce more flexible results at runtime.

- `imagefile`, `imageid` and `imagemask`—These determine what image the note will display. In the **ConfirmationNote** example, the standard

Chapter 6 Notes

confirmation note icon is not displayed. For simplicity, another image from `avkon.mbm` was chosen, but you can define your own image. The `imagefile` specifies the location of the .mbm file containing the image to be displayed. The `imageid` is the position of the image in that file, and the `imagemask` is the position of the image mask in the file.

Constructing and Executing a Customized Note

To create a customized note, derive the note dialog class from `CAknNoteDialog`. It should be constructed using standard dialog construction and execution methods. In the **ConfirmationNote** example, this is performed in the container's `ConstructL()` method.

```
void CConfirmationNoteContainer::ConstructL(const TRect& aRect)
    {
    CreateWindowL();
    CAknNoteDialog* note = new (ELeave) CAknNoteDialog(CAknNoteDialog::EConfirmationTone,
CAknNoteDialog::EShortTimeout);

    note->PrepareLC(R_CONFIRMATIONNOTE_LOADED_GAMES_NOTE);
    TInt numberOfLoadedGames = LoadGames();
    note->SetTextPluralityL(numberOfLoadedGames > 1);
    note->SetTextNumberL(numberOfLoadedGames);
    note->RunLD();

    SetRect(aRect);
    ActivateL();
    }
```

The constructor requires arguments for the tone and timeout value. As the **ConfirmationNote** example customizes a confirmation note, it passes in `EConfirmationTone` for the tone and `EShortTimeout` for the timeout. This is essentially what the confirmation wrapper class does for you. The note can be executed by calling `ExecuteLD()`, passing in the `DIALOG` resource created in the previous step.

Alternatively, if it is required to call other methods before execution, you must call `PrepareLC()` first, then call any other methods, and then call `RunLD()`. In the **ConfirmationNote** example, the dynamic element of the message label must be set, so the `PrepareLC()`, `RunLD()` route was chosen. The dynamic text for the **ConfirmationNote** example relates to the number of games loaded. A dummy method, `LoadGames()`, has been implemented to determine this value. If more than one game has been loaded, `SetTextPluralityL()` is called, passing `ETrue`, and the `plural_label` will then be used. If only one game has been loaded, `SetTextPluralityL()` is called with `EFalse` and the `singular_label` used.

Wait Notes

A wait note is used to inform the user that processing is taking place—no indication is given of the time needed to complete the processing. This subsection looks at the **WaitNote** example, which shows how to create a wait note. This involves defining the note in resource, constructing and executing a wait note wrapper object, and creating a `MAknBackgroundProcess` derived class.

In the **WaitNote** application, selecting **Save Game** from the **Options** menu causes a wait note to display while the game saves, as shown in Figure 6-21. If the user cancels the note, the application cancels the saving of the game. The methods for saving the game and canceling saving are not actually implemented—the dummy methods, `SaveGamePartToFile()`, `CompleteGameSave()`, and `CancelGameSave()` are used to indicate where this functionality would be performed.

Figure 6-21 Wait note.

Defining a Wait Note in Resource

A wait note is defined in a resource structure using a `DIALOG` in a similar way to a customized note.

```
RESOURCE DIALOG r_waitnote_saving_game_note
    {
    flags = EAknWaitNoteFlags;

    items =
        {
        DLG_LINE
            {
            type = EAknCtNote;
            id = EConfirmationNoteDlgCIdSavingNote;
            control = AVKON_NOTE
                {
                layout = EWaitLayout;
                singular_label = SAVING_GAME_TEXT;
                imagefile = "z:\\system\data\avkon.mbm";
                imageid = EMbmAvkonQgn_note_progress;
                imagemask = EMbmAvkonQgn_note_progress_mask;
                animation = R_QGN_GRAF_WAIT_BAR_ANIM;
                };
            }
        };
    }
```

Set the dialog `flags` to `EAknWaitNoteFlags`.

The `AVKON_NOTE` should have:

Chapter 6 Notes

- `layout` set to `EWaitLayout`.
- `animation` field to animate the wait bar. Set this to `R_QGN_GRAF_WAIT_BAR_ANIM` to obtain the standard animation.
- `imagefile`, `imageid` and `imagemask` appropriate for a wait note. This is set to standard wait note values in the **WaitNote** example.

Constructing and Executing a Wait Note Wrapper Object

Wait note wrappers are constructed and executed in a different way to other dialogs. A wait note wrapper is actually a wait note wrapped up in an Active Object, so it does not behave in the same way as other dialogs. The same functionality can be achieved using the `CAknWaitDialog`, but this involves writing an Active Object (`CActive`-derived class) and is therefore slightly more complex. The progress note in the next subsection shows how this is done.

A wait note wrapper is constructed and executed using its `NewL()` and `ExecuteL()` methods. In the **WaitNote** example, this is performed in response to the user selecting **Saved Game** from the **Options** menu, in `CWaitNoteContainer::SaveGameL()`.

```
void CWaitNoteContainer::SaveGameL()
    {
    CAknWaitNoteWrapper* waitNoteWrapper =
CAknWaitNoteWrapper::NewL();

    // reinterpret_cast is required as CAknWaitNoteWrapper
inherits privately from CActive!
        CleanupStack::PushL(
reinterpret_cast<CBase*>(waitNoteWrapper));

    if (!waitNoteWrapper->ExecuteL(R_WAITNOTE_SAVING_GAME_NOTE,
*this))
        {
        CancelGameSave();
        }
    CleanupStack::PopAndDestroy(waitNoteWrapper);
    }
```

The `ExecuteL()` function must be supplied with a `DIALOG` resource and a reference to a `MAknBackgroundProcess` object (implementation of this mixin class is discussed in the next subsection). In the **WaitNote** example, the container class, `CWaitNoteContainer`, implements `MAknBackgroundProcess`, so a dereferenced `this` pointer is passed through. `ExecuteL()` blocks until the processing is complete, and a `TBool` is returned indicating whether the processing completed (`ETrue`) or the user cancelled the note (`EFalse`). In the **WaitNote** example, the game save is cancelled if the method returns `EFalse`.

Unlike other dialogs, wait notes do not delete when they finish executing; they must be deleted explicitly. In the **WaitNote** example, the note is a local variable, so deletion is achieved through the call to `CleanupStack::PopAndDestroy()`.

TIP If the note wrapper is not member data of a class, use a reinterpret cast to cast it to a `CBase` pointer in order to place it onto the `CleanupStack`. It cannot be placed directly onto the `CleanupStack`, as it *privately* inherits from `CActive`, which means that it has no (implicit) public converter to `CBase*`!

Creating an MAknBackgroundProcess-Derived class

A class that implements `MAknBackgroundProcess` is required in order to be able to instantiate the wrapper. In the **WaitNote** example, the container class does this job.

```
class CWaitNoteContainer : public CCoeControl, public
MAknBackgroundProcess
    {
private: // from MAknBackgroundProcess
    void DialogDismissedL(TInt /*aButtonId*/);
    TBool IsProcessDone() const;
    void ProcessFinished();
    void StepL();
    …
```

`MAknBackgroundProcess` encapsulates the processing that occurs while the wait note displays. In **WaitNote**, the processing performed is the saving of the game—the wait note wrapper encapsulates an Active Object, which calls the methods in `MAknBackgroundProcess` as it runs.

The processing should occur in a number of small steps. In the **WaitNote** example, the saving of the game is split into steps, which execute in the `SaveGamePartToFile()` method (this is a dummy method and nothing is actually saved to a file). The number of steps completed so far is indicated by `iStepsCompleted`. Active Objects are covered in more detail in Chapter 3.

`CAknWaitNoteWrapper` is an Active Object, and so calling `ExecuteL()` results in a service request being made to the Active Scheduler. When the request has been serviced, `MAknBackgroundProcess::IsProcessDone()` is called to determine whether all of the steps of the processing have been carried out. If they have not, `EFalse` will be returned and `MAknBackgroundProcess::StepL()` will be called, causing the next step of the processing to be carried out.

If the processing is allowed to complete—that is, the user does not cancel the dialog—the framework calls `MAknBackgroundProcess::ProcessFinished()` and then `MAknBackgroundProcess::DialogDismissedL()`. If the user does end the processing, `DialogDismissedL()` is called first by the framework. It then calls `IsProcessDone()` to check whether the processing has finished, followed by `ProcessFinished()`.

The method to determine whether processing is complete is `IsProcessDone()`.

```
TBool CWaitNoteContainer::IsProcessDone() const
    {
    return (iStepsCompleted == KNumberOfStepsToSaveGame);
    }
```

If the processing is complete, `ETrue` is returned, and `EFalse` if not. In the **WaitNote** example, this means checking that all the required steps in the procedure have been carried out, and if so, returning `ETrue`. In a real application, this would mean implementing some kind of state machine, or at the very least using a member flag to indicate when processing is complete.

The `StepL()` method is where small steps of the processing should be completed.

```
void CWaitNoteContainer::StepL()
    {
    SaveGamePartToFile();
    iStepsCompleted++;
    }
```

In the **WaitNote** example, the `SaveGamePartToFile()` dummy method simply wastes some time using an `RTimer` object.

TIP The `StepL()` method executes synchronously. It should therefore process quickly—otherwise it will not be possible to cancel the note. The thread will block, and there will be a large delay in displaying the note, as it does not display until one step completes.

Use the `ProcessFinished()` or the `DialogDismissedL()` methods to do any postprocessing work required, such as closing files, resetting state machines, and the like.

```
void CWaitNoteContainer::ProcessFinished()
    {
    CompleteGameSave();
    iStepsCompleted = 0;
    }
```

In the **WaitNote** example, the implementation of `ProcessFinished()` just calls a dummy method to complete the game save and to reset the steps completed counter.

Progress Notes

Progress Notes are similar to wait notes, but they give the user a visual indication of what portion of the processing is still left to do. In this subsection, the application **ProgressNote** is used to demonstrate the techniques you should use to create a progress note. The basic steps are as follows:

1. Declare the progress note.
2. Create the progress note for a variable-length process.
3. Execute the progress note.
4. Update the progress note as a process executes.
5. Handle user cancels and completion of the progress note.

Figure 6-22 Progress note.

In **ProgressNote**, selecting **Save Game** from the **Options** menu displays a progress note while the game saves, as shown in Figure 6-22. If the user cancels the note, the application cancels the game saving. The functions to save the game and cancel saving have not been fully implemented—the dummy methods `SaveGamePartToFile()`, `CompleteGameSave()`, and `CancelGameSave()` are used to indicate where this functionality would be performed.

Declaring a Progress Note

A progress note contains a progress bar. At regular intervals during the execution of the process it is necessary to increment and redraw the progress bar. The process needs to run asynchronously in small steps, preferably using an Active Object. Otherwise, the thread will block and the note will not update or may not even display until the processing is complete.

An Active Object is created in the usual way by deriving from `CActive`. In **ProgressNote**, `CProgressNoteGameSaverAO` is declared as an Active Object:

```
class CProgressNoteGameSaverAO : public CActive,
MProgressDialogCallback
    {
    ...
private: // from CActive
    void RunL();
    void DoCancel();

private: //data
    CAknProgressDialog* iProgressDialog;
    ...
```

A progress note is a nonwaiting dialog. That is, while the dialog is executing, the application can continue processing in the background. This is essential if the note is to update correctly. A progress note should be declared as member data, and then deleted in the class destructor. If execution completes normally—that is, no leave occurs—the progress note deletes itself and sets the pointer to it to NULL. However, if a leave occurs during the background processing, the note has no knowledge of this and cannot delete itself. Therefore, its deletion must be handled in the destructor. (You will see how it is possible for the dialog to null the pointer to itself in the next step of the **ProgressNote** example.)

Creating a Progress Note for a Variable-Length Process

A progress note is constructed just before a long-running process is about to begin, usually within an Active Object. In **ProgressNote**, the note is constructed in `CProgressNoteGameSaverAO::SaveGameL()`.

```
void CProgressNoteGameSaverAO::SaveGameL()
    {
    iProgressDialog = new (ELeave) CAknProgressDialog(
reinterpret_cast<CEikDialog**>(&iProgressDialog));
    ...
```

There are two constructors that can be used to create a variable-length progress note. The one used here creates a progress note that will display only after 1.5 seconds of processing and will display for a minimum of 1.5 seconds. If the processing completes within the initial 1.5 seconds, the note will not be displayed. This stops the note from flashing on screen and then disappearing in the case of short processes. The other constructor has a TBool parameter, which can be set to ETrue to prevent this delay. Use this only if you are certain that processing will take at least 1.5 seconds.

Both constructors have a parameter referred to in the documentation as a self-pointer. It is actually the address of the member data pointer for the note, so in the **ProgressNote** example it is the address of `iProgressDialog`. The method expects a CEikDialog**, so a reinterpret cast is required, as shown in `SaveGameL()`. The note sets the self-pointer to NULL when it deletes itself. As discussed in the previous step, this is to allow the safe deletion of the note in the destructor if necessary, without any possibility of a double deletion occurring.

Executing a Progress Note

Once the progress note is constructed, it can be prepared and run. This is performed in `CProgressNoteGameSaverAO::SaveGameL()`:

```
    ...
    iProgressDialog
->PrepareLC(R_PROGRESSNOTE_SAVING_GAME_NOTE);
    iProgressDialog->SetCallback(this);
    CEikProgressInfo* progressBar = iProgressDialog
->GetProgressInfoL();
```

```
progressBar->SetFinalValue(KFinalValue);
iProgressDialog->RunLD();
SaveGamePartToFile();
progressBar->IncrementAndDraw(KIncrement);
SetActive();
}
```

As methods will be called on the progress note, the dialog is prepared for use by calling `PrepareLC()` and passing through the resource. (The definition of the `DIALOG` resource, `r_progressnote_saving_game_note`, can be found in `ProgressNote.rss`.)

As with all dialogs, progress notes are, by default, nonwaiting. Nonwaiting progress notes do not return a value when they finish executing, so this cannot be used to determine whether the user cancelled the note or the process completed normally. Instead, a callback is set on the note by calling `CAknProgressDialog::SetCallback()`, passing in an `MProgressDialogCallback` object. `DialogDismissedL()` is called on the callback object when the dialog has been dismissed. The mechanics of this function is discussed in detail later—see *Completion of a Progress Note and User Cancellation.*

As the progress bar will represent the proportion of the task completed, it is necessary to set the number of individual units for the bar. The progress bar, of type `CEikProgressInfo`, can be retrieved by calling `GetProgressInfoL()` on the dialog. A call to `CEikProgressInfo::SetFinalValue()` is used to set the number of units the bar represents. Note that, although `GetProgressInfoL()` returns a pointer to the progress bar, the application does not take ownership of it.

The note is instantiated using `RunLD()`—this returns immediately, allowing processing to begin. The `RunLD()` method has a "D" suffix, which usually indicates that the object deletes itself just before the method returns. In common with other nonwaiting dialogs, however, the note does not delete when the `RunLD()` method returns. Instead it is deleted when `ProcessFinishedL()` is called, or, if the user dismisses the note, after a call to the `DialogDismissedL()` method (called by `OkToExitL()`).

The processing in the **ProgressNote** example is performed in the `SaveGamePartToFile()` function. This encapsulates an asynchronous request, hence the call to `SetActive()`.

Updating a Progress Note as a Process Executes

It is necessary to update the progress bar at regular stages during the execution of the process and dismiss it when processing finishes. The `RunL()` of the Active Object is the natural place to do this.

```
void CProgressNoteGameSaverAO::RunL()
    {
    if (!IsProcessDone())
        {
        SaveGamePartToFile();
        iProgressDialog->GetProgressInfoL()
->IncrementAndDraw(KIncrement);
        SetActive();
        }
    else
        {
        iProgressDialog->ProcessFinishedL();
        }
    }
```

The `RunL()` method should first check whether all processing is complete. In the **ProgressNote** example, the local method `IsProcessDone()` is used to determine this—it simply checks that a counter has not reached its limit. If processing is complete, the progress note should be informed by calling its `ProcessFinishedL()` method. The note will then call the `DialogDismissedL()` method and delete itself, setting its self-pointer to null.

If processing is not yet complete, then the following steps should be continued until it is:

1. Perform a step in the processing. The **ProgressNote** example simulates saving the game in the `SaveGamePartToFile()` method—this is the asynchronous request for the Active Object.

2. Update the progress bar accordingly, using the `IncrementAndDraw()` method. In the **ProgressNote** example, the steps are of even length, so the progress bar can be incremented by a constant value, `KIncrement`. If the process has uneven steps, you may use a variable for this. The progress bar, instead of being incremented and drawn, can be set and drawn using the `SetAndDraw()` method. This disregards any previous value for the progress bar and sets its position to that specified in the arguments to the method.

3. Set the Active Object to be active. This is so that `RunL()` is called again to complete the next step, if there is one. The progress note will dismiss automatically, once it has been incremented sufficiently that it has reached its final value. The final value should be selected and incremented with care, to ensure that it does not dismiss before the processing is complete.

Completion of a Progress Note and User Cancellation

`MProgressDialogCallback::DialogDismissedL()` is called when the progress dialog is dismissed. Perform any cleanup in this method—in the case of a user cancellation—or do further processing after a successful completion.

Chapter 6 Dialogs

In the **ProgressNote** application, `IsProcessDone()` is called to determine whether or not the processing was complete when the dialog was dismissed. If it was not, then the user must have chosen to cancel the processing, so the saving of the game will be canceled.

```
void CProgressNoteGameSaverAO::DialogDismissedL(TInt
/*aButtonId*/)
  {
  if (!IsProcessDone())
      {
      CancelGameSave();
      }
  else
      {
      CompleteGameSave();
      }

  iStepsCompleted = 0;
  }
```

Fixed-Period Progress Notes

If the length of time a process will take is known, a progress dialog can be instantiated and its display can be updated automatically. Again, this is performed within an Active Object. However, the note object itself increments and redraws the progress bar automatically, so there is no need for this to be done in the code.

```
  const TInt KFinalValue = 5;
  const TInt KInterval = 100;
  const TInt KIncrement = 1;

  SaveGamePartToFile();
  SetActive();

  iProgressDialog = new (ELeave)
CAknProgressDialog(KFinalValue, KIncrement, KInterval,
reinterpret_cast<CEikDialog**>(&iProgressDialog));

  iProgressDialog
->ExecuteLD(R_PROGRESSNOTE_SAVING_GAME_NOTE);
```

The following three parameters are required. They determine how the progress bar increments, and the length of the process:

- Final Value—As in the **ProgressNote** example, this determines the number of units in the progress bar.

Chapter 6 Notes

- Increment—The number of units the progress will move when it is incremented.
- Interval—How often the progress bar increments in hundredths of a second.

For example, for a process that takes 5 seconds the bar could be 5 units long, with 1-unit increments, and increments occurring every second. The progress note would display for 5 seconds. The following formula can be used to calculate approximate values for an application: `duration = interval * final value / increment`.

You may need to experiment with the values to make a progress bar update smoothly and to give it the appearance of doing something all the time. An interval of less than 100 can cause the total time the note displays to increase dramatically from what might be expected. Dismissing the dialog as soon as the processing is finished is possible, but it might be confusing for the user if the progress bar does not reach its final value.

Global Notes

A global note dialog is displayed on screen even if the application that owns it is not in focus. The code used to display a global note is quite simple, but should be used sparingly. It does not require the definition of a note in a resource structure, although this can be done for a customized note.

Construction and execution of a global note is performed using the `NewLC()` and `ShowNoteL()` methods. A note type is supplied to the `ShowNoteL()` method, plus the text to be displayed. The `aknnotifystd.h` header file defines the possible note types in the `TAknGlobalNoteType` enumeration.

```
HBufC* noteText;
noteText = StringLoader::LoadLC(R_GLOBAL_NOTE_TEXT);
CAknGlobalNote* globalNote = CAknGlobalNote::NewLC();
iNoteId = globalNote->ShowNoteL(EAknGlobalPermanentNote,
*noteText);
CleanupStack::PopAndDestroy(globalNote);
CleanupStack::PopAndDestroy(noteText);
```

All global notes have a `CancelNoteL()` method that can be called to dismiss them. This method requires the ID of the note to be cancelled—the ID returned when calling the `ShowNoteL()` method. Permanent notes do not timeout and cannot be dismissed by the user, they must be cancelled programmatically using `CancelNoteL()`.

WARNING Global notes do not have menus or soft keys defined and so will lock the UI of a device until cancelled.

Queries

Series 60 applications use various queries to ask the user a question, which may determine how the application then proceeds. A response to a query may require the user to enter some data, select one or more items from a list or simply confirm an action. The basic types of query are a confirmation query, a data query, a list query and a global query.

A **confirmation query** poses a question to which there is only a positive or a negative response—for example, confirming whether a saved game should be deleted. It is possible for a confirmation query to have a single answer. A **data query** requests that a user enter some text, for example, a player's name. The text can be plain or formatted as a password, PIN, phone number, date, time, duration, or floating-point number. Table 6–3 provides a more comprehensive list of the different types. A multiline data query can have two lines of text, each plain or formatted as per single-line data queries.

Table 6–3 Data Query Types

Type	Class	Layout	Control
Plain Text	CAknTextQueryDialog	EDataLayout	EDWIN
Phone Number	CAknTextQueryDialog	EPhoneLayout	EDWIN
PIN	CAknTextQueryDialog	EPinLayout	SECRETED
Password	CAknTextQueryDialog	ECodeLayout	SECRETED
Date	CAknTimeQueryDialog	EDateLayout	DATE_EDITOR
Time	CAknTimeQueryDialog	ETimeLayout	TIME_EDITOR
Duration	CAknDurationQueryDialog	EDurationLayout	DURATION_EDITOR
Float	CAknFloatingPointQueryDialog	EFloatingPointLayout	FLPTED
Multiline	CAknMultiLineDataQueryDialog	Each AVKON_DATA_QUERY has a value from EAknMultilineData Layout depending on the type of control	Varies depending on usage. Up to two lines.
Number	CAknNumberQueryDialog	ENumberLayout	AVKON_INTEGER_EDWIN

Chapter 6 Queries

A **list query** contains a menu list, from which one or more items can be selected. Items in the list can contain one or two lines of text, with or without an icon or heading. As with all lists, items must have the same type of content.

Queries can be local to the application or global. The three types mentioned earlier are all local. Series 60 hides local queries when another application switches to the foreground, but global queries are not hidden.

NOTE   Global queries are not dialogs, strictly speaking, as they do not inherit from `CEikDialog`.

You can use **global queries** for servers or applications that are currently in the background. Do this only if the information to be conveyed to the user is very important. If you display a global query from your application, the user cannot switch away to another application until they dismiss it. Global queries can be used to confirm an action, display a message or request an item be selected from a list. The different types are listed in Table 6–4.

Table 6–4 Global Queries

Type of Global Query	Description
Confirmation	Poses a question in a global manner, such as confirming whether to delete *all* games. You can customize a global confirmation query so that it plays a tone, dismisses with any key rather than just the soft keys, or shows a different icon and animation.
Message	Displays a message with a header—for example, informing that the current game will close. You can customize a global message query so that it plays a tone and/or shows a small icon in the top right-hand corner.
List	Contains a selection list from which you can select an item. The items can contain a single line of text only. The query focuses the first item of the list by default, but you can change this.

The following section discusses three example applications that deal with data queries (**DataQuery**), list queries (**ListQuery**), and global queries (**GlobalQuery**). A confirmation query is very similar to a data query and therefore is not covered explicitly here, although sample code can be found in **DataQuery**.

Data Queries

The example application **DataQuery** provides sample implementations of both a confirmation query and a data query. As you would expect, the techniques involved in creating these dialogs are similar, and therefore only the data query will be analyzed in detail here.

DataQuery has just one menu option, **Exit**. If the user chooses to exit the application, a dialog is shown asking if they would like to save the game. This dialog is a confirmation query with soft key options of **Yes** and **No**, as shown in Figure 6–23.

If the user decides to save the game, a data query is displayed, requesting a name for the saved game, as shown in Figure 6–24. This subsection describes how to define and execute the data query.

Figure 6–23 Confirmation query.

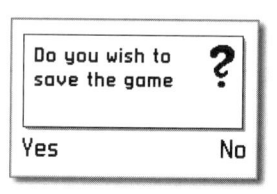

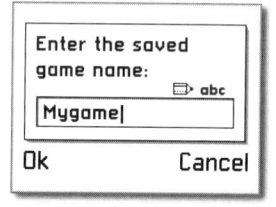

Figure 6–24 Data query.

Defining a Data Query in Resource

As a data query is a type of dialog, it should be defined in a `DIALOG` resource.

```
RESOURCE DIALOG r_dataquery_data_query
   {
   flags = EGeneralQueryFlags;
   buttons = R_AVKON_SOFTKEYS_OK_CANCEL;

   items =
      {
      DLG_LINE
         {
         type = EAknCtQuery;
         id = EDataQueryDlgCIdDataQuery;
         control = AVKON_DATA_QUERY
            {
            layout = EDataLayout;
            control =
               EDWIN
                  {
                  width = KMaxGameNameLength;
                  lines = 1;
                  maxlength = KMaxGameNameLength;
```

```
            };
         };
      }
   };
}
```

The flags should be set to `EGeneralQueryFlags` or `EAknGeneralQueryFlags`—both equate to the same thing. The dialog should contain a single dialog line, which contains the query control. An appropriate layout for the **DataQuery** example is `EDataLayout`—other layouts are defined in `avkon.hrh`. An `EDWIN` control is used for the data entry. Optionally, it is possible to set the text for the control's label, but in **DataQuery,** the label text is set dynamically.

> **NOTE** A heading can be set for a query, and this should be constructed as a dialog line. The exception to this is a multiline data query. This has two dialog lines, each of which must contain a query.

Table 6–3 shows the type of control to use for each type of query. Chapter 8 gives more details on `EDWIN` and other editor resources.

Constructing and Executing a Data Query

The general method for executing a data query is to construct it using the `NewL()` method and execute it using `ExecuteLD()`. In **DataQuery** this is performed in `CDataQueryContainer::SaveGameL()`. A descriptor representing the game's name is passed into the query on construction. This will be modified using the contents of the editor control.

```
...
    TBuf<KMaxGameNameLength> gameName;
    CAknTextQueryDialog* gameNameQuery =
CAknTextQueryDialog::NewL(gameName);
    CleanupStack::PushL(gameNameQuery);

    HBufC* prompt = StringLoader::LoadLC(R_DATA_QUERY_PROMPT);
    gameNameQuery->SetPromptL(*prompt);
    CleanupStack::PopAndDestroy(prompt);
    CleanupStack::Pop(gameNameQuery);

    if (gameNameQuery->ExecuteLD(R_DATAQUERY_DATA_QUERY))
       {
       SaveGameToFileL(gameName);
       }
    }
...
```

TIP CAknQueryDialog is the base class for all local queries. It provides overloaded `NewL()` methods, which construct an appropriate derived class based on the type of the argument.

After constructing the query, a prompt can be set that will be displayed with the control. This is done by calling the `SetPromptL()` method, passing in a constant reference to a descriptor containing the prompt. In **DataQuery**, a string is loaded from resource into the `prompt` descriptor. Because the method is not leave-safe, the query should be put onto the Cleanup Stack before it is called. However, before `ExecuteLD()` is called, it must be popped off the stack, because the query will put itself onto the Cleanup Stack as part of the execution.

List Queries

A **list query** contains a menu list, from which the user can select one or more items. Items in the list can contain one or two lines of text with or without an icon or heading. As with all lists, items must have the same type of content. This subsection gives details of how to construct a list query, with the application **ListQuery** providing the basis of the discussion. This application is very similar to the previous example, **DataQuery**, but instead of asking the user to enter a name for the saved game, it presents a list of names for the user to choose from, as shown in Figure 6–25.

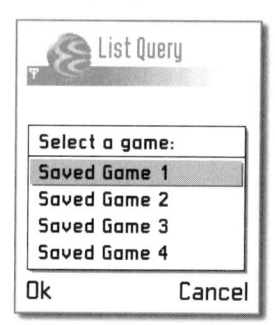

Figure 6–25 List query.

List Query Resources

The resource type for a list query is an AVKON_LIST_QUERY resource. The resource should have a single dialog line, AVKON_LIST_QUERY_DLG_LINE, containing an AVKON_LIST_QUERY_CONTROL. The control should define the `listtype` field to indicate the type of list the query will contain. This could be one of the pop-up menu list types, but in **ListQuery** a single line of text is chosen. The `listbox` refers to an AVKON_LIST_QUERY_LIST resource. To define the list items statically, an array identifier can be specified. The `array_id` specified in the AVKON_LIST_QUERY_LIST resource structure can refer to another resource containing an array of static strings representing the list items. In the **ListQuery** example, the list items are going to be defined dynamically, so the resource is left blank.

```
RESOURCE AVKON_LIST_QUERY r_listquery_list_query
    {
    items =
```

Chapter 6 Queries

```
      {
   AVKON_LIST_QUERY_DLG_LINE
      {
      control = AVKON_LIST_QUERY_CONTROL
         {
         listtype = EAknCtSinglePopupMenuListBox;
         heading = SELECT_GAME_TEXT;
         listbox = AVKON_LIST_QUERY_LIST
            {
            // array of items will be defined dynamically
            };
         };
      }
   };
}
```

> **TIP** The AVKON_LIST_QUERY has a flags field, which defaults to EGeneralQueryFlags, and a softkeys field, which defaults to **Ok** and **Cancel**. These values can be overridden for a custom list query.

Creating and Executing a List Query

In **ListQuery**, the query list is created in CListQueryContainer::SaveGameL(). The CAknListQueryDialog constructor takes a reference to a TInt. This represents the index of the selected item and can be set to a nonzero value to preselect a particular item in the list. The TInt passed in is updated by the control with the user selection.

```
TInt index(0);
CAknListQueryDialog* query = new (ELeave)
CAknListQueryDialog(&index);
```

As you might expect, the list query contains a list, which in turn contains an item array. In **ListQuery**, an item array is a set up in the local method SetupListQueryItemArrayL(), which is called from CListQueryContainer::ConstructL(). The item array is a standard list item array, which is covered further in Chapter 7.

Setting the item array is done once the construction of the query is completed by PrepareLC(). In **ListQuery** this is performed in CListQueryContainer::SaveGameL() by calling the query's SetItemTextArray() method.

```
query->PrepareLC(R_LISTQUERY_LIST_QUERY);
query->SetItemTextArray(iListQueryItemArray);
query->SetOwnershipType(ELbmDoesNotOwnItemArray);
```

To see which item was selected by the user, rather than just its index, the application should take ownership of the item array by calling `SetOwnershipType()`. The query destroys itself after execution, so if the application does not take ownership of the array, it will be destroyed too.

A list query is invoked in the same manner as other queries—that is, using `PrepareLC()` and `RunLD()` or using `ExecuteLD()`. In **ListQuery**, `PrepareLC()` is used, so `RunLD()` is used for execution.

```
if (query->RunLD())
    {
    SaveGameToFileL((*iListQueryItemArray)[index]);
    }
```

The execution methods return a `TBool` indicating whether the user closed the query by pressing **Ok** (`ETrue`) or **Cancel** (`EFalse`). Assuming the application owns the item array, it can obtain the selected item, by using the `TInt` passed to the query on construction, to index into the array of items. Here the resulting list item is passed to a local method called `SaveGameToFileL()`.

Multiselection List Queries

A multiselection list allows the user to choose more than one list item. A multiselection list query is created in the same way as a list query, with the following exceptions:

- An `AVKON_MULTISELECTION_LIST_QUERY_DLG_LINE` is used rather than a `AVKON_LIST_QUERY_DLG_LINE` in the `AVKON_LIST_QUERY`.

- The `listbox` in the `AVKON_LIST_QUERY_LIST` is set to be an `AVKON_MULTISELECTION_LIST_QUERY_LIST`.

- The list items will need to contain an icon index to refer to the check box—for example, "1\tMy Item". The list class supplies the bitmaps for the check boxes automatically, so unlike in multiselection lists, adding them to the list's icon array is not required.

- The list query is constructed, passing in a `TInt` array (`CSelectionIndexArray`). This will contain the indices of the selected items, once the query is dismissed.

Using Global Queries

A global query always takes focus, even when the application that owns it is running in the background. The user cannot task away from a global query, they must dismiss it, and so this type of query should be used sparingly—applications should not dominate the screen if the user has tasked away.

Global queries are slightly more complex than the queries seen so far. Their functionality relies on both the notifier framework and the Active Object mechanism. The notifier framework is an operating system facility that allows messages or dialogs to be displayed.

To produce a global query, an Active Object has to be constructed to handle the dismissal of the dialog.

The example application, **GlobalQuery**, implements a global query to inform the user that a game will be closed, as shown in Figure 6–26. This note is displayed when the application is opened.

Figure 6–26 Global query.

Creating a Global Message Query

GlobalQuery does nothing more than create a global query and then handle its dismissal—no other processing takes place. The query is created by the container class, `CGlobalQueryContainer`, during its construction. An Active Object, `CGlobalQueryQueryHandlerAO`, handles the notification of the dialog's dismissal from the framework. Because the **GlobalQuery** example is trivial, the Active Object simply informs an observer class that the dialog has been closed.

Since the container launches the query, it is also defined as the observer. The container class therefore implements the interface, `MGlobalQueryCloseGameObserver`, which has one method, `CloseGameL()`. A reduced class definition for `CGlobalQueryContainer`, presenting only the parts involved in the query construction, is shown below:

```
class CGlobalQueryContainer : public CCoeControl,
MGlobalQueryCloseGameObserver
   {
public: // from MGlobalQueryCloseGameObserver
   void CloseGameL();

private: // constructor
   void ConstructL(const TRect& aRect);

private: //data
   CGlobalQueryQueryHandlerAO* iQueryHandlerAO;
   CAknGlobalMsgQuery* iGlobalMsgQuery;
   };
```

Before a global query can be executed, the Active Object (`CActive` derived class) must be constructed and set active. As the query is shown upon startup of **GlobalQuery**, this is performed in `CGlobalQueryContainer::ConstructL()`.

```
iQueryHandlerAO = CGlobalQueryQueryHandlerAO::NewL(*this);
iQueryHandlerAO->Start();
```

The `NewL()` function of the Active Object takes a reference to a `MGlobalQueryCloseGameObserver` object. As the container class itself is the observer, a dereferenced `this` pointer is passed in.

In the **GlobalQuery** example, the Active Object's `Start()` method initiates the process. It simply sets it to be active using the `SetActive()` method.

The query itself is created in the container's `ConstructL()` method:

```
  HBufC* header =
StringLoader::LoadLC(R_GLOBALQUERY_HEADER_TEXT);
  HBufC* message =
StringLoader::LoadLC(R_GLOBALQUERY_MESSAGE_TEXT);

  iGlobalMsgQuery = CAknGlobalMsgQuery::NewL();
  iGlobalMsgQuery->ShowMsgQueryL(iQueryHandlerAO->iStatus,
*message, R_AVKON_SOFTKEYS_OK_CANCEL, *header, KNullDesC);

  CleanupStack::PopAndDestroy(message);
  CleanupStack::PopAndDestroy(header);
```

A global message query is constructed via `NewL()` and is executed using the `ShowMsgQueryL()` method. `ShowMsgQueryL()` has a number of parameters, some of which have default values and so are optional:

- `TRequestStatus& aStatus`—The `iStatus` of the Active Object that will handle the query's dismissal. The message query will set this to the value of the soft key used to dismiss it, which in turn will cause the Active Object's `RunL()` method to run.
- `const TDesC& aMsgText`—The text the query will display.
- `TInt aSoftkeys`—The resource identity for the soft keys.
- `const TDesC& aHeaderText`—A heading for the query.
- `const TDesC& aHeaderImageFile`—The name of the file containing the icon to be displayed with the query. The file should be an `.mbg` file. Set this to an empty string using `KNullDesC` if an icon is not required.
- `TInt aImageId`—The ID of the icon to be used for the query. This is optional, but must be specified if a file name is given in the previous parameter.
- `TInt aImageMaskId`—The ID of the icon mask. This is optional, but it must be specified if a file name is given in the previous parameter.
- `CAknQueryDialog::TTone aTone`—Tone to play when the query displays (optional).

Handling Dismissal of a Global Message Query

When the query is dismissed by the user, it causes the Active Object's `RunL()` to be called:

```
void CGlobalQueryQueryHandlerAO::RunL()
    {
    if (iStatus == EAknSoftkeyOk)
        {
        iCloseGameObserver.CloseGameL();
        }
    Cancel();
    }
```

Testing the value of `iStatus` will reveal which soft key dismissed the query, and appropriate action can be taken. (The soft key values are defined in `avkon.hrh`.) In the **GlobalQuery** example, if **Ok** is pressed, `MGlobalQueryCloseGameObserver::CloseGameL()` is called, which is implemented in the container. The Active Object is then cancelled, using its `Cancel()` method.

It is very important that the `RunL()` method should complete quickly.

Other Global Queries

In addition to global message queries, global confirmation queries and global list queries can be created. Creating these types of queries is very similar to creating the global message query. A global confirmation query should be created using `CAknGlobalConfirmationQuery::NewL()` and executed using `ShowConfirmationQueryL()`, while a global list query is created using `CAknGlobalListQuery::NewL()` and executed with `ShowListQueryL()`.

List Dialogs

Series 60 provides two types of list dialogs: a selection list dialog, as shown in Figure 6–27, and a markable list dialog, as shown in Figure 6–28. The lists behave in the same way as standard lists, but are contained within a dialog. A selection list dialog can optionally have a find pane—this allows the user to search for items in the list by entering matching characters from the start of the item name.

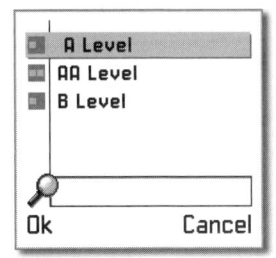

Figure 6–27 Selection list dialog with find pane.

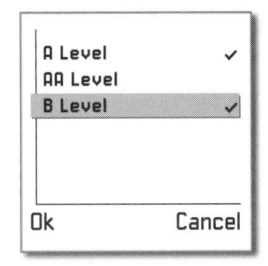

Figure 6–28 Markable list dialog.

Both types of list are constructed using the same techniques. The focus in this subsection is on the selection list dialog—the application **ListDlg** is used to illustrate the main principles involved.

On startup, the **ListDlg** application displays a list of saved games. Each has a small icon representing the number of players and a single line of text representing the name. There is a find pane at the bottom of the screen, which allows the user to search for items in the list. As characters are typed in, a filter is applied to the list so that only those items that match the entered text are listed, as shown in Figure 6–29. In the **ListDlg** example, selection of an item from the list results in a call to the empty method PlaySelectedGame()—no game-playing functionality has been implemented.

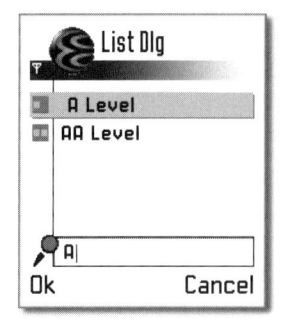

Figure 6–29 List filtering.

Defining a Selection List Dialog in a Resource

A selection list dialog should be defined in a DIALOG resource structure. A find pane is implemented in the resource file, so the DIALOG resource structure in this example has two DLG_LINE sections. If the filtering facility is not required, the second line can simply be omitted. The list items could be defined dynamically in the code, but here they are defined statically in the resource r_listdlg_list_array:

```
RESOURCE DIALOG r_listdlg_dialog
    {
    flags = EAknDialogSelectionList;
    buttons = R_AVKON_SOFTKEYS_OK_CANCEL;

    items =
       {
       DLG_LINE
          {
          type = EAknCtSingleGraphicListBox;
          id = ESelectionListControl;
          control = LISTBOX
             {
             flags = EAknListBoxSelectionList;
             array_id = r_listdlg_list_array;
             };
          },
       DLG_LINE
          {
          itemflags = EEikDlgItemNonFocusing;
          id = EFindControl;
          type = EAknCtSelectionListFixedFind;
          }
       };
    }
```

Chapter 6 List Dialogs

Further details of list types, item formats, and their appearance are provided in Chapter 7.

Constructing a Selection List Dialog

Construction of a selection list dialog is performed using its `NewL()` method. This function has three mandatory parameters. The first, a `TInt&`, is used to return the selected item in the list. The second parameter is the list item array. This is a standard list item array, containing descriptors representing the list items. In **ListDlg**, the list items are defined statically in the resource, so this parameter is set to `NULL`. However, if a dynamic array of items is required, an array of descriptors can be created and passed through instead. The list does not take ownership of the array, so its subsequent deletion must be performed externally. The third parameter should be the value of the resource for any menu bar required. In the **ListDlg** example, the dialog is defined in the resource file with **Ok** and **Cancel** soft keys, so there is no need for a menu bar, and so zero is passed in. If an **Options** menu is needed, an appropriate `MENU_BAR` resource should be passed in here.

In **ListDlg**, the dialog is constructed in `CListDlgContainer::ConstructL()`:

```
TInt openedItem(0);

// Construct and prepare the dialog
CAknSelectionListDialog* dialog =
CAknSelectionListDialog::NewL(openedItem, NULL, 0);
```

Adding Icons to a Selection List Dialog

For a selection list to contain graphics items, they must be added to the dialog's icon array.

```
dialog->PrepareLC(R_LISTDLG_DIALOG);
HBufC* iconFileName =
StringLoader::LoadLC(R_ICON_FILE_NAME);

CArrayPtr<CGulIcon>* icons = new (ELeave)
CAknIconArray(KNumberOfIcons);
CleanupStack::PushL(icons);

icons->AppendL(iEikonEnv->CreateIconL(*iconFileName,
EMbmDialoglistex11player, EMbmDialoglistex11player_mask));

icons->AppendL(iEikonEnv->CreateIconL(*iconFileName,
EMbmDialoglistex12player, EMbmDialoglistex12player_mask));

dialog->SetIconArrayL(icons); // transferring ownership of icons
CleanupStack::Pop(icons);
...
CleanupStack::PopAndDestroy(iconFileName);
```

The list's item drawer owns the icons. Icons cannot be added to the list before it exists, so the dialog must be prepared first. This is performed using the standard dialog `PrepareLC()` method, passing in the dialog resource. A standard array of list icons could then be created and set into the dialog using its `SetIconArrayL()` method. See Chapter 8 for further details on lists and icon arrays.

Executing a Selection List Dialog

After the dialog has been prepared, it is executed using the standard `RunLD()` method. If dialog preparation is not required, then the standard `ExecuteLD()` method can be used instead.

```
if (dialog->RunLD ())
    {
    PlaySelectedGame(openedItem);
    }
```

As usual, this returns a `TBool` indicating whether the dialog closed with a positive or negative action. Once closed, the `TInt`, supplied on construction, in this case `openedItem`, will contain the index value of the selected item.

Markable List Dialogs

To create a markable list dialog that allows the user to select more than one item, a few alterations are necessary to the **ListDlg** application.

The `DIALOG` resource should be defined as for the selection list, but the flags should be set to `EAknDialogMarkableList` and the `LISTBOX` flags should be `EAknListBoxMarkableList`. A markable list dialog is defined by the `CAknMarkableListDialog` class. It is instantiated by passing in an array (`CArrayFix<TInt>`) by reference, which contains the selected indices. A reference to a context-sensitive menu can also be passed in.

For a markable list without graphics, default marking icons will be supplied automatically. For a markable list with graphics, they are supplied as in the selection list example **ListDlg**. The marking graphics are added as the first element of the array. (The IDs for the standard marking graphic can be accessed from `avkon.mbg`.)

The array is executed using the usual `ExecuteLD()`, or `PrepareLC()` and `RunLD()`.

Summary

This chapter has described the different types of dialogs available in Series 60. Using example applications, it has demonstrated how to create the most commonly used types. Dialog types range from providing simple notifications to the user through to sophisticated data display and data entry capabilities.

Some guidance has been given about when each type of dialog should be used. It is recommended that you also consult the Series 60 UI Style Guide in the SDK documentation for further detailed information.

Dialogs are commonly used UI component that can be employed to convey information to the user or to query information from them. In the following chapters, you will learn about other controls that can be used to communicate with the user, such as lists and editor controls.

chapter 7

Lists

Lists are user interface controls for presenting a collection of items, in a one- or two-dimensional display

A list is a user interface control containing a number of items, each of which can be individually selected and activated. Series 60 provides a range of list types that are in common use throughout the platform. This chapter describes the various list types available, and discusses example applications to demonstrate how each type is used.

Series 60 lists are divided into three broad categories: **vertical lists**, **grids** and **settings lists**. Vertical lists are used for laying out a one-dimensional set of items—each occupying the full width of the screen, as shown in Figure 7–1.

Grids are used for two-dimensional arrangements of items, as shown in Figure 7–2.

Settings lists are a special type of vertical list, used for modifying an application's configuration settings, as shown in Figure 7–3.

Figure 7–1 Vertical list. 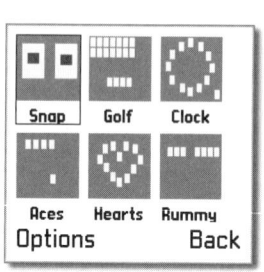 **Figure 7–2** Grid. 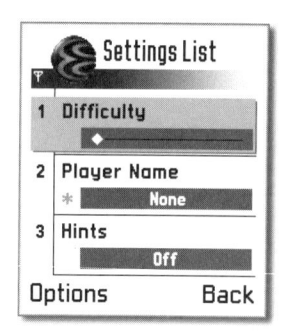 **Figure 7–3** Settings list.

Within these three broad categories, there are many subtypes of list, each designed to address a particular type of requirement. This chapter covers all of these subtypes and helps you to understand the typical use case for each. It includes the following sections:

- **List Basics**—A short section covering the fundamental features shared by all Series 60 lists.

- **Vertical Lists**—Description of the various types of vertical list in Series 60: selection lists, menu lists, markable lists, and multiselection lists, and information about when each type would be used.

Chapter 7   Lists Basics

- **Using Vertical Lists**—Full coverage of the example applications that illustrate each kind of vertical list.
- **Grids**—Explanation of the different types of grids available—selection grids, menu grids, and markable grids—and how to use them.
- **Using Grids**—Detailed coverage of the example code in detail, demonstrating how to create and use each kind of grid.
- **Settings Lists**—Description of settings lists and the associated settings controls.
- **Using Settings Lists**—Explanation of the example code used for creating settings lists and settings pages, plus a reference section with detailed information about individual settings list items.

Throughout this chapter, code snippets from example applications are used to illustrate the ideas under discussion. In each section the examples expand on the previous example, adding some new functionality. Since only new features in the code will be discussed, you are advised to work through each section in the order it is presented. The example applications will be introduced at the relevant point in the text.

A basic understanding of the Series 60 UI framework, and the concept of UI controls, will be useful before reading this chapter—further information on these topics is available in Chapter 4 and Chapter 5, respectively.

List Basics

The specific characteristics of each type of list will be discussed throughout this section. However, all lists share some basic characteristics, and is useful to establish them before progressing:

- List items must be brought into focus before they can be selected.
- Only one item can be in focus at once. The way focus is indicated depends on the list type, current UI scheme, and so on. Typically, a focused list item will have a different background color to other items and/or be surrounded by a border.
- Focus is moved using the standard scroll keys or joystick. Note that **Left** and **Right** are available only for grids.
- Once in focus, a list item is selected by using the **Selection** key, or by using an item from the **Options** menu. (This item will usually be the uppermost on the **Options** menu—its text will depend on the context.)

Selection behavior (that is, the action performed when an item is selected) depends on the individual list's implementation.

Vertical Lists

Vertical lists display to a user a one-dimensional set of items, each occupying one line in the list. Various types of vertical list exist—Table 7–1 provides an overview of their functionality, with each type described in detail in this section. An example application illustrating each type provides a practical guide to using vertical lists.

Selection Lists

A selection list allows one item at a time to be selected. If you need to enable the user to select multiple items simultaneously, you should use a multiselection list. The items in a selection list represent persistent data—if the user switches away from the application or closes it, the items will all be the same when the list is next displayed. An example use for a selection list would be for a game to display a list of saved games, as shown in Figure 7–4. Selecting an item allows you to do several other things as well, like opening a more detailed view to show details of the game, which the user could edit. You can also execute a command, such as to play the game or open a context-sensitive menu.

Figure 7–4 Selection list.

Table 7–1 List Types

Type	Example	LISTBOX Flags in Resource File
Selection	**SimpleList**	EAknListBoxSelectionList
Menu	**PopUpList**	EAknListBoxMenuList
Markable	**MarkableList**	EAknListBoxMarkableList
Multiselection	**MarkableList** can be easily changed to create this functionality	EAknListBoxMultiselectionList

Menu Lists

Menu lists appear in a pop-up window in place of the **Options** menu and are cancelled by high-priority events (for example, incoming calls). Note that the **Options** menu itself is a menu list.

The usual way to invoke a menu list is by choosing an item from a selection list—the menu list then contains further options for the selected item. For example, selecting a saved game from a selection list could invoke a menu list showing levels from a previously saved game, as shown in Figure 7–5. Selecting an item from the pop-up menu invokes a command, such as playing a game at the selected level.

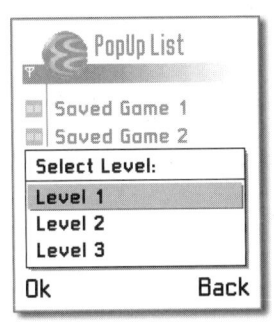

Figure 7–5 Menu list.

Menu lists are very similar to context-sensitive menus invoked by the **Selection** key. You should use a menu list when you need to provide the same set of actions for a number of items (for example, to play a game at a specific level). A context-sensitive menu should be used when you need to provide alternative actions for a selected item (for example, to play or delete a selected game).

	Keys		
Scroll Up/Down	**Selection Key**	**Left Soft Key**	**Right Soft Key**
Moves focus	Selects item	Opens options menu	Back-steps
Moves focus	Selects item	Selects item	Cancels list and returns to previous state
With Edit key: Marks or unmarks all items scrolled over Without Edit key: Moves focus	With Edit key: Marks or unmarks currently focused item Without Edit key: Selects item	Selects item	Unmarks all items in list and back-steps
Moves focus	Check or uncheck item	Accept list checks	Cancels list and return to previous state

Markable Lists

Markable lists allow a user to mark multiple items and perform an action on all of them. For example, in a markable list of saved games, the user may choose to delete all of the marked games, as shown in Figure 7–6.

Selecting **Mark**, **Mark All**, **Unmark** or **Unmark All** from the **Options** menu, or using certain keys, marks and unmarks the items graphically. Note that nothing is displayed by a markable list (such as an empty checkbox) to indicate explicitly that an item can be marked. If you need to show explicitly that items should be marked, you should use a multiselection list instead.

Figure 7–6 Markable list.

When a command is invoked from the **Options** menu (for example, **Delete**), it will be applied to the marked items only—if an item is focused but not marked, it will not be affected by the command. Pressing the **Selection** key could invoke a command (for example, **Delete**) on the marked items or open a context-sensitive menu. Further details of context-sensitive menus can be found in the **Menus** section of Chapter 5.

Multiselection Lists

Multiselection lists, as shown in Figure 7–7, allow you to mark a number of items and perform an action on all of them.

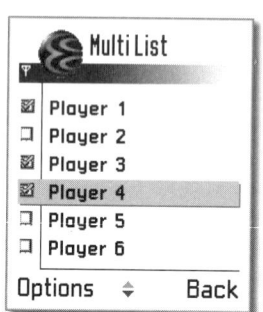

In contrast to markable lists, the ability to mark items in a multiselection list is transparent—usually it is shown by means of a checkbox. Multiselection lists should be used in preference to markable lists if you need to show that there is a *requirement*, rather than an *option*, to perform an action on multiple items.

Figure 7–7 Multiselection list.

Items are checked and unchecked using the **Selection** key. Pressing the left soft key accepts the list of checked items and invokes a command (for example, to play a multi-player game against the checked players). Note that the **Options** menu is not available in a multiselection list.

Markable and multiselection lists are very useful for speeding up repetitive actions for a user. For example, deleting multiple items from a long list of saved games in a selection list would take a long time and many key presses to perform, but less time and key presses in a markable or multiselection list.

List Items and Fields

As you have seen, a list contains a number of items, which may be individually selected and activated. Items may in turn be made up of smaller components

Chapter 7 Using Vertical Lists

referred to as *fields*. Fields cannot be individually selected or activated—they are just the building blocks used to construct list items. Series 60 supports the following types of field:

- Icons
- Headings
- Numbers, or times
- One or more lines of text

For example, each item in a list of saved games might consist of an icon representing the number of players, and two lines of text—one for the saved game name, and one for a short description. An item in a list of high scores may have a number indicating the position within the high-score table, and a single line of text indicating the player's name.

Note that, in any given list, every item has exactly the same type and layout of fields.

Finding Items in a List

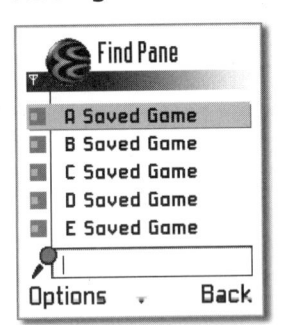

You can search for items in a list using a search field (also known as a **find pane**), as shown in Figure 7–8. This is a particularly useful feature in long lists. You type into the search field part of the text of the item you wish to find. As you type, the list filters to display only those items that start with text you have typed.

Figure 7–8 Search field.

Using Vertical Lists

This section walks you through four example applications—**SimpleList**, **DynamicList**, **MarkableList**, and **PopUpList**—that explore the list features described.

SimpleList just makes very simple use of lists, in order to show you the basics. Specifically, it demonstrates how to use resources to define a list and list items (including icons), and how to instantiate the list. It also shows how to ensure scrollbars will be displayed if your list requires them and how to handle list events.

DynamicList covers the use of dynamic list items. In particular, it describes how to define a dynamic list resource and to set and change items dynamically in a list.

MarkableList builds on the **DynamicList** example to demonstrate how to produce a list that can be marked (by a tick or some other icon). It shows you how to define the icons to be used for marking and how to determine which items have been marked. Finally, **PopUpList** explains how to create a pop-up list, set the title for a pop-up list, display a pop-up list and handle selections.

Basic Lists

The application **SimpleList** displays a list of saved games—each has a small icon representing the number of players, and a single line of text representing the name, as shown in Figure 7–9. When you select an item from the list, the appropriate game opens. The code for playing the game was not implemented, but instead an empty method, `PlaySelectedGame()`, is used to indicate where the game-playing functionality should be.

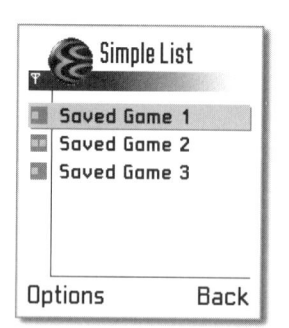

Figure 7–9 SimpleList screen.

Defining a List Using Resources

The first step when creating a list is to define it a resource. This is achieved using a `LISTBOX` resource, as in `SimpleList.rss`:

```
RESOURCE LISTBOX r_simplelist_saved_games_listbox
    {
    array_id = r_simplelist_saved_games_items;
    flags = EAknListBoxSelectionList;
    }
```

The `flags` field determines the type of list—in **SimpleList**, a selection list is chosen. The `array_id` field refers to an array of `LBUF` strings, which defines the items contained in the list. You will see how to define the actual strings representing each item in the list later in the **SimpleList** example.

> **NOTE** Series 60 ignores the `height` and `width` flags in the `LISTBOX` resource. In addition, you should not change the `version` field.

A list can be made to loop (that is, when the top item is in focus and the up key is pressed, focus moves to the bottom item, and vice versa). To do this you need to add the flag `EAknListBoxLoopScrolling` to the flags list. For example:

```
flags = EAknListBoxSelectionList | EAknListBoxLoopScrolling;
```

Instantiating a List Using Resources

In the **SimpleList** example, the list is declared as member data in the header file of the container class, `SimpleListContainer.h`:

```
CAknColumnListBox* iSavedGamesListBox;
```

The list is instantiated in the `ConstructL()` method of the container by calling a private convenience method `CreateListL()`:

```
void CSimpleListContainer::CreateListL()
    {
    iSavedGamesListBox = new (ELeave) CAknSingleGraphicStyleListBox();
    iSavedGamesListBox->SetContainerWindowL(*this);

    TResourceReader reader;
    iEikonEnv->CreateResourceReaderLC(reader,
R_SIMPLELIST_SAVED_GAMES_LISTBOX);
    iSavedGamesListBox->ConstructFromResourceL(reader);
    CleanupStack::PopAndDestroy(); // reader
    }
```

A list of appropriate type is constructed, in this case `CAknSingleGraphicStyleListBox`, and a list resource (in this example, `R_SIMPLELIST_SAVED_GAMES_LISTBOX`) is passed in during the second phase of construction. The name of the list class indicates the fields contained in each item of the list. `CAknSingleGraphicStyleListBox` indicates a single line of text and a small graphic. (Table 7–2 provides a graphical representation of each of the available list types to help you decide which is appropriate.)

Adding Icons to a List

For the list defined in the **SimpleList** example (`CAknSingleGraphicStyleListBox`), an icon of size 13 by 13-pixels is displayed on the left-hand side. Two bitmaps are required for each icon—the graphic itself and a mask for the icon, and these are packaged into an `.mbm` file. The `.mbg` file, generated when creating the `.mbm` file, should be included in the file implementing the container class, as this contains the enumerated indices for each bitmap stored within the `.mbm` file.

> **TIP** Lists will crop any icons supplied that are too large. If you do not know the correct size for an icon, supply an icon of any size to the list, run up the debugger, and look at the output window. It will say that the supplied icon is not of the expected size and give the size that it should be. For example, "`Error: Icon size (50, 50) in listbox different from expected size (13, 13)`."

Table 7–2 List Item Definitions

Class	Example
CAknSingleStyleListBox	Label
CAknSingleGraphicStyleListBox	▪ Label
CAknSingleNumberStyleListBox	1 Label
CAknSingleHeadingStyleListBox	Head Label
CAknSingleGraphicHeadingStyleListBox	▪ Head Label
CAknSingleLargeStyleListBox	Big Label
CAknSingleNumberHeadingStyleListBox	1Head Label
CAknDoubleStyleListBox	Main Label / Secondary Label
CAknDoubleStyle2ListBox	Main Label / Secondary Label
CAknDoubleNumberStyleListBox	1 Main Label / Secondary Label
CAknDoubleTimeStyleListBox	14.00 Main Label / Secondary Label
CAknDoubleLargeStyleListBox	Big Main Label / Secondary Label

In **SimpleList**, the icons are added to the list in the SetupListIconsL() method, which is called by ConstructL():

```
void CSimpleListContainer::SetupListIconsL()
  {
  HBufC* iconFileName;
  iconFileName = StringLoader::LoadLC(R_ICON_FILE_NAME);

  CArrayPtr<CGulIcon>* icons = new(ELeave)
```

Table 7–2 continued

Col A	Col B	Col C	Example Item String
Main text			"\tLabel"
13 by 13 icon	Main text		"1\tLabel"
Number	Main text		"1\tLabel"
Heading		Main text	"Head\tLabel"
13 by 13 icon	Heading	Main Text	"0\tHead\tLabel"
42 by 29 icon		Main text	"1\tLabel"
Number	Heading	Main text	"1\tHead\tLabel"
Main text Secondary text			"\tMain Label\tSecondary Label"
Main text Secondary text			"\tMain Label\tSecondary Label"
Number	Main text Secondary text		"1\tMain Label\tSecondary Label"
Time	AM or PM	Main text Secondary text	"13.00\tPM\tMain Label\tSecondary Label"
42 by 36 icon	Main text Secondary text		"1\tMain Label\tSecondary Label"

```
CAknIconArray(KNumberOfIcons);

  CleanupStack::PushL(icons);
  icons->AppendL(iEikonEnv->CreateIconL(*iconFileName,
EMbmSimplelist1player, EMbmSimplelist1player_mask));

  icons->AppendL(iEikonEnv->CreateIconL(*iconFileName,
EMbmSimplelist2player, EMbmSimplelist2player_mask));

  CleanupStack::Pop(icons);
```

```
CleanupStack::PopAndDestroy(iconFileName);

iSavedGamesListBox->ItemDrawer()->ColumnData()
->SetIconArray(icons);
    }
```

To create the icons, you need the name of the .mbm file containing the bitmaps for the icons. In **SimpleList**, this filename is loaded from the R_ICON_FILE_NAME resource, and equates to \system\apps\simplelist\simplelist.mbm.

The icons are stored in an array, and this is populated using CEikonEnv::CreateIconL(), which uses the name of the .mbm file, and two enumerated indices from the .mbg file (for the icon and its mask) to create each CGulIcon.

The list has a drawer object, which is responsible for drawing the list on screen. You can get a handle to this by using the ItemDrawer() method. The drawer contains a data object, which determines how each item will appear in the list. In the **SimpleList** example, a handle to the data object is acquired using the drawer's ColumnData() method.

The data object owns the icon array, so you should give it a pointer to the icon array you have just created and populated, using the SetIconArray() method. As you are passing ownership of the array, there is no need to handle its deletion.

> **NOTE**
> **SimpleList** uses a list derived from CAknColumnListBox. This has a drawer object of type CEikColumnListDrawer, which uses the ColumnData() method to get a handle on its data object. If the list derived from CEikFormattedCellListBox, which has a CEikFormattedCellDrawer, then you would get a handle on its data object using the FormattedCellData() method.

Defining List Items Using Resources

When you define the list in a resource, the list items are defined as an array_id field in the list structure. This refers to an array of LBUF strings, one for each list item. In the **SimpleList** example, the list items are defined statically in the resource file. In a real-world application, list items are often dynamically inserted; this is covered in the *Dynamic Lists* subsection later in this chapter.

The string for a list item is separated into columns by the tab escape sequence ("\t"). Each string defined for the **SimpleList** example contains one field for the icon and one for the game name—for example,
#define SAVED_GAME1_TEXT "0\tSaved Game 1". Here, the number "0" denotes the icon to use; it specifies the position of an icon in the icon array (created previously in *Adding Icons to a List*). The game name is defined as "Saved Game 1".

Lists have three virtual columns, referred to as A, B and C, as shown in Figure 7–10. Each field within a list item occupies one or more columns. Column A can

Chapter 7   Using Vertical Lists   353

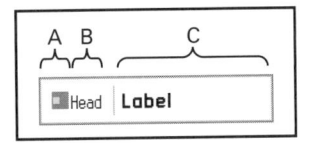

Figure 7–10 List virtual columns.

contain a graphic or a number, column B a heading, and column C the main text for the item.

If required, a field can span more than one column. The span of columns AB can contain a heading or a large graphic, as shown in Figure 7–11. The span of columns BC or ABC can contain the main text of an item.

The list type determines the columns in which fields will reside. In the **SimpleList** application, there is a small icon in column A, and the main text in columns BC, as shown in Figure 7–9.

Table 7–2 also contains string formats for each list type.

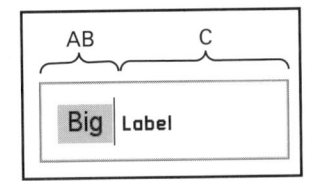

Figure 7–11 List virtual columns with span across columns A and B.

Column C can contain additional indicator icons. These are small icons, which need not appear in each item of the list (hence, they are part of column C, rather than columns in their own right). To specify these additional indicators, you will need to add \t and the number of the icon to the end of the item strings for each indicator. For example, if a third icon had been defined, it could be used as an indicator in SAVED_GAME1_TEXT using "0\tLevel 3\t2".

WARNING Incorrect item strings are a common cause of panics when constructing lists. If your application crashes as soon as you attempt to launch it, check that you have formatted the strings correctly, and that the values for icons are numeric and within the bounds of the icon array.

Allowing the List to Scroll

The **SimpleList** example list contains only three items and so has no reason to scroll. However, it is good practice to ensure that any list has the capability to scroll. This means that if further items are added to the list later, then the list will still operate correctly.

In the **SimpleList** example, the scroll bars are created in the ConstructL() function by calling the private convenience method SetupScrollBarsL().

```
void CSimpleListContainer::SetupScrollBarsL()
    {
    iSavedGamesListBox->CreateScrollBarFrameL(ETrue);
    iSavedGamesListBox->ScrollBarFrame()
->SetScrollBarVisibilityL(
CEikScrollBarFrame::EOff, CEikScrollBarFrame::EAuto);
    }
```

Scroll bars are automatic for all controls apart from lists and editors, where the developer needs to enable scrollbars explicitly. (For editors, to enable the

Chapter 7 Lists

scrollbars, a different API is used than is used for lists—see Chapter 8 for details.) To enable the scroll bars for a list, you use the list's `CreateScrollBarFrameL()` method. Note that, although `CreateScrollBarFrameL()` can have an argument to indicate that scrollbars should be preallocated, this has no effect.

The scroll bars are made visible using the scroll bar frame's `SetVisibility()` method. The arguments specify visibility for horizontal and vertical scrollbars, respectively. In the **SimpleList** example, horizontal scroll bars do not appear, and vertical scroll bars appear only when necessary.

Handling List Events

In order to perform an action on a selected item in the list, you will need to generate and handle list events. In **SimpleList**, the user can choose to open a selected game.

The list will generate events only if you offer key presses to it. You do this by overriding `OfferKeyEventL()` in the container such that it passes key events to the list.

```
TKeyResponse CSimpleListContainer::OfferKeyEventL(
const TKeyEvent& aKeyEvent, TEventCode aType)
    {
    if (iSavedGamesListBox)
        {
        return iSavedGamesListBox->OfferKeyEventL(aKeyEvent,
aType);
        }
    else
        {
        return EKeyWasNotConsumed;
        }
    }
```

To handle list events you need to implement the `MEikListboxObserver` mixin and register with the list to receive events. In **SimpleList**, the mixin is implemented by the container class, and its pure virtual method `HandleListBoxEventL()` is overridden.

```
void CSimpleListContainer::HandleListBoxEventL(
sCEikListBox* /*aListBox*/, TListBoxEvent aListBoxEvent)
    {
    if (aListBoxEvent ==
MEikListBoxObserver::EEventEnterKeyPressed)
        {
        PlaySelectedGame();
        }
    }
```

Pressing the **Selection** key creates an `EEventEnterKeyPressed` event. In **SimpleList**, the `PlaySelectedGame()` method is called when this occurs.

Chapter 7 Using Vertical Lists   355

In order for the observer to receive events, you register it with the list, using the line `SetListBoxObserver()` method. This is performed in the `ConstructL()` method, using the line `iSavedGamesListBox->SetListBoxObserver(this)`.

> **TIP** List box observers can register themselves with more than one list in an application. The `HandleListBoxEventL()` method has a list box parameter that indicates which list generated the event.

Dynamic Lists

To demonstrate how you can alter the contents of a list dynamically, the application **DynamicList** is considered. This example displays a list of saved games, as shown in Figure 7–12, the names of which are set at runtime. Selecting **Delete** from the **Options** menu, as shown in Figure 7–13, will delete the currently focused item.

Figure 7-12 DynamicList screen.

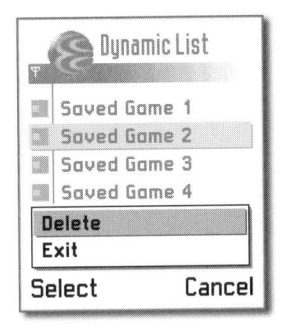

Figure 7-13 DynamicList options menu.

Defining a Dynamic List Resource

When defining the `LISTBOX` resource for a dynamic list, you do not need the `array_id` field, as you will create the item array at runtime. The list resource definition is therefore very simple, as shown in `DynamicList.rss`:

```
RESOURCE LISTBOX r_dynamiclist_saved_games_listbox
    {
    flags = EAknListBoxSelectionList;
    }
```

Setting Items Dynamically in a List

Once the list is constructed, you can add items to it. This is performed in the `ConstructL()` method by calling the private convenience method `SetupListItemsL()`:

```
void CDynamicListContainer::SetupListItemsL()
    {
    CTextListBoxModel* model = iSavedGamesListBox->Model();
    model->SetOwnershipType(ELbmOwnsItemArray);

    CDesCArray* savedGamesArray = STATIC_CAST(CDesCArray*,
model->ItemTextArray());
    LoadSavedGamesL(*savedGamesArray);
    }
```

You will need to get a pointer to the list's model (which manages the list items) using the `Model()` method. Note that, although the `Model()` method returns a pointer to the model, it does not transfer ownership of the model. In other words, the returned pointer must not be deleted or pushed onto the Cleanup Stack.

The list items are defined by descriptors, which are stored in an array of type `CDesCArray`. You can choose whether the model owns the array, using the `SetOwnershipType()` method. The model owns the item array by default—although here the model is explicitly set as the owner, both to demonstrate the method and for maintainability.

If the model owns the array, as in the **DynamicList** example, you can get a pointer to it using the `ItemTextArray()` method. You will need to cast it up to a `CDesCArray` pointer before you can populate it, as the returned `MDesCArray` pointer does not give you functionality to do this. Further information on descriptor arrays, such as `CDesCArray`, can be found in Chapter 3.

You can then populate the array with item strings, which you can create in whatever way you choose. In **DynamicList** this is performed using the `LoadSavedGamesL()` method, which simply creates a list of saved game names of the form "`Saved Game X`", where X is a value initially set to 1, which is incremented as each item is created in a loop.

TIP If you need to own the array, you should set the ownership type accordingly and then call the model's `SetItemTextArray()` method to set it.

Changing Items Dynamically in a List

It is possible to alter the items in your list dynamically. To demonstrate this, **DynamicList** provides an option to delete an item—selecting **Delete** from the **Options** menu results in a call to the container method `DeleteSelectedL()`:

```
void CDynamicListContainer::DeleteSelectedL()
    {
    if (iSavedGamesListBox)
        {
```

```
            CTextListBoxModel* model = iSavedGamesListBox->Model();

            if (model->NumberOfItems() > 0)
                {
                CDesCArray* itemArray = STATIC_CAST(CDesCArray*,
model->ItemTextArray());

                TInt currentItem = iSavedGamesListBox
->CurrentItemIndex();

                itemArray->Delete(currentItem);

    AknListBoxUtils::
    HandleItemRemovalAndPositionHighlightL(iSavedGamesListBox,
    currentItem, ETrue);
                }
            }
        }
```

To access the item text array, you will need to get a pointer to the model first. You should then check that there is actually something to delete. The **DynamicList** example ensures that the number of items in the model is greater than zero. The list item that you want to delete is the one that has been selected by the user. To get the index of the currently selected item, call the list's `CurrentItemIndex()` method. This will return the position within the item text array of the required item. The item can then be removed from the list using the array's `Delete()` method, passing in the item's index as an argument. To effect the change on screen, use the static method `AknListBoxUtils::HandleItemRemovalAndPositionHighlightL()`, passing in the list, the index of the removed item, and `ETrue` to indicate that you have removed the currently selected item.

> **NOTE** For portability, note that the `HandleItemRemovalAndPositionHighlightL()` method is available only in Series 60, not in other Symbian OS platforms. You should not use `CEikListBox::HandleItemRemovalL()` in Series 60, as this causes indexing problems for the current and top item.

Markable Lists

The application **MarkableList** demonstrates the implementation of a list where the individual items can be marked. On opening, the application displays a list of saved games. Any item from this list can be marked by selecting **Mark** from the menu, as shown in Figure 7–14. When an item in the list is marked, a small tick displays next to it, as shown in Figure 7–15. Selecting **Delete** from the **Options** menu deletes the marked items.

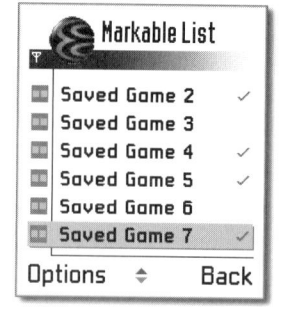

Figure 7–14 MarkableList Options menu.

Figure 7–15 MarkableList screen.

Defining a Markable List Using Resources

The `LISTBOX` resource in this case indicates that the list is markable, as in `MarkableList.rss`:

```
RESOURCE LISTBOX r_markablelist_saved_games_listbox
    {
    flags = EAknListBoxMarkableList;
    }
```

Add Marking Icons to a List

To indicate that a list item has been marked, a specified icon is displayed next to it. Therefore, you will need to create an icon for this purpose. In the **MarkableList** example, the icon chosen is a tick. Adding the icon to the list is performed in exactly the same way as in the **SimpleList** example, with the marking icon being defined as the first element in the array.

Handling Marking Commands

The user needs to be able to mark and unmark items via the **Options** menu, so a submenu for marking is defined with options for the user to mark and unmark specific items, or mark and unmark all items. These menu items have specific system-defined commands associated with them: `EAknCmdMark`, `EAknCmdUnmark`, `EAknMarkAll`, and `EAknCmdUnmarkAll`, respectively.

The list handles marking commands by calling the static method `AknSelectionService::HandleMarkableListProcessCommandL()`. In the **MarkableList** example, this is performed in the container class method `HandleMarkCommandL()`, which is called by the AppUi whenever it receives a marking command from the menu.

```
void CMarkableListContainer::HandleMarkCommandL(TInt aCommand)
    {
    if (iSavedGamesListBox)
```

Chapter 7 Using Vertical Lists

```
        {
        AknSelectionService::
        HandleMarkableListProcessCommandL(
        aCommand, iSavedGamesListBox);
        }
    }
```

The marking command is simply passed on to `HandleMarkableListProcessCommandL()`, along with a pointer to the list that is to be marked. Note that the framework provides functionality for marking items via key presses, rather than the **Options** menu, so you do not need to handle this in code.

Dynamically Changing Marked List Items

As with **DynamicList**, the deletion of an item will be used to demonstrate dynamically changing a marked list. (Remember, you need to add the **Delete** command to the **Options** menu, and handle it in the usual way.) In **MarkableList**, the container method `DeleteSelectedL()` is called to handle the deletion.

```
void CMarkableListContainer::DeleteSelectedL()
    {
    if (iSavedGamesListBox)
        {
        CTextListBoxModel* model = iSavedGamesListBox
->Model();

        if (model->NumberOfItems() > 0)
            {
            // Create a copy of the currently selected
            // items (copyIndices) in numeric order.
            const CListBoxView::CSelectionIndexArray*
                selectionIndices = iSavedGamesListBox
                ->SelectionIndexes();

            RArray<TInt> copyIndices;
            CleanupClosePushL(copyIndices);
            TInt numberSelectedGames = selectionIndices
->Count();

            for (TInt i = 0; i < numberSelectedGames; i++)
                {
                copyIndices.InsertInOrder (
(*selectionIndices)[i]);
                }

            // Iterate through the copyIndices in reverse order,
            // deleting them from the list's item array
            TInt currentItem(0);
```

```
                CDesCArray* itemArray = STATIC_CAST(
                  CDesCArray*, model->ItemTextArray());

                for (TInt j = numberSelectedGames-1; j >= 0; j--)
                  {
                  currentItem = copyIndices[j];
                  itemArray->Delete(currentItem);
                  }

                // Close copyIndices and free memory
                CleanupStack::PopAndDestroy();

                // Redraw the list following removal
                AknListBoxUtils::
                HandleItemRemovalAndPositionHighlightL(
                  iSavedGamesListBox,
                  currentItem,
                  ETrue);
                }
        }
    }
```

The marked items are stored in an array in the list, which you access using the `SelectionIndexes()` method. As with the **DynamicList** example, you delete them from the item text array using the `Delete()` method. The array contains the indices in the order in which they were marked, not numerical order.

Because you have only the indices of the array, and not the objects in it, you will have to be careful about the order in which you delete—otherwise you could end up deleting items that are not marked. Consider a list in which items 3 and 4 have been marked. If you first delete item 3, when you delete item 4, it will delete what is currently in position 4, in other words, what was originally item 5. To get around this, you should copy the indices from the array into an `RArray` in numerical order. You can then iterate through this array in reverse order, deleting the items in turn.

You will need to call `HandleItemRemovalAndPositionHighlightL()` to update the screen. Take care, however, when using this method, as it resets the marks and the selection indices. Call it after you have deleted all marked items from the array; otherwise it will only delete the first item.

To use a multiselection list in an application, rather than a markable list, is straightforward. You change the `LISTBOX` resource flags to `EAknMultiselectionList` and supply two icons only: one for checked and one for unchecked. The checked icon must again be the first in the icon array. The unchecked icon should be shown in the A column of each menu item—for example, "`1\tSaved Game 1`". If you still wish to show the number-of-player's icon, supply those icons as well, but since the checking icon is in column A, put the number-of-player's icon in column C—for example, "`1\tSaved Game 1\t2`".

Chapter 7 Using Vertical Lists   361

Pop-up Menu Lists

The example application, **PopUpList** displays a list of saved games, as shown in Figure 7–16. When you select a saved game, it displays a pop-up menu list showing a number of levels at which you can restart the game, as shown in Figure 7–17. If the code had been implemented, selecting a level would play the game at the appropriate point.

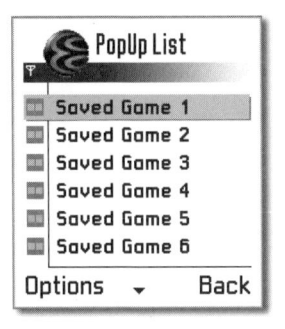

Figure 7–16 PopUpList screen.

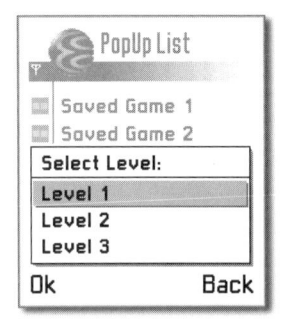

Figure 7–17 PopUpList menu list.

Creating a Pop-up List

The usual way to invoke a pop-up menu list is to select an item from a selection list. In the **PopUpList** example, the pop-up list is displayed when a saved game is selected. It is common to construct it in response to an `EEventEnterKeyPressed` event in `HandleListBoxEventL()`. In **PopUpList**, the list is displayed through a call to `OpenMenuListForSavedGameL()`, with the name of the currently selected item passed as a parameter.

```
void CPopUpListContainer::OpenMenuListForSavedGameL(
TDesC& aSavedGameName)
    {
    CAknSinglePopupMenuStyleListBox* savedGameMenuList = new
(ELeave) CAknSinglePopupMenuStyleListBox;

    CleanupStack::PushL(savedGameMenuList);

    CAknPopupList* popupList =
CAknPopupList::NewL(savedGameMenuList,
R_AVKON_SOFTKEYS_OK_BACK);

    CleanupStack::PushL(popupList);
    savedGameMenuList->ConstructL(popupList,
EAknListBoxMenuList);
    ...
```

You need to construct the pop-up list using an appropriate list class. Here `CAknSinglePopupMenuStyleListBox` is used, to enable a single line of text to

Table 7–3 Pop-up Menu List Classes and Item Definitions

Class	Example
CAknSinglePopupMenuStyleListBox	Label
CAknSingleGraphicBtPopupMenuStyleListBox	■ Label
CAknSingleHeadingPopupMenuStyleListBox	Head Label
CAknSingleGraphicPopupMenuStyleListBox	■ Label
CAknDoublePopupMenuStyleListBox	Main Text / Secondary Text
CAknSingleGraphicHeadingPopupMenuStyleListBox	■Head Label
CAknDoubleLargeGraphicPopupMenuStyleListBox	Big Main Text / Secondary Text

be displayed in a menu list. As with the other types of list, it is also possible to display items containing small or large graphics, two lines of text, or headings by using a different class. Table 7–3 contains details of the pop-up list classes.

At this point, you do not continue construction in resource, but instead construct a `CAknPopupList` object, passing a pointer to the list, and a soft key resource. The `CAknPopupList` is a helper class, which displays the list in a pop-up window and provides a set of soft keys for dismissing it. You then complete the menu list's construction by calling its `ConstructL()` method, passing in the pop-up list as its parent, and a type. In the **PopUpList** example, a type of `EAknListBoxMenuList` is passed in, indicating that the list is a menu.

Setting the Title for a Pop-up List

The title of a pop-up list should describe the list's contents or give the user an indication of what they should do—"`Select Level:`" is used in the **PopUpList** example. As with other strings, you should load the title from a resource file. You then set it in the `CAknPopupList`, using the `SetTitleL()` method:

```
…
HBufC* title;
title = StringLoader::LoadLC(R_SAVED_GAME_MENU_LIST_TITLE);
popupList->SetTitleL(*title);
CleanupStack::PopAndDestroy(title);
…
```

Chapter 7 Using Vertical Lists

Table 7-3 continued

Col A	Col B	Col C	Example Item String
Main text			"Label"
13 by 13 icon	Main text		"1\tLabel"
Heading	Main text		"Head\tLabel"
13 by 13 icon	Main text		"1\tLabel"
Main text Secondary text			"MainText\tSecondaryText"
13 by 13 icon	Head	Main text	"1\tHead\tLabel"
42 by 32 icon	Main text Secondary text		"1\tMainText\tSecondaryText"

Displaying a Pop-up List and Handling Selections

Once constructed, displaying the pop-up list is simply a matter of calling its `ExecuteLD()` method. In the **PopUpList** example, this is performed in the continuation of the `OpenMenuListForSavedGameL()` method:

```
...
TInt popupOk = popupList->ExecuteLD();
CleanupStack::Pop(popupList);

if (popupOk)
    {
    TDesC level =
(*savedGameMenuListArray)[savedGameMenuList
->CurrentItemIndex()];
    CleanupStack::PopAndDestroy(savedGameMenuList);

    PlayGame(aSavedGameName, level);
    }
else
    {
    CleanupStack::PopAndDestroy(savedGameMenuList);
    }
...
```

The method does not return until you dismiss the pop-up menu. If you select an item and accept the selection, in this case by pressing the **Ok** soft key, it returns `ETrue`; otherwise, it returns `EFalse`. You can handle an accepted selection using the usual list methods—in this case, the index of the selected item is used to play the chosen level of a game.

TIP The suffix `D` on the `ExecuteLD()` method indicates that the method destroys the pop-up list. Because of this, it should only be popped it from the Cleanup Stack rather than using `PopAndDestroy()`.

List Dialogs

You can implement selection and markable lists using predefined dialog classes, such as `CAknSelectionListDialog` or `CAknMarkableListDialog`. These simplify the use of lists, in particular when using a search field. They are covered in Chapter 6.

Grids

Grids are UI controls similar to the lists seen earlier in this chapter, in that they display a number of items that can be individually focused and selected. The difference between grids and other kinds of list is that a grid may contain more than one item on a single line. In other words, a grid displays a two-dimensional set of items, a list only a one-dimensional set. Figure 7–18 shows an example of a grid control.

Each item in a grid occupies one *cell*, and each *row* contains multiple cells. As with lists, Series 60 provides a number of different types of grid. Table 7–4 shows the available grid types.

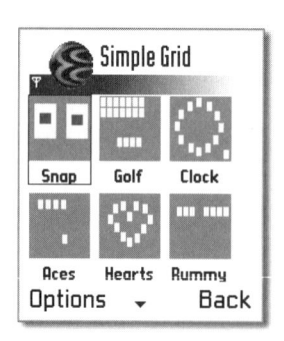

Figure 7–18 Grid control.

The functionality of each grid type is analogous with that of the list types previously described.

Table 7–4 Grid Types

Type	Flags to Use in the `GRID` Resource
Selection	`EAknListBoxSelectionGrid`
Markable	`EAknListBoxMarkableGrid`
Multiselection	`EAknListBoxMarkableGrid`
Menu	`EAknListBoxMenuGrid`

Chapter 7 Grids

Each cell within a grid can contain one or more graphics or pieces of text. Each is referred to as a *sub-cell*. As with list item fields, a sub-cell is not individually selectable: it is just a component part of the cell, which displays a particular type of data. For example, a grid whose cells contain an icon and a text caption would consist of two sub-cells: one for the icon and one for the text.

The grid's *orientation* refers to the plane in which items append to it—this can be either horizontal or vertical. You can control the order in which items focus programmatically, such that the focus loops around rows and columns, snakes from row to row or column to column, or stops at particular points. The focus scrolls along rows or columns. When reaching an edge, it can loop, snake or stop. Looping means that if the user attempts to scroll past an edge, it returns to the opposite end of the row or column. Snake means that it moves to the opposite end of the next row or column. Stop means that it will stay at the edge.

As with lists, grid items can be defined dynamic or statically. Unlike lists, you usually determine a grid's layout by calculating exact values, such as the size of the cells and the sub-cells within them, in your own concrete grid class.

The following subsections contain information you may find useful about some partially complete concrete grid classes that are available.

Monthly Calendar Grid

This type of grid, detailed in Table 7–5, is used in calendar applications. It displays the days in a single month, as shown in Figure 7–19. The first row shows the days, and the first column the week numbers. The remaining cells show six rows of weeks (to allow for days falling in the previous and next month).

Table 7–5 Calendar Grid

Class	`CAknCaleMonthlyStyleGrid`
Dimensions	8 by 7 cells with 1 pixel between cells
Example Item String	`"0\t1\t2"` where: 0 = 21 by 19 icon occupying the whole cell—use this for outlining a cell, for example, with a border or colored box 1 = 5 by 5 icon occupying bottom right of cell—use this for marking a day 2 = day number. This should be a maximum of 2 characters wide. Any more and it will show the first character followed by an ellipsis, for example, `"123"` shows as `"1..."`
Orientation	Vertical

Figure 7–19 Monthly calendar grid.

Figure 7–20 Pin-board grid.

Pin-Board Grid

This grid type, detailed in Table 7–6, represents a pin board used to display shortcuts. A possible example use would be in a favorites application, which stores shortcuts to a user's preferred applications and files, as shown in Figure 7–20.

Table 7–6 Pin-Board Grid

Class	CAknPinbStyleGrid
Dimensions	5 by 5
Example Item String	"0\t1\t2", where: 0 = 29 by 25 icon occupying center of cell, showing the application for which this is a shortcut, or an icon representing an application instance 1 = 13 by 13 icon occupying top right of cell, used for marking the item 2 = 13 by 13 icon occupying bottom left of cell, used to show the application for which this is a shortcut, if this is a instance of an application
Orientation	Vertical

Chapter 7 Using Grids

GMS Grid

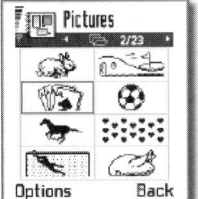

This type of grid, detailed in Table 7–7, is a markable graphic grid. It could be used, for example, by a graphics messaging application to show a grid of picture messages that can be marked, as shown in Figure 7–21.

Figure 7–21 GMS grid.

Table 7–7 GMS Grid

Class	CAknGMSStyleGrid
Dimensions	2 by 4 as default, but can be changed in ConstructL()
Example Item String	"0\t1", where: 0 = 74 by 28 icon occupying background of whole cell 1 = 13 by 13 icon occupying foreground in top right of cell, used for marking the item N.B.: A 13 by 13 unmarked item icon should also be supplied and will be shown by default until the item is marked
Orientation	Vertical

Using Grids

This section guides you through two example applications: **SimpleGrid** and **MarkableGrid**. **SimpleGrid** covers the basics of using grids. Specifically, it shows you how to define a grid using resources, create a concrete grid class, and then draw the grid. **MarkableGrid** looks at markable grids, which are used much as markable lists are. It covers creating a markable grid, setting the icons for a markable grid, and drawing the grid.

As in the **Using Vertical Lists** section, each of the grid examples employs a very simple user interface, with each example incrementally building on the previous example. This section will only discuss features in the code that are new to each example. Similarly, many aspects of the code are the same as those in the vertical list examples discussed in the **Using Vertical Lists** section of this chapter—duplicated code is not discussed here.

Grid Basics

The example **SimpleGrid** displays a number of card games that a player can choose from, as shown in Figure 7–22. Each item represents one game and consists of an icon and a caption. Selecting an item starts the selected game. As this is not a full application, no code is included to actually play the game—instead an empty method is called (`PlaySelectedGame()`).

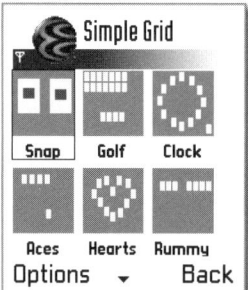

Figure 7–22 SimpleGrid screen.

Defining a Grid Using Resources

You define the grid in a `GRID` resource, as shown in `SimpleGrid.rss` for the **SimpleGrid** example:

```
RESOURCE GRID R_SIMPLEGRID_GAMES_GRID
    {
    array_id = r_simplegrid_games_grid_items;
    flags = EAknListBoxSelectionGrid;
    style = r_simplegrid_games_grid_style;
    }
```

The `flags` field determines the type of grid, in this case a selection grid. (Refer back to Table 7–4 for other options.) The `array_id` field refers to an array of `LBUF` strings, which defines the items contained in the grid. As usual, you should localize the text for the items in an appropriate resource file. Each grid item comprises of a number of fields separated by a tab character (`\t`)—for example:

```
#define GAMES_GRID_ITEM1_TEXT "0\tSnap"
```

As with lists, the number indicates an icon in the icon array. However, the tab character only separates the fields rather than indicating their position in the grid cell. You will see how the position is determined shortly when subcells are discussed. See *Adding Icons to a List* and the **SimpleList** example in the **Using Vertical Lists** section for details on setting the icon array.

Series 60 ignores the `height` and `width` flags in the `GRID` resource, but these values can be specified in the `GRID_STYLE` resource. The `emptytext` field can be overridden to supply a string which displays when the grid is empty. You should not change the `version` field.

The `style` value refers to a `GRID_STYLE` resource:

Chapter 7 Using Grids

```
RESOURCE GRID_STYLE r_simplegrid_games_grid_style
    {
    layoutflags = EAknGridHorizontalOrientation|
EAknGridLeftToRight | EAknGridTopToBottom;
    // horizontal scroll
    primaryscroll = EAknGridFollowsItemsAndLoops;
    // vertical scroll
    secondaryscroll = EAknGridFollowsItemsAndLoops;
    // number of items horizontally per screen
    itemsinprimaryorient = 3;
    // number of items vertically per screen
    itemsinsecondaryorient = 2;
    gapwidth = 5;
    gapheight = 5;
    height = 68;
    width = 50;
    }
```

The `GRID_STYLE` resource determines the layout of the grid, but not the layout of individual cells.

The `layout_flags` field indicates how the grid appends items. You need to specify an orientation flag. In this case, `EAknGridHorizontalOrientation` is used to indicate that the grid will append items to rows and will append additional rows as necessary. If you choose a vertical orientation, using `EAknVerticalOrientation`, the grid will append items to columns and append additional columns as necessary.

You also specify flags to indicate how the grid should add rows, and how it should add items to rows. In **SimpleGrid**, `EAknGridTopToBottom` indicates that the grid should add new rows from top to bottom and `EAknGridLeftToRight` that it should add items to rows from left to right.

In a horizontal orientation, `primaryscroll` indicates how horizontal scrolling will behave, and `secondaryscroll` relates to vertical scrolling. In a vertical orientation, these are reversed: `primaryscroll` indicates vertical scrolling and `secondaryscroll` indicates horizontal scrolling.

In this case, `primaryscroll` and `secondaryscroll` are set to `EAknGridFollowsItemsAndLoops`. This flag means that as you scroll in one direction, the focus moves from one item to the next until you reach the end of the row or column. At this point, if you scroll in the same direction again, it will either: loop round to the other end of the same row or column, or snake to the other end of the next row or column. Whether it loops or snakes depends on the orientation of the grid. When scrolling in the same plane as the orientation—for example, scrolling to the right in a horizontally oriented grid—it will snake. When scrolling in the other plane—for example, scrolling down in a horizontally oriented grid—it will loop.

So, when scrolling to the right in a horizontally oriented grid, the focus moves along the items in a row from left to right. If, when the focus reaches the last item in the row, you scroll right again, it then moves to the first item on the **next** row. When you reach the last item of the last row, scrolling will continue at the first item of the first row. Figure 7–23 depicts this scenario.

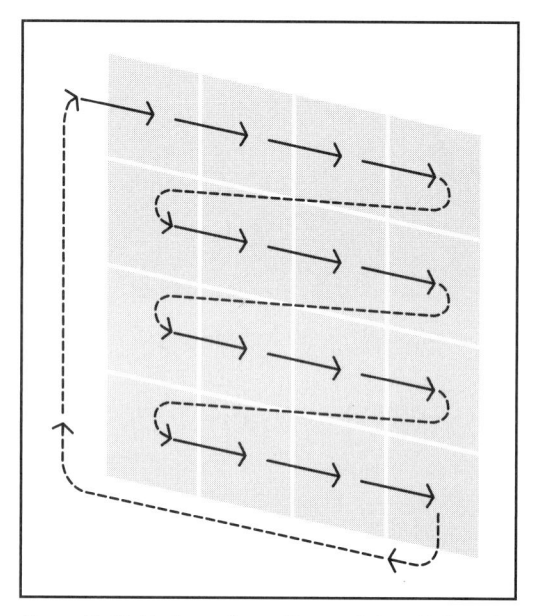

Figure 7–23 Horizontal scrolling to right with loop in a horizontally oriented grid.

If you are scrolling down in a horizontally oriented grid and the focus reaches the last item in the column, then, if you scroll down again, the focus will move to the first item in the same column. Table 7–8 provides the details of possible types of scrolling.

The values for `itemsinprimaryorient` and `itemsinsecondaryorient` determine the horizontal and vertical dimensions of the grid. In the **SimpleGrid** example there are three items in each row and two in each column. (In a vertical orientation, `itemsinprimaryorient` indicates the number of items in each column and `itemsinsecondaryorient` the number in each row.)

TIP When working out the values for `itemsinprimaryorient` and `itemsinsecondaryorient`, you should select values that will fit comfortably on one screen. If you select more items in an orientation than will fit on the screen, the scrolling might not work as you would expect. Items will only scroll once you pass the final item in the orient, regardless of whether it is visible on screen.

The `gapwidth` and `gapheight` fields determine the amount of space between each cell in pixels.

Table 7-8 Grid Primary and Secondary Scrolling

Type	Scrolling Behavior at Edge in Vertically Oriented Grids	Scrolling Behavior at Edge in Horizontally Oriented Grids
`EAknGridFollowsItemsAndStops`	*Horizontal:* **Stops**. *Vertical:* **Snakes**. Except when moving up from the first cell in last the grid or down from the last cell in the grid where it **stops**.	*Horizontal:* **Snakes**. Except when moving left from the first cell in the grid or right from the cell in the grid where it **stops**. *Vertical:* **Stops**.
`EAknGridFollowsItemsAndLoops`	*Horizontal:* **Loops**. *Vertical:* **Snakes**. Additionally, moves to the first cell when moving down from the last cell in the grid, and to the last cell when moving up from the first cell in the grid.	*Horizontal:* **Snakes**. Additionally, moves to the first cell when moving right from the last cell, and to the last cell when moving left from the first cell. *Vertical:* **Loops**.
`EAknGridFollowsGrid`	*Horizontal:* **Loops**. *Vertical:* **Loops**.	*Horizontal:* **Loops**. *Vertical:* **Loops**.
`EAknGridStops`	*Horizontal:* **Stops**. *Vertical:* **Stops**.	*Horizontal:* **Stops**. *Vertical:* **Stops**.
`EAknGridIncrementLineAndLoops`	*Horizontal:* **Loops**. *Vertical:* **Snakes**. Additionally, moves to the first cell when moving down from the last cell in the grid, and to the last cell when moving up from the first cell in the grid.	*Horizontal:* **Snakes**. Additionally, moves to the first cell when moving right from the last cell, and to the last cell when moving left from the first cell. *Vertical:* **Loops**.
`EAknGridIncrementLineAndStops`	*Horizontal:* **Snakes**. Except when moving left from the first cell in the grid or right from the last cell in the grid, where it **stops**. *Horizontal:* **Snakes**. Except when moving left from the first cell in the grid or right from the last cell in the grid, where it **stops**.	*Vertical:* **Snakes**. Except when moving up from the first cell in the grid or down from the last cell in the grid, where it **stops**. *Vertical:* **Snakes**. Except when moving up from the first cell in the grid or down from the last cell in the grid, where it **stops**.

The `height` and `width` fields determine the size of each cell in pixels. You set the size according to the exact layout you require. For **SimpleGrid**, a 50 by 50-pixel bitmap was created for the icons; the graphics should appear above the text and take up approximately 18 pixels. A width of 50 and a height of 68 were therefore set.

Possible values of `flags`, `primaryscroll` and `secondaryscroll` are provided in Table 7-8.

Creating a Concrete Grid Class

You will need to implement your grid as a class derived from `CAknGrid`. In the **SimpleGrid** example, the grid class is `CSimpleGridGrid`. This is contained within a standard Symbian OS container `CSimpleGridContainer`. The code for the AppUi and Container will not be analyzed in depth here, as there is little difference between it and the code in the vertical list examples. There are, however, a few points worthy of note:

- In the AppUi, the container is constructed using `new(Eleave)`, and the MOP parent is set before calling `ConstructL()`, rather than calling `NewL()` and then setting the MOP parent. This is because the scroll bars will appear only if the MOP parent is set before the call to `ConstructL()`.

- In the container's `SizeChanged()` method, the extent of the grid is set using `Rect().Size()`, rather than the grid's minimum size. This is because using the minimum size affects the redrawing of the grid when scrolling.

The grid needs to construct itself from resource, set up its icons and then display itself. This is performed in `CSimpleGridGamesGrid::ConstructL()`, which is called from the container. The container passes the resources for the icon file name (R_ICON_FILE_NAME) and the grid (R_SIMPLEGRID_GAMES_GRID), which was defined previously in *Defining a Grid Using Resources*.

```
void CSimpleGridGamesGrid::ConstructL(TInt aGridResource,
TInt aIconFileResource)
    {
    TResourceReader reader;
    CEikonEnv::Static()->CreateResourceReaderLC(reader,
aGridResource);

    ConstructFromResourceL(reader);
    CleanupStack::PopAndDestroy();

    SetupGridIconsL(aIconFileResource);
    SizeChanged();
    }
```

The base class has a `ConstructFromResourceL()` method, which you should call with a resource reader created using the GRID resource. Once constructed, you can set up the icons for the grid. This is performed by calling the `SetupGridIconsL()` method. Apart from the difference in the actual icons, the code for adding icons is the same as for lists. See the **SimpleList** example in the **Using Vertical Lists** section for details of how to add icons.

> **TIP** You do not need to set up the scroll bars for the grid as you did with lists. `CAknGrid` takes care of them for you.

Chapter 7 Using Grids

The last thing you need to do is draw the grid. You do this by overriding its `SizeChanged()` method.

```
void CSimpleGridGamesGrid::SizeChanged()
   {
   CAknGrid::SizeChanged();
   SetupGrid();
   }


void CSimpleGridGamesGrid::SetupGrid()
   {
   // Setup text foreground and background colours to default
   AknListBoxLayouts::SetupStandardGrid(*this);

   // Get local copies of data we will need
   CFormattedCellListBoxItemDrawer* itemDrawer =
      this->ItemDrawer();

   TInt cellWidth = ColumnWidth();
   TInt cellHeight = ItemHeight();

   // Set up graphics subcells
   AknListBoxLayouts::SetupFormGfxCell(
       *this,        //the grid
       itemDrawer,   // the grid's drawer
       0,            // index of the graphic within item strings
       0,            // left position
       0,            // top position
       0,            // right - unused
       0,            // bottom - unused
       cellWidth,    // width of graphic
       KGraphicsHeight,            // height of graphic
       TPoint(0, 0), // start position
       TPoint(cellWidth , KGraphicsHeight)); // end position

   // Set up text subcells
   const CFont* KFont = LatinBold12();
   TInt baseline = cellHeight - KFont->DescentInPixels() - 1;

   // N.B. although color is commented out
   // in the header file, it is still used!

   AknListBoxLayouts::SetupFormTextCell(
       *this,        // the grid
       itemDrawer    // the grid's drawer
       1,            // index of text within item strings
       KFont,        // the font for the text
       KTextColor,   // the color of the text
       0,            // left margin
       0,            // right margin - unused
       baseline,     // Baseline
```

```
        cellWidth,      // text width (take margin into account
if set)
        CGraphicsContext::ECenter      // Alignment
        TPoint(0, KGraphicsHeight)     // start position
        TPoint(cellWidth, cellHeight)  // end position
    }
```

You should call the base class `SizeChanged()` method, so that you retain important functionality in your overridden method. Then you set up the cell layout using a number of static methods defined in `AknListBoxLayouts`.

The default foreground and background colors are set using the `SetupStandardGrid()` method. You can determine how the graphics and text will appear in each cell by setting up the subcells using `SetupFormGfxCell()` for the graphics subcell and `SetupFormTextCell()` for the text subcell. Each method has an `index` parameter, which refers to the position in the item strings of the graphic or text. All item strings used here are of the form "n\tText", where n is an icon number and `Text` is the text that will appear beneath that icon. Therefore, the index provided for the graphics subcell is 0 (the first position in the string), and the index for the text subcell is 1 (second position). When setting up the graphics subcell you must provide the following parameters:

- A reference to the grid itself.
- The grid's drawer.
- The index, as already discussed.
- Left position—the position in pixels relative to the left-hand side of the cell.
- Top position—the position in pixels relative to the top side of the cell.
- The width of the graphic.
- The height of the graphic.
- Start position—the coordinates of the start of the graphic.
- End position—the coordinates of the end of the graphic.

When setting up the text subcell you must provide the following parameters:

- The grid itself.
- The grid's drawer.
- The index, as already discussed.
- The font.
- The color of the text as a `TInt`—this is an index into the color palette, the values of which can be found in the Appendix of the Series 60 UI guide, part of the SDK documentation.

Chapter 7 Using Grids

- Left margin position—position in pixels relative to the left-hand side of the cell.
- Baseline position—the position between the ascent and descent of the font. See Chapter 11 or the SDK documentation (*Character ascent, descent, height and baseline*) for further information.
- The width of the text.
- The alignment of the text.
- Start Position—the coordinates of the start of the text area.
- End Position—the coordinates of the start of the text area.

> **TIP** The header file comments out the color parameter of `AknListBoxLayouts::SetupFormTextCell()`, but it is actually used.

Markable Grids

The **MarkableGrid** example displays a list of saved games. Each list item has an icon representing the game, and a single line of text representing the name, as shown in Figure 7–24. When you select an item from the list, a tick appears beneath it. You can delete marked items by selecting **Delete** from the **Options** menu.

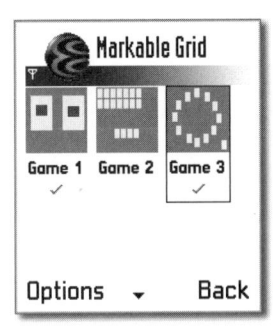

Figure 7–24 MarkableGrid screen.

Creating a Markable Grid

You need to define a grid resource and set the flags to `EAknListBoxMarkableGrid` to indicate a markable grid. The item strings will need to include the unmarked icon. For example:

```
#define GAMES_GRID_ITEM1_TEXT "2\tGame 1\t1"
```

The first field defines the index of the icon representing the game, the second defines the name, and the third is the index of the unmarked item icon.

Setting Markable Grid Icons

You will need two additional icons, to denote the marked and unmarked states, which you should add to the icon array.

In code, the markable grid is more akin to a multiselection list than to a markable list. That is, you provide two marking icons: one to represent a marked item and one to represent an unmarked one. In the **MarkableGrid** example, a blank icon is provided to represent an unmarked item, meaning that the markable grid will have the same behavioral characteristics as a markable list.

In the **MarkableList** example, it was necessary for the marking icon to be the first element in the icon array. With markable grids, it is up to you to decide which element the icon appears in. In order to remain consistent with markable list behavior, the **MarkableGrid** example uses the first element in the array.

Drawing a Markable Grid

As in the **SimpleGrid** example, you set the layout of the grid in the `SizeChanged()` method. This time, you need to change the layout of the cells to accommodate the marking item. You will need to adjust height, width, start and end positions accordingly, and you can add another graphics subcell, by calling `SetupFormGfxCell()`. Then you need to call methods on the item drawer to allow marking to occur. In **MarkableList** these steps are performed in `CMarkableGridGamesGrid::SetupGrid()`:

```
...
itemDrawer->SetItemMarkPosition(2);

if (!iMarkOnValue)
    {
    iMarkOnValue =
StringLoader::LoadL(R_MARKABLEGRID_MARK_ON_VALUE);
    }

// the index of the mark on icon
itemDrawer->SetItemMarkReplacement(*iMarkOnValue);
// Don't display all items as marked initially
itemDrawer->SetItemMarkReverse(ETrue);
...
```

The `SetItemMarkPosition()` method sets the position in the item string of the unmarked item.

When an item is marked, the framework alters the item string such that the marking icon number replaces the nonmarking icon number. In the **MarkableGrid** example, "2\tGame 1\t1" would become "2\tGame 1\t0" when marked. You use the `SetItemMarkReplacement()` method to specify what replaces the unmarked icon in the string. This is a descriptor, which you should define in a resource file and set to the position of the marking icon in the icon array (in this case "0").

Finally, you should call `SetItemMarkReverse()`. This determines whether the icon replacement occurs when an item is marked or unmarked. Setting it to `ETrue` ensures that the marking icon will appear when the item is marked.

Settings Lists

Settings lists group together an application's user-configurable settings. They contain one or more settings items, each having a title and a value. Settings items may also have a sequence number and an indicator to show that they are compulsory, but this is optional. Figure 7–25 shows an example settings list with a number of items—note that the player name is compulsory.

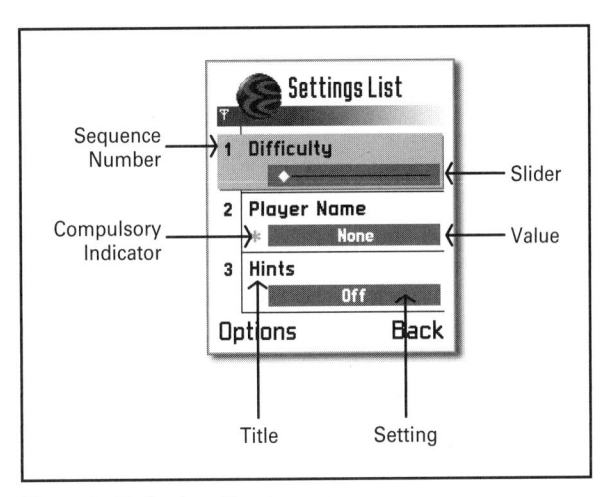

Figure 7–25 Settings list elements.

The types of settings item that can appear in a list are as follows:

- Volume control
- Text editor
- Slider control
- Enumerated text (pop-up)
- Time editor
- Date editor
- IP field editor
- Binary switch
- Password editor (numeric or alphabetic)

Unlike other lists, you can mix the types of item in a settings list. For example, a settings list for a game could contain a slider control to represent the difficulty level, a text editor to enter the player name and a binary switch to determine whether hints are shown or not.

You can adjust the value of an item by selecting it with the **Selection** key, or by invoking **Change** from the **Options** menu. On selection of the item, a settings page will usually display to allow the value to be changed. You will see how the framework determines whether to display a settings page when you go through the examples. The exact appearance of the settings page will depend on the type of settings item you are adjusting. It will have a title and a control, which you can use to adjust the value of the item. Optionally, it may have a number and some hint text, describing what the user should do to adjust the value. An example settings page is shown in Figure 7–26.

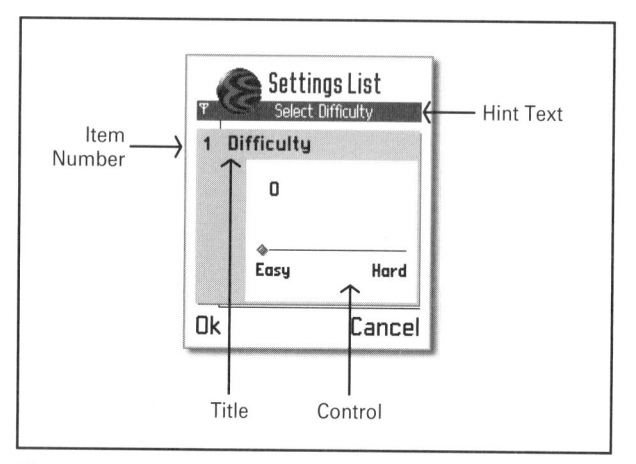

Figure 7–26 Settings page.

The soft keys for the settings page are usually labeled **Ok** and **Back**. However, it is possible to change them programmatically—for example, to allow an **Options** menu. Once you have adjusted the value of the settings page item, if you select a positive soft key, such as **Ok**, it saves the value.

For some items, such as the binary switch, the settings list will not always use a settings page for editing the item. In this case, the value will be edited in-place—when the **Selection** key is pressed, the binary switch will toggle between the on and off values.

Using Settings Lists

This section guides you through the **SettingsList** example application, which illustrates various types of settings list items. This example demonstrates the basic techniques for producing a settings list, including defining the list and

Chapter 7   Using Settings Lists   379

settings list items in resource, creating a settings list subclass and constructing the list. It also covers the method used to change the value of a settings list item and also how to retrieve the values set by the user.

Settings List Basics

SettingsList displays the settings list for a game, as shown in Figure 7–27. The list contains three items: a slider to set the difficulty level, a player name editor, and a binary switch to enable or disable game hints. The items are numbered, and player name is marked as compulsory. Figure 7–28 shows the settings page screen for the difficulty level.

Figure 7–27 SettingsList screen.

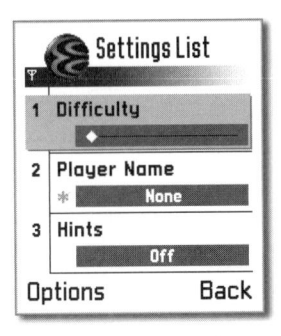

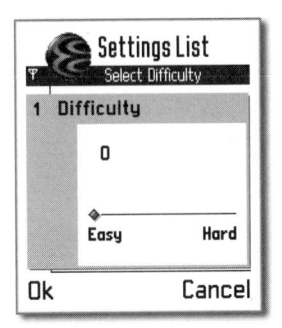

Figure 7–28 SettingsList settings page screen.

Defining a Settings List Using Resources

Settings lists are defined in a resource file using an AVKON_SETTING_ITEM_LIST resource structure. The resource structure defined in `SettingsList.rss` from the **SettingsList** example is shown here:

```
RESOURCE AVKON_SETTING_ITEM_LIST
r_settingslist_setting_item_list
    {
    flags = EAknSettingItemNumberedStyle;
    title = SETTING_ITEM_LIST_TITLE;

    items =
        {
        AVKON_SETTING_ITEM
            {
            identifier = ESettingsListDifficultySettingItem;
            setting_page_resource =
r_settingslist_difficulty_setting_page;
            name = DIFFICULTY_SETTING_ITEM_TITLE;
            },
        AVKON_SETTING_ITEM
            {
            identifier = ESettingsListPlayerNameSettingItem;
```

```
                    setting_page_resource =
r_settingslist_player_name_setting_page;
            name = PLAYER_NAME_SETTING_ITEM_TITLE;
            compulsory_ind_string = "*";
            },
        AVKON_SETTING_ITEM
            {
            identifier = ESettingsListHintsSettingItem;
            setting_page_resource =
r_settingslist_hints_setting_page;
            associated_resource =
r_settingslist_hints_popup_setting_texts;
            name = HINTS_SETTING_ITEM_TITLE;
            }
        };
    }
```

The `AVKON_SETTING_ITEM_LIST` structure has a number of fields, summarized here:

- `flags`—This allows you to specify if the list is numbered using `EAknSettingItemNumberedStyle`, as in the **SettingsList** example. If it is numbered, you can add "| `EAknSettingItemIncludeHiddenInOrdinal`" to this value to specify that hidden items are numbered.

- `title`—This is a string, displayed at the top of the list.

- `items`—This refers to an array of `AVKON_SETTING_ITEM` resources.

> **TIP** The default starting point for the sequence of numbers is one. To change this, you can override a default field (`initial_number`) in the `AVKON_SETTING_ITEM_LIST` and set it to a value of your choice.

Each settings item (`AVKON_SETTING_ITEM`) must have the following fields:

- `identifier`—This uniquely identifies the settings item in code. You define identifiers in an enum in the `.hrh` file.

- `setting_page_resource`—This refers to an `AVKON_SETTING_PAGE` resource.

In the **SettingsList** example, optional fields are defined in the settings items:

- `name`—The title for the item, displayed above its value.

- `compulsory_ind_string`—Marks a field as compulsory with the specified string.

- `associated_resource`—Links to another control used by the settings page. In the **SettingsList** example, the "hints" settings item uses a pop-up list to switch hints on or off. In this case, the settings item refers to an

Chapter 7 Using Settings Lists

AVKON_POPUP_SETTING_TEXTS resource, which the list uses for strings to display in both its pop-up and normal state.

> **TIP** A binary switch can have only two enumerated text items for its pop-up settings texts. One should have a value of zero and the other a value of one. Any deviation from this will cause a panic on construction.

```
RESOURCE AVKON_SETTING_PAGE
r_settingslist_difficulty_setting_page
    {
    number = 1;
    hint_text = DIFFICULTY_HINT_TEXT;
    label = DIFFICULTY_SETTING_ITEM_TITLE;
    type = EAknCtSlider;
    editor_resource_id = r_settingslist_difficulty_slider;
    }
```

As a minimum, a settings page resource needs a `type` and an `editor_resource_id`. The `type` specifies the type of control used to edit the item, in this case a slider control. The `editor_resource_id` refers to the resource definition for the `SLIDER` control.

Some optional fields are defined in the **SettingsList** example. The `number` is displayed in the top left corner of the settings page, if set to a value other than zero. This overrides any numbering on the settings item itself. The `hint_text` appears in the navigation pane, if set. The `label` is the title for the item, which appears above its value.

> **SEE ALSO** The `SLIDER` resource is a standard control, well documented in the Series 60 SDK "Slider Control Example." Table 7–9 provides brief details, but you are advised to consult the SDK documentation if you require in-depth information.

Deriving a Settings List

You will need to create a subclass of `CAknSettingItemList` and override some of its methods. In the **SettingsList** example, this role is performed by `CSettingsListSettingItemList`. It is usual to pass in data for the list to modify—this is performed in the **SettingsList** example by the class `TSettingsListSettings`.

Note that `TSettingsListSettings` has public accessor methods which obtain references to private data. This is not normally advisable, as it breaks encapsulation. However, in this case the list's settings items require non-const references to manipulate. Data is passed by reference into the first-phase constructor, which then stores it.

Chapter 7 Lists

```
CSettingsListSettingItemList(TSettingsListSettings& aSettings)
: iSettings(aSettings)
   {
   }
```

Creating the Settings Items

Although you define the settings items in resource, you need to construct them dynamically in your code; otherwise they will not display. You do this in the `CreateSettingItemL()` method, which you override in the list subclass.

```
CAknSettingItem*
CSettingsListSettingItemList::CreateSettingItemL(
TInt aIdentifier)
    {
    CAknSettingItem* settingItem = NULL;

    switch (aIdentifier)
        {
        case ESettingsListDifficultySettingItem:
            settingItem = new (ELeave)
CAknSliderSettingItem(aIdentifier, iDifficultyLevel);
            break;

        case ESettingsListPlayerNameSettingItem:
            settingItem = new (ELeave)
CAknTextSettingItem(aIdentifier, iPlayerName);
            break;

        case ESettingsListHintsSettingItem:
            settingItem = new (ELeave)
CAknBinaryPopupSettingItem(aIdentifier,
iSettings.HintsEnabled());
            break;
        }
    return settingItem; // passing ownership
    }
```

The base class calls the method for each item defined in resource, passing in the identifier for the item (`aIdentifier`). You then construct the appropriate type of settings item for the identifier, passing in a piece of data for it to alter. The base class owns the settings items and holds them in an array. You can access this using the `SettingItemArray()` method.

Note that the base class calls `CreateSettingItemL()` when the settings list is constructed. If the application fails when it first shows the list, it is likely that there is a problem with this method. For example, a panic would occur if `iDifficulty` were out of range for the slider.

Changing Settings List Values

Whenever you select an item using the **Selection** key, the base class calls the `EditItemL()` method.

```
void CSettingsListSettingItemList::EditItemL(TInt aIndex,
TBool aCalledFromMenu)
    {
    CAknSettingItemList::EditItemL(aIndex, aCalledFromMenu);
    (*SettingItemArray())[aIndex]->StoreL();
    }
```

You should first call the base-class version of `EditItemL()`, as this allows you to edit the item either via the settings page or in-place. Which method is used depends on the value of `aCalledFromMenu`. If it is set to `EFalse`, editing is in place, otherwise it is edited via the settings page. The framework calls `EditItemL()` whenever you select an item. It sets `aCalledFromMenu` to `EFalse` for all components apart from the binary switch, which it sets to `ETrue`.

> **TIP** You should add a **Change** item to the **Options** menu to allow editing. This should call the `EditItemL()` method. It is common to have the binary switch open the settings page when called from the menu, but edit in-place when the **Selection** key is pressed. To do this, just have `EditItemL()` called with a value of `ETrue` for all controls.

When you change settings item values, they do not automatically update the data items supplied to them on construction. Settings items have a method `StoreL()` that writes the value in the control to the data. In the **SettingsList** example, the `EditItemL()` method is overridden to provide a convenient location for this storage to occur. The settings items are stored in an array in the base class, which can be accessed using the `SettingItemArray()` method. The appropriate item is selected from the array and its `StoreL()` method called to store the value that has just been edited.

The settings list provides another method for storing items: `StoreSettingsL()`. This calls `StoreL()` on each of the settings items in the list. You could use this method to store items when you remove the list from the display, instead of storing each item individually after editing.

Constructing a Settings List

You will need to construct your list using the resource defined for it. This can be performed in the container's `ConstructL()` method using standard two-phase construction from resource.

Chapter 7 Lists

```
iSettingItemList = new CSettingsListSettingItemList(
iSettings);
    iSettingItemList->SetMopParent(this);
    iSettingItemList
->ConstructFromResourceL(R_SETTINGSLIST_SETTING_ITEM_LIST);
```

Settings Item Types

Tables 7-9 through 7-16 are provided to assist you in adding items to your settings lists. Each table describes a particular settings item, illustrated with an example screen shot, giving sample resource definitions and sample code for creating the item.

Table 7-9 Slider Settings Item Reference

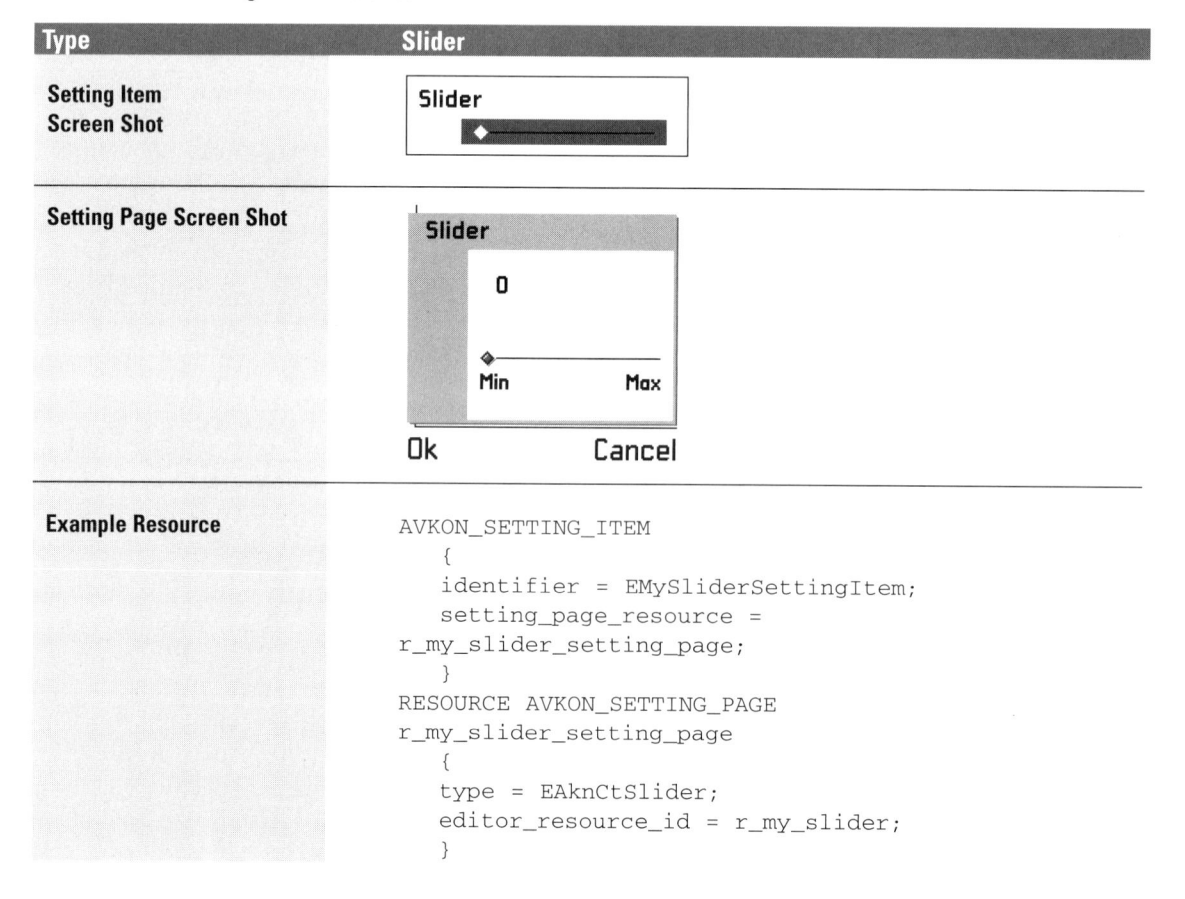

Type	Slider
Setting Item Screen Shot	Slider
Setting Page Screen Shot	Slider — 0 — Min Max — Ok Cancel

Example Resource	```AVKON_SETTING_ITEM
	    {
	    identifier = EMySliderSettingItem;
	    setting_page_resource =
	r_my_slider_setting_page;
	    }
	RESOURCE AVKON_SETTING_PAGE
	r_my_slider_setting_page
	    {
	    type = EAknCtSlider;
	    editor_resource_id = r_my_slider;
	    }```

Chapter 7 Using Settings Lists   385

Table 7-9 continued

Type	Slider
| **Example Resource** | ```
RESOURCE SLIDER r_my_slider
    {
    layout = EAknSettingsItemSliderLayout;
    minvalue = 0;
    maxvalue = 2;
    step = 1;
    valuetype = EAknSliderValueBareFigure;
    minlabel = "Min";
    maxlabel = "Max";
    }
``` |
| **Setting Item Class** | CAknSliderSettingItem |
| **Example** `CreateSettingItemL()` | ```
TInt iSlider;
CAknSettingItem* CMySettingItemList::
CreateSettingItemL(TInt aIdentifier)
    {
    return new CAknSliderSettingItem
    (aIdentifier, iSlider);
    }
``` |
| **Further Information** | `layout`—determines whether to show the header, current value and labels. `minvalue` and `maxvalue`—determine the minimum and maximum values represented by the slider. `step`—determines how much the slider increments by. `valuetype`—determines how the value will be displayed (figure, percent, fraction or decimal). `minlabel` and `maxlabel`—are the minimum and maximum labels. If you were to use a slider outside a settings list, you would use a `CAknSlider` control. Methods are available to alter the values and labels of the slider dynamically and to change the step size and minimum and maximum values. Further information can be found in the SDK documentation. |

Table 7–10 Volume Setting Item Reference

Type	Volume
Setting Item Screen Shot	Volume
Setting Page Screen Shot	Volume (with bar chart, Ok / Cancel)
| **Example Resource** | ```
AVKON_SETTING_ITEM
    {
    identifier = EMyVolumeSettingItem;
    setting_page_resource =
r_my_volume_setting_page;
    }
RESOURCE AVKON_SETTING_PAGE r_my_volume_setting_page
    {
    type = EAknCtVolumeControl;
    editor_resource_id = r_my_volume;
    }
RESOURCE VOLUME r_my_volume
    {
    flags = ESettingsVolumeControl;
    value = 1;
    }
``` |
| **Setting Item Class** | CAknVolumeSettingItem |
| **Example** CreateSettingItemL() | ```
TInt iVolume;
CAknSettingItem* CMySettingItemList::
CreateSettingItemL(TInt aIdentifier)
    {
    return new CAknVolumeSettingItem(
aIdentifier, iVolume);
    }
``` |
| **Further Information** | The header file for the CAknVolumeSettings class indicates that the volume can be in the range 0 to 10. It can actually be in the range 1 to 10—a panic occurs if you set it to zero! |

Table 7-11 Text Editor Setting Item Reference

Type	Text Editor
Setting Item Screen Shot	Text Editor — None
Setting Page Screen Shot	2 Text Editor ... Cancel
| Example Resource | ```
AVKON_SETTING_ITEM
    {
    identifier = EMyTextEditorSettingItem;
    setting_page_resource =
r_my_text_editor_setting_page;
    }
RESOURCE AVKON_SETTING_PAGE
r_my_text_editor_setting_page
    {
    type = EEikCtEdwin;
    editor_resource_id = r_my_text_editor;
    }
``` |
| Setting Item Class | CAknTextSettingItem |
| Example CreateSettingItemL() | ```
TBuf<10> iText;
CAknSettingItem* CMySettingItemList::
CreateSettingItemL(TInt aIdentifier)
    {
    return new CAknTextSettingItem(aIdentifier,
iText);
    }
``` |
| Further Information | When empty, the text editor displays a placeholder string. By default this is "None", but it can be changed using CAknTextSettingItem::SetEmptyItemTextL(). The **Ok** soft key is not visible on the settings page screen, until the editor contains one character. |

Table 7-12 Enumerated Text Setting Item Reference

Type	Enumerated Text
Setting Item Screen Shot	Enumerated Text / First
Setting Page Screen Shot	Enumerated Text / ● First / ○ Second / Ok Cancel
| Example Resource | ```
AVKON_SETTING_ITEM
    {
    identifier = EMyEnumeratedTextSettingItem;
    setting_page_resource = r_my_setting_page;
    associated_resource =
r_my_popup_setting_texts;
    }
RESOURCE AVKON_SETTING_PAGE r_my_setting_page
    {
    type = EAknCtPopupSettingList;
    editor_resource_id = r_my_popup_setting_list;
    }
``` |
| Setting Item Class | CAknEnumeratedTextPopupSettingItem |
| Example CreateSettingItemL() | ```
TTime iEnumeration;
CAknSettingItem* CMySettingItemList::
CreateSettingItemL(TInt aIdentifier)
    {
    return new
CAknEnumeratedTextPopupSettingItem(aIdentifier,
iEnumeration);
    }
``` |

Chapter 7 Using Settings Lists   389

Table 7-13 Time or Date Editor Setting Item Reference

Type	Time or Date Editor
Setting Item Screen Shot	Time 8:06 PM
Setting Page Screen Shot	Time 08:06 PM Ok Cancel

Example Resource	```
	AVKON_SETTING_ITEM
	    {
	    identifier = EMyTimeSettingItem;
	    setting_page_resource =
	r_my_time_setting _page;
	    }
	RESOURCE AVKON_SETTING_PAGE
	r_setlistex2_time_setting_page
	    {
	    type = EEikCtTimeEditor;
	    editor_resource_id = r_my_time_editor;
	    }
	``

Setting Item Class	CAknTimeOrDateSettingItem

Example `CreateSettingItemL()`	```
	TTime iTime;
	CAknSettingItem* CMySettingItemList::
	CreateSettingItemL(TInt aIdentifier)
	    {
	    return new CAknTimeOrDateSettingItem(
	aIdentifier, CAknTimeOrDateSettingItem::
	ETime, iTime);
	    }
	``

Further Information	To make this a date, rather than time editor, change the second parameter in the `CAknTimeOrDateSettingItem` constructor to `EDate`, change the type in the settings page to `EEikCtDateEditor`, and set the resource's `editor_resource_id` to a `DATE_EDITOR`.

Table 7–14 IP Address Editor Setting Item Reference

Type	IP Address Editor
Setting Item Screen Shot	IP Address 0.0.0.0
Setting Page Screen Shot	IP Address 0.0.0.0 Ok Cancel
| Example Resource | ```
AVKON_SETTING_ITEM
  {
  identifier = EMyIpSettingItem;
  setting_page_resource = r_my_ip_setting_page;
  }
RESOURCE AVKON_SETTING_PAGE = r_my_ip_setting_page
  {
  type = EAknCtIpFieldEditor;
  editor_resource_id = r_my_ip_field_editor;
  }
``` |
| Setting Item Class | CAknIpFieldSettingItem |
| Example CreateSettingItemL() | ```
TInetAddr iIp;
CAknSettingItem* CMySettingItemList::
CreateSettingItemL(TInt aIdentifier)
  {
  return new
CAknIpFieldSettingItem(aIdentifier, iIP);
  }
``` |

Chapter 7 Using Settings Lists   391

Table 7-15 Binary Switch Setting Item Reference

Type	Binary Switch
Setting Item Screen Shot	Binary Switch On
Setting Page Screen Shot	Binary Switch ● Enabled ○ Disabled Ok Cancel

Example Resource	```
	AVKON_SETTING_ITEM
	    {
	    identifier = EMyBinarySwitchSettingItem;
	    setting_page_resource = r_my_setting_page;
	    associated_resource =
	r_my_popup_setting_texts;
	    }
	RESOURCE AVKON_SETTING_PAGE r_my_setting_page
	    {
	    type = EAknCtPopupSettingList;
	    editor_resource_id = r_my_popup_setting_list;
	    }
	``

Setting Item Class	CAknBinaryPopupSettingItem

Example `CreateSettingItemL()`	```
	TBool iBinarySwitch
	CAknSettingItem* CMySettingItemList::
	CreateSettingItemL()ItemL(TInt aIdentifier)
	    {
	    return new CAknBinaryPopupSettingItem(
	aIdentifier, iBinarySwitch);
	    }
	``

Further Information	There can be only two items in the `associated_resource` and `editor_resource_id` resource. See the **SettingsList** example for further details.

Table 7–16 Password Editor Setting Item Reference

Type	Password Editor
Setting Item Screen Shot	Password / None
Setting Page Screen Shot	Password / **** / Ok Cancel
| Example Resource | ```
AVKON_SETTING_ITEM
    {
    identifier = EMyPasswordSettingItem;
    setting_page_resource =
r_my_password_setting_page;
    }
RESOURCE AVKON_SETTING_PAGE
r_my_password_setting_page
    {
    type = EEikCtSecretEd;
    editor_resource_id = r_my_password_secreted;
    }
``` |
| Setting Item Class | CAknPasswordSettingItem |
| Example CreateSettingItemL() | ```
TBuf <10> iPassword;
CAknSettingItem* CMySettingItemList::
CreateSettingItemL(TInt aIdentifier)
    {
    return new CAknPasswordSettingItem(
aIdentifier, CAknPasswordSettingItem::EAlpha,
iPassword);
    }
``` |
| Further Information | To make this a numeric password, or PIN, change the second parameter in the CAknPasswordSettingItem constructor to ENumeric. |

Summary

This chapter has introduced lists, the Series 60 UI controls used to organize and display collections of items. It described the three main types of list available in Series 60: vertical lists, grids, and settings lists. Vertical lists can be used to present data in a one-dimensional array, while grids allow you to produce a two-dimensional display. User interaction is achieved using techniques to mark or select items from a list. Settings lists are a specific type of list that allow the user to modify configuration settings for your application. A range of example applications were presented to demonstrate how to construct the different types of list and illustrate how they could be used.

When producing applications for a small device, presentation is particularly important. One crucial aspect is consistency of UIs across applications. You are therefore encouraged to consult the Series 60 UI guide (provided with the SDK documentation) to help you decide which UI control is most suitable for a particular situation.

chapter 8

Editors

Using User Interface components for entering, displaying and editing data

Almost all applications display data and allow users to add new, or edit existing data items. Editors are a family of controls that Series 60 provides in order to handle these common requirements.

Editors, by necessity, offer a diverse range of behavior—primarily because they must be able to handle numerous data types and formats. For example, they need to be able to handle phone numbers, text entry, floating-point numbers and so on. For each specific data type, editors may need to restrict certain input—numbers should not be accepted, for example, if the editor has been configured to receive only text.

Also, some data entry formats may require other special characters, apart from the standard letters and numbers that users can intuitively access from the keypad—for example, emails would require the at ("@") character to complete an address. Therefore, editor controls must also provide access to relevant additional characters. This chapter will show you how to configure editors in order to handle these various requirements.

Apart from handling data entry, editors also provide sophisticated text representations. Editors must be able to display, and allow input of, formatted alphanumeric data—formatting that parallels the capabilities of contemporary word processor applications (bold, italic and underlined text; a variety of fonts; and so on). The Series 60 family of editors provides all of these facilities, and this chapter will explain and illustrate the techniques required to achieve these aims.

This chapter assumes a good grounding in the fundamentals of Symbian OS, and an understanding of how to create an application UI, using standard Series 60 architectures. Of particular importance is the use of UI component controls and resource files. If necessary, you should familiarize yourself with the material in Chapters 3, 4 and particularly 5.

The family of editors is divided into the following categories, each of which will be covered in turn:

- **Text Editors**—These allow single or multiline text. The most basic accept and display only unformatted (**plain**) **text**, but more specialized ones accommodate highly formatted (**rich**) **text**.

Chapter 8 Text Editors

- **Numeric Editors**—These specialized text editors restrict input to allow integer, fixed-point and floating-point number editing.
- **Secret Editors**—These hide text as it is typed to provide secure entry of passwords and PINs (Personal Identification Numbers).
- **Multi-Field Numeric Editors (MFNEs)**—These are numeric editors that have one or more fields separated by data-specific characters. There are specialized MFNEs that allow the user to enter IP addresses, ranges, times, dates, durations and so on.

All of the topics covered in this chapter are illustrated using code excerpts, largely taken from example applications. Details of how to download the full buildable source for these, along with the other example applications, are given in the Preface. Some of the following topics can be amply explained without a comprehensive example—in these cases the concepts are conveyed using illustrative code excerpts.

For clarity, the examples have been kept very simple—their purpose is solely to illustrate how to use the editors programmatically. All examples use the Traditional Symbian OS Control-Based Architecture, rather than Dialog-Based or Avkon View-Switching Architectures.

The first example, **PlainTextEditor**, shows how to create and use a plain text editor in an application. Attributes are set on the editor such that the user could use it to enter a URL (Uniform Resource Locator) in an application.

The **RichTextEditor** example shows how to create and use a rich text editor, which has attributes set such that you could use it to display a help screen for a game application.

The final example, **NumericEditor**, shows how to create and use numeric editors in an application.

Text Editors

Text editors are feature-rich editors that can comprise single or multiple lines of text. They have numerous configurable attributes, such as: their dimensions, the case of the characters they display and so on.

There are three main types of text editors:

- **Plain Editor**—This is the most basic text editor, which contains unformatted text as shown in Figure 8–1.
- **Global Editor**—This allows global formatting of paragraphs and characters but does not allow you to format characters and paragraphs individually— it applies the formatting to all of the text.

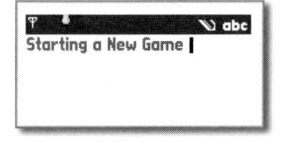

Figure 8–1 A plain text editor.

Chapter 8 Editors

- **Rich Text Editor**—This allows individual formatting of paragraphs and characters, as shown in Figure 8–2. It encompasses all of the functionality of a global editor, and because it is a functional superset, only the rich text editor will be illustrated in this chapter.

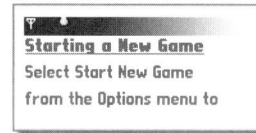

Figure 8–2 A rich text editor.

Table 8–1 details the classes, resource and control types for each of the text editors. You will see how to use the plain and rich text editors in the examples that follow.

Table 8–1 Text Editor Resource and Classes

Text Editor	Class	Resource	Control
Plain	CEikEdwin	EDWIN	EEikCtEdwin
Global	CEikGlobalTextEditor	GTXTED	EEikCtGlobalTextEditor
Rich Text	CEikRichTextEditor	RICHTEXTED	EEikCtRichTextEditor

Text editors have features that you can define, either statically (defining a resource in a resource file), or dynamically (using the text editor's API in code at runtime). Further information on these two general types of control interface can be found in Chapter 5. More specific information on the dynamic features of text editors is available in the SDK documentation for CEikEdwin. Additionally, static (resource file) features of text editors are covered in the SDK documentation for the appropriate text editor resource (such as EDWIN, GTXTED and so on).

The common features that can be defined for all text editors are:

- **Dimensions**—For example, width and number of lines.
- **Input Mode**—Controls what type of characters the keypad can generate (such as all alphabetical, or all numeric), and by doing so, controls the type of input the control accepts.
- **Input Case**—Governs the case convention for text input by controlling what the keypad can generate, such as all capitals or all lowercase characters.
- **Numeric Key Map**—Provides a means of conveniently accessing commonly required nonnumeric characters in numeric input mode—for example, the asterisk (or "star") character: "*".
- **Special Character Table**—Provides a means of conveniently accessing groups of special characters that do not map directly to the keypad.
- **Properties**—Special characteristics that can be applied.

Chapter 8   Text Editors   399

The following subsections discuss each of these features in more detail. In addition, there are a number of features that cannot be set in a resource and must be accessed using the text editor's API. These features are shown in Table 8–2.

Table 8–2 Dynamic Text Editor Features

Feature	Example Use
Read-Only	To check and set whether an editor is read-only.
Text	To get and set the text, and count the words in it.
Selected Text	To get and set the selected text, to select all text, and to clear all selected text.
Cursor Position	To get and set the cursor position.
Infrared	Use infrared to send or receive the editor text.
Clipboard	To cut, copy and paste.
Fields	To insert new fields (for example, a field could mark a table of contents in a rich text editor) and to update fields.
Sizes and Colors	To set margins and to set the background color.
Zoom	To set a zoom factor.
Search and Replace	To search for text and replace it.
Scroll Bars	To display and update the scroll bars.
Utility Functions	For example, to remove non-ASCII characters, calculate the average number of characters per line and so on.
Miscellaneous Properties	To word wrap, and to determine whether the editor owns the text it is editing.
Formatting	To format characters and paragraphs (in global and rich text editors only).
Objects	To insert pictures (in rich text editors only).

Dimensions and Input Capacity

All text editors have three basic configurable characteristics: height (or number of lines of text to display), width and input capacity. These are detailed in Table 8–3.

Table 8–3 Characteristics That Determine Dimensions and Input Capacity for a Text Editor

Characteristic	Description	Additional Information
Width	This is the physical width of the control. The width can be specified either by the number of characters to accommodate, or by pixels.	Editors have a default font, and this font's characters are proportional. So if the width is set in characters rather than pixels, the editor sets the width based on the font's widest character. If the user enters narrower characters, it is possible that they will see more characters per line than expected. For example, with a width of five, they would get more than five "i" characters per line. Global and rich text editors always have their width defined in pixels.
Number of Lines	This is the number of lines of text that the editor can display at a time. The default value for this is one.	Rich and global text editors also define a height, in pixels, in addition to the number of lines. You can use either of these characteristics to define a multiline editor.
Maximum Length	This is the maximum number of characters that can be stored in the editor. It is different from the above two characteristics because it specifies capacity instead of a physical dimension.	The editor will scroll if the maximum length is greater than the number of characters it can display in the width. Rich and global text editors use the term "text limit" in their API methods and resource definition, to refer to the maximum length.

Filtering Keypad Input

The Series 60 keypad allows users to input more characters than there are keys—each key on the keypad maps to multiple characters, and users can, for example, scroll through a key's mappings by pressing that key rapidly. Series 60 editors can reduce the inefficiency of having to make numerous key presses by filtering out those characters that are not relevant to the editor's corresponding data format. For example, editors that exclusively require numeric characters (such as an editor field for inputting a person's age) can be configured to filter out all alphabetic characters from keypad input. The subsection that follows explains the three primary input modes that are used to accomplish this type of filtering.

Chapter 8   Text Editors   401

Input Modes

Essentially, input modes are filters that make data entry more efficient for the user. At the same time, they also prevent editors from receiving input that is invalid for their purposes. The three main modes available are text, numeric and secret, as described in Table 8–4.

Table 8–4 Text Editor Input Modes

Input Mode	Description
Text	Key presses can produce letters, numbers and special characters, as normal.
Numeric	Key presses produce only numbers (and characters defined by the numeric key map).
Secret	Text is replaced by an asterisk (*) on screen as the user types.

Editors have a default input mode, but it is also possible to specify additional allowed input modes. For example, an MMS address may be purely numeric if sending to another phone, or alphanumeric if sending to an email account. An editor can be set up to allow both modes, and users can manually switch modes (if more than one is attributed to an editor) by pressing the hash (#) key.

NOTE   To switch between text and numeric input modes, users must press and hold down the hash (#) key.

You must take care not to choose conflicting values—the list of allowed input modes must include the default input mode. By default, the input mode is text and the editor allows all input modes.

Input Case

Input case settings also filter input, but only when the editor's input mode is set to text. Text input can be set so that it is uppercase only, lowercase only, or follows text rules—for example, setting the first letter of the first word in a sentence to uppercase. You can set the editor so that it has a default case, and a set of allowed cases—a combination of cases the editor will allow you to change to.

Again, you must take care not to choose conflicting default and allowed values. For example, you cannot set the allowed cases to be uppercase only and then set the default case to lowercase only. By default, the case follows text rules, and the editor allows all cases. Table 8–5 shows the enumeration values allowed:

Chapter 8 Editors

Table 8–5 Text Editor Input Cases

Input Case	Use	Screenshot
`EAknEditorUpperCase`	Uppercase only.	THE CAT
`EAknEditorLowerCase`	Lowercase only.	the cat sat. the|
`EAknEditorTextCase`	Automatic. First letter is uppercase.	The cat sat. The|
`EAknEditorAllCaseModes`	Upper, lower and text cases.	Any of the above.

Providing Mappings to Additional Characters

Certain data types require a subset of special characters. URLs, for example, typically comprise alphanumeric characters interspersed with colons, forward slashes and full stops (such as **http://www.emccsoft.com**). For efficiency, it is necessary to group together these additional characters and to provide a convenient and consistent mechanism for accessing them. The asterisk (*) key provides the necessary extension—pressing it will bring up a visible array of all additional characters available for a particular editor. Users can then use the direction and **Selection** keys to acquire characters from this map (see Table 8–6 for examples of various context-dependent mappings).

A single control can deploy more than one mapping, as detailed in the next two subsections—it can provide different mappings according to its current input mode.

Numeric Key Maps

When the input mode is numeric, the asterisk (*) and hash (#) keys can be used to input numerical operators and signs. Generally, the asterisk key brings up a mapping of multiple characters.

Figure 8–3 The standard numeric key map for the asterisk key.

Editors in numeric mode default to a mapping suitable for phone numbers (that is, "*", "+", "p" or "w" for the asterisk key and "#" for the hash key), as shown in Figure 8–3.

NOTE In a phone number, the "p" and "w" characters are used to respectively pause and wait before dialing a number. The user can employ this when dialing automated services. For example, a voicemail system may require the user to dial a number and then, on answer, enter their mailbox number. In this case, they could specify the number to dial followed by w, and then their mailbox number. This would dial the number, and then wait. When the voicemail answered, the user could then press the **Send** key to send the remaining mailbox number digits.

Chapter 8   Text Editors

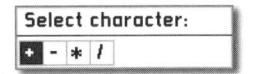

Figure 8-4 The calculator numeric key map for the asterisk key.

Alternative settings bring up mappings suitable for other entry types, such as calculations ("+", "-", "*" and "/" for the asterisk key and "." for the hash key), as shown in Figure 8-4.

Providing Specific Mappings for Text-Based Entries with Special Character Tables

When the input mode is text, pressing and holding the asterisk (*) key displays a special character table. This allows the user to select punctuation and other special characters to input into the editor.

There are different types of special character tables, which contain different characters. The Standard, Email Address, URL and Currency character tables are shown in Table 8-6.

Table 8-6 Text Editor Special Character Tables

Use	Resource Identifier	Screenshot
Default	R_AVKON_SPECIAL_CHARACTER_TABLE_DIALOG	
URL Addresses	R_AVKON_URL_SPECIAL_CHARACTER_TABLE_DIALOG	
Email Addresses	R_AVKON_EMAIL_ADDR_SPECIAL_CHARACTER_TABLE_DIALOG	
Currency	R_AVKON_CURRENCY_NAME_SPECIAL_CHARACTER_TABLE_DIALOG	

Note that these settings may not give you exactly what you would expect—the currency table has no pound (£), yen (¥), euro (€) or general currency (¤) characters, only the dollar ($) character!

Properties

Text editors also allow a number of special properties to be set, which you can use to perform tasks such as:

- Preventing the case from being changed.
- Preventing predictive text (T9) entry.
- Removing the edit indicators from the navigation pane.
- Preventing left and right scrolling.
- Preventing the special character table from displaying.
- Enabling the scroll bar indicator.
- Enabling the text to overwrite rather than insert.
- Enabling the numeric key map, and preventing anything other than Latin characters from being entered.
- Specifying the width as being measured in pixels.
- Making the editor read-only.
- Disabling the cursor.

These properties are set by flags, as will be illustrated in the **PlainTextEditor** example. By default, no properties are set.

Configuring a Plain Text Editor

The **PlainTextEditor** example, as shown in Figure 8-5, illustrates all of the attributes and features discussed so far. The example control is configured to facilitate URL entries. It demonstrates the following concepts:

- Configuring a plain text editor in a resource file.
- Providing multiple input modes (text and numeric).
- Providing multiple case modes (upper or lowercase).
- Providing a URL-specific special character mapping for text input mode.
- Instantiating the editor from its resource definition.

The editor does not accept predictive text entry, and it has no numeric key map—this means that when the input mode is set to numeric, the asterisk (*) and hash (#) keys have no effect. If the input mode is set to text, then pressing the asterisk key will display a special character table suitable for URLs, as shown in Figure 8-6.

A short press of the hash key switches case in text mode, and a long press switches between text and numeric input modes.

Figure 8-5 The PlainTextEditor main screen.

Figure 8-6 The PlainTextEditor special character table.

Defining a Plain Text Editor Resource

The resource structure for the plain text editor is given below. Note that the resource type for a plain text editor is `EDWIN`.

```
RESOURCE EDWIN r_plaintexteditor_urleditor
    {
    width = KUrlWidth;
    lines = KUrlNumberLines;
    maxlength = KMaxUrlLength;
    flags = EEikEdwinWidthInPixels;
    avkon_flags = EAknEditorFlagNoT9;
    default_case = EAknEditorLowerCase;
    allowed_case_modes = EAknEditorUpperCase |
EAknEditorLowerCase;
    numeric_keymap = EAknEditorPlainNumberModeKeymap;
    allowed_input_modes = EAknEditorTextInputMode |
EAknEditorNumericInputMode;
    default_input_mode = EAknEditorTextInputMode;
    special_character_table =
R_AVKON_URL_SPECIAL_CHARACTER_TABLE_DIALOG;
    }
```

This structure complies with the minimum required definition by specifying `width`, `lines` and `maxlength`—these set the dimensions of the editor, as described previously.

Note that constants are used to set the values of the dimensions, and these are defined by macro (rather than using `const TInt` as you might expect), because they occur within a resource file:

```
#define KUrlWidth 176
#define KUrlNumberLines 1
#define KMaxUrlLength 256
```

Note also that Series 60 has expanded on the Symbian OS implementations and defined new fields in the text editor resources. For example, in `EDWIN` the new fields are: `avkon_flags`, `default_case`, `allowed_case_modes`, `numeric_keymap`, `allowed_input_modes`, `default_input_mode`, `special_character_table`, `max_view_height_in_lines` and `base_line_delta`. There are also some Series 60-specific values defined for `flags`—these all contain the word `Avkon` in their name.

The configurable features of the text editor in the **PlainTextEditor** example span six main categories. Table 8–7 provides a breakdown of the resource definition, in terms of the category each resource line addresses, and its purpose.

Table 8–8 provides a breakdown of the properties applied by specific `avkon_flags`.

Table 8–7 Explanation of Resource Definitions for the PlainTextEditor Example

General Feature Affected	Line Item	Purpose in Resource Definition	Reference Information
Dimensions	`width` `lines` `maxlength`	Specifies the dimensions of the control.	See Table 8–3
Input Mode	`allowed_input_modes` `default_input_mode`	The default input mode is set to text. However, numeric input is specified as an additional allowed mode, because URLs can sometimes require numeric input.	See Table 8–10
Input Case	`default_case` `allowed_case_modes`	These specify which filter to apply when the editor is in text mode. The resource states that lowercase should be the default mode, but uppercase is included as an additional allowed mode.	See `uikon.hrh`
Numeric Key Map	`numeric_keymap`	This line sets the numeric key map to be used when the editor is in numeric entry mode. The chosen value actually prevents a map from being presented.	See Table 8–9
Special Character Table	`special_character_table`	This line sets the mapping (for when the editor is in text mode) to a collection of characters that are often required in URL entries.	See Table 8–6

Table 8–7 continued

General Feature Affected	Line Item	Purpose in Resource Definition	Reference Information
Properties	flags	EEikEdwinWidthInPixels sets the control so that its width is determined in pixels instead of characters.	All flags are defined within uikon.hrh. Additionally, the SDK documentation for CEikEdwin describes a number of flags (TFlags) that you can set when constructing an editor dynamically using ConstructL().
Properties	avkon_flags	EAknEditorFlagNoT9 disables predictive text, which may confuse users when entering URLs.	See Table 8–8 for a list of avkon_flags properties.

All files referenced in this table can be found in the directory \epoc32\include *in the root of your SDK.*

Table 8–8 Text Editor avkon_flags Properties

Value of avkon_flags	Use
EAknEditorFlagDefault	Default. Resets all flags.
EAknEditorFlagFixedCase	Prevents the case from being changed.
EAknEditorFlagNoT9	Prevents predictive text entry.
EAknEditorFlagNoEditIndicators	Removes edit indicators from the navigation pane. Particularly useful in tabbed views, such as with multipage dialogs.
EAknEditorFlagNoLRNavigation	Prevents left and right scrolling.
EAknEditorFlagSupressShiftMenu	Prevents the special character table from displaying.
EAknEditorFlagEnableScrollBars	Enables the scroll bar indicator.
EAknEditorFlagMTAutoOverwrite	Enables the text to overwrite rather than insert.
EAknEditorFlagUseSCTNumericCharmap	In numeric mode, the editor displays the special character table rather than the numeric key map when the user presses and holds down the asterisk (*) key. It does this only if you have defined a special character table in the editor using the special_character_table field.

Table 8-8 continued

Value of `avkon_flags`	Use
EAknEditorFlagLatinInputModesOnly	Prevents anything other than Latin characters from being entered.
EAknEditorFlagForceTransparentFepModes	Forces front-end-processor (FEP) modes to be transparent. For example, in the find pane in Chinese variants, a pop-up window would be displayed to allow the user to choose between Stroke and Zhuyin text entry. If this flag is set, the entry will be transparent—no pop-up window will be displayed.

Table 8-9 provides a breakdown of the numeric key map modes.

Table 8-9 Text Editor Numeric Key Maps

Value of `numeric_keymap`	* Key Enters	# Key Events
EAknEditorStandardNumberModeKeyMap	*, +, p, w	#
EAknEditorPlainNumberModeKeyMap	Nothing	Nothing
EAknEditorCalculatorNumberModeKeyMap	+, -, *, /	.
EAknEditorConverterNumberModeKeyMap	+, -, E	.
EAknEditorToFieldNumberModeKeyMap	+	;
EAknEditorFixedDiallingNumberModeKeyMap	*, +, p, w	#
EAknEditorSATNumberModeKeymap	*, +	#
EAknEditorSATHiddenNumberModeKeymap	*	#

Table 8-10 provides a breakdown of the available input modes.

Table 8-10 Text Editor Input Modes

Value of `default_input_mode` and `allowed_input_modes`	Use
EAknEditorTextInputMode	Text
EAknEditorNumericInputMode	Numeric
EAknEditorSecretAlphaInputMode	Secret
EAknEditorAllInputModes	All of the above

Instantiating a Plain Text Editor from a Resource

The plain text editor is constructed using an instance of `CEikEdwin`. **PlainTextEditor** performs this construction in the container's second-phase constructor:

```
void CPlainTextEditorContainer::ConstructL(const TRect& aRect)
    {
    CreateWindowL();

    iEditor = new (ELeave) CEikEdwin;
    iEditor->SetContainerWindowL(*this);
    TResourceReader reader;
    iCoeEnv->CreateResourceReaderLC(reader,
R_PLAINTEXTEDITOR_URLEDITOR);
    iEditor->ConstructFromResourceL(reader);
    CleanupStack::PopAndDestroy(); // reader
    iEditor->SetFocus(ETrue);

    SetRect(aRect);
    ActivateL();
    }
```

As shown above, it is necessary to call the first-phase constructor for `CEikEdwin`, construct it from a resource using `ConstructFromResourceL()`, and set the editor so that the user can type into it, by calling `SetFocus()` with a value of `ETrue`.

Key events must be offered to the control so that it receives keypad input from special keys (such as the **Clear** key). For this purpose, `CPlainTextEditorContainer` overrides `OfferKeyEventL()`:

```
TKeyResponse CPlainTextEditorContainer::OfferKeyEventL(const TKeyEvent& aKeyEvent,TEventCode aType)
    {
    if (iEditor)
        {
        return iEditor->OfferKeyEventL(aKeyEvent, aType);
        }
    else
        {
        return CCoeControl::OfferKeyEventL(aKeyEvent, aType);
        }
    }
```

Note that, as this is a Traditional Symbian OS Architecture example (as described in Chapter 4), the container class derives from `CCoeControl`, and so is itself offered key events by the application framework. It passes these first to its child editor control here.

Configuring a Rich Text Editor

The main purpose of **PlainTextEditor** was to show how you can configure editors to facilitate different types of data entry. **RichTextEditor** focuses instead on using an editor to display formatted text. The example emulates a help document browser by displaying a body of formatted sample help text, as shown in Figure 8-7. It allows the user to scroll up and down by page increments. In keeping with text browser characteristics, the editor is read-only and the cursor is hidden. These aims are accomplished by using a rich text editor.

Figure 8-7 The RichTextEditor main screen.

RichTextEditor demonstrates the basic programming tasks associated with rich text editors. These tasks include defining the editor within a resource file and instantiating the editor from resource.

In addition, **RichTextEditor** demonstrates how to develop a customized control by deriving from `CRichTextEditor`. The derived class in the example conveniently encapsulates some of the `CRichTextEditor` text formatting capabilities in intuitively-named public member functions (such as `SetUnderlineOn()` and `SetBoldOn()`). The custom control also enables additional help browser behavior, such as scrolling by page.

Defining a Rich Text Editor Resource

A rich text editor is defined in an `RTXTED` resource structure, as shown below:

```
RESOURCE RTXTED r_richtexteditor_rich_text_editor
   {
   width = KWidth;
   height = KHeight;
   textlimit = KMaxLength;
   flags = EEikEdwinReadOnly | EEikEdwinAvkonDisableCursor;
   avkon_flags = EAknEditorFlagEnableScrollBars;
   }
```

The minimum specification of the above structure comprises the width and height of the editor in pixels, and the maximum number of characters that the editor will hold (`textlimit`).

The **RichTextEditor** example also specifies values for `flags` and `avkon_flags`. Please refer to Table 8-7 for a general explanation of how these and other optional fields can be used. (Note, though, that values in Table 8-7 refer specifically to the **PlainTextEditor** example.)

Because the example is designed only to display text, and not to accept user input, the value of `flags` is set to make the editor read-only

(`EEikEdwinReadOnly`) and hide the cursor from display (`EEikEdwinAvkonDisableCursor`). To enable scrolling, the value of `avkon_flags` is set to `EAknEditorFlagEnableScrollBars`. This causes scroll bars to appear when the length of the editor's text exceeds its physical dimensions.

Creating a Custom Rich Text Editor Control

Although it is possible to use a rich text editor object (`CEikRichTextEditor`) directly, **RichTextEditor** uses a custom class that inherits from `CEikRichTextEditor`. The derived control provides help browser behavior (such as scrolling in page increments) and presents a simplified interface for formatting text. Consider the custom control's class definition below:

```
class CRichTextEditorRichTextEditor : public CEikRichTextEditor
    {
public: // Constructors and destructor
    static CRichTextEditorRichTextEditor* NewL();
    static CRichTextEditorRichTextEditor* NewLC();

private: // Constructor
    void ConstructL();

public: // from CoeControl
    TKeyResponse OfferKeyEventL(const TKeyEvent& aKeyEvent,
TEventCode aType);

public: // members
    void SetUnderlineOn(TBool aOn);
    void SetBoldOn(TBool aOn);
    void AddCarriageReturnL (),
    void AddTextL(const TDesC& aText);
    void DisplayStartOfHelpTextL();

private: // data
    TCharFormatMask iCharFormatMask;      // Current formatting,
e.g. bold, etc.
    TCharFormat iCharFormat;              // Current formatting,
e.g. bold, etc.
    };
```

Notable member functions of this class include:

- An overridden version of `OfferKeyEventL()` to implement scrolling a page at a time.

- Members that simplify the rich text editor interface, such as `SetBoldOn()`.

- Member data for character formatting.

Chapter 8 Editors

You will see the implementation of these methods and use of the member data as you work through this example.

Setting Formatting Attributes

Text formatting in a rich text editor can be applied to either individual characters or entire paragraphs. The **RichTextEditor** example uses character (rather than paragraph) formatting and encapsulates this formatting functionality via member functions that set up individual attributes.

The following is the code for CRichTextEditorRichTextEditor::SetBoldOn():

```
void CRichTextEditorRichTextEditor::SetBoldOn(TBool aOn)
    {
    iCharFormatMask.SetAttrib(EAttFontStrokeWeight);

    if (aOn)
        {
    iCharFormat.iFontSpec.iFontStyle.SetStrokeWeight(
EStrokeWeightBold);
        }
    else
        {
    iCharFormat.iFontSpec.iFontStyle.SetStrokeWeight(
EStrokeWeightNormal);
        }
```

To set a particular attribute, you use a character format mask (TCharFormatMask) to identify the attribute (such as the font stroke weight) that you wish to set, calling the SetAttrib() method to set the actual attribute. In **RichTextEditor**, the attribute used is EAttFontStrokeWeight.

Next, you need to use a character format (TCharFormat) to set the value of this attribute (such as bold or normal). The character format has two (public) member variables that you can set—the font specification (iFontSpec) and the font presentation (iFontPresentation). The font specification gives you access to the typeface, height and font style. The font presentation gives you access to the highlighting and alignment of the characters. **RichTextEditor** deploys the font specification to set the stroke weight, setting it to EStrokeWeightBold or EStrokeWeightNormal as appropriate.

Further information on the full range of specifiable attributes and values is available in the SDK documentation.

> **NOTE** If you wish to set paragraph attributes rather than character attributes, you should use a TParaFormatMask format mask and a CParaFormat format object instead.

Setting the Text in a Rich Text Editor and Applying Character Formatting

A rich text editor contains a `CRichText` object, and this object is used to set and update the editor's text and apply formatting to it. The **RichTextEditor** example illustrates the use of this object in its `AddCarriageReturnL()` and `AddTextL()` functions:

```
void CRichTextEditorRichTextEditor::AddTextL(const TDesC& aText)
  {
  CRichText* richText = RichText();
  TInt documentLength = richText->DocumentLength();
  richText->InsertL(documentLength, aText);
  richText->ApplyCharFormatL(iCharFormat, iCharFormatMask, documentLength, aText.Length());
  }
```

As shown here, a handle on the rich text object is obtained by calling the `RichText()` method of `CEikRichTextEditor`. Despite `RichText()` returning a pointer to the object, there is no transfer of ownership. Therefore, you must not put it on the Cleanup Stack or destroy it.

NOTE Like the rich text editor, other text editors also contain text objects, although of different types. A plain text editor (`CEikEdwin`) contains a `CPlainText` object, and a global text editor (`CEikGlobalTextEditor`) contains a `CGlobalText` object. You can obtain pointers to these objects by using `CEikEdwin::Text()` and `CEikGlobalTextEditor::GlobalText()`, respectively. All text objects derive from `CPlainText`.

The rich text object does not have a method for appending text. However, it does have one for inserting text. If you wish to append text to the end of the existing text and apply character formatting to it, you will need to:

- Calculate the length of the existing text using the `DocumentLength()` method.

- Insert the new text at the calculated position using the `InsertL()` method. This requires two arguments: the position and the text. The **RichTextEditor** example code shown previously passes in the length of the existing text (`documentLength`) as the position so that the text appends onto the existing text.

- Apply the formatting to the text by calling `ApplyCharFormatL()`. This requires four arguments: the character format, the character format mask, the position where the formatting should begin and the number of characters to format. In the **RichTextEditor** example, the recently appended text gets formatted by setting the start position to the length of the previously existing text (`documentLength`), and the number of characters to be the length of the text that was just added (`aText.Length()`).

> **NOTE** If you are applying paragraph formatting, rather than character formatting, then call `ApplyParaFormatL()` instead.

A carriage return is appended to the text in `AddCarriageReturnL()` by inserting `CEditableText::ELineBreak` at the end of the text:

```
void CRichTextEditorRichTextEditor::AddCarriageReturnL()
    {
    CRichText* richText = RichText();
    richText->InsertL(richText->DocumentLength(),
CEditableText::ELineBreak);
    }
```

Positioning the Editor's Cursor

To make the text scroll a page at a time, you need to handle up and down scroll key events. As with all key events, you can do this by overriding the `OfferKeyEventL()` method:

```
TKeyResponse CRichTextEditorRichTextEditor::OfferKeyEventL(
const TKeyEvent& aKeyEvent, TEventCode aType)
    {
    if (aType == EEventKey)
        {
        if (aKeyEvent.iCode == EKeyDownArrow)
            {
            MoveCursorL(TCursorPosition::EFPageDown, EFalse);
            return EKeyWasConsumed;
            }
        else if (aKeyEvent.iCode == EKeyUpArrow)
            {
            MoveCursorL(TCursorPosition::EFPageUp, EFalse);
            return EKeyWasConsumed;
            }
        else
            {
            return CEikRichTextEditor::OfferKeyEventL(aKeyEvent,
aType);
            }
        }
    return EKeyWasNotConsumed;
    }
```

To scroll the text, you simply need to move the cursor to the appropriate position using the `MoveCursorL()` method. This requires two arguments: a desired logical position and a flag indicating whether to select all text delimited by the cursor's previous and new positions. If the selection flag is set to `ETrue`, then the text between the current cursor position and the position you are moving to is selected. In the **RichTextEditor** example, selection of text is not appropriate, so the value passed in is `EFalse`.

The `TCursorPosition` enumeration defines the possible logical positions to move to. In **RichTextEditor**, the application moves up and down "pages" using the values `EFPageUp` and `EFPageDown`.

If, as in **RichTextEditor**, you have appended to the existing text, you will find that the cursor is positioned at the end of that text. If there is more text than will fit on the screen, you will also find that the last page of text is displayed rather than the first. In order to make the first page display, you will need to move the cursor to the start of the text. This is accomplished in `CRichTextEditorRichTextEditor::DisplayStartOfHelpTextL()`:

```
void CRichTextEditorRichTextEditor::DisplayStartOfHelpTextL()
    {
    SetCursorPosL(0, EFalse);
    }
```

This simply calls the `SetCursorPosL()` method, with the new cursor position (`0`) and a flag (`EFalse`) indicating whether to select the text between the current position and the new one.

It is important to call this method **after you have activated the containing window**. In **RichTextEditor** this occurs in `CRichTextEditorContainer::ConstructL()`:

```
void CRichTextEditorContainer::ConstructL(const TRect& aRect)
    {
    CreateWindowL();

    iEditor = CRichTextEditorRichTextEditor::NewL();
    iEditor->SetContainerWindowL(*this);
    AddTextToEditorL();
    SetRect(aRect);
    ActivateL();
    iEditor->DisplayStartOfHelpTextL();

    }
```

As you can see, this method creates the container's window and the rich text editor object, sets the editor's container window, adds text to the editor, activates the container and then moves the cursor to the start of the text.

Constructing a Rich Text Editor from a Resource

The following excerpt shows how to construct a rich text editor from a resource file, using `ConstructFromResourceL()`. The

Chapter 8 Editors

`CRichTextEditorRichTextEditor` class wraps up its construction into standard `NewL()`, and `NewLC()` factory methods. These factory methods call the first-phase constructor and an overridden `ConstructL()`, as shown below:

```
void CRichTextEditorRichTextEditor::ConstructL()
    {
    TResourceReader reader;
    iCoeEnv->CreateResourceReaderLC(reader,
R_RICHTEXTEDITOR_RICH_TEXT_EDITOR);
    ConstructFromResourceL(reader);
    CleanupStack::PopAndDestroy(); // reader
    SetFocus(ETrue);
    }
```

As you can see, this completes the construction using `ConstructFromResourceL()` with the `RTXTED` resource that was defined in the resource file. After construction, it then sets the editor's input focus by calling `SetFocus()` with a value of `ETrue`. This is an important step, without which the editor will not receive keypad input.

Using the Custom Rich Text Editor Control

To complete this example, the following code shows how formatted text is added to the custom rich text editor control. It is taken from the `AddTextToEditorL()` function and uses the `SetBoldOn()`, `SetUnderlineOn()`, `AddCarriageReturnL()` and `AddTextL()` methods discussed previously in order to create and format the text:

```
    ...
    HBufC* header1 =
StringLoader::LoadLC(R_RICHTEXTEDITOR_HELP_HEADER1);
    iEditor->SetUnderlineOn(ETrue);
    iEditor->SetBoldOn(ETrue);
    iEditor->AddTextL(*header1);
    CleanupStack::PopAndDestroy(header1);
    iEditor->AddCarriageReturnL();
    iEditor->SetUnderlineOn(EFalse);
    iEditor->SetBoldOn(EFalse);
    ...
```

The results of this code can be seen in the first line of text presented in Figure 8–7.

Using Styles

In addition to formatting characters as shown in the **RichTextEditor** example, it is possible to specify named styles that can be applied to whole paragraphs. You could, for example, have a named style, `Heading`, which you could use to apply formatting for all headings in your text. This could, for example, specify that font underlining is on and that the stroke weight is set to bold.

All the styles for an editor are stored in a style list. You can pass this to the `CRichText` object on construction or set it afterward. The style list is a `CStyleList` object, which encapsulates a fixed-length array of pointers to `RParagraphStyleInfo` objects. Each `RParagraphStyleInfo` encapsulates a `CParagraphStyle`, which contains the formatting for a paragraph, and an optional reference to the next paragraph style. This optional reference allows you to know which style to apply to the next paragraph. For example, you may want the `Normal` style to always follow the `Heading` style, so in the `Heading`'s `RParagraphStyleInfo` you would set the next paragraph style to be `Normal`.

For each defined style you need to construct a `CParagraphStyle`, passing in a `CParaFormatLayer`, and a `CCharFormatLayer`—these are constructed using masks and formats, as in the **RichTextEditor** example. You then need to construct an instance of `RParagraphStyleInfo`, passing in the `CParagraphStyle` on construction, and add this `RParagraphStyleInfo` to the `CStyleList` using the `AppendL()` method.

Styles are applied to paragraphs much as character formatting is applied, but this time using the `ApplyParagraphStyleL()` method, rather than the `ApplyCharFormatL()` method. You can obtain the particular style required using the `CStyleList` API methods.

For further information on using styles, refer to the SDK documentation.

Numeric Editors

Numeric editors are single-line editors that limit the characters that users can enter into them to numeric types. Series 60 supplies three different editors that can be used to enter numeric data:

- Integer editor.
- Floating-Point Number editor.
- Fixed-Point Number editor.

All numeric editors have characteristics that determine their behavior, as shown in Table 8–11.

In addition to these, the fixed-point number editor can specify of the number of places to keep after the decimal.

The resource structures and classes used for the numeric editors are shown in Table 8–12.

Table 8–11 Numeric Editor Characteristics

Characteristic	Description
Minimum and Maximum Values	This sets the valid range of values for the editor. As you will see in the **NumericEditor** example, each editor deals with out-of-range values differently.
Default Value	This specifies the editor's initial value. All editors specify a default, although the integer editor's default is referred to as the "unset_value".
Maximum Length	This is the maximum number of characters that the editor can hold. Nonnumeric characters, such as decimal points, count as well—take this into account when specifying a maximum length.

Table 8–12 Numeric Editor Resources and Classes

Numeric Editor	Class	Resource	Control
Floating-Point	CEikFloatingPointEditor	FLPTED	EEikCtFlPtEd
Integer	CAknIntegerEdwin	AVKON_INTEGER_EDWIN	EAknCtIntegerEdwin
Fixed-Point	CEikFixedPointEditor	FIXPTED	EEikCtFxPtEd

Configuring Numeric Editors

The example application **NumericEditor** shows you how to define resources for an integer and fixed-point editor, and how to set, get and validate values in any numeric editor. It allows the user to enter and edit minimal details for the employees of a company, and displays brief details of them in a list. The main screen of **NumericEditor** is shown in Figure 8–8.

Selecting **New** from the **Options** menu allows the user to add a new employee by filling in a form. Selecting **Open** allows the user to edit an existing employee's details. The form for creating and editing entries is shown in Figure 8–9.

Figure 8–8 The NumericEditor list screen.

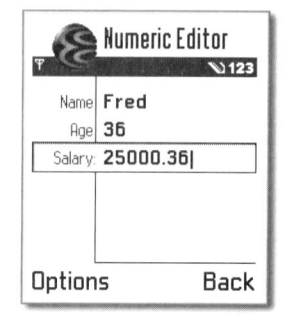

Figure 8–9 The NumericEditor form containing numeric editors.

Chapter 8 Numeric Editors

This form contains two numeric editors: age (an integer editor) and salary (a fixed-point editor).

SEE ALSO

NumericEditor is based on the form example, **OpponentForm**, and so details of the form or the list will not be covered here. See Chapter 6 for more information on forms, and Chapter 7 for more information on lists.

Defining an Integer Editor Resource

You define an integer editor in an AVKON_INTEGER_EDWIN resource, as shown in Table 8–12. In **NumericEditor**, a form contains the editors, so the resource is defined in one of the dialog lines of the r_numericeditor_form resource. The form uses the AVKON_INTEGER_EDWIN to construct a CAknIntegerEdwin. Part of the form resource is shown below:

```
…
control = AVKON_INTEGER_EDWIN
    {
    maxlength = 2;
    min = 18;
    max = 70;
    unset_value = 18;
    };
…
```

The editor definition specifies default values for all of its fields. The maxlength value of 2 provides for a two-digit number—note that for floating-point editors, this value must be incremented by 1 to accommodate the decimal point. The minimum value is set to 18 and the maximum is set to 70—note that you can specify negative numbers if this is appropriate to your application. unset_value is the default value for the editor, and here this is also set to 18.

WARNING

Unlike other numeric editors, there are no checks in an integer editor (CAknIntegerEdwin) that prevent you from setting the maximum to be less than the minimum!

Defining a Fixed-Point Editor Resource

The **NumericEditor** example's fixed-point editor is defined in a FIXPTED resource, as shown in Table 8–12. As before, this resource is defined in one of the dialog lines of the r_numericeditor_form resource. The form uses the FIXPTED resource to construct a CEikFixedPointEditor, and the relevant part of the form resource is shown here:

```
    ...
    control = FIXPTED
        {
        decimalplaces = 2;
        min = 0;
        max = 10000000;        // 100000.00 with the decimal point
shifted 2 decimal places.
        default = 0;
        };
    ...
```

As with the integer editor, each field has a default value. Unlike the integer editor, however, if you specify a maximum value less than the minimum, then the `CEikFixedPointEditor` editor object will panic on construction.

Obtaining and Validating the Value in a Numeric Editor

Integer editors provide access to their values via the `GetTextAsInteger()` method, floating-point editors use the `GetValueAsReal()` method, and fixed-point editors use the `GetValueAsInteger()` method.

Note that fixed-point editors actually store and manipulate integer values—to obtain the actual numeric value held, you need to divide by 10^N, where N is the number of decimal places set for the editor. Floating-point editors work with `TReal` values, as you would expect.

In the **NumericEditor** example, values are retrieved from the editors at the point where the form data is saved in `CNumericEditorForm::SaveFormDataL()`:

```
    ...
    CAknIntegerEdwin* ageEditor =
static_cast<CAknIntegerEdwin*>(
ControlOrNull(ENumericEditorDlgCIdIntegerEdwin));

    if (ageEditor)
        {
        TInt age = 0;
        CAknNumericEdwin::TValidationStatus ageStatus =
ageEditor->GetTextAsInteger(age);

        switch (ageStatus)
            {
            case CAknNumericEdwin::EValueValid:
                {
                iEmployee.SetAge(age);
                break;
                }
            ...
            }
        }
```

Chapter 8 Numeric Editors

```
    CEikFixedPointEditor* salaryEditor =
static_cast<CEikFixedPointEditor*>(
ControlOrNull(ENumericEditorDlgCIdFxPtEd));

    if (salaryEditor)
       {
       TInt salary = 0;
       CAknNumericEdwin::TValidationStatus salaryStatus =
salaryEditor->GetValueAsInteger(salary);

       switch (salaryStatus)
          {
          case CAknNumericEdwin::EValueValid:
             {
             iEmployee.SetSalary(salary);
             break;
             }
          ...
          }
       }
    ...
```

`GetTextAsInteger()` is called to obtain and subsequently validate the age in the integer editor. The method is passed a reference to a `TInt`, which it sets to the value contained in the editor, and returns an enumerated `TValidationStatus` value. This represents the validity of the numeric value set, and it can be any of the following values:

- `EValueValid`—if the value is valid.
- `EValueTooLarge`—if the value is greater than the maximum.
- `EValueTooSmall`—if the value is less than the minimum.
- `EValueNotParsed`—if the value was not a valid number.

`GetValueAsInteger()` is similarly called to obtain and validate the salary in the fixed-point editor. Note that fixed-point and floating-point editors actually validate their value whenever they lose focus. If the current value is outside of the minimum and maximum value constraints, then it is set to the closest legal value. This means that the **NumericEditor** example does not actually need to validate the value obtained from the fixed-point salary editor.

Setting the Value in a Numeric Editor

You can set the value in a numeric editor using the `SetValueL()` method. In **NumericEditor**, this occurs when the form is loaded in the `LoadFormValuesFromDataL()` method, part of which is shown below:

```
...
CAknIntegerEdwin* ageEditor =
static_cast<CAknIntegerEdwin*>(
ControlOrNull(ENumericEditorDlgCIdIntegerEdwin));
    if (ageEditor)
        {
        ageEditor->SetValueL(iEmployee.Age());
        }

    CEikFixedPointEditor* salaryEditor =
static_cast<CEikFixedPointEditor*>(
ControlOrNull(ENumericEditorDlgCIdFxPtEd));

    if (salaryEditor)
        {
        TInt salary = iEmployee.Salary();
        salaryEditor->SetValueL(&salary);
        }
...
```

The parameter required by `SetValueL()` depends on the type of editor. The integer editor's `SetValueL()` requires a `TInt` passed by value, the floating-point editor requires a pointer to a `TReal`, and the fixed-point editor requires a pointer to a `TInt`. Passing in `NULL` clears a fixed- or floating-point editor.

Secret Editors

You can use secret editors to securely obtain PINs or passwords from a user. As the user types into a secret editor, it represents each character with an asterisk on the display, as shown in Figure 8–10.

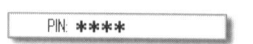

Figure 8–10 A secret editor in a form.

Series 60 provides two types of secret editor: a numeric one that can be used for PINs, and an alphanumeric one that can be used for passwords. Unlike the numeric secret editor, the alphanumeric one displays the character briefly before converting it into an asterisk. This is because in alphanumeric mode, each key on the keypad can produce different characters.

Using a Secret Editor

The code for using secret editors is quite simple and so does not require a dedicated example. This section shows how to define a secret editor in a resource, instantiate it and get and set its value.

Defining a Secret Editor in a Resource

Numeric secret editors are defined using a `NUMSECRETED` resource structure, and alphanumeric secret editors are defined using a `SECRETED` structure:

```
RESOURCE NUMSECRETED r_my_numeric_secret_editor
    {
    num_code_chars = 4;
    }

RESOURCE SECRETED r_my_secret_editor
    {
    num_letters = 10;
    }
```

Each resource has a field representing the maximum length of the editor. In the NUMSECRETED resource, this is the num_code_chars field, and in the SECRETED resource, the num_letters field. The value must be less than a predefined constant, EMaxSecEdLength, otherwise the editor will panic on construction with the error code EEikPanicSecretEditorTooLong.

Instantiating a Secret Editor

Secret editors can be instantiated using any of the standard methods previously seen—for example, using ConstructFromResourceL(), or creating one as part of a dialog, in which case the dialog constructs, and owns the editor.

The alphanumeric secret editor class is CEikSecretEditor and the numeric secret editor class is CAknNumericSecretEditor, which is derived from CEikSecretEditor. Table 8–13 summarizes the values that you should use for each type of secret editor resource, including the control value that you should use to create the editor in a dialog.

Table 8–13 Secret Editor Resource and Classes

Secret Editor	Class	Resource	Control
Alphanumeric	CEikSecretEditor	SECRETED	EEikCtSecretEd
Numeric	CAknNumericSecretEditor	NUMSECRETED	EAknCtNumericSecretEditor

Getting and Setting a Secret Editor's Value

Once constructed, you can simply get a reference to the editor's text using GetText() and initialize the contents of the editor with SetText().

Multi-Field Numeric Editors

As their name suggests, multi-field numeric editors (**MFNE**s) are used for entry of multiple numeric fields into a single editor. Note that the simplest MFNE, namely the number editor, actually contains only one field.

Figure 8-11 An IP address editor.

Characters appropriate to the type of editor separate the fields, for example, a dot (".") is used to separate fields in an IP (Internet Protocol) address editor, as shown in Figure 8-11.

As well as IP addresses, Series 60 provides MFNEs for editing single-integer numbers, ranges, times, dates and durations. Each editor has a minimum, maximum and a current value. The minimum and maximum values limit the range of values that the user can enter. Depending on the type of editor, this limit is applied either to each field or to the editor as a whole. For example, in an IP address editor, the minimum and maximum values determine the range of the values that could be entered into each field of the address, whereas in a date editor, the minimum and maximum values limit the range of dates that could be entered in the editor as a whole.

This section provides details of the available MFNEs and describes how to define a resource structure and instantiate one. It also explains how to get and set the value of an MFNE.

IP Address Editor

The IP address editor allows the user to enter IPv4 addresses in decimal format—that is, four decimal numbers between 0 and 255, separated by a dot ("."). An example of an IP address editor is shown in Figure 8-11.

Number Editor

The number editor is very similar to the integer editor, in that it can be used for editing integers. However, its behavior is more like that of a floating-point editor—when values are out of range, they are set to the nearest valid value. Also, when the maximum value of the number editor is set to be greater than its minimum value, the editor will panic during its construction. An example of a number editor is shown in Figure 8-12.

Figure 8-12 A number editor.

Note that the integer editor derives from `CEikEdwin`, whereas the number editor derives from `CEikMfne`, and this provides them with different configuration options.

Range Editor

The range editor allows the user to enter two numbers representing a range. The second number must be greater than or equal to the first, and by default the numbers are separated by a dash ("-"). However, you can change the separator used in the editor's resource definition. An example of a range editor is shown in Figure 8-13.

Figure 8-13 A range editor.

Time Editor

The time editor has fields for hours, minutes and seconds, and you can use flags to determine which of these fields are displayed. It is available in either 12-hour or 24-hour format—the default is 12-hour format. The locale settings are used to determine formatting, and which characters are used to separate the fields. An example of a time editor is shown in Figure 8–14.

Figure 8–14 A time editor shown in 12-hour format with no seconds field.

Date Editor

The date editor provides day, month and year fields. Again, the locale settings are used to determine formatting and the characters that are used to separate the fields. An example of a date editor is shown in Figure 8–15.

Figure 8–15 A date editor.

Time and Date Editor

The time and date editor has the properties of both the time and date editors. Yet again, the locale settings are used to determine formatting and the characters that are used to separate the fields. An example of a time and date editor is shown in Figure 8–16.

Figure 8–16 A time and date editor.

Duration Editor

The duration editor displays a length of time in hours, minutes and seconds. As with the time editor, you can use flags to determine which of these fields are shown. The locale settings are used to determine formatting and the characters that are used to separate the fields. An example duration editor is shown in Figure 8–17.

Figure 8–17 A duration editor.

Time Offset Editor

The time offset editor allows the user to edit a signed time offset in hours, minutes and seconds. The locale settings are used to determine formatting and the characters that are used to separate the fields. An example time offset editor is shown in Figure 8–18.

Figure 8–18 A time offset editor.

Using an MFNE

As with the secret editor, the code for MFNEs is quite simple. Therefore, a comprehensive example is not required. Instead, this section focuses on defining an MFNE in a resource, instantiating an MFNE and getting and setting an MFNE's value.

Defining an MFNE in a Resource

The type of resource you define depends on the MFNE that you are using. Table 8–14 shows which resource to use when instantiating an MFNE, and which control to use if using the MFNE as part of a dialog:

For example, you can define a number editor in a `NUMBER_EDITOR` resource:

```
RESOURCE NUMBER_EDITOR my_number_editor
    {
    min = 0;
    max = 200;
    }
```

Each resource has a field representing the maximum and minimum values for the editor. The type and name of these will depend on the particular MFNE that you are using. As shown, the number editor has integer values for the "`min`" and "`max`" fields. Further information on these resource structures can be found in the file `\epoc32\include\eikon.rh` in the root directory of your SDK. The values of any flags used can be found in `\epoc32\include\eikon.hrh`, also in this directory—their names all contain the string "`Mfne`".

Instantiating an MFNE

You can instantiate the editor using any of the standard methods. For example, you can use the first-phase constructor of its class and then call `ConstructFromResourceL()`, or you can instantiate one as part of a dialog line—in which case the dialog constructs and owns the editor.

Table 8–14 shows which class you should use when instantiating an MFNE.

Table 8–14 MFNE Editor Resources and Classes

MFNE Editor	Class	Resource	Control
Range	`CEikRangeEditor`	`RANGE_EDITOR`	`EEikCtRangeEditor`
Time	`CEikTimeEditor`	`TIME_EDITOR`	`EEikCtTimeEditor`
Date	`CEikDateEditor`	`DATE_EDITOR`	`EEikCtDateEditor`
Duration	`CEikDurationEditor`	`DURATION_EDITOR`	`EEikCtDurationEditor`
Time and Date	`CEikTimeAndDateEditor`	`TIME_AND_DATE_EDITOR`	`EEikCtTimeAndDateEditor`
IP Address	`CAknIpFieldEditor`	`IP_FIELD_EDITOR`	`EAknCtIpFieldEditor`
Number	`CEikNumberEditor`	`NUMBER_EDITOR`	`EEikCtNumberEditor`
Time Offset	`CEikTimeOffsetEditor`	`TIME_OFFSET_EDITOR`	`EEikCtTimeOffsetEditor`

Getting and Setting the Value of an MFNE

Once constructed, you can get and set the value of the editor using the API methods for the particular class of object that you are using—these are typically functions of the form `GetXXX()` and `SetXXX()`. Refer to the class definitions in the header files `\epoc32\include\eikmfne.h` and `\epoc32\include\aknipfed.h` in the root directory of your SDK.

Also, it is possible to dynamically get and set the minimum and maximum values of an MFNE using the applicable class overloads of the methods `SetMinimumAndMaximum()` and `GetMinimumAndMaximum()`.

Summary

This chapter has shown how you may use editors in a Series 60 application to enter, edit and view numerous data types, in various formats. It has also described how input restrictions can be enforced, and how access to extended character sets can be configured.

Editors can be created statically (using a resource file) or dynamically (using API construction) and may exist as stand-alone UI components. However, more typically they are used as part of a dialog, or form—in which case they are usually constructed and owned as part of that control.

The first example application, **PlainTextEditor**, showed you how to create and use a plain text editor, applying attributes that made it into an editor suitable for the insertion of URLs. The second example application, **RichTextEditor**, showed you how to use a rich text editor and how to apply complex character formatting, also giving an overview of paragraph formatting, or "styles." The third example, **NumericEditor**, showed you how to use numeric editors and, in particular, how to configure them for entry of integer, or fixed-point, data.

Additionally, this chapter gave you direction on how to create and use secret editors—in other words, editors that hide their input values, and also multi-field numeric editors (MFNEs). These provide specialized entry for specific numeric data types, such as times, dates and IP addresses.

Further information on these and other editor controls can be found in the SDK documentation.

chapter 9

Communications Fundamentals

Basic communication APIs for Series 60

Chapter 9 Communications Fundamentals

Communication is central to a Series 60 phone. Of course, voice communication is the core feature of all phones, but a Series 60 device offers you far more ways to communicate than just making a call. Series 60 can enable services as diverse as browsing and downloading from the Web, playing a game over Bluetooth between two phones, sending a document to your office printer over infrared, receiving the latest score of a football match by text message or connecting to your bank to get a secure update of your bank balance.

Each of these actions involves forming some kind of connection between your Series 60 phone and another device. The "other device" may be anything from a phone in the same room to a network server on the other side of the world. Series 60 provides support for a range of communication technologies to enable communication with different types of devices, at different distances and for different purposes.

This chapter introduces some of the more basic-level communications APIs provided by Series 60. Many of these APIs operate at a relatively low level—you may often need to implement extra layers of code on top of them in order to achieve the kind of communication your application requires. Chapter 10 covers some higher-level APIs, which often provide all you need to establish communication with another device.

The main topics of this chapter are as follows:

- **Serial Communication**—Serial communication is a simple, low-level communication between two devices at close range. Series 60 enables this type of communication over infrared or Bluetooth. This section discusses the techniques involved in creating a serial connection.

- **Sockets**—This section covers sockets in Series 60. The Sockets API offers a higher-level point-to-point communication between two participants. It can be implemented over a range of physical and logical network types including infrared, Bluetooth and TCP/IP.

- **TCP/IP**—TCP/IP makes use of the Sockets APIs for accessing standard Internet services. In this section, Series 60 support for TCP/IP is demonstrated with a simple application. Series 60 2.x adds support for multihoming, which allows multiple network connections to be open concurrently.

- **Infrared**—Infrared (or IR) is a simple, point-to-point communication between two devices at close range with a direct "line of sight" between them. This section discusses the various IR protocols supported in Series 60.

- **Bluetooth**—Often used for similar purposes as infrared, Bluetooth is a radio technology used for close-range communication between two or more devices. This section introduces the Bluetooth capabilities of Series 60, and the APIs provided to make use of them.

The last three sections—**TCP/IP, Infrared** and **Bluetooth**—all build on information given earlier in the chapter. It is therefore recommended that you read the sections on **Serial Communication** and **Sockets** before moving on to the latter parts of the chapter.

The following example applications illustrate these technologies and provide the source code used throughout this chapter:

- **IrSerial**—Illustrates serial communication in a simple application that sends data to an IR-capable printer.

- **IrSockets**—Demonstrates sending small text-based chat messages between two Series 60 devices using an infrared sockets interface.

- **TcpipMultihomingEx**—Performs simple transactions with a Daytime server, demonstrating the multihoming APIs present in Series 60 2.x.

- **TcpIpImpEx**—Demonstrates how developers who wish to avoid the added complexity of multihoming APIs can use TCP/IP via an implicit connection. This example also works on Series 60 1.x.

- **BluetoothChat**—Implements a simple instant chat messaging application, sending text messages between two Series 60 devices using Bluetooth sockets.

Details of how to download the full buildable source of these example applications are given in the Preface.

Serial Communication

Serial communication is a low-level, point-to-point technology used to transfer data between two devices, typically at close range—Series 60 supports serial communication over infrared and Bluetooth. Central to the Series 60 implementation is the **Serial Communications Server** (also referred to as the **Comms Server** or **C32**). This uses the familiar Symbian OS Client/Server framework to provide access to serial hardware, and it is *generic* (in other words, the same API is used for both infrared and Bluetooth serial communication—note that other hardware can be supported by the addition of plug-ins) and *shared* (in other words, several client threads can safely use the same serial port concurrently).

Chapter 9 Communications Fundamentals

All serial communication on Series 60 is implemented using the following basic steps:

1. Load the serial device drivers.
2. Start the Comms Server.
3. Connect to the Comms Server.
4. Load a **comms module** (otherwise known as a **CSY**)—this is the Comms Server plug-in that determines which type of serial port you wish to use (for example, infrared or Bluetooth).
5. Open a serial port.
6. Configure the serial port.
7. Write data to (and/or read data from) the port.
8. Finally, close the port.

Using Serial Communication

To illustrate how to implement the basic steps that have been outlined, an example application is provided. This application is called **IrSerial**, and it allows the user to enter some text, which it then sends to an IR-enabled printer using serial communication. A screenshot of **IrSerial** is shown in Figure 9–1.

Figure 9–1 The IrSerial application.

All the example code shown here comes from the `CIrSerialEngine` class. Here are the relevant member variables from the class, which are featured throughout this section:

```
RCommServ iCommServer;   // serial comms
RComm iCommPort;         // comm port
// data to transmit
TBuf8<EIrSerialFormEdwinMaxLength> iDataBuf;
```

You should also note that `CIrSerialEngine` is an Active Object—in other words, it is derived from `CActive`.

The following subsections show how to implement each of the basic steps needed for serial communication.

Loading the Serial Device Drivers

Your first step is to load the serial communication drivers. There are two parts you need to load: a **physical driver** (which interacts with the hardware directly), and a **logical driver** (which provides an API onto the physical driver). The names

Chapter 9 Serial Communication

for these are fixed and are defined here globally. Notice that the physical device driver for emulator builds (WINS) is different from that for target builds, as you would expect when dealing with different hardware.

```
    // Physical device driver names
#if defined (__WINS__)
    _LIT(KPddName, "ECDRV");
#else
    _LIT(KPddName, "EUART1");
#endif

    // Logical device driver name
    _LIT(KLddName, "ECOMM");
```

These literals are then used to load the drivers in the `CIrSerialEngine::InitialiseL()` method:

```
    TInt err;

#if defined (__WINS__) // File Server required in WINS to
                       //enable loading of device drivers
    RFs fileServer;
    User::LeaveIfError(fileServer.Connect());
    fileServer.Close();
#endif

    // Load the physical device driver.
    err = User::LoadPhysicalDevice(KPddName);

    if (err != KErrNone && err != KErrAlreadyExists)
       {
       User::Leave(err);
       }

    // Load the logical device driver.
    err = User::LoadLogicalDevice(KLddName);

    if (err != KErrNone && err != KErrAlreadyExists)
       {
       User::Leave(err);
       }
```

Note that a return value of `KErrAlreadyExists` from the loading functions indicates that the driver is already loaded. This is not considered an error, so you can call this code "blindly," without needing to explicitly check in advance whether the driver is already loaded. If your code is part of a UI application, these drivers will already have been loaded by the Series 60 UI framework.

Also, at the top of this example, note the unusual WINS-only lines, to connect an `RFs` file session and then immediately close it. The purpose is to ensure that the File Server is loaded before the calls to load the drivers. Opening a session will start the server if it is not already started, but since you do not use the file session itself, you should then just close it and discard it.

Starting the Comms Server

Next, you need to start the Comms Server itself. The code for this is very similar to the code you used to load the drivers. In particular, note that this code can also be safely called even if the Comms Server has already been started. Also, note that `StartC32()` is a global function (defined in `c32comm.h`).

```
// Start the comms server process
err = StartC32();

if (err != KErrNone && err != KErrAlreadyExists)
   {
   User::Leave(err);
   }
```

Connecting to the Comms Server

`RCommServ` is a client-side handle on the Comms Server and is used by all subsequent code to communicate with it. First you need to connect the handle to the server:

```
// Connect to the Serial comms server.
User::LeaveIfError(iCommServer.Connect());
```

Note that here, unlike in the previous code snippets, any return value other than `KErrNone` is considered an error.

Loading a CSY

A CSY is a DLL plug-in to the Comms Server that needs to be explicitly loaded before data transfer can take place. As with the drivers, CSYs are loaded by name: "`IRCOMM`" for infrared and "`BTCOMM`" for Bluetooth. The **IrSerial** example illustrates how to load the infrared CSY:

```
// Comms modules
_LIT (KIrComm, "IRCOMM");
...
// Load the CSY module.
User::LeaveIfError(iCommServer.LoadCommModule(KIrComm));
```

Again, note that any return value other than `KErrNone` should be treated as an error. To change your code to use Bluetooth instead of infrared, you would have to load the Bluetooth CSY (`BTCOMM`) here.

Opening a Serial Port

Your next step is to open a serial port. Again, this is opened by name. Note, however, that port names are not fixed and may need to be obtained dynamically. In the case of infrared, however, the port name is well known, so the **IrSerial** example defines this as a string literal and uses it directly:

Chapter 9 Serial Communication

```
// Comm Port Name
_LIT(KPortName, "IRCOMM::0");
…
User::LeaveIfError(iCommPort.Open(iCommServer, KPortName,
ECommShared));
```

The third parameter to `RComm::Open()` indicates a mode in which the port should be opened. The possible values are as follows:

- `ECommExclusive`—Once open, the port cannot be used by any other `RComm` clients.
- `ECommShared`—The port can be shared by other `RComm` clients who open in the same mode.
- `ECommPreemptable`—The port will be lost if another clients tries to open it.

In cases where the port name is not well known, the following code (not taken from the **IrSerial** example) shows how you would obtain it dynamically:

```
TSerialInfo portInfo;

// Get the port information for your chosen CSY
User::LeaveIfError(iCommServer.GetPortInfo(KIrComm,
portInfo));

// Construct a descriptor to contain the full name of
// the lowest port this CSY supports. This needs to be
// big enough to contain the value of TSerialInfo.iName
// (maximum: 16 characters, defined as KMaxPortName),
// plus two colons and one or two digits.
// Example: "IRCOMM::0"

_LIT(KColons, "::");

TBuf<KMaxPortName + 4> portName;
portName.Append(portInfo.iName);
portName.Append(KColons);
portName.AppendNum(portInfo.iLowUnit);

// Now open the first serial port - here in exclusive mode
User::LeaveIfError(iCommPort.Open(iCommServer, portName,
ECommExclusive));
```

Configuring the Serial Port

Before you finally make use of the serial port you have just opened, you may need to configure it. This will depend on the requirements of the device you are connecting to; devices offering serial communication services will usually be accompanied by some form of documentation detailing the configuration required (such as maximum port speed and use of data format elements such

Chapter 9 Communications Fundamentals

as stop bit and parity bit). You can also query its capabilities to ensure that it supports the configuration you require.

> **SEE ALSO**
>
> The API Guide section of the Series 60 SDK help includes a section entitled "Serial Comms", which provides detailed information on configuring serial ports.

The port's capabilities are queried as follows:

```
TCommCaps portCapabilities;
iCommPort.Caps(portCapabilities);

if (((portCapabilities().iRate & KCapsBps19200) == 0) ||
    ((portCapabilities().iDataBits & KCapsData8) == 0) ||
    ((portCapabilities().iStopBits & KCapsStop1) == 0) ||
    ((portCapabilities().iParity & KCapsParityNone) == 0))
    {
    User::Leave(KErrNotSupported);
    }
```

The constant values used here (such as `KCapsBps19200`) are all defined in the system header file `d32comm.h`.

Note that the reason for the odd syntax is that `TCommCaps` is defined as a package buffer—`operator()` returns the underlying `TCommCapsV01` object that the package buffer contains.

> **SEE ALSO**
>
> Package buffers are discussed in Chapter 3.

Note also that you do not need to leave if the port's capabilities do not match your requirements. You may choose to handle this condition differently—for example, by prompting the user to select a different port or CSY.

Next, you can set the configuration of the port. You should start by obtaining the current configuration so that values you do not wish to set explicitly stay the same.

```
TCommConfig portSettings;
// Get current configuration
iCommPort.Config(portSettings);

// Set port characteristics
portSettings().iRate = EBps19200;
portSettings().iParity = EParityNone;
portSettings().iDataBits = EData8;
portSettings().iStopBits = EStop1;
portSettings().iFifo = EFifoEnable;
```

Chapter 9 Serial Communication

```
    portSettings().iHandshake = KConfigObeyXoff |
KConfigSendXoff;
    // line feed character
    portSettings().iTerminator[0] = 10;
    portSettings().iTerminatorCount = 1;

    User::LeaveIfError(iCommPort.SetConfig(portSettings));
```

Note again that `TCommConfig` is a package buffer and so requires slightly unusual syntax.

Finally, there are some miscellaneous configuration details that cannot be set using a `TCommConfig` object. For example, flow control signals such as DTR (Data Terminal Ready) and RTS (Ready To Send) are set using the `RComm::SetSignals()` function, as follows:

```
    // Turn on DTR and RTS
    iCommPort.SetSignals(KSignalDTR, 0);
    iCommPort.SetSignals(KSignalRTS, 0);

    // Set buffer size
    const TInt KBufferLength = 4096;
    iCommPort.SetReceiveBufferLength(KBufferLength);
```

Once configuration is complete, the serial port is ready to use. The next subsection shows how data is sent.

Transfering Data over the Port

The `CIrSerialEngine::Transmit()` method is used to send data over the port you have just opened. The user provides data to send by typing text into a dialog, which is passed to this function as the `aTxData` parameter.

```
void CIrSerialEngine::Transmit(const TDesC& aTxData)
    {
    // Timeout Value
    const TTimeIntervalMicroSeconds32 KTimeOut(4000000);
    Cancel();

    iDataBuf.Copy(aTxData);
    iCommPort.Write(iStatus, KTimeOut, iDataBuf);
    SetActive();
    }
```

There are some interesting points to note here:

- `Cancel()` is called before calling the `Write()` function. As `Write()` is asynchronous, this ensures that `CIrSerialEngine` does not attempt to make two write operations simultaneously, which would cause a panic.

- `aTxData` is a 16-bit descriptor, as Series 60 is a Unicode platform. As such it cannot be passed directly to `RComm::Write()`, which requires an explicit 8-bit (`const TDesC8&`) descriptor—so it is copied into an 8-bit descriptor (`iDataBuf`), and this is passed to the `Write()` function. Note that any characters with a Unicode value greater than `255` will be set to an 8-bit (ASCII) value of "`1`"—this is covered further in the **Descriptors** section of Chapter 3.

- The second parameter to `Write()` is a timeout value, defined here as four seconds. If the `Write()` request has not completed normally within this time, it will be forced to complete, posting a value of `KErrTimedOut` into the `iStatus` member of `CIrSerialEngine`.

Closing the Port

When you have finished sending and receiving data over the serial port, you must close the port (or more specifically, close the handle on the port), and likewise close the server. Fortunately, this is the simplest step of the whole process!

```
void CIrSerialEngine::Close()
    {
    iCommPort.Close();
    iCommServer.Close();
    }
```

Sockets

Sockets provide higher-level access to communications protocols than seen in the **Serial Communication** section. The Sockets API was originally designed to facilitate TCP/IP connections in BSD UNIX, and it has become a standard API for doing so across a range of platforms, including UNIX, Windows and now Series 60. If you have used sockets before on other platforms, then many of the concepts explained here will be quite familiar already.

In addition to TCP/IP connections, the Sockets API is sufficiently generic to allow for other underlying network types and protocols. This fact has been exploited in Series 60, where the Sockets API can be used to make connections over protocols such as infrared, Bluetooth and WAP.

This section outlines the key essentials of sockets programming and provides an example application discussed in a step-by-step fashion. The Series 60 APIs for secure sockets programming are also discussed.

The subsections that follow present the basics you will need to understand in order to use sockets on Series 60.

Sockets on Series 60

The structure of the Series 60 Sockets API is very similar to that already seen for serial communication. The key components are:

Chapter 9 Sockets   439

- A Socket Server, to provide shared, generic access to sockets protocols.
- A client-side API, providing controlled access to this server.
- A system of server plug-ins (here known as **protocol modules**, or **PRTs**, analogous to the CSYs used by the Serial Comms Server).

The internal name of the Symbian OS Socket Server is "ESOCK", so you may see this used in the SDK documentation and elsewhere to refer to this server.

The plug-in protocol modules are just DLLs with a .prt file extension. You do not use these directly—they are used only by the Socket Server itself. Each protocol module specifies its capabilities in an .esk text file, which can be found in the \system\data folder on a device or in the equivalent PC emulator directory. The .esk files replace an earlier architecture which used a single configuration file, named esock.ini. Again, you may see this file referred to in other documentation—you should understand that it no longer exists and has been replaced by a set of .esk files, one per protocol.

Client and Server

A socket represents one end of a connection between two participants. In any sockets scenario, one participant is referred to as the **server** and the other as the **client**. This distinction is really meaningful only at connection time—a server passively waits for incoming connections, whereas the client actively requests a connection with a server. Once connected, both client and server may write data to and/or read data from the other participant.

A single program can be written to perform either role. As you will see, the **IrSockets** example application can behave as either client or server, depending on a menu command chosen by the user.

> **NOTE** It is important not to confuse the terms client and server described here with the Symbian OS Client/Server framework. A server can indeed be implemented in a Symbian OS client process and, for example, connect to the Symbian OS Socket Server.

Connectionless and Connected Sockets

A socket is said to be either **connectionless** or **connected** (the latter is also referred to as **connection-oriented**). Connectionless sockets simply ensure delivery of datagrams (data packets) from one participant to another. There is no concept of an ongoing "session" between the two participants, and this entails a number of important consequences:

- Each time you write data to a connectionless socket, you must specify the target participant's address.

- Likewise, when reading from a connectionless socket, you must specify the sender's address at each read.

- Connectionless sockets cannot guarantee that the datagrams will arrive in the same order as they were sent.

You can think of using a connectionless socket as being something like sending a series of letters to a friend—each letter needs to be individually addressed and posted, and there is no guarantee that they will arrive in the exact order in which they were sent (or, indeed, that they will arrive at all!).

Connected sockets, on the other hand, *do* maintain the concept of a session between the two participants—so the address needs to be specified only once, at first connection, and datagrams will arrive in an ordered fashion. You might think of this as loosely analogous to a telephone call—you dial the person's number only once, at the start of the conversation, and you can guarantee that they will hear everything you say in exactly the order in which you say it! (You should not take this analogy too far, however—there are many respects in which a connected socket session is nothing at all like a phone call!) The disadvantage of connected sockets is that they are less efficient to set up and use.

Both connected and connectionless sockets are supported in Series 60, and both use exactly the same classes (for example, RSocket, RSocketServ, and so on). The difference is just in the way the sockets are set up and configured, and in the member functions of RSocket that are subsequently used to send and receive data. The rest of this section focuses on connected sockets, as these are far more frequently used on Series 60. For details of the differences between connected and connectionless sockets programming, refer to the SDK documentation.

Connected Sockets

Connected sockets are much more commonly used than connectionless sockets. For a small overhead in terms of efficiency, connected sockets provide you with a reliable stream of data between two endpoints, freeing the developer to concentrate on the data itself rather than the mechanics of how it is transmitted.

This subsection details how to use connected sockets in Series 60, using example code taken from the **IrSockets** application shown in Figure 9-2. This is a simple application which sends short text-based chat messages between two devices using infrared sockets. Each device can choose whether to send or receive, so data can be sent in both directions.

The example code in this subsection is drawn from the CIrSocketsEngine class in the **IrSockets** application. The following code segment shows the member data of CIrSocketsEngine that is used in this subsection:

Figure 9-2 The IrSockets example.

Chapter 9 Sockets

```
               // data to transmit
               TBuf8<KFormEdwinMaxSize16>      iSendBuffer;
               // data to receive
               TBuf8<KFormEdwinMaxSize16>      iRecvBuffer;

               /*! @var the actual sockets */
               RSocket iSocket;
               RSocket iServiceSocket;

               /*! @var the Symbian OS socket server */
               RSocketServ iSocketServer;

               /*! @var DNS name resolver */
               RHostResolver iHostResolver;

               /*! @var The result from the name resolver */
               TNameEntry iNameEntry;

               /*! @var The name record found by the resolver */
               TNameRecord iNameRecord;

               TProtocolDesc iProtocolInfo;
               TIrdaSockAddr iIrdaSockAddr;
               TCommsStatus iState;
```

`TCommsStatus` is a bespoke enumerated type defined for this application; it will be discussed in more detail below.

The basic steps performed by the application are as follows:

1. Connect to the Symbian OS Socket Server.
2. Open a socket.
3. Connect the server socket.
4. Connect the client socket.
5. Transfer data.
6. Close the socket and other handles.

These steps are described in detail here:

Connecting to the Symbian OS Socket Server

The first step of any sockets programming for Series 60, whether client or server, is to connect to the Symbian OS Socket Server. This is a simple step compared to the initialization required for serial communication—with sockets, there is no need to preload drivers, nor to start the server explicitly.

```
               // Connect to the Symbian OS Socket Server.
               TInt err = iSocketServer.Connect();
```

```
if (err != KErrNone && err != KErrAlreadyExists)
    {
    User::Leave(err);
    }
```

Note that all applications that wish to use the Sockets API must connect to the Symbian OS Socket Server—this is not the same thing as making a connection to a server socket. Applications that contain a client socket must connect to the Symbian OS Socket Server to create the socket, and this client socket would then connect to a server socket, which will typically exist on another device. The terminology can be a little confusing at first!

Now there are just a couple of further steps required. First, as a sanity check, you should make sure that the Socket Server supports at least one protocol (in other words, that one or more PRT plug-in modules are present):

```
TUint numProtocols;
// Number of protocols the Symbian OS
// Socket Server is aware of
err = iSocketServer.NumProtocols(numProtocols);
User::LeaveIfError(err);
```

Finally, you need to obtain information about the protocol you intend to use—in this case, **IrTinyTP**, which is a transport-layer protocol for infrared.

```
_LIT(KProtocol, "IrTinyTP");
...
// Find protocol by name and get description
TProtocolName protocolName(KProtocol);
err = iSocketServer.FindProtocol(protocolName,
iProtocolInfo);
User::LeaveIfError(err);
```

Opening a Socket

Having connected to the Symbian OS Socket Server, you are ready to open a socket. This is done using `RSocket::Open()`, which takes the following parameters:

- `RSocketServ& aServer`—a reference to an open socket server session
- `TUint addrFamily`—a constant value representing an address family (see Table 9–1)
- `TUint sockType`—a constant value representing a socket type (see Table 9–2)
- `TUint protocol`—a constant value representing a protocol (see Table 9–3)

Tables 9–1 through 9–3 list the constant values that may be passed to `RSocket::Open()`. All header files are located in the `\epoc32\include` folder, relative to the root of your SDK installation.

Chapter 9 Sockets

Table 9-1 Socket Address Family Constants

Constant	Defined in	Description
KAfInet	in_sock.h	Internet sockets
KIrdaAddrFamily	ir_sock.h	Infrared sockets
KBTAddrFamily	bt_sock.h	Bluetooth sockets

Table 9-2 Socket Type Constants

Constant	Defined in	Description
KSockStream	es_sock.h	Reliable (connection-oriented) socket
KSockDatagram	es_sock.h	Unreliable (connectionless) socket
KSockSeqPacket	es_sock.h	(Currently unused)
KSockRaw	es_sock.h	Raw socket

Table 9-3 Socket Protocol Constants

Constant	Defined in	Description
KProtocolInetIcmp	in_sock.h	ICMP—Internet Message Control Protocol (For information on this and other TCP/IP protocols, see the TCP/IP section later in this chapter.)
KProtocolInetTcp	in_sock.h	TCP—Transfer Control Protocol
KProtocolInetUdp	in_sock.h	UDP—User Datagram Protocol
KProtocolInetIp	in_sock.h	IP—Internet Protocol
KIrmux	ir_sock.h	IrMUX (For information on this and other IR protocols, see the Infrared section later in this chapter.)
KIrTinyTP	ir_sock.h	IrTinyTP
KBTLinkManager	bt_sock.h	Bluetooth Link Manager Protocol (For information on this and other Bluetooth protocols, see the Bluetooth section later in this chapter.)

Table 9-3 continued

Constant	Defined in	Description
KL2CAP	bt_sock.h	Bluetooth Logical Link Control and Adaptation Protocol
KRFCOMM	bt_sock.h	Bluetooth RFCOMM Protocol
KSDP	bt_sock.h	Bluetooth Service Discovery Protocol

These constants may be passed directly to `RSocket::Open()`. However, as in the **IrSockets** application, it is best practice to find a protocol by name (as shown in the previous code segment) and get these values from the Sockets Server. The following segment shows how a socket is opened using values obtained from the Sockets Server.

```
TInt err;

// Open socket
err = iSocket.Open(iSocketServer,
iProtocolInfo.iAddrFamily, iProtocolInfo.iSockType,
iProtocolInfo.iProtocol);

if (err != KErrNone && err != KErrAlreadyExists)
    {
    User::Leave(err);
    }
```

Again, this step is the same for both client and server sockets code. Note that *opening* a socket is a slightly confusing term—it does not *connect* it. All you have done here is to initialize a socket object. Connection is shown in the following text, and this is where the code starts to differ between client and server usage.

Connecting the Server Socket

To initialize your socket as a server you need to carry out three steps. First you need to **bind** it to an address and port number which the client socket will specify in order to connect to it. In **IrSockets**, this takes place in the `CIrSocketsEngine::ConfigureReceiverL()` function:

```
// Port number
const TUint KPortNumber = 0x02;
…
// Set port
iIrdaSockAddr.SetPort(KPortNumber);

// Bind socket locally and set up listening queue
iSocket.Bind(iIrdaSockAddr);
```

Chapter 9 Sockets

After the socket is bound to an address and port number, you use the `Listen()` method to make it **listen** for incoming connections from elsewhere. The `KSizeOfListenQueue` parameter passed to the `Listen()` method in the **IrSockets** example specifies the length of the **listen queue**—in other words, the number of outstanding connection requests that the socket can queue up before further requests are rejected.

In addition, you will need a second socket to service any incoming requests, as your listening socket must stay free to monitor for further incoming connection requests. This is also done in `CIrSocketsEngine::ConfigureReceiverL()`:

```
// Size of Listening Queue
const TUint KSizeOfListenQueue = 1;
...
// Listen for connection from sender
iSocket.Listen(KSizeOfListenQueue);

// Create blank socket
iServiceSocket.Open(iSocketServer);
```

Finally, you need to call `Accept()` to **accept** the first incoming request. Note that this request may not yet have been made. However, `Accept()` is an asynchronous call, so it will eventually complete when a request is received. (Alternatively, it can also be cancelled before this ever happens, using `CancelAccept()`.) If, on the other hand, a request *has* already been issued by another device, this will currently be pending in the listen queue that was set up by `Listen()`, with the result that `Accept()` will complete very quickly.

In any case, the call to `Accept()` takes exactly the same form, as follows:

```
iSocket.Accept(iServiceSocket, iStatus);
iState = EAccepting;
SetActive();
```

Note the following interesting points about the preceding code segment:

- You pass your second socket (`iServiceSocket`) through as a parameter to `Accept()`, as this is the socket that will actually be used to transfer data, once a connection has been established.

- The `CIrSocketsEngine` class is an Active Object, so it passes its own `iStatus` member to `AcceptL()` and subsequently calls `SetActive()` on itself. Its `RunL()` method will be called when a connection request occurs.

- The `iState` member is used to implement a basic state machine within the `CIrSocketsEngine` class. This is essential, owing to the asynchronous nature of the Sockets API—when the `RunL()` method is called, you must be able to determine what state you are currently in (for example, accepting, reading, writing and so on). `iState` is a part of the implementation of such a state machine in the `CIrSocketsEngine` class and is not a part of the

Sockets API. Be aware that you may not see exactly this same variable name (or type) in other Series 60 sockets code, although generally it will always implement some kind of state machine along these lines.

While the server just waits for any incoming connection, the client actively pursues a connection. The following discussion shows what is required on the client side in order to make a connection request.

Connecting the Client Socket

At the simplest level, connecting a client socket just involves calling `Connect()`, specifying the required address and port information. This would be the case, for example, in connecting a TCP/IP socket to an Internet server at a known address and port.

However, for an infrared connection, the address of the target machine is typically *not* known in advance. Connections are not being made over a permanent, physical network, and so are more ad hoc in nature. The same is true of Bluetooth. In both cases, connection is usually preceded by a phase called **discovery**, where the device looks for other machines that are in range and offering services (listening on a socket) for a given protocol.

Discovery

The class used to discover other devices is `RHostResolver`. This is also used with TCP/IP sockets to resolve host names (such as **www.emccsoft.com**) into numerical IP addresses. This is a quite different process from discovering IR or Bluetooth devices, but it performs a similar role in the process of establishing a socket, and the `RHostResolver` class ensures a consistent set of APIs across all socket types. An `RHostResolver` object is opened in a similar way to an `RSocket`: with an `Open()` method which takes an open Socket Server session, an address family and a protocol.

As discovery could potentially take a relatively long time, the corresponding API (`RHostResolver::GetByName()`) is asynchronous. As in the example of `Accept()` above, the `CIrSocketsEngine` class is itself an Active Object that will issue such asynchronous requests and handle their completion. The code required to start the discovery phase is shown here:

```
// Number of devices to search for during discovery
const TUint KNumOfDevicesToDiscover = 1;

…

TInt err;
THostName hostname;

// Set up socket for device discovery.
// Use one slot, i.e. search for one device.
TPckgBuf<TUint> buf(KNumOfDevicesToDiscover);
iSocket.SetOpt(KDiscoverySlotsOpt, KLevelIrlap, buf);
```

```
// Open host resolver for device discovery
err = iHostResolver.Open(iSocketServer,
iProtocolInfo.iAddrFamily, iProtocolInfo.iProtocol);

User::LeaveIfError(err);

iState = EDiscovering;
iHostResolver.GetByName(hostname, iNameEntry, iStatus);
SetActive();
```

This code finishes by calling the asynchronous function `iHostResolver.GetByName()` to start the discovery process, passing in its own `iStatus` member and setting itself active. This will result in the `RunL()` method getting called when discovery is complete. Note that the `iState` variable is also set accordingly, so that `RunL()` can correctly determine which request it is handling.

NOTE `GetByName()`, and other asynchronous calls on `RHostResolver`, can be cancelled by calling `RHostResolver::Cancel()`.

Connection

When the Active Object handling the call to `GetByName()` completes successfully, the `iNameEntry` member will contain the address of a valid server machine that was discovered. This is then used to connect, as follows:

```
iHostResolver.Close();  // As it's no longer needed.

// Store remote device address
iIrdaSockAddr = TIrdaSockAddr(iNameEntry().iAddr);

...

iIrdaSockAddr.SetPort(KPortNumber);
iState = EConnecting;
// Request connection.
iSocket.Connect(iIrdaSockAddr, iStatus);
SetActive();
```

This code is taken from the `CIrSocketsEngine::RunL()` and `CIrSocketsEngine::ConnectL()` methods, but brought together here for the sake of clarity. Note that the `KPortNumber` passed to `SetPort()` is the same value used when binding the server socket—this ensures that the client will attempt to connect with the correct port number.

As you might expect, `Connect()` is another asynchronous function. Note that this may fail even if the discovery phase succeeded—for example, if the server has gone out of range since it was discovered.

Transferring Data

Once a socket connection is established between two devices, data transfer is achieved by means of the `Read()` and `Write()` methods; note that both functions are available to each socket. In the **IrSockets** example, only the client writes and only the server reads, but this need not be the case—data may flow either way, or in both directions. (Remember that on the server side, it is a separate socket, `iServiceSocket`, which performs any reading or writing. The main socket, `iSocket`, is busy listening for any further incoming connection requests.)

The implication is that there must be an agreement or "contract" between the client and server implementations to ensure that a write on one side is matched by a read on the other. In the case of the **IrSockets** example, this "contract" is very simple but is specific to the example. It can be summarized as follows:

- Client writes one descriptor.
- Server reads one descriptor.
- Both sockets disconnect.

Of course, in the real world most sockets "contracts" are more complex than this, and in order to be useful they need to be widely understood and implemented. Such contracts become specified as higher-level protocols, such as HTTP and FTP.

The **IrSockets** application sends a Unicode descriptor from one participant to the other. This is complicated slightly by the fact that sockets can handle only 8-bit descriptors. Hence the application actually externalizes the 16-bit descriptor to an 8-bit one, which it then transfers.

The `CIrSocketsEngine::SendL()` function shows how the data is externalized to an 8-bit descriptor and then sent over the socket:

```
void CIrSocketsEngine::SendL(const TDesC& aData)
    {
    if (iState == ESending)
        {
        // Create a TDesBuf referring to iSendBuffer.
        // This will allow us to create a stream to stream data
        // into iSendBuffer.
        TDesBuf buffer;
        buffer.Set(iSendBuffer);

        // Create a write stream
        RWriteStream stream(&buffer);
        CleanupClosePushL(stream);

        // Stream aData (the 16-bit descriptor argument)
        // into iSendBuffer
        stream << aData;
```

Chapter 9 Sockets

```
        // Clean up the stream
        CleanupStack::PopAndDestroy();

        // Now write the 8-bit descriptor iSendBuffer to the
        // socket.
        // On the reading side this can then be read in, then
        // the 16-bit descriptor internalized from the 8-bit
        // buffer.
        iSocket.Write(iSendBuffer, iStatus);
        SetActive();
        }
    }
```

Note that `Write()` is an asynchronous function, which is handled using an Active Object in the usual way.

That is all that is required on the writing side—the next thing that happens will be a call to `RunL()` with a status value that indicates either success or failure.

On the receiving side, the process is split into two steps:

- An asynchronous read occurs on the socket which will receive the 8-bit data into the member `iRecvBuffer`.

- Once this has completed, code in the `RunL()` method, which streams the 16-bit descriptor from `iRecvBuffer`.

First, here is the `CIrSocketsEngine::ReceiveL()` function which handles reading the 8-bit data from the socket:

```
void CIrSocketsEngine::ReceiveL()
    {
    if (iState == EAccepted)
        {
        iRecvBuffer.Zero();
        iState = EReceiving;
        iServiceSocket.Read(iRecvBuffer, iStatus);
        SetActive();
        }
    }
```

Note that the `Read()` function is called on `iServiceSocket`, not `iSocket`. Remember that `iSocket` is in permanent use listening for new connection requests, and `iServiceSocket` is used to service those requests.

When the `Read()` request completes, it will result in a call to `CIrSocketsEngine::RunL()`, with `iState` set to `EReceiving`. This is handled as follows—you should be able to see a clear correlation between the steps of this process and those used above in `SendL()`:

```
    // Create a TDesBuf object referring to iRecvBuffer
    TDesBuf desBuf;
```

```
    desBuf.Set(iRecvBuffer);

    // Open a stream on iRecvBuffer using desBuf
    RReadStream stream(&desBuf);
    CleanupClosePushL(stream);

    // Stream the data into a Unicode buffer
    TBuf<EFormEdwinMaxLength> unicodeBuffer;
    stream >> unicodeBuffer;

    // Clean up the stream
    CleanupStack::PopAndDestroy();

    // Pass the Unicode descriptor to the UI
    iIrSocketsEngineObserver.DisplayReceivedDataL(unicodeBuffer);
```

Closing Sockets

In order to clean up neatly, you will need to ensure that all your R-class handles are closed. CIrSocketsEngine cancels and closes all sockets and the host resolver in a CloseSubSessions() function, which in turn is called from its DoCancel() method. This ensures that in the event of a cancel, any outstanding asynchronous requests will be correctly cancelled. In its destructor, the class calls Cancel() (following good practice for Active Objects)—which will result in a call to DoCancel(), and therefore CloseSubSessions()—and also closes the Socket Server session.

These functions are shown below.

```
CIrSocketsEngine::~CIrSocketsEngine()
    {
    Cancel(); // This will close all subsessions

    // Close session on Sockets Server
    if (iSocketServer.Handle() != 0)
        {
        iSocketServer.Close();
        }
    }


void CIrSocketsEngine::DoCancel()
    {
    CloseSubSessions();
    }


void CIrSocketsEngine::CloseSubSessions()
    {
    // Cancel and close host resolver
    if (iHostResolver.SubSessionHandle() != 0)
        {
```

```
        iHostResolver.Cancel();
        iHostResolver.Close();
        }

    // Cancel and close service socket
    if (iServiceSocket.SubSessionHandle() != 0)
        {
        iServiceSocket.CancelAll();
        iServiceSocket.Close();
        }

    // Cancel and close main socket
    if (iSocket.SubSessionHandle() != 0)
        {
        iSocket.CancelAll();
        iSocket.Close();
        }

    // Set state to idle
    iState = EIdle;
    }
```

It is important that before you make a call to cancel or close any handles, you ensure that the handles have been initialized (using `Handle()` and `SubSessionHandle()`). Failure to do this can result in a panic.

Before closing a socket, you should call `RSocket::CancelAll()`. This is a convenient function that will cancel any outstanding asynchronous request on a socket, regardless of the original function involved (`Read()`, `Write()`, `Accept()` and so on) — there is no need to check the value of `iState` here, which makes the `CloseSubSessions()` function a lot simpler.

Secure Sockets

Secure sockets encrypt the data that passes between the client and server sockets. Series 60 provides support for two secure sockets standards: **SSL** (Secure Sockets Layer) v3.0, and **TLS** (Transport Layer Security) v1.0.

SSL is a thin protocol layer, which provides almost completely transparent encryption over a standard connection-oriented socket. After some minimal initialization, an SSL-enabled socket is used in exactly the same way as a nonencrypted socket. In Series 60, SSL can even be enabled on a standard socket after it has been connected. SSL is the mechanism used to implement secure Web pages — any page with a URL beginning **https://** has been encrypted using SSL.

TLS operates at the transport layer (usually TCP/IP) and requires slightly more initialization than SSL. TLS is intended as a successor to SSL and can be regarded as "SSL 3.1." However, as SSL is still the prevalent standard on the Internet, this is the default option for secure sockets in Series 60 — TLS must be enabled explicitly.

It should be noted that SSL and TLS are not interoperable. Also, Series 60 provides support only for client-side secure sockets—you cannot implement an SSL-enabled server using native Series 60 APIs.

Finally, note that in Series 60, only a TCP/IP socket can be secured.

Series 60 1.x Secure Sockets

The APIs used for secure sockets have changed between Series 60 1.x and Series 60 2.x. Both APIs are described here, starting with Series 60 1.x.

Securing a New Socket

In Series 60 1.x, secure sockets programming is performed using exactly the same classes as seen previously in this section: `RSocketServ`, and `RSocket`. To enable SSL, all you need to do is call `RSocket::SetOpt()` with the parameters `KSoSecureSocket` and `KSolInetSSL` (defined in `in_sock.h`), as shown in the following code segment:

```
// RSocketServ iSocketServ; defined in class definition.

// Connect iSocketServ...

// RSocket iSocket; defined in class definition.

// First, open a socket.
// This socket must be opened with the KAfInet,
// KSockStream and KProtocolInetTcp options
// to allow SSL.
User::LeaveIfError(iSocket.Open(iSocketServ, KAfInet,
KSockStream, KProtocolInetTcp));

// Then, to enable SSL on that socket, call SetOpt()
// with the option name of KSoSecureSocket, and the
// option level of KSolInetSSL.
User::LeaveIfError(iSocket.SetOpt(KSoSecureSocket,
KSolInetSSL));

// SSL is now enabled on the socket, and no
// further actions are required.

// Now the socket can attempt to connect to the
// remote host.
iSocket.Connect(iInetAddr, iStatus);
```

If you wish to set other secure options on the socket, additional calls to `SetOpt()` can be made before calling `Connect()`. The various options available for secure sockets are listed on Symbian's Web site (**http://www.symbian.com/developer/techlib/papers/sslapi/sslapi.html**).

Chapter 9 Sockets

Securing an Already Connected Socket

You can also enable encryption on a socket that has already been connected, provided it has been opened with the same parameters as shown in the code above (that is, for TCP/IP). To do this, simply call SetOpt() with the KSoSecureSocket and KSolInetSSL parameters. If you need to set any other security options (also with SetOpt()), be sure to do this first and call socket.SetOpt(KSoSecureSocket, KSolInetSSL) last of all.

Using a Secure Socket — the SSL Handshake

There is an important point to note on the use of secure sockets, regardless of whether you enable security before connecting or vice versa. In both cases you do **not** receive explicit notification that SSL has been successfully enabled on the socket. Instead, the following happens:

- Where you enable SSL before connecting the socket, the call to RSocket::Connect() completes once a TCP/IP connection has been made. Only **after** this does the process of establishing the SSL connection — known as the *SSL handshake* — begin.

- When enabling SSL on an already connected socket, the SSL handshake starts when SetOpt(KSoSecureSocket, KSolInetSSL) is called.

In each case, any attempts to read from, or write to, the socket while the handshake is still in progress will be buffered. If the SSL handshake fails, the first read/write operation after the handshake was initiated will complete with an error code. This is your only way of determining if the SSL handshake completed successfully.

Using TLS

TLS is used in a very similar way to SSL, but it requires a little more configuration. Unfortunately, there is no explicitly named function to enable TLS. Instead, it is again enabled by passing particular parameters to SetOpt(), as described below.

A secure socket encrypts data using a mechanism known as a **cipher suite**. A number of cipher suites are defined within the SSL/TLS standards, and during the SSL handshake the client runs through a list of cipher suites, enquiring whether the server implements them, until one is found which both sides support.

The cipher suites listed in Table 9-4 are available to use with the weak version of the cryptography library; this is included with the public SDK. Those in Table 9-5 are available for use with the strong version of the cryptography library, in addition to those available in the weak library. The strong library is not available in the public SDK, as it is subject to export restrictions under UK law. It may be present in licensee products, depending on legal arrangements. Such arrangements are beyond the scope of this book.

Table 9-4 Weak Cryptography Cipher Suites

Name	Value
SSL_RSA_WITH_DES_CBC_SHA	[0x00][0x09]
SSL_DHE_DSS_EXPORT_WITH_DES40_CBC_SHA	[0x00][0x11]
SSL_DHE_RSA_EXPORT_WITH_DES40_CBC_SHA	[0x00][0x14]

Table 9-5 Strong Cryptography Cipher Suites

Name	Value
SSL_RSA_EXPORT_WITH_RC4_40_MD5	[0x00][0x03]
SSL_RSA_WITH_RC4_128_MD5	[0x00][0x04]
SSL_RSA_WITH_RC4_128_SHA	[0x00][0x05]
SSL_RSA_EXPORT_WITH_DES40_CBC_SHA	[0x00][0x08]
SSL_RSA_WITH_3DES_EDE_CBC_SHA	[0x00][0x0a]
SSL_DHE_DSS_WITH_DES_CBC_SHA	[0x00][0x12]
SSL_DHE_DSS_WITH_3DES_EDE_CBC_SHA	[0x00][0x13]
SSL_DHE_RSA_WITH_DES_CBC_SHA	[0x00][0x15]
SSL_DHE_RSA_WITH_3DES_EDE_CBC_SHA	[0x00][0x16]

The default ordering of the supported cipher suites is shown in Table 9-6. Note that this is the order when using the strong version of the cryptography library. If the weak version is used, the order remains the same, but suites from the strong cryptography library are not included.

Calling RSocket::SetOpt() with a first parameter of KSoCurrentCipherSuite can change the order of the list of cipher suites which the client sends to the server. Each cipher suite is referred to by a two-byte value, and a list of these is passed to RSocket::SetOpt(). In order to enable TLS, you must make this call and set the last value in the list to [0xff][0xff].

Chapter 9 Sockets

Table 9–6 Default Order of Cipher Suites

Name	Value
SSL_RSA_WITH_RC4_128_MD5	[0x00][0x04]
SSL_RSA_WITH_RC4_128_SHA	[0x00][0x05]
SSL_RSA_WITH_3DES_EDE_CBC_SHA	[0x00][0x0a]
SSL_DHE_RSA_WITH_3DES_EDE_CBC_SHA	[0x00][0x16]
SSL_DHE_DSS_WITH_3DES_EDE_CBC_SHA	[0x00][0x13]
SSL_DHE_DSS_WITH_DES_CBC_SHA	[0x00][0x12]
SSL_RSA_WITH_DES_CBC_SHA	[0x00][0x09]
SSL_DHE_DSS_EXPORT_WITH_DES40_CBC_SHA	[0x00][0x11]
SSL_DHE_RSA_EXPORT_WITH_DES40_CBC_SHA	[0x00][0x14]
SSL_RSA_EXPORT_WITH_DES40_CBC_SHA	[0x00][0x08]
SSL_RSA_EXPORT_WITH_RC4_40_MD5	[0x00][0x03]

The following code segment provides an example. Note how the list of values is passed in as a descriptor.

```
// Call SetOpt()to enable security on the socket
  User::LeaveIfError(iSocket.SetOpt(KSoSecureSocket,
KSolInetSSL));

// Now call again to set TLS and a preferred
// list of cipher suites

const TInt KNoOfCiphers = 4;
// 2 bytes per cipher suite
const TInt KArraySize = KNoOfCiphers * 2;
const TUint8 array[KArraySize] =
    {
    0x00, 0x0a,   // SSL_RSA_WITH_3DES_EDE_CBC_SHA
    0x00, 0x04,   // SSL_RSA_WITH_RC4_128_MD5
    0x00, 0x16,   // SSL_DHE_RSA_WITH_3DES_EDE_CBC_SHA
    0xff, 0xff    // to specify TLS
    };
```

```
// Create a descriptor version of the array
TPtrC8 ptr(array, KArraySize);

// Call SetOpt(). Note that the second parameter is
// still KSolInetSSL, even though we want to use TLS.
User::LeaveIfError(iSocket.SetOpt(KSoCurrentCipherSuite,
KSolInetSSL, ptr));

// Now the socket can attempt to connect to the
// remote host
iSocket.Connect(iInetAddr, iStatus);
```

NOTE As with SSL, `Connect()` will return as soon as a TCP/IP connection is established. The completion value of the first read or write operation should be checked to determine whether the TLS handshake was successful.

Series 60 2.x Secure Sockets

Series 60 2.x adds some new classes to handle secure sockets. The main one is `CSecureSocket`, which is instantiated from an `RSocket` handle to represent a secure socket. Its API is derived from the new mixin interface `MSecureSocket`.

The main shortcoming from Series 60 1.x that is addressed by the new API is the lack of notification when the secure handshake completes. In Series 60 2.x, the handshake is initiated explicitly by means of an asynchronous function, and the corresponding Active Object is notified when the handshake completes.

Another key difference from Series 60 1.x is that the `RSocket` handle *must* be connected before it is secured. So now the sequence of events is always as follows:

- Open a TCP/IP socket exactly as before, using `RSocket::Open()`.
- Connect the socket, again just as in Series 60 1.x, using `RSocket::Connect()`.
- Once `RSocket::Connect()` has completed, create a `CSecureSocket` object, passing the connected `RSocket` handle into its `NewL()` function.
- Call `StartClientHandshake()` on your secure socket, and wait for this to complete.
- Transfer data securely using the `Send()` and `Recv()` methods of `CSecureSocket`.

The following code segment shows construction of a `CSecureSocket` object and the initiation of the SSL handshake. Assume this is within a class derived from `CActive`, with member data `iSocket` (`RSocket`) and `iSecureSocket` (`CSecureSocket*`). `iSocket` has been opened and connected as described above.

```
// Construct the secure socket object
_LIT(KSSL3, "SSL3.0"); // or "TLS1.0" for TLS
iSecureSocket = CSecureSocket::NewL(iSocket, KSSL3);

// Start the SSL handshake process
iSecureSocket->StartClientHandshake(iStatus);
SetActive();
```

When the handshake completes, the Active Object's `RunL()` method will be called, with the value of `iStatus` indicating whether the handshake was successful. If the handshake succeeded, you can then go on to transfer data as follows:

```
// Class has the following members:
// - iRequestBuffer, a TBuf8 big enough to contain what we
// want to send
// - iBytesSent, a TInt

// initialize iRequestBuffer here

iSecureSocket->Send(iRequestBuffer, iStatus, iBytesSent);
SetActive();
```

When the Active Object's `RunL()` is called, `iBytesSent` will contain the number of bytes successfully sent from `iRequestBuffer`. If not all bytes have been sent, you will need to issue further `Send()` requests until the entire buffer has been transmitted.

TCP/IP

Serial communication is sufficient for point-to-point transmission of data, but the physical constraints imposed, and the relative slowness of the transport, prohibits its use for multipoint communication. Networking technologies support communication between multiple devices and employ an addressing mechanism for identification of each machine. Very fast speeds can be achieved through the use of mediums such as Ethernet.

The most common network protocol suite is **TCP/IP**, which consists of several protocols over two layers. At the network layer, the Internet Protocol (**IP**) and Internet Control Message Protocol (**ICMP**) protocols are employed, while at the transport layer, Transmission Control Protocol (**TCP**) and User Datagram Protocol (**UDP**) operate.

UDP is an unreliable protocol that provides a datagram (connectionless) service and is suited to simple query and response applications. TCP is reliable and provides stream-oriented connections using data packet sequence numbers to coordinate transmission. Services such as FTP, telnet, email delivery by SMTP and Web services via HTTP are defined at this level, with each service being assigned a specific port number. (A port is an endpoint to a logical connection. Numbers are used to distinguish between different logical channels on the same network interface.)

At the network level, the ICMP protocol detects and announces network errors and congestion. The IP protocol is responsible for routing the data between networks and employs an addressing system to achieve this. IPv4 is supported, and Series 60 2.x adds support for IPv6 with a dual stack implementation provided by Symbian OS 7.0s.

IPv6

A primary motivation for the introduction of IPv6 was the need for more IP addresses. An increase in address size from 32-bit to 128-bit means that it supports more levels of addressing hierarchy and ultimately more unique addresses. There are further differences between IPv4 and IPv6, in particular the format of the headers, but in general, applications do not need to be concerned about the IP version. However, if an address is to be stored, displayed or manipulated, then you will need to check whether it is an IPv4 or IPv6 address, as the size and format of the addresses differ. This can be achieved through a call to `TInetAddr::Family()`.

IPv6 also offers increased performance and enhanced security and is mandated by 3GPP standards.

TCP/IP Programming for Series 60

To communicate over TCP/IP you use the Sockets API, specifying the TCP/IP family as the protocol. The general principles of sockets programming, are followed, as described in the **Sockets** section. An important point to note is that a network interface needs to be started to provide the transport for the connection. There can be many Internet Access Points (**IAPs**) defined (the device may know many different ways to connect to the Internet), and you can choose whichever one you wish to use, or just use the default setup.

Besides the Sockets API, other important components include the connection management API, `RConnection`, which allows an application to use a specific Internet Access Point; **NIFMAN**, a server to control network interfaces; and the communications database, **CommDB**, which stores all of the configuration settings. This section describes the Series 60 implementation of TCP/IP and shows you the steps required to create a TCP/IP connection. The differences between Series 60 1.x and Series 60 2.x are also discussed.

CommDB

Information relating to communications settings (IAPs, ISPs, GPRS, modems, locations, charge cards, proxies and WAP) is held in a Symbian OS database. A series of relational tables store the data, with the database being managed by the communications database server, CommDB. Applications can access and

Chapter 9 TCP/IP

modify the settings through CommDB, with the underlying DBMS ensuring the integrity of the database. Users can use the Settings application to view and manipulate the details stored. Note that the database not only stores connection details but also maintains user preferences.

Series 60 1.x defines the `CApDataHandler` class to enable easy reading and updating of IAP settings. In Series 60 2.x this is replaced by `CApUtils`. The `CApAccessPointItem` class is used to represent an individual access point.

Multihoming

Series 60 2.x supports multihoming, which means that several network interfaces can be active at one time. For users, this means that they can access many services, such as MMS, Internet and WAP, simultaneously, with each one using a different technology or setting to connect to the Internet. So a dial-up connection along with several Packet Switched Data (**PSD**) connections could be active, each with its own IP address and each potentially on a different network.

NOTE You may encounter the term **PDP** (Packet Data Protocol) **context** in discussions about multihoming. Essentially a PDP context refers to a set of information relating to, for example, security, billing, Quality of Service and routing, which describes a mobile wireless service call or session.

Series 60 1.x is effectively "single-homed"—that is, it allows only one network interface to be active at a time, and although applications can use the `RGenericAgent` API to request a connection through a specific IAP, the request will not be successful if another connection is already active. A single-homed system can also lead to problems relating to security and billing, as the TCP/IP stack may not route `RSocket` and `RHostResolver` requests through the correct interface. The `RConnection` API introduced in Series 60 2.x allows an application to specify which IAP is used, and so prevents these problems.

RConnection

The `RConnection` API, exported by `ESOCK`, offers full support for multihoming and overcomes the security and billing problems mentioned earlier. Essentially the `RConnection` API can be used to provide the following functionality:

- Starting and stopping interfaces—with or without overriding the preferences specified in `CommDb`.
- Monitoring progress during start-up of an interface.
- Indicating when interfaces start and stop.
- Enumerating active connections.

- Retrieving statistics on traffic sent and received on an interface.
- Indicating when subconnections (PDP contexts) start and stop.
- Reading `CommDb` fields specific to an active connection.

`RConnection` objects are implemented as Symbian OS Socket Server subsessions, so they need to be opened on an existing `RSocketServ` session. Multiple `RConnection` objects can be defined on a single `RSocketServ` session, but an individual `RConnection` object must be used by only a single client. However, it is possible for several `RConnection` objects to refer to the same underlying interface.

Explicit Connections

If you want to use `RConnection` to create an explicit interface connection for your TCP/IP sockets, the general procedure is:

- Connect to the Symbian OS Socket Server.
- Open an `RConnection` on the Socket Server.
- Create IAP preferences using `TCommDbConnPref` (this is optional).
- Start the outgoing connection—with preference overrides if required, using `RConnection::Start()`.
- Optionally open a host resolver associated with the connection for DNS look-ups using `RHostResolver::Open()`.
- Open a socket associated with the connection using `RSocket::Open()`.
- Request the socket to connect to the destination `RSocket::Connect()`.

Most of these steps have been explained in the **Sockets** section, but an example, **TcpipMultihomingEx**, is provided to show the extra stages required. Note once more that `RConnection`, explicit connections and multihoming are not available in Series 60 1.x, and so this application is Series 60 2.x-specific.

The purpose of the **TcpipMultihomingEx** example is to read the current date and time from a Daytime Server. To do this, you simply need to connect to an appropriate server on port 13.

NOTE The Internet Assigned Numbers Authority (IANA) maintains a list of well-known ports. Port 13 is defined as the port number for the Daytime service. Connection to this port results in the current date and time being sent out.

The following code segment shows the data members of the class `CTcpipMultiHomingExEngine`:

Chapter 9 TCP/IP

```
TEngineState                         iEngineStatus;
CTcpipMultiHomingExAppUi&            iAppUi;
RSocket                              iSocket;
RConnection                          iRConn;
RSocketServ                          iSocketServ;
RHostResolver                        iResolver;
TNameEntry                           iNameEntry;
TNameRecord                          iNameRecord;
CTcpipMultiHomingExTimer*            iTimer;
TInetAddr                            iAddress;
TInt                                 iPort;
TBuf<KMaxServerNameLength>           iServerName;
TBuf8<KReadBufferSize>               iBuffer;
TSockXfrLength                       iLength;
```

The first step, as always, is connecting to the Symbian OS Socket Server. This is done in the `CTcpipMultiHomingExEngine::ConstructL()` function, as shown in the earlier **Sockets** section. You then create an `RConnection` object on the Socket Server, leaving if there are any errors:

```
//Open an RConnection on the Socket Server
User::LeaveIfError(iRConn.Open(iSocketServ));
```

You can now decide which IAP you want your application to use. By default, the user will be presented with a list of the available IAPs from which to select. Alternatively you can specify a particular access point in your code, either by selecting one from CommDb or by creating a new one. The class `TCommDbConnPref` can be used to set the preference. If you wish to define an alternative IAP, to be used in case of failure of the preferred connection, `TCommDbMultiConnPref` is available. You need to consider whether you want the user to be notified of the connection being used. The enumeration `TCommDbDialogPref` defines the available options that can be passed to `TCommDbConnPref::SetDialogPreference()`. If you have specified more than one IAP, then it is recommended that you do warn the user if it is the second choice that is being used (`ECommDbDialogPrefWarn`). The use of `TCommDbConnPref` is shown in the code below. (Note that this code is not taken from **TcpipMultihomingEx**, as this example allows the user to select the IAP from the list available.)

```
// Set the required IAP
TCommDbConnPref prefs;
prefs.SetDialogPreference(ECommDbDialogPrefDoNotPrompt);
prefs.SetDirection(ECommDbConnectionDirectionOutgoing);
prefs.SetIapId(2);
```

You then call one of the implementations of `RConnection::Start()`. If you have defined your IAP preference(s) then they should be passed through; although, if you decide to allow the user to select an IAP, this is not necessary. In the example **TcpipMultihomingEx**, the user chooses the IAP.

There is a synchronous form of `RConnection::Start()`, but the asynchronous form (as always) is preferred. In the **TcpipMultihomingEx** example, `CTcpipMultiHomingExEngine` is an Active Object and issues the request passing through its `iStatus`. It uses a simple state machine, represented by `iState`, to manage the connection process.

```
// Start the connection
iRConn.Start(iStatus);
ChangeStatus(ERConnStarting);
iTimer->After(KTimeOut);
SetActive();
```

`CTcpipMultiHomingExEngine::RunL()` will be called when the Active Object completes. To determine whether or not the connection was started successfully, `iStatus` should be tested.

If DNS lookup is required, you can open an `RHostResolver` object and associate it with the `RConnection`. The socket should also be opened and associated with the `RConnection`. This ensures that the data flows over the same underlying interface. Note that `KAfInet` should always be specified as the address family argument for both `RSocket::Open()` and `RHostResolver::Open()`:

```
   User::LeaveIfError(iResolver.Open(iSocketServ, KAfInet,
KProtocolInetUdp, iRConn));
   ...
   User::LeaveIfError(iSocket.Open(iSocketServ, KAfInet,
KSockStream, KProtocolInetTcp, iRConn));
```

The destination you wish to connect to is specified using the `TInetAddr` class. `TInetAddr` is the generic Socket Server address class for TCP/IP and can store either an IPv4 or IPv6 address. The `Input()` method takes the IP address as a `TDesC&` which can be of either format—this is particularly useful if the IP address has been entered into a dialog box by the user. If you have an IPv4 address in the form of a `TUint32`, or an IPv6 address in the form `TIp6Addr`, then `SetAddress()` can be called instead. In this example, the connection is to port 13.

`RSocket::Connect()` is asynchronous, so the `iStatus` member of `CTcpipMultiHomingExEngine` is also passed in, and the state machine is updated:

```
// Initiate socket connection
iSocket.Connect(iAddress, iStatus);
ChangeStatus(EConnecting);

// Start a timeout
iTimer->After(KTimeOut);

SetActive();
```

Chapter 9 TCP/IP 463

> **TIP** It is strongly recommended that you set up a timer, so that if there is a problem connecting to the specified address, you can time-out the connection. Remember that if your timer does expire before your connection completes, you must cleanup correctly, canceling the connection attempt and closing the socket.

Transferring Data

Once the socket is connected, data can be transferred using the techniques described in the **Sockets** section. In this example, the time is read from the `DAYTIME` server (connecting on port 13) using `RSocket::RcvOneOrMore()`. This is an asynchronous function that completes as soon as any data is available. The data is written to the descriptor, `iBuffer`, and the length of the data received is written to `iLength`.

```
iSocket.RcvOneOrMore(iBuffer, 0, iStatus, iLength);
ChangeStatus(EReading);
SetActive();
```

Closing the Connections

It is important that once all of the required reading and writing has completed, all of the handles are closed. As was mentioned in the **Sockets** section, it is good practice to make the `DoCancel()` function responsible for the clean-up of sockets and other subsessions, and for `Cancel()` to be called from the destructor.

```
void CTcpipMultiHomingExEngine::DoCancel()
    {
    iTimer->Cancel();

    TEngineState state = ENotConnected;
    // Cancel appropriate request to socket
    switch (iEngineStatus)
        {
        case EDestroyed:
            state = EDestroyed;
        case ENotConnected:
        case ERConnStarting:
        case EConnecting:
        case EConnected:
        case ELookingUp:
        case EReading:
            if (iRConn.SubSessionHandle() != 0)
                {
                iRConn.Close();
                }
            if (iResolver.SubSessionHandle() != 0)
                {
```

```
            iResolver.Cancel();
            iResolver.Close();
            }
        if (iSocket.SubSessionHandle() != 0)
            {
            iSocket.CancelAll();
            iSocket.Close();
            }
        break;
    default:
        User::Panic(KPanicTcpipMultiHomingExEngine,
ETcpipMultiHomingExBadStatus);
        break;
        }
    ChangeStatus(state);
    }
```

Note that closing the `RConnection` will not necessarily stop the underlying connection, as other clients may still be using it—however, the system will take care of this.

> **WARNING** There is an `RConnection::Stop()` function, but this will cause the connection to be terminated—even if it is being used elsewhere. Generally you should not use this!

Finally, the session to the Socket Server itself is closed in the destructor:

```
CTcpipMultiHomingExEngine::~CTcpipMultiHomingExEngine()
    {
    ...
    if (iSocketServ.Handle() != 0)
        {
        iSocketServ.Close();
        }
    }
```

Implicit Connections

Although `RConnection` can be used to start an interface, it is not necessary for an application to complete this step. `RSocket::Connect()` or `RHostResolver::GetByName()` will automatically cause an interface to be started, with the Symbian OS Socket Server opening a default connection. The availability of this implicit connection provides compatibility across all versions of Series 60, as the code to produce the implicit connection in Series 60 2.x is the same as the code required to start a TCP/IP (implicit) connection in Series 60 1.x. Note, however, that applications that used `RNif` and `RGenericAgent` to explicitly specify a connection in Series 60 1.x will still have to be modified to work with Series 60 2.x.

Chapter 9 Infrared

The application **TcpIpImpEx** provides an example implementation of an implicit connection. Again, the purpose of the application, apart from demonstrating TCP/IP techniques, is to simply connect to a `DAYTIME` server (port 13) and display the date and time values returned. As an implicit connection is used, this example will work for any version of Series 60.

The example code is taken from the class `CTcpipImpExEngine`. The important members of this class are shown here:

```
RSocketServ iSocketServ;
RSocket iSocket;
```

As with the explicit connection, a call to `RSocketServ::Connect()` is made during construction. The implicit connection is made in the engine's `ConnectL()` function:

```
void CTcpipImpExEngine::ConnectL(TUint32 aAddr)
    {
    // Initiate attempt to connect to a socket by IP address
    if (iEngineStatus == ENotConnected)
        {
        // Open a TCP socket
        User::LeaveIfError(iSocket.Open(iSocketServ, KAfInet,
KSockStream, KProtocolInetTcp));

        // Set up address information
        iAddress.SetPort(iPort);
        iAddress.SetAddress(aAddr);

        // Initiate socket connection
        iSocket.Connect(iAddress, iStatus);
        ChangeStatus(EConnecting);

        // Start a timeout
        iTimer->After(KTimeOut);

        SetActive();
        }
    }
```

Once the implicit connection has been made, the method to read the data from the server is identical to that explained in the **TcpipMultihomingEx** example.

Infrared

Infrared is a wireless data connection that uses infrared radiation as the physical bearer. Series 60 support for infrared is provided through an implementation of the IrDA protocol stack. IrDA is an example of "directed" infrared—meaning that it is a point-to-point communications system, offering secure transmission and receipt of data and requiring the communicating devices to have a line-of-sight connection. The connection range is limited—the IrDA standard specifies that a range of up to 1m must be supported, but many devices do exceed this capability.

IrDA Stack

Before looking in detail at the Series 60 implementation of IrDA, this subsection briefly outlines the role of the various layers of the IrDA stack. You should note that although implementation of some of the protocols is optional, Series 60 does support all of them. The layers of the IrDA stack are shown in Figure 9–3.

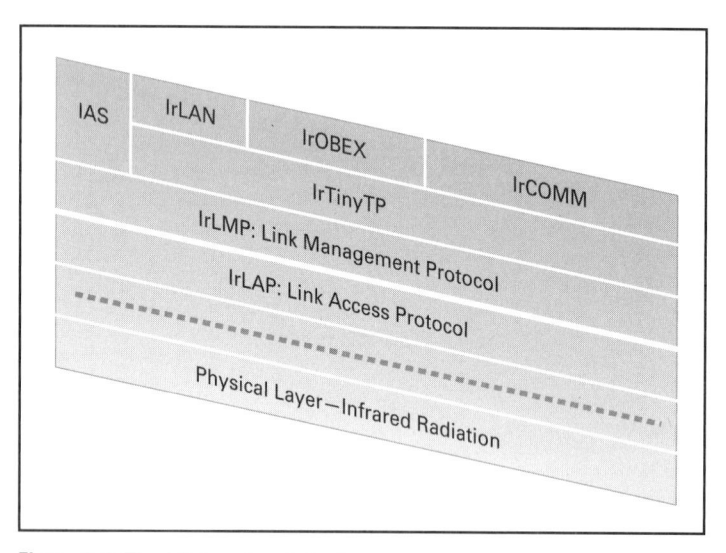

Figure 9–3 The IrDA protocol stack.

The physical layer of the stack specifies the optical characteristics and deals with the encoding of data. The **Link Access Protocol (IrLAP)** layer provides reliable data transfer over the physical layer. It ensures that either data is safely delivered or else, if not, the protocols higher up the stack are informed so that they can take appropriate action. Note that as the success of a transaction is dependent upon the physical line of sight between devices being maintained, failure can be fairly common. The **Link Management Protocol (IrLMP)** provides multiple-session support, enabling multiple clients to run over a single IrLAP link. It is also responsible for high-level device discovery and provides the **Information Access Service (IAS)**, often termed the "yellow pages," of services available for incoming connections. Conflicts in addressing are also resolved at this level.

The two main jobs carried out by the **Tiny Transport Protocol (IrTinyTP)** layer are flow control, at the IrLMP level, and data stream segmentation and reassembly. Flow control is important to allow the multiplexing capabilities of the IrLMP layer to be used, as it allows each IrLMP connection to stop its data stream without interfering with other clients. Sending large streams of data across a connection is not desirable, as a failure means the whole lot will have to be resent. By breaking the data into smaller chunks before sending, efficiency can be improved. Error checking is performed on each segment, with segments being resent if necessary. The receiving device is able to reassemble the constituent parts to produce the original stream—the technical term for this is **Segmentation And Reassembly (SAR)**.

At the top of the stack are protocols to deal with specific situations. The **Infrared OBject EXchange (IrOBEX)** protocol is designed to facilitate the exchange of simple objects via IR. It is commonly used to exchange vCard and vCal files (open file formats for transferring business cards and calendar entries, respectively, between diverse devices). This standard is well defined and details what a packet should contain to enable two devices to communicate. The **Infrared Communications (IrCOMM)** protocol emulates both the serial and parallel ports and was designed as a legacy protocol for use by existing applications. Its use is not encouraged for new applications, as it hides some of the useful features of the lower protocols, such as negotiation of parameters. **Infrared Local Area Network (IrLAN)**, as the name suggests, enables a device to connect to a local network using infrared.

Further information on the IrDA protocol stack and infrared communication, in general, can be found at **http://www.irda.org**.

Programming Infrared on a Series 60 Device

The earlier sections of this chapter concerning **Serial Communication** and **Sockets** programming cover most of the practical basics of infrared programming. It is recommended that you read them before continuing with this section.

Whether you use the IrDA Sockets API or the IrDA Serial API, the general concepts are the same—namely: loading a protocol, discovery, connection, transfer of data and disconnection. Examples of using both APIs were covered in the sections mentioned above, so rather than repeat this information here, this subsection just summarizes the important points and adds further details as required.

Other methods of infrared communication, specifically using the protocols IrTranP and IrOBEX, will also be described in this section.

IrDA Sockets API

The **Sockets** section in this chapter describes the Sockets API for Series 60, using the IrTinyTP protocol in its example application—refer to that section for more detailed coverage of infrared sockets.

IrTinyTP is a connection-oriented protocol which is commonly used for infrared sockets programming. Series 60 provides sockets support for another infrared protocol, **IrMUX**. This is a connectionless protocol—in other words, it does not provide guaranteed or ordered delivery of data packets. This is much less useful over infrared, and so is far less commonly used. Its use is not fully covered here. However, if you wish to use IrMUX, you need to start by loading the appropriate protocol module instead of IrTinyTP, as follows:

```
_LIT(KProtocol, "Irmux");
TProtocolDesc protocolInfo;
User::LeaveIfError(socketServ.FindProtocol(KProtocol,
protocolInfo));
```

NOTE Pay close attention to the string being passed in here to specify the IrMUX protocol. Specifying "IrMUX" (with the last three letters in uppercase) will **not** work—it has to be specified exactly as shown here.

IAS Queries

One important component of the IrDA API that has not yet been discussed is the Information Access Service (IAS) database. This can be viewed as an advertising service for all the IrDA-capable applications on a device. Applications register themselves with the IAS, and then, by interrogating this database, other devices can determine whether they are interested in what is on offer.

In Series 60, the `RNetDatabase` class implements the IAS database, and entries take the form of a `TIASDatabaseEntry`. The following code sets the attributes for a particular entry and then adds it to the IAS database:

```
_LIT8(KIrClassName, "com:emcc:ir:IRTest");
_LIT8(KIrAttributeName, "IrDA:TinyTP:LsapSel");

iIASEntry.SetClassName(KIrClassName);
iIASEntry.SetAttributeName(KIrAttributeName);
iIASEntry.SetToInteger(iListenSocket.LocalPort());
TInt ret = iNetDatabase.Add(iIASEntry);

if (ret != KErrNone && ret != KErrAlreadyExists)
    {
    User::Leave(ret);
    }
```

There are a few points you should note here. First, the class name must be unique within the database, so you should bear that in mind when you are choosing it. `SetToInteger()` sets the response type as an integer and should be set to the port number of your listening socket. Also, you should open the database before you try to add your entry to it, and you must also check whether the attempted addition was successful.

To query the database of another device, you need to construct a query using the `TIASQuery` class and then interrogate the database.

```
// TIASResponse iResponse; defined in member data.
TIASQuery query(KIrClassName, KIrAttributeName,
GetRemoteDevAddr());
   iNetDatabase.Query(query, iResponse, iStatus);
```

The `TIASQuery` takes the class name and attribute parameters set when constructing the `TIASEntry`, and also the address of the remote device (a `TUInt`). Querying the database is performed asynchronously. You pass through a `TIASResponse` object that, on return, will have been populated. If the query completes successfully, the value of `iStatus` will be `KErrNone`. A value of `KErrBadName` means the class name was not present. `KErrUnknown` signifies that the class was present but not the attribute.

The value stored in `TIASResponse` will depend upon the response type set in the database entry. In the above example, `response.Type()` would return `EIASDataInteger`. The port number could then be retrieved using `response.GetInteger()`.

IrDA Serial API

Series 60 allows you to treat the IR port as a serial port and use the APIs that relate to serial programming to perform transactions over infrared. The **Serial Communication** section of this chapter describes in detail the steps to take, so you should read that information and look at the **IrSerial** application.

IrTranP

The IrTranP protocol was introduced to handle the transfer of digital images and as such is obviously relevant to Series 60. As yet, however, only the ability to receive digital images is supported by Series 60.

Implementation of the protocol is through the `CTranpSession` class, with the abstract class `MTranpNotification` providing an observer mechanism. To use the API, you need to create your own notification class, derived from `MTranpNotification`, and implement all of the callback functions.

The general sequence of events when receiving a picture is as follows:

1. The notification object is created.
2. The `CTranpSession` object is created and the notification object registered.
3. The device is put into receive mode through a call to `CTranpSession::Get()`.

4. The remote device makes a connection. You will be notified through the callback function `MTranpNotification::Connected()`.

5. The remote device sends the picture. You will be updated on the progress of the transaction through repeated calls to `MTranpNotification::ProgressIndication()` and then, when the picture has been fully received, a call to `MTranpNotification::GetComplete()`.

6. The connection is broken. You will be notified through a call to `MTranpNotification::Disconnected()`.

7. The appropriate clean-up is performed.

IrOBEX

The IrOBEX protocol is designed for the exchange of objects—for example, vCard and vCal objects. This protocol is supported in Series 60 by the OBEX Client/Server framework. Although a detailed description of IrOBEX is beyond the scope of the book, some of the implementation details are introduced here.

The classes `CObexServer` and `CObexClient` represent the OBEX server and OBEX client, respectively. OBEX is not exclusive to infrared communication, so when you construct these objects you need to specify the protocol you wish to use. The class `TObexIrProtocolInfo` should be used for this purpose, specifying the class name and the attribute name of the service you are interested in. For the default OBEX server use "`OBEX`" and "`IrDA:TinyTP:LsapSel`".

The interface `MObexServerNotify` should also be implemented, so that notification of server events, such as connection and error occurrence, is received. Functions on the client and server are called to perform the transactions, with the data being transported using sockets.

The objects that can be transported using this protocol are of type `CObexBaseObject`. Specialized classes are also defined—namely, `CObexNullObject` (for use if only headers are required), `CObexFileObject` and `CObexBufObject`.

Bluetooth

Bluetooth is a short-range wireless communication protocol, transmitting on a frequency of around 2.45 GHz. Connections can be point-to-point (between two devices) or multipoint (between many devices), over a range of up to 10 meters. Further information about Bluetooth, including technical specifications, can be found at the Official Bluetooth Membership Site, **http://www.bluetooth.org**. Consumer-oriented information on Bluetooth products is available on the website of the Bluetooth Special Interest Group (SIG), **http://www.bluetooth.com**.

Chapter 9 Bluetooth

Series 60 provides support for a range of Bluetooth technologies. Most usefully, it provides a Bluetooth plug-in to the Symbian OS Socket Server, which allows developers to form Bluetooth connections using the standard Sockets API. This API is covered in detail in the **Sockets** section earlier in this chapter, and you should ensure that you are familiar with that first.

This section presents a brief overview of Bluetooth technology before providing a detailed discussion of Bluetooth socket programming for Series 60. This is supported by an example application, **BluetoothChat**, which shows the key APIs in use.

Bluetooth Overview

This subsection provides a brief overview of Bluetooth essentials, in order to place the material that follows into context.

The Bluetooth Stack

The combination of hardware and software required on a device for it to support Bluetooth is known as the Bluetooth stack. It comprises the following parts:

- **Physical Layer**—The Bluetooth radio transceiver.

- **Baseband**—A low-level layer concerned with establishing connections and controlling the transmission of data bits over the physical layer. Typically this is also implemented in hardware.

- **Bluetooth Protocols**—A set of software components, similar in concept to network protocols such as IP and TCP, which provide support for Bluetooth communication at varying levels of capability and reliability.

- **Bluetooth Profiles**—Higher-level software components that implement well-defined services on Bluetooth-capable devices and ensure correct interoperability with other devices.

Bluetooth Protocols

The following are some of the most commonly used Bluetooth protocols:

- **Link Manager Protocol (LMP)**—Manages the behavior of the wireless link, controlling the baseband device and allowing service discovery (which will be covered later in this section). It also addresses security.

- **Logical Link Control and Adaptation Protocol (L2CAP)**—Offers connectionless and connection-oriented data services between the baseband layer and the upper protocols. It also supports data stream segmentation and reassembly, allowing the upper layers to use data packets larger than the baseband can handle.

- **Radio Frequency Communications (RFCOMM)**—A reliable transport protocol that provides emulation of RS-232 serial ports over the L2CAP protocol.

- **Service Discovery Protocol (SDP)**—Allows Bluetooth devices to discover services offered by other devices. It is responsible for broadcasting queries and fielding the responses to determine which devices offer particular services.

Bluetooth Profiles

A number of different profiles are defined by the Bluetooth specification, and each Series 60 device supports a subset of these. The exact set of profiles supported is determined by the handset manufacturer—there is no single set of Series 60-supported profiles. However, commonly implemented profiles include:

- **Object Push Profile**—Allows a Bluetooth device to send and receive OBEX objects, such as vCard and vCal items.

- **File Transfer Profile**—Depending on the profile implementation on the device, this profile may allow you to browse, delete, send files to and receive files from a Series 60 device. File Transfer Profile can also be implemented as a server only.

- **Fax Profile**—The Fax profile will allow you to send faxes to, and receive faxes from, a Bluetooth device (typically a PC) using the phone as a fax modem.

- **DUN (Dial-Up Networking) Profile**—Allows a device such as a laptop to use a phone as a data modem.

- **Serial Port Profile**—Provides legacy support for older applications and protocols using the RFCOMM protocol. It provides a virtual COM port, similar to an RS-232 port on a PC. This profile is often used as the data carrier for other profiles and applications.

- **Headset Profile, Hands Free Profile**—Enable phones to connect to wireless headsets and Bluetooth car kits.

Service Advertisement and Discovery

A Bluetooth device may offer services for other devices to connect to, and/or connect to other devices itself and make use of their services. A Bluetooth device that offers services to others is known as a server, and the user is known as a client.

Chapter 9 Bluetooth 473

NOTE Do not confuse the terms client and server used here with the Symbian OS Client/Server framework. The two are not the same—for example, a Bluetooth server can be implemented in a Symbian OS client process. Throughout the **Bluetooth** section, unless otherwise specified, the terms client and server are used in the Bluetooth sense.

A Bluetooth server makes services available by means of a process known as **Service Advertisement**—information about the new service is provided to any other Bluetooth devices that wish to query it. It does not necessarily guarantee that an attempted connection will be successful. When an application advertises its ability to accept connections, an entry is placed in the device's Bluetooth Service Discovery Database (**SDD**).

Regardless of the type of service provided, a Bluetooth service advertisement will consist of the following information:

- **Port**—The port number an incoming connection will be accepted on. Many applications can advertise their services on a device, with each accepting a connection on a specified port.

- **Class ID**—A class ID that will provide information on the type of service provided.

- **Availability**—The availability of this service. When an application accepts an incoming connection, it will mark its entry in the Service Discovery Database as not available. This will inform any other devices that may wish to connect that it is currently occupied.

Most entries will also contain user-readable text to explain the type of service they offer, and it is even possible to provide a graphical icon for a service.

On the client side, an attempt to make a Bluetooth connection is typically preceded by a process of searching for local Bluetooth-enabled devices and querying the services they provide. This process is known as **Service Discovery**.

The Series 60 APIs for service advertisement and discovery are covered later in this section.

Example Bluetooth Application

The rest of this section looks at the example application **BluetoothChat**, which allows two users to exchange short instant chat messages using Bluetooth as the communication medium. The application contains both client and server code and requires one user to act as a server and one as a client. This section examines the Bluetooth-specific APIs used by the application, in both the server and client code—starting with the service advertisment and discovery phase, before moving on to detail the socket-based communication between the devices.

Service Advertisement

In the **BluetoothChat** application the user can choose whether they wish their device to act as the client or the server. If they choose **Start Receiver** from the menu, as shown in Figure 9–4, then their phone will act as the server and listen for incoming connections. After the appropriate port number is obtained from the Socket Server, the first step in this process is to advertise the **BluetoothChat** service.

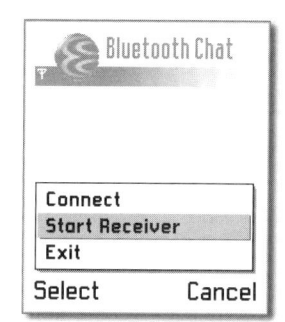

Figure 9–4 The BluetoothChat menu options.

Service advertisement of the **BluetoothChat** service is performed by the `CBluetoothAdvertiser` class, in the `StartAdvertisingL()` function. This class's member data is shown below:

```
RSdp                    iSdpSession;
RSdpDatabase            iSdpDatabase;
TSdpServRecordHandle    iRecord;
TInt                    iRecordState;
```

The first step is to connect to the Service Discovery Database (SDD). This is performed via an `RSdp` session. You may then use this `RSdp` session to open an `RSdpDatabase` subsession for access and manipulation of the database.

```
// Connect to the Service Discovery Database.
if (!iIsConnected)
  {
  User::LeaveIfError(iSdpSession.Connect());
  User::LeaveIfError(iSdpDatabase.Open(iSdpSession));
  iIsConnected = ETrue;
  }
```

You are now ready to create entries within the Service Discovery Database. First you must create a UUID (Universally Unique Identifier) that uniquely defines your service. In the **BluetoothChat** example, the UID of the application itself is used, as this is guaranteed to be unique within Series 60.

The `CreateServiceRecordL()` function is used to create service records within the database.

```
// UID of the application.
const TUint KServiceClass = 0x101F6148;
iSdpDatabase.CreateServiceRecordL(KServiceClass, iRecord);
```

The `iRecord` member is an instance of `TSdpServRecordHandle`. This is passed by reference, and when the function returns, it will contain the service handle of the newly created service record.

Chapter 9 Bluetooth

There are two overloaded versions of `CreateServiceRecordL()`. The example above assigns the record to one service class, whereas the overloaded version will assign the record to multiple service classes. (An example of a service with multiple classes is a color printer that can advertise its services as being a printing device *and* a *color* printing device.)

Now that a record has been entered into the database, you can assign some attributes to it.

Creating Service Record Attributes

Attributes are items of additional information that can be added to a service record. The **BluetoothChat** application adds the following attributes to its service record:

- Protocol Description.
- Service Name.
- Service Description.

Service Name and Service Description are just extra textual information that may prove to be useful:

```
_LIT(KServiceName, "EMCC Chat");
_LIT(KServiceDescription, "Bluetooth Chat Application");
```

The Protocol Description contains information that is needed to establish a successful connection. Creating the protocol description will involve building a Data Element Sequence (**DES**) using the `CSdpAttrValueDES` class.

The `CSdpAttrValueDES::StartListL()` function opens and returns a new DES list within the current list sequence, and by using the `BuildxxxxL()` functions it is possible to populate this list with actual data. When the current DES list has been appropriately populated, calling the `EndListL()` function will close it and return its parent list. In the **BluetoothChat** application this is performed in the function `CBluetoothAdvertiser::BuildProtocolDescriptionL()`.

Notice that the bulk of this function is a single line of code, split over a number of lines for readability. As the `CSdpAttrValueDES` APIs used here all modify the DES list and then return a pointer to it, this is a standard way of building a list, which you will see used in Bluetooth code throughout Series 60.

```
void CBluetoothAdvertiser::BuildProtocolDescriptionL(
CSdpAttrValueDES* aProtocolDescriptor, TInt aPort)
    {
    TBuf8<1> channel;
    channel.Append((TChar)aPort);
```

```
        aProtocolDescriptor
        ->StartListL()    // List of protocols required
          ->BuildDESL()
          ->StartListL()        // Details of lowest level protocol
            ->BuildUUIDL(KL2CAP)
          ->EndListL()

          ->BuildDESL()
          ->StartListL()
            ->BuildUUIDL(KRFCOMM)
            ->BuildUintL(channel)
          ->EndListL()
        ->EndListL();
    }
```

`KL2CAP` and `KRFCOMM` are constant values that represent parts of the Bluetooth stack. Their actual values are defined within the Bluetooth specification.

The parameter `channel` represents the port that this service will be advertised on. This is retrieved via a call to obtain an available RFCOMM channel on the listening socket that the server will advertise on, and which the client should bind to.

```
    // Get a channel to listen on
    // same as the socket's port number
    TInt channel;
    User::LeaveIfError(iListeningSocket.GetOpt(
    KRFCOMMGetAvailableServerChannel, KSolBtRFCOMM, channel));
```

Now the attributes are added in the following manner:

```
    // Add protocol information
    iSdpDatabase.UpdateAttributeL(iRecord,
KSdpAttrIdProtocolDescriptorList, *vProtocolDescriptor);
```

The `KSdpAttrIdProtocolDescriptorList` constant value is part of the Bluetooth specification and defines attributes key to every service record.

```
    // Add a name to the record
    iSdpDatabase.UpdateAttributeL(iRecord,
KSdpAttrIdBasePrimaryLanguage + KSdpAttrIdOffsetServiceName,
KServiceName);

    // Add a description to the record
    iSdpDatabase.UpdateAttributeL(iRecord,
KSdpAttrIdBasePrimaryLanguage +
KSdpAttrIdOffsetServiceDescription, KServiceDescription);
```

Removing Service Records

Once the service record has been added to the database, it will become available to other devices and their applications. When a connection is made, you should stop advertising the service. You can do this by marking the service as busy, using `UpdateAttributeL()` to modify the `KSdpAttrIdServiceAvailability` parameter in the service record. Alternatively, you could delete the record from the database, as shown in this code snippet from `CBluetoothAdvertiser::StopAdvertisingL()`:

```
iSdpDatabase.DeleteRecordL(iRecord);
```

Closing the Service Discovery Database

When a client application has finished with its Bluetooth services, it should disconnect from the Service Discovery Database. In **BluetoothChat** this is performed in the destructor of `CBluetoothAdvertiser`. The SDP session is closed at this point as well.

```
iSdpDatabase.Close();
iSdpSession.Close();
```

Series 60 keeps track of all sessions that are connected to servers. Once a server no longer has any sessions accessing it, the operating system will shutdown the server. This means that the SDD will be shutdown once there are no longer any applications with open sessions onto it, even if there are other Bluetooth devices accessing it. It is therefore recommended that the client application keep a session to the database open during the period it wishes to publicize its services.

Bluetooth Security

Typically a receiving device will stipulate a collection of security settings for the Bluetooth connections it offers. Series 60 provides a Bluetooth Security Manager API, and this allows applications to specify a set of security settings in order to determine whether user authorization, authentication or encryption are required when establishing a connection. These security settings are applied to a specific server, protocol and channel.

The Bluetooth Security Manager ensures that incoming connections adhere to the application settings; it is implemented as a server within the Series 60 Bluetooth API. The class `RBTMan` encapsulates a session to this server, and subsessions are used to access specific functionality. The subsession required for a service to register its security settings with the security manager is `RBTSecuritySettings`.

The `TBTServiceSecurity` class encapsulates the security settings of a Bluetooth service—containing information regarding the service UID, the protocol and channel IDs and access requirements. Access requirements are set either on or off by passing a `TBool` value to the functions `SetAuthentication()`, `SetAuthorisation()` and `SetEncryption()`. It is also possible to deny all access using the `SetDenied()` function.

Once the security settings within `TBTServiceSecurity` have been set, the `RBTSecuritySettings::RegisterService()` function allows the application to register its security settings with the Security Server.

> **NOTE** The `UnregisterService()` function allows an application to remove any security requirements.

Once the security settings are registered with the Security Server, the application no longer requires access to the server, so its session and connection can be terminated. In the **BluetoothChat** application these steps are carried out by the `CBluetoothServer` class using the following member data:

```
RBTMan                   iSecManager;
RBTSecuritySettings      iSecSettingsSession;
TState                   iState;
```

The function `CBluetoothServer::SetSecurityOnChannelL()` is responsible for configuring security settings:

```
void CBluetoothServer::SetSecurityOnChannelL(TBool
aAuthentication, TBool aEncryption, TBool aAuthorisation)
    {
    // connect to the security manager
    // and open a settings session.
    User::LeaveIfError(iSecManager.Connect());
    User::LeaveIfError(iSecSettingsSession.Open(iSecManager));

    // the security settings
    TBTServiceSecurity serviceSecurity(KUidBluetoothChat,
KSolBtRFCOMM, 0);

    //Define security requirements
    serviceSecurity.SetAuthentication(aAuthentication);
    serviceSecurity.SetEncryption(aEncryption);
    serviceSecurity.SetAuthorisation(aAuthorisation);

    serviceSecurity.SetChannelID(iChannel);

    // make an asynchronous request
    iSecSettingsSession.RegisterService(serviceSecurity,
iStatus);
    iState = ESettingSecurity;
    SetActive();
    }
```

Typically it is the receiving device that stipulates the security settings for any given Bluetooth connection. When a requesting device discovers that a receiving device stipulates a set of security requirements, the requesting device

Chapter 9 Bluetooth

will automatically implement these requirements, and this process will appear transparent to the application.

Device and Service Discovery

The preceding subsection demonstrated how to advertise services in a Bluetooth server application. This subsection covers the API for searching for devices and querying their SDD for the services they provide.

There are two methods of searching for remote devices:

- **RNotifier**—The `RNotifier` class presents the user with a UI dialog listing all Bluetooth devices currently in range and advertising appropriate services. The address of the device selected by the user is returned to the application. This is sufficient for most client applications, but its major drawback is that it does not filter the list of devices it displays. Therefore it is not possible for the client application to use this API to display a list of devices providing a specific service.

- **RHostResolver**—The `RHostResolver` API provides a way for the client application to programmatically search for local devices. This allows the author of the application to filter the list of locally available devices before displaying them on the screen. Although this method provides more control over the list of displayed devices, it is more complicated than using `RNotifier`.

As the use of `RHostResolver` has already been documented in the **Infrared** section, the **BluetoothChat** application uses the `RNotifier` API.

Device Discovery Using RNotifier

`RNotifier` represents a session to the Notifier Server, which is used to display system dialogs. As with any other session class, the first step in using it is to create and connect an instance. In this case, the `CBluetoothDeviceSearcher` class has an `RNotifier` member variable (`iNotifier`) which is connected in its `SelectDeviceL()` method by calling `iNotifier.Connect()`.

The `RNotifier` class is responsible for displaying many different types of dialog. Each type is represented by a unique ID. The UID representing the Bluetooth Device Selection dialog is `KDeviceSelectionNotifierUid`, and this is passed to `RNotifier::StartNotifierAndGetResponse()` to invoke the dialog.

The full list of parameters passed to this function is as follows:

- A `TRequestStatus&`, as this is an asynchronous function.
- The `RNotifier` UID.

- A `TBTDeviceSelectionParamsPckg`—a package descriptor containing an instance of `TBTDeviceSelectionParams`, which specifies a way of filtering the number of devices that are returned. This parameter is currently ignored and could be left uninitialized. However, it is hoped that it will be supported in future releases of the platform, so its correct use is shown in the **BluetoothChat** application.

- A `TBTDeviceResponseParamsPckg`, which will be populated with device information after a device successful selection.

The call to the function can be seen here:

```
void CBluetoothDeviceSearcher::SelectDeviceL(
TBTDeviceResponseParamsPckg& aResponse)
    {
    // store a pointer to the response buffer
    iResponse = &aResponse;

    TUUID targetServiceClass(KServiceClass);
    TBTDeviceClass deviceClass(KServiceClass);
    TBTDeviceSelectionParams selectionFilter;
    selectionFilter.SetUUID(targetServiceClass);
    selectionFilter.SetDeviceClass(deviceClass);

    // These filtering methods do not work at
    // the time of writing, and ALL Bluetooth
    // devices in range are returned

TBTDeviceSelectionParamsPckg selectionParams(selectionFilter);

    User::LeaveIfError(iNotifier.Connect());

    iNotifier.StartNotifierAndGetResponse(iStatus,
KDeviceSelectionNotifierUid, selectionParams, aResponse);
    SetActive();
    }
```

As you can see, `CBluetoothDeviceSearcher` is an Active Object which will handle completion of the request in its `RunL()` method. The following code is used to verify successful device retrieval and close the handle to the notifier:

```
void CBluetoothDeviceSearcher::RunL()
    {
    TInt retVal = iStatus.Int();

    if (retVal == KErrNone)
        {
        if ((*iResponse)().IsValidDeviceName())
            {
            retVal = KErrNone;
```

Chapter 9 Bluetooth

```
        }
    }

    iNotifier.CancelNotifier(KDeviceSelectionNotifierUid);
    iNotifier.Close();

    iObserver.DeviceFoundL(retVal);
}
```

The `TBTDeviceResponseParamsPckg` variable now contains the 48-bit Bluetooth device address (`TBTDevAddr`) of the device you have chosen to communicate with. This may be retrieved via a call to `TBTDevAddr::BDAddr()`.

Service Discovery

Once you have the address of a Bluetooth device, you can query it for the services it provides. This is essential, as you will need to discover whether the remote device provides the service you require—in this case, **BluetoothChat**. You will also need to discover on what port number the service is advertised, and whether it is presently available.

Service discovery involves querying the remote device's SDD to see whether the service you requested is advertised. The process involves finding records that correspond to the required service and then viewing attributes on those records.

Service discovery is an asynchronous process, and a number of callback functions need to be implemented in order to handle information retrieval from it. The Bluetooth API provides the `MSdpAgentNotifier` mixin class in order to handle callbacks, and this contains the following three pure virtual functions that need to be implemented:

```
    virtual void NextRecordRequestComplete(TInt aError,
TSdpServRecordHandle aHandle, TInt aTotalRecordsCount) = 0;

    virtual void AttributeRequestResult(TSdpServRecordHandle
aHandle, TSdpAttributeID aAttrID, CSdpAttrValue* aAttrValue)
= 0;

    virtual void AttributeRequestComplete(TSdpServRecordHandle,
TInt aError) = 0;
```

A `CSdpAgent` object is used to perform the query. The `CSdpAgent` must be created specifying an `MSdpAgentNotifier` observer object, and the Bluetooth device address of the device the client application wishes to query. In the **BluetoothChat** application, the `CBluetoothServiceSearcher` class implements the `MSdpAgentNotifier` mixin.

The `CSdpSearchPattern` class is used to specify the SDD records that you are interested in. Once a search pattern list has been built you should pass it through to `CSdpAgent::SetRecordFilterL()`.

The following code shows these APIs in action:

```
iSdpSearchPattern = CSdpSearchPattern::NewL();
iSdpSearchPattern->AddL(KServiceClass);
…
iAgent = CSdpAgent::NewL(*this, aDeviceAddress);
iAgent->SetRecordFilterL(*iSdpSearchPattern);
iAgent->NextRecordRequestL();
```

When the record request completes, the function `NextRecordRequestComplete()` will be called on the `CBluetoothServiceSearcher` class. The parameters of this function are an error code, a service record handle and a count of matching records.

Within the implementation of this function you must first check that the `aError` variable is set to `KErrNone`, indicating that the record request was successful. Next check the value of `aTotalRecordsCount` to see how many matching records were found. If there was no error and the number of records is greater than zero, then `aHandle` will be a valid handle on the first record matching the search filter parameters. Its attributes can be queried via a call to `CSdpAgent::AttributeRequestL()`, as shown below:

```
iMatchList = CSdpAttrIdMatchList::NewL();
// Availability
iMatchList->AddL(KSdpAttrIdServiceAvailability);
// Port Number, RFCOMM, L2CAP
iMatchList->AddL(KSdpAttrIdProtocolDescriptorList);
…
  TRAPD(error, iAgent->AttributeRequestL(aHandle,
*iMatchList));
```

> **NOTE** You can also search for a single attribute using an overloaded version of `AttributeRequestL()`. See the SDK documentation for further details.

When performing an attribute search, the `AttributeRequestResult()` function will be called once for every attribute that matches the search criteria. The parameters passed to this function are a record handle, an attribute ID and a pointer to an attribute value (`CSdpAttrValue`) object. The attribute ID can be queried to see which attribute's values are being passed back, and the `aAttrValue` parameter will contain a pointer to the relevant attributes values, as shown below:

```
void CBluetoothServiceSearcher::AttributeRequestResult(
   TSdpServRecordHandle aHandle,
   TSdpAttributeID aAttrID,
   CSdpAttrValue* aAttrValue)
   {
   TRAPD(error, AttributeRequestResultL(aHandle, aAttrID,
```

Chapter 9 Bluetooth

```
   aAttrValue));
   }

void CBluetoothServiceSearcher::AttributeRequestResultL(
   TSdpServRecordHandle /*aHandle*/,
   TSdpAttributeID aAttrID,
   CSdpAttrValue* aAttrValue)
   {
   if (aAttrID == KSdpAttrIdProtocolDescriptorList)
      {
      // Validate the attribute value, and extract
      // the RFCOMM channel
      // iContinueSearching flag will show ETrue if this
      // parse is successful
      TBluetoothAttributeParser parser(*this,
iContinueSearching);
      aAttrValue->AcceptVisitorL(parser);
      }
   else if (iContinueSearching && aAttrID ==
KSdpAttrIdServiceAvailability)
      {
      if (aAttrValue->Type() == ETypeUint)
         {
         iAvailable = static_cast<TBool>(aAttrValue->Uint());
         ...
         }
      }
   delete aAttrValue;      // Ownership has been transferred
   }
```

> **NOTE** Ownership of the `aAttrValue` parameter is passed into the function. It is your responsibility to delete this, once you have finished with it, otherwise a memory leak will occur.

It is important to know the type of each value passed into this function, as the `CSdpAttrValue` class provides retrieval functions for many different types. You can call `aAttrValue->Type()` to find the value's type, and then you must call the appropriate getter for that type. The code above shows how to get an unsigned integer value in this way.

An attribute may also be a list of values; such lists are known as Data Element Alternatives (DEA) or Data Element Sequences (DES). If an attribute is of one of these types, it needs to be parsed in order to extract individual values. The mixin class `MSdpAttributeValueVisitor` is used for this purpose. In the **BluetoothChat** application, this mixin is implemented by the `TBluetoothAttributeParser` class and is used for extracting values from the `KSdpAttrIdProtocolDescriptorList` element.
`TBluetoothAttributeParser` is passed through as a parameter to the `aAttrValue->AcceptVisitorL()` function, and `VisitAttributeValueL()`

is called for each member of the list. Below is the implementation of this function in the **BluetoothChat** application:

```
void TBluetoothAttributeParser::VisitAttributeValueL(
CSdpAttrValue& aValue, TSdpElementType /*aType*/)
    {
    switch (iProcessingState)
        {
        case EStartOfDesOne:
        case EStartOfDesTwo:
        case EL2Cap:
        case EEndOfDesTwo:
        case EStartDesThree:
            break; // check nothing

        case ERFComm:
            CompareRFCOMM(aValue);
            break;

        case ERFCommPort:
            GetPort(aValue);
            break;

        case EEndOfDesThree:
        case EEndOfDesOne:
            break; // check nothing

        default: // unexpected output - panic
            Panic(EBadAttributeValue);
            break;
        }

    Next();
    }
```

This enables you to extract whether RFCOMM is an available protocol and to obtain the port number of the remote device that this service is advertised on.

Once all attributes matching the search parameters have been returned to the client application, the observer function `MSdpAgentNotifier::AttributeRequestComplete()` will be invoked. At this point, if the service you are interested in has not been found, then it is normal practice to issue another `iAgent->NextRecordRequestL()`, unless the error code parameter to this function is equal to `KErrEOF`, which highlights that all records have been processed.

Bluetooth Socket Communication

Provided that the device and service discovery phase was successful, you now have an address and port number identifying a service on a remote device. This information can be used to open a socket session with that device.

Chapter 9 Bluetooth

Note that the general steps involved in opening a socket in Series 60 are covered in detail in the **Sockets** section earlier in this chapter—you should read through that section to familiarize yourself with it before continuing with this section. The generic steps that have been described previously will not be covered in detail again here.

The Bluetooth Server

Just as before, socket communication starts by connecting to the Symbian OS Socket Server and then opening a socket. The only difference in this case is the protocol specified when opening the socket, which is set to "RFCOMM" (here represented by the constant `KServerTransportName`).

```
User::LeaveIfError(iSocketServer.Connect());
    User::LeaveIfError(iListeningSocket.Open(iSocketServer,
KServerTransportName));
```

Next you need to obtain a channel (a port number) to listen on. This is achieved with a call to `RSocket::GetOpt()`. The parameters define the type of option request (`KRFCOMMGetAvailableServerChannel`) and the protocol (`KSolBtRFCOMM`). These constants are defined in the system header file `bt_sock.h`. The third parameter represents the channel to listen on, and it is populated if this function returns without error.

```
User::LeaveIfError(iListeningSocket.GetOpt(
    KRFCOMMGetAvailableServerChannel, KSolBtRFCOMM, iChannel));
```

You then create a Bluetooth socket address object (`TBTSockAddr`), set its port to the channel just obtained and bind the socket to this address. Note you do not have to set the local Bluetooth device address in this case.

```
TBTSockAddr listeningAddress;
    listeningAddress.SetPort(iChannel);
    User::LeaveIfError(iListeningSocket.Bind(
listeningAddress));
```

Now you are ready to listen for incoming connections. As before, you need to specify a listening queue size—in other words, the number of outstanding requests you want to queue before further requests are rejected. Presently devices are limited to only one Bluetooth connection at a time, so the value of `KListeningQueSize` is set to 1.

```
// Listen for incoming connections
    User::LeaveIfError(iListeningSocket.Listen(
KListeningQueSize));
```

Finally, you should make a call to the asynchronous function `Accept()`, which will complete when the first connection attempt is successful.

```
iListeningSocket.Accept(iAcceptedSocket, iStatus);
SetActive();
```

Once the socket has been set up successfully, you need to configure the security manager. In the **BluetoothChat** example application this is performed in the function `CBluetoothServer::SetSecurityOnChannelL()`, which is shown in the **Bluetooth Security** subsection above. Once this is done, you are ready to advertise the new Bluetooth service you have created. In the **BluetoothChat** application this is achieved using the `CBluetoothAdvertiser` methods `StartAdvertisingL()` and `UpdateAvailabilityL()`. The server must also ensure that when an incoming connection is accepted, the appropriate entry in the SDP database is marked as being used (if the record has not been deleted). This ensures that a second client will not attempt to connect on the same port. This is performed in the `CBluetoothAdvertiser::StopAdvertisingL()` function, detailed above.

> **NOTE** You should also ensure that you call `StopAdvertisingL()` when your application is shut down, in order to remove your entry from the SDD.

Once these steps are complete, you can start to send and receive data over `iAcceptedSocket` in the generic way described in the **Sockets** section.

The Bluetooth Client

As with the server code, the first step on the client side is to connect to the Symbian OS Socket Server and then open a socket, specifying RFCOMM as the required protocol.

As shown previously in this section, the service discovery phase is used to obtain an address and port number, which are then used to initialize a `TBTSockAddr` object. This object is then passed to `RSocket::Connect()`:

```
iSendingSocket.Connect(iSocketAddress, iStatus);
SetActive();
```

Once the connection succeeds, client and server can communicate using standard `RSocket` function calls.

Summary

This chapter has covered the fundamental communication technologies available in Series 60.

The chapter started by introducing the Serial Comms Server and the client-side APIs used to carry out **Serial Communication**. Sample code was provided to show the steps required to set up a successful serial connection over infrared.

The Symbian OS Socket Server offers a higher-level API for many types of connection, in keeping with familiar **Sockets** APIs on other platforms. The chapter showed how to construct a typical socket session, from both the client and the server side, again using infrared as the bearer. It also covered the Series 60 APIs for secure sockets, using the SSL and TLS protocols.

TCP/IP is a common use of sockets. This chapter covered both the TCP/IP APIs from Series 60 1.x and the new multihoming APIs added in Series 60 2.x.

The **Infrared** section provided a short overview of the various IR technologies supported by Series 60. Example code for the most common uses of IR, serial and socket communications, was provided in the earlier sections of the chapter.

Finally, the **Bluetooth** section explained the concepts of Bluetooth service advertisement and discovery, as well as Bluetooth-specific security considerations, and showed these in use in an example application.

The technologies covered in this chapter offer developers a high degree of flexibility, although, since this is at a relatively low level, configuration can often be complicated. Chapter 10 covers the higher-level communications APIs available in Series 60, which provide quicker, more convenient access to common communications functionality.

chapter 10

Advanced Communication Technologies

Sophisticated communication APIs for Series 60 developers

Chapter 9 introduced you to a range of fundamental communications capabilities available to Series 60 C++ developers. This chapter introduces some powerful technologies that build on those basics to provide you with a more sophisticated set of high-level APIs. Although these technologies are built on top of the fundamental functionality covered in the previous chapter, well-encapsulated high-level APIs usually mean that you do not need to understand the underlying technology before reading this chapter. An exception is HTTP on Series 60 1.x, which requires you to make explicit use of the Sockets APIs covered in the previous chapter. Note that in Series 60 2.x, a high-level HTTP API is introduced, which provides a level of abstraction above the raw sockets interface.

Specifically, this chapter covers the following:

- **HTTP**—The HyperText Transfer Protocol is used for communicating with machines such as Web servers. Typically it is used by a browser application to retrieve Web pages, but it may also be used to download any kind of file, and also to upload data to the server.

- **WAP**—The Wireless Application Protocol defines a set of network layers designed for use by mobile phones and other small, resource-constrained devices. As with HTTP, its typical use is by browsers to download data from a server, but it has a range of other uses which are discussed in this chapter.

- **Messaging**—Series 60 contains a powerful "messaging architecture" for sending, receiving and processing messages such as email, SMS and MMS. This chapter shows you how to use the various messaging architecture APIs to send and receive messages from your own application.

- **Telephony services**—This section covers the APIs used to access a Series 60 phone's voice call capabilities. It shows you how to make and receive calls, as well as accessing the last calling number.

The following example applications are provided with this chapter:

- **HTTPExample**—A simple HTTP application which demonstrates how to perform GET and POST requests using the Series 60 2.x HTTP API.

- **MtmsExample**—Demonstrates use of the messaging architecture's Client MTM API to create and send messages of different types.

Chapter 10   HTTP

- **SendAsExample**—Shows the messaging Send-As API, a simple, restricted API to perform common messaging tasks.

- **SendAppUiExample**—Demonstrates use of the `CSendAppUi` class to add message-sending capabilities to an application's user interface.

- **MsgObserver**—An application which monitors the *Inbox* and receives notification when a new message is received.

- **AnsPhone**—A simple telephony application which shows how to use Series 60 APIs to make and receive calls and to get the last calling number on the phone.

The details for downloading the full buildable source for these examples can be found in the Preface.

HTTP

The Hypertext Transfer Protocol (**HTTP**) is an Application Layer protocol used for information transport over the Internet. HTTP servers are able to host many types of data, including HTML and XML pages, multimedia files or images. A client can retrieve data by opening a connection to the server (usually on TCP port 80) and sending a request. A Uniform Resource Identifier (URI) specifies a file on the server. The URI includes the server name, the path on the server and the name of the file.

Series 60 1.x does not have a public API that offers direct support for HTTP. However, the TCP/IP section in the previous chapter shows how you can implement a simple HTTP layer yourself using the Sockets API.

Series 60 2.x introduces the HTTP Client API. This provides a client interface enabling applications to communicate with HTTP servers on the Internet. The API supports HTTP 1.0 by default, and it also provides conditional HTTP 1.1 compliance, as defined in RFC 2616. The API allows the Client to specify features such as the encoding and transport protocol required. The Client stack supports persistent connections, pipelining and chunked transfer encoding. Filters provide support for HTTP Redirection and HTTP Basic and Digest Authentication.

To use the Client API correctly, you need to understand the main ideas behind conducting a transaction over HTTP: sessions, transactions, headers, data suppliers, and filters. This section briefly explains the function of each component before walking through an example use of the API.

Sessions

A session describes one or more transactions sharing the same connection. A session has a series of settings and filters defined that are applicable to all transactions. It is possible to have more than one session active concurrently, although this is not usual.

In the HTTP Client API, the `RHTTPSession` class represents the session. The class has functions to open and close the session and to create a transaction. To be notified of session events you need to call `RHTTPSession::SetSessionEventCallback()`; otherwise all events are handled by the session and will not be passed to your application.

The default protocol for the `RHTTPSession` class is HTTP over TCP, but the API provides support for choosing other combinations. If you wish to use a different protocol, you can obtain a list of alternatives by calling `RHTTPSession::ListAvailableProtocols()`. The required protocol can then be passed as a parameter to an overloaded version of `RHTTPSession::OpenL()`. Note, however, that in the vast majority of cases the default protocol is what you want, so you can usually just call the no-parameter version of `RHTTPSession::OpenL()`.

Note that if you are using the default protocols, the `Connect()` and `Disconnect()` methods of `RHTTPSession` should not be used—these are for use with the WSP protocol.

Transactions

The simple way to think of a transaction is as a single exchange of messages between an HTTP client and HTTP origin server—a request from the client and a response from the server. (In practice, filters set for the transaction may complicate the exchange. Filters are discussed later in this section.)

HTTP defines a number of methods such as GET, to ask for a specific document, and HEAD, to retrieve information about a document. When making a request, you need to specify which method you want to use and the URI of the resource. Both the request and response consist of a header and, optionally, a body. The response received from the server contains an HTTP status code and message, which indicate the success of the method or the state of the resource following the method.

The `RHTTPTransaction` class represents an HTTP transaction.

Headers

Each request and response message has a header section containing one or more fields. The purpose of these headers is to allow the client and server to communicate information to each other. This could relate to the connection or maybe to properties of the client or server. For example, the client may specify the acceptable content type that can be received.

The class `RHTTPHeaders` should be used to create and modify the headers in the HTTP client API, with `THTTPHdrVal` describing each field.

Data Suppliers

A data supplier object represents the body of a message. The HTTP client API defines the mixin class MHTTPDataSupplier for this purpose. Clients should implement this class to receive response message data from the framework and to supply the request message body to the framework.

The classes RHTTPRequest and RHTTPResponse are defined to represent the whole message. The request or response can be retrieved from the transaction using the functions RHTTPTransaction::Request() and RHTTPTransaction::Response(), respectively.

Filters

Filters are applied at session level and can modify transactions as they travel to or are retrieved from the server. Typically they can cause headers to be added or removed, terminate or a cancel a transaction, or transform the body.

You can write your own filter by deriving from MHTTPFilter. A filter is registered using RHTTPFilterCollection::AddFilterL().

Example Application

The **HTTPExample** application is provided to illustrate the concepts outlined above. This example provides the user with options to perform simple GET and POST requests, as shown in Figure 10–1.

Implementation

Use of the HTTP APIs in the **HTTPExample** application is encapsulated in the CHTTPExampleEngine class. As well as deriving from CBase, this class also derives from MHTTPAuthenticationCallback to enable it to receive progress events during an outstanding transaction. The application provides a further observer mixin class, MHTTPExampleEngineObserver, for the purpose of passing these events up to the application UI.

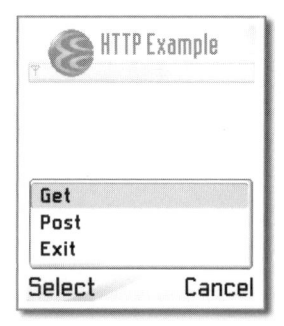

Figure 10–1 The HTTPExample application.

The important functions and variables of the MHTTPExampleEngineObserver and CHTTPExampleEngine classes are shown in the abridged class declarations below.

```
class MHTTPExampleEngineObserver
    {
public:
    virtual void ResponseStatusL(
TInt aStatusCode, const TDesC& aStatusText) = 0;
```

```
    virtual void ResponseReceivedL(
const TDesC& aResponseBuffer) = 0;
    };


class CHTTPExampleEngine : public CBase, public
MHTTPTransactionCallback
    {
public:
    static CHTTPExampleEngine* NewL(MHTTPExampleEngineObserver&
aObserver);

    ~CHTTPExampleEngine();

    void GetRequestL(const TDesC& aUri);
    void PostRequestL(const TDesC& aName);

    // Cancel an outstanding transaction
    void Cancel();

private: // from MHTTPTransactionCallback
    virtual void MHFRunL(RHTTPTransaction aTransaction, const
THTTPEvent& aEvent);

    virtual TInt MHFRunError(TInt aError, RHTTPTransaction
aTransaction, const THTTPEvent& aEvent);

private:
    CHTTPExampleEngine(MHTTPExampleEngineObserver& aObserver);
    void ConstructL();

    void ParseUriL(const TDesC& aUri);
    void AddHeaderL(RHTTPHeaders aHeaders, TInt aHeaderField,
const TDesC8& aHeaderValue);

private:
    RHTTPSession iSession;
    RHTTPTransaction iTransaction;

    HBufC* iResponseBuffer;
    HBufC8* iUri;
    TUriParser8 iUriParser;
    CHTTPFormEncoder* iFormEncoder;

    MHTTPExampleEngineObserver& iObserver;
    };
```

Opening a Session

The first step an application must perform in order to use the HTTP APIs is to open an HTTP session. In the **HTTPExample** example, this is performed in the engine's `ConstructL()` method. No parameters are specified, because the default protocol, HTTP over TCP/IP, is required.

Chapter 10 HTTP

```
void CHTTPExampleEngine::ConstructL()
    {
    // Open the RHTTPSession
    iSession.OpenL();
    …
    }
```

Making a GET Request

Once the `RHTTPSession` object is open, you are ready to make a HTTP request. This application demonstrates both GET and POST requests — GET requests are discussed first.

A GET request is initiated by calling `CHTTPExampleEngine::GetRequestL()`. This takes a single descriptor parameter, which should contain the URI to be requested. In order to be passed to the HTTP request transaction, this first needs to be converted into a `TUriC8` object — `TUriC8` is the base class of a family of classes provided by Series 60 for handling URIs.

So the first step taken by `CHTTPExampleEngine::GetRequestL()` is to call `CHTTPExampleEngine::ParseUriL()`, which takes a descriptor and parses it using the `TUriParser8` data member `iUriParser`. The function will leave if the URI is invalid. This `iUriParser` member is later passed into the request function (as `TUriParser8` is derived from `TUriC8`). Here is the `ParseUriL()` function in full:

```
void CHTTPExampleEngine::ParseUriL(const TDesC& aUri)
    {
    // Convert the URI to an 8-bit descriptor
    // then set iUriParser to point at it
    delete iUri;
    iUri = NULL;
    iUri = HBufC8::NewL(aUri.Length());
    iUri->Des().Copy(aUri);
    User::LeaveIfError(iUriParser.Parse(*iUri));
    }
```

Notice that the descriptor has to be converted from 16-bit to 8-bit before it can be parsed. This is achieved using the `TDes::Copy()` method.

Once the URI has been successfully parsed, you need to create a transaction object. This will encapsulate the request to and response from the server. It is represented by the `RHTTPTransaction` class and opened using the `RHTTPSession::OpenTransactionL()` method. The parameters to this method are:

- A URI — in this case, `iUriParser`.

- A reference to an `MHTTPTransactionCallback` object. As `CHTTPExampleEngine` implements this mixin, it passes a reference to itself here.

- A string pool entry identifying the type of request to make. Strings are defined for each of the standard HTTP request types.

SEE ALSO The String Pool section of the SDK documentation for further information on using string pools.

The following section of code shows the function call in full, and the way a request string is retrieved from the `RHTTPSession` object.

```
iTransaction = iSession.OpenTransactionL(iUriParser, *this,
    iSession.StringPool().StringF(HTTP::EGET,
    RHTTPSession::GetTable()));
```

The `RStringF` type is used for a string pool string—F is short for "folded," meaning that comparisons performed on this string are case-insensitive.

Adding Request Headers

After constructing the transaction object, one more stage is required before a request can be submitted—to add a set of request headers. Transaction headers are represented by the `RHTTPHeaders` class, and a transaction's request headers can be obtained by calling `iTransaction.Request().GetHeaderCollection()`. The **HTTPExample** example explicitly adds *Accept* and *User-Agent* headers, as shown below:

```
// Set transaction headers
RHTTPHeaders headers =
iTransaction.Request().GetHeaderCollection();
  AddHeaderL(headers, HTTP::EUserAgent, KUserAgent);
  AddHeaderL(headers, HTTP::EAccept, KAccept);
```

The Accept header defines the content type (or types) which will be accepted in the response message body. The User-Agent header provides a string identifying the client application. The constants used here are defined as follows at the top of the file:

```
// CONSTANTS
// HTTP header values
// Name of this client app
_LIT8(KUserAgent, "HTTPExample (1.0)");
// Accept any (but only) text content
_LIT8(KAccept, "text/*");
```

The Accept value of `text/*` specifies that any text-based content type is acceptable in the response body.

`AddHeaderL()` is a utility function in `CHTTPExampleEngine` which encapsulates the code required to add a header. It uses the `THTTPHdrVal` class to represent each individual header and uses the Series 60 HTTP string pool to ensure the correct strings are specified:

```
void CHTTPExampleEngine::AddHeaderL(RHTTPHeaders aHeaders,
TInt aHeaderField, const TDesC8& aHeaderValue)
    {
    RStringPool stringPool = iSession.StringPool();
    RStringF valStr = stringPool.OpenFStringL(aHeaderValue);
    THTTPHdrVal headerVal(valStr);
    aHeaders.SetFieldL(stringPool.StringF(aHeaderField,
RHTTPSession::GetTable()), headerVal);
    valStr.Close();
    }
```

Headers that are defined by RFC 2616 as being mandatory for HTTP 1.1 are automatically added to a request. So your application code does **not** need to add the following to a request:

- **Content-Length**—This is derived from the client request's `MHTTPDataSupplier`.
- **Connection**—This is handled automatically to manage the HTTP/1.1 persistent connection. Only if a client wants to indicate that the persistent connection is to close should the client add a Connection header.
- **Host**—This is taken from the client's request URL. Only if a relative URL is supplied should the client add a Host header.
- **Transfer-Encoding**—This is added automatically if the client's request `MHTTPDataSupplier` specifies an unknown overall data size.
- **Authorization**—This is added automatically by the Authentication filter.

Submitting the Request and Receiving the Response

Once the request headers have been set, the request can be submitted. This is achieved by calling the `RHTTPTransaction::SubmitL()` method:

```
// Submit the request
iTransaction.SubmitL();
```

`RHTTPTransaction::SubmitL()` is an asynchronous function, but you do not need to explicitly create an active object in order to handle it. (This is performed by the HTTP stack for you.) During the course of the request transaction, progress notification events are sent to the `MHFRunL()` method of the `MHTTPTransactionCallback` object that was passed by reference to `RHTTPSession::OpenTransactionL()`. In this case, this is the `CHTTPExampleEngine` class itself. This callback function is defined in `MHTTPTransactionCallback` as follows:

```
virtual void MHFRunL(RHTTPTransaction aTransaction,
const THTTPEvent& aEvent) = 0;
```

The `aTransaction` parameter identifies the transaction that has generated the progress event, and aEvent identifies the type of event. `CHTTPExampleEngine` overrides this to handle the following event types:

- `EGotResponseHeaders`—Generated when the response headers are received. When this event is received, the engine notifies its observer with a call to `MHTTPExampleEngineObserver::ResponseStatusL()`.

- `EGotResponseBodyData`—Generated when part (or all) of the response body is received. The engine handles this by appending the currently received part to a buffer used for storing the complete response.

- `EResponseComplete`—Generated when the entire response has been received. On receipt of this event the engine passes the response to its observer by calling `MHTTPExampleEngineObserver::ResponseReceivedL()`.

> **NOTE**
> If a leave occurs in your `MHFRunL()` method, you will get the chance to handle it in the `MHFRunError()` method, also inherited from `MHTTPTransactionCallback`. The leave code is passed into this function as a parameter, along with the transaction and event that were being handled when the leave occurred. The function should return `KErrNone` if it handles the error. If it returns any other error code, the application will panic (with `HTTP-CORE 6`).

The following code shows how the engine obtains the status code and status text returned with the `EGotResponseHeaders` event. An `RHTTPResponse` object is used to achieve this. Notice also that the response text is converted from 8-bit text back into a 16-bit descriptor.

```
// HTTP response headers have been received.
// Pass status information to observer.
RHTTPResponse resp = aTransaction.Response();

// Get status code
TInt statusCode = resp.StatusCode();

// Get status text
RStringF statusStr = resp.StatusText();
HBufC* statusBuf = HBufC::NewLC(statusStr.DesC().Length());
statusBuf->Des().Copy(statusStr.DesC());

// Inform observer
iObserver.ResponseStatusL(statusCode, *statusBuf);

CleanupStack::PopAndDestroy(statusBuf);
```

> **TIP**
> You can check whether a request succeeded by verifying that the status code returned is in the range `200` to `299`, indicating a successful request. See RFC 2616 for a full list of HTTP response status codes.

Chapter 10 HTTP

Provided the request was successful, the next event the engine handles is `EGotResponseBodyData`, indicating that at least part of the body data has been received. This event will occur one or more times until all body data has been received, at which point the `EResponseComplete` event is sent. Because data may arrive in parts, the engine handles `EGotResponseBodyData` by appending the current chunk to the end of a buffer, and handles `EResponseComplete` by passing a reference to this buffer to the observer. Each body part is obtained in the `EGotResponseBodyData` handler code from a data supplier object by calling `MHTTPDataSupplier::GetNextDataPart()`. As with the header data, the engine converts the data received from 8-bit text to a 16-bit descriptor. (Remember that the response body is guaranteed to be text rather than binary content, as the Accept header was set to `text/*`.)

The full code of the handler is shown below:

```
// Get text of response body
MHTTPDataSupplier* dataSupplier =
aTransaction.Response().Body();

TPtrC8 ptr;
dataSupplier->GetNextDataPart(ptr);

// Convert to 16-bit descriptor
HBufC* buf = HBufC::NewLC(ptr.Length());
buf->Des().Copy(ptr);

// Append to iResponseBuffer
if (!iResponseBuffer)
    {
    iResponseBuffer = buf->AllocL();
    }
else
    {
    iResponseBuffer = iResponseBuffer
->ReAllocL(iResponseBuffer->Length() + buf->Length());
    iResponseBuffer->Des().Append(*buf);
    }

// Release buf
CleanupStack::PopAndDestroy(buf);

// Release the body data
dataSupplier->ReleaseData();
```

The handler for the `EResponseComplete` event simply passes the complete buffer to the observer:

```
// Pass the response buffer by reference to the observer
iObserver.ResponseReceivedL(*iResponseBuffer);
```

Chapter 10 Advanced Communication Technologies

In the **HTTPExample** application, the observer is the `CHTTPExampleAppUi` class, which implements `ResponseReceivedL()` by inserting the text into its rich text view.

Making a POST Request

Another common kind of HTTP request is the POST request. This differs from the GET request mainly in that the request itself contains a body as well as headers. It is most commonly used for transmitting data entered in an HTML form to some kind of program on the server, such as a CGI script or Java servlet, which then returns a dynamically generated response. The **HTTPExample** application makes a POST request to a CGI script, passing the user's name as a parameter. The script returns the text "Hello <name>!".

The only difference between the GET and POST requests in this application is in the way the request is issued—the application handles responses from both kinds of request identically, using the code shown above. Since a POST request has a body, a data supplier is required to provide the body data. You can provide your own data supplier by implementing a class which derives from `HTTPDataSupplier`, or you can use an existing data supplier class. As **HTTPExample** just sends HTTP form data, it uses the HTTP API class `CHTTPFormEncoder` for its data supplier. The function `CHTTPExampleEngine::PostRequestL()` starts by initializing this form encoder object with the user's name, passed in from the UI:

```
_LIT8(KPostParamName, "NAME"); // Name of the parameter sent
in a POST request
...

void CHTTPExampleEngine::PostRequestL(const TDesC& aName)
    {
    // Build form encoder
    // Start by removing any previous content
    delete iFormEncoder;
    iFormEncoder = NULL;
    iFormEncoder = CHTTPFormEncoder::NewL();
    TBuf8<EMaxNameLength> buf8;
    buf8.Copy(aName);
    iFormEncoder->AddFieldL(KPostParamName, buf8);
    ...
    }
```

To complete the request, the application needs to set the request headers (just as it did for a GET request) and set the `iFormEncoder` object to be the request's data supplier. This time it sets one extra header, Content-Type, to indicate the content type of the request body:

```
_LIT8(KPostContentType, "text/plain");  // Content type sent
in a POST request
...
```

Chapter 10 HTTP

```
// Set transaction headers
RHTTPHeaders headers =
iTransaction.Request().GetHeaderCollection();
  AddHeaderL(headers, HTTP::EUserAgent, KUserAgent);
  AddHeaderL(headers, HTTP::EAccept, KAccept);
  AddHeaderL(headers, HTTP::EContentType, KPostContentType);
```

The `iFormEncoder` object is associated with the request by passing it to the `RHTTPRequest::SetBody()` function. Remember that this will cause the transaction to call back into the data supplier asynchronously when the body data is required, so your data supplier must not be a local variable that will go out of scope. This is why `CHTTPExampleEngine` has `iFormEncoder` as member data. The concluding part of the request is shown here:

```
// Set the form encoder as the data supplier
iTransaction.Request().SetBody(*iFormEncoder);

// Submit the request
iTransaction.SubmitL();
```

Canceling and Aborting Transactions

An outstanding transaction can be canceled at any time using `RHTTPTransaction::Cancel()`. This means that you will not receive any more events for that transaction, unless you choose to resubmit it. This is called in `CHTTPExampleEngine::Cancel()`, which allows the UI to cancel the request if necessary:

```
void CHTTPExampleEngine::Cancel()
  {
  iTransaction.Cancel();
  }
```

You can also abort a transaction using `RHTTPTransaction::Close()`, in which case there is no need to call `Cancel()` first.

> **NOTE** Calling `Close()` causes all the resources associated with a transaction to be cleaned up, so you must not try to use a transaction once it has been closed.

Terminating the Session

When you have finished all your transactions, the session should be closed. Note that closing the session will cause any outstanding transactions to be cancelled. This is performed in the destructor of the `CHTTPExampleEngine` class:

```
// Close session
iSession.Close();// Will also close any open transactions
```

WAP

The Wireless Application Protocol (WAP) is described as "the de-facto world standard for the presentation and delivery of wireless information and telephony services on mobile phones and other wireless terminals."[1] The standard encompasses transport and delivery protocols as well as content technology. Series 60 has a WAP stack implementation and provides developers with the APIs required to use it. These APIs are the main focus of this section, after a brief introduction to the WAP architecture.

WAP Architecture

The hierarchy of the stack can be seen in Figure 10–2. Note that the bearer layer is not specified as part of the WAP standard—the existence of suitable bearer, such as GSM, SMS, or GPRS, and appropriate APIs are assumed.

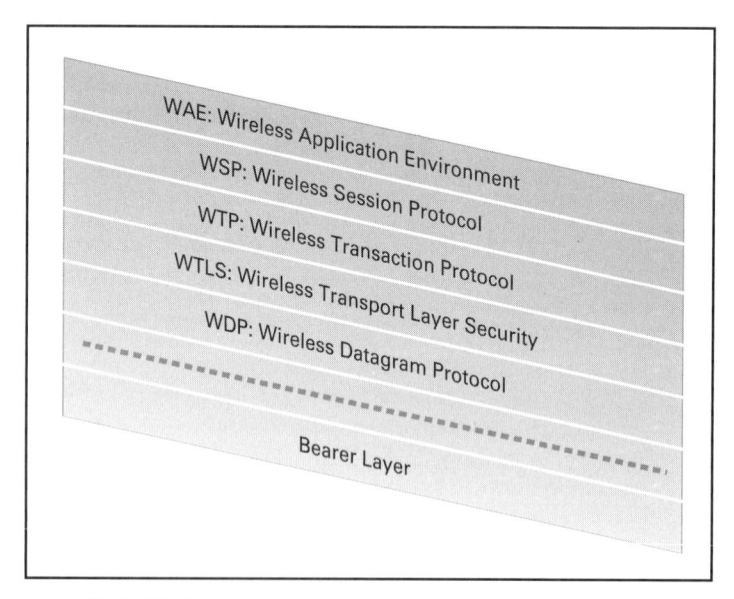

Figure 10–2 WAP stack.

WDP is an unreliable transport protocol, analogous to UDP, which is essentially responsible for moving data between the sender and receiver and back again. As the base of the WAP stack it provides an interface to the different bearers. The WTLS protocol provides a mechanism for secure connections that is based on the industry-standard Transport Secure Layer (TSL). It can perform encryption of data, has methods for authentication, and ensures privacy of data. The WSP layer of the stack is concerned with the delivery of data and offers both a connected and connectionless service. If a connected service is requested, the WTP layer manages the transaction by demanding an

---

[1] http://www.wapforum.org/what/WAP_white_pages.pdf.

acknowledgement of the receipt of data and resending any packets that are not acknowledged. The top layer of the WAP stack, WAE, defines the development environment for applications and specifies the mark-up and scripting languages and the telephony interface.

WAP supports two modes of transaction between client and server: **client pull** and **server push**. In a client pull transaction, a client sends a request for data to a server, and the server then responds. In a server push transaction, data is sent from server to client without having been requested by the client.

Series 60 Implementation

WAP was supported in early releases of Series 60, but Series 60 2.x introduces a new API and, unfortunately, provides no backward compatibility. The new API supports the multihoming framework discussed in the previous chapter.

Series 60 1.x WAP APIs

In Series 60 1.x, a WAP stack API allows the client access to the WSP, WTP and WDP layers, providing a means of performing connected and connectionless transactions. A Client/Server architecture is used, with the WAP server (this is not the remote server that provides WAP content) managing client access to WAP services. The client can access the WAP server through the RWapServ class. A session with this server must be opened (RWapServ::Connect()) before any of the other protocol session is opened—this is analogous to opening a session with RSocketServ in sockets programming.

The various layers of the WAP stack were outlined earlier in this chapter. In Series 60 there are classes to represent each layer, and, as many of the functions that must be available at each layer are the same, they are all derived from the base class, RWAPConn. An Open() function is defined in RWAPConn that takes the parameters necessary to open a WAP protocol session; these include a reference to the WAP server, the address of the remote host, port numbers, bearer and whether the connection should be secure. This Open() function can be described as fully specified, meaning that the address for each end of the connection is defined. Methods to cancel pending operations and close the session are, of course, provided, as well as functions to retrieve the remote address, port and the like. If a secure connection was requested, the WTLS layer is created, and the function Wtls() will return a handle to this layer.

Sending WAP Datagrams

It is usual, when sending WAP content, to do so using the upper levels of the protocol stack, creating a WSP session to ensure a reliable transport. However, it is also possible to use the WDP protocol to send WAP datagrams—you would use the RWDPConn class for this. This class implements the base class, fully specified, Open() function, but also defines an Open() method that just takes a reference to the Symbian OS WAP server and local port number as parameters. When sending data there are two options: the fully specified version of `Open()` can be used in conjunction with `Send()`, or `SendTo()`, which requires a destination address to be passed through, and can be used with the overloaded `Open()` implementation. `SendTo()` allows data to be sent to multiple destinations, whereas with `Send()` you are restricted to the one address.

The asynchronous `Recv()` function also takes a remote address as a parameter, meaning that, as long as the fully specified `Open()` was not used, datagrams can be received from multiple destinations, too.

WSP Sessions

Since the most common way to perform WAP transactions is to create a WSP session, the related APIs will be examined in more detail and supported with example code. The idea of connected and connectionless sessions is common in communications programming, and WSP sessions can be of either variety. Connected sessions are appropriate if a reliable medium for request and response is required; connectionless can be used if a conversational protocol is not needed. Both session types support the push model of delivery as well as the usual pull mode.

Connected WSP Session

The class designed to create a connected WSP session is `RWSPCOConn`. To use the class, you must first create a session with the WAP server. As it is used only for connected sessions, `RWSPCOConn` defines just the fully specified version of the `Open()` function.

```
User::LeaveIfError(iWapServ->Connect());
User::LeaveIfError(iWspConn->Open(*iWapServ, remoteHost,
remotePort, iLocalPort, iBearer, EFalse));
```

The local port number can be left as zero to indicate that the device should allocate a suitable port (`GetLocalPort()` can be used later to retrieve the port number). Suitable bearers are defined by the enumeration `TBearer` and include `EIP` (Internet Protocol) and `ESMS` (8-bit SMS). As secure connections are not supported, the `aSecureConn` parameter should be set to `EFalse`.

To start an open WSP session, you use the `Connect()` method. Session headers are sent to the remote server, and, optionally, information about the size of buffers and data packets is communicated through the codec, `CCapCodec`. You can set the desired client parameters—for example, SDU size (set by default to 1400)—using the functions on `CCapCodec`, and the WAP server will also set server parameters, which you can retrieve.

```
iCapCodec = CCapCodec::NewL();
iCapCodec->SetClientSDUSize(KClientSDUSize);
User::LeaveIfError(iWSPConn->Connect(clientHeaders,
iCapCodec));
```

Chapter 10 WAP

NOTE Besides the systemwide error codes that can be returned, WAP-specific errors are defined by `RWSPCOConn::TReturnCodes` and `RWapConn::TReturnCodes`.

`CreateTransaction()`, as the name suggests, is used to kick-off the data exchange process. `TMethod` describes the type of transaction—examples include `EGet`, `ETrace`, and `EPost`. Other parameters to be passed are the URI of the destination, the body and WAP headers. You must also create an `RWSPCOTrans` object and pass a reference to it as a parameter. On return of the function, you can use this object to manage the transaction. For example, you could call `GetState()` to determine what stage the transaction is at.

```
iWSPTransaction = new (ELeave) RWSPCOTrans();
User::LeaveIfError(iWSPConn->CreateTransaction(
RWapConn::EGet, uri, headers, body, *iWSPTransaction));
```

During one transaction, a number of different events can occur at both session level and transaction level. You can monitor the events by use of the `GetEvent()` function, which has a synchronous and asynchronous implementation—as usual, the asynchronous version is preferred. The types of event are described by the enumeration `TEventType` and include notification of exceptions and disconnection. If data is associated with the event, then you can retrieve it by calling `GetSessionData()` with the appropriate `TSessionDataType` parameter. Note that if the buffer you pass in is too small, `EMoreData` will be returned.

To retrieve the data from the transaction, once it has completed, you need to call the `GetData()` function of the `RWSPCOTrans` object, passing in a buffer for the data and specifying the data type as `RWSSPCOTrans::EResultBody`. As with `GetSessionData()`, `EMoreData` will be returned if the buffer is too small.

```
TInt error = iWSPTransaction->GetData(iData,
RWSPCOTrans::EResultBody);
```

When all transactions are complete, the WTP session should be disconnected and the WAP server session closed.

```
iWSPConn->Disconnect();
iWapServ->Close();
```

Connectionless WSP

If you require a connectionless WSP session, `RWSPCLConn` should be used. The session is created in much the same way as for a connected session, the difference being the availability of both fully specified and not fully specified `Open()` functions. The connectionless session does not afford the same degree of transaction management as the connected session, relying on just an ID for control—this means that you do not need to create a transaction object.

To retrieve content, the method you select depends upon whether a push or pull model is being used. For the pull method, you call `UnitInvoke()` to initiate the transaction. Note that the parameters for this function are the same as those in the `CreateTransaction()` function used for connected sessions, except that an ID is passed in place of the transaction object. You do not need to set the ID yourself; an ID will be allocated, and on return of the function the `aID` parameter will have been updated. To retrieve the response from this initial request, you call `UnitWaitResult()` (available as either a synchronous or an asynchronous implementation). If the ID returned matches that from the `UnitInvoke()` function, then you can be sure it is a response to your request. The body and WAP headers are returned.

If the data is to be pushed to the client, then `UnitWaitPush()` can be called to wait for the incoming message—synchronous and asynchronous versions are available. On return, the body and headers can be retrieved.

Series 60 2.x WAP APIs

The WAP stack API that was described for Series 60 1.x has been withdrawn in Series 60 2.x and replaced with a WAP messaging API. The differentiation between a fully specified connection and one that does not define a specific endpoint for all transactions leads to the definition of four interfaces (ECom plug-in interface).

- CWAPBoundDatagramService
- CWAPFullySpecDatagramService
- CWAPBoundCLPushService
- CWAPFullySpecCLPushService

Sending WAP Datagrams

The classes `CWAPBoundDatagramService` and `CWAPFullySpecDatagramService` offer functionality akin to that of the `RWDPConn` class that was previously supported. One major difference is that, in 2.x, the client can specify which network interface should be used for the connection, employing the multihoming framework. The difference between `CWAPBoundDatagramService` and `CWAPFullySpecDatagramService` is that the latter should be used if you want to restrict all transactions to a specific remote host.

CWAPBoundDatagramService

If you are not just concerned with one specific remote host, `CWAPBoundDatagramService` can be used to send and receive datagrams—instantiation of an object is through the `NewL()` function. To connect to the WAP server, you obviously use a version of the `Connect()` function. There are two implementations, one that takes the IP address of the network interface to be used, and another that does not—you need to decide whether or not you wish to use the default interface. The other parameters you need to pass through relate to the bearer you want to use (you can choose `TBearer::EAll` for listening and then define a particular bearer when you send data) and the local port number to listen on (a value of 0 leaves the selection to the system).

When you want to send data, you will need to use the `SendTo()` function and nominate the recipient—the format of the address of the remote host will depend upon the bearer that you are using. You will need to select a bearer type for the transaction—this is necessary even if you specified a particular bearer during connection, but will be ignored unless you specified the bearer value `EAll` initially. The remote port number needs to be set and the data to be sent passed through in an 8-bit descriptor. The `SendTo()` function is synchronous and returns `KErrNone` if successful.

The asynchronous `AwaitRecvDataSize()` should be called when you want to listen for a datagram, as this will allow you to retrieve the size of the data. When you have done this, you can then call `RecvFrom()` with a buffer of the appropriate size to get hold of the data. Note that you can set a timeout value using the `aTimeout` parameter—it is probably a good idea to do so, but as you have been notified that there is data waiting for you (through `AwaitRcvData()`), it should be OK to set an infinite timeout using `aTimeout = 0`.

There is no specific `Disconnect()` function; destruction of the object closes the connection, but you should always make sure you have canceled any outstanding asynchronous listening functions using `CancelRecv()`.

CWAPFullySpecDatagramService

If you want to exchange datagrams with a particular remote host, `CWAPFullySpecDatagramService` can be used. Use of this class is very similar to that of `CWAPBoundDatagramService` that has just been described, but its connection function is fully specified. As both endpoints have been nominated, there is a `Send()` function with just one parameter, the descriptor to be sent, instead of `SendTo()`. Similarly this class has a `Recv()` function rather than a `RecvFrom()`, because you have already set up the remote address.

CWAPBoundCLPushService and CWAPFullySpecCLPushService

The `CWAPBoundCLPushService` class can be used to listen for push messages from any sender. Again, the class is used in much the same way as the datagram classes, but because it operates above the WDP layer, there is the option of creating a secure connection. This is simply achieved by setting the Boolean parameter, `aSecure`. Once you have made a connection to the WAP stack, you can listen for messages by calling the asynchronous `AwaitPush()` method, passing through buffers for the headers and body. If the buffers you have allocated will not hold all of the data received, then the `iStatus` returned will be `RWAPConn::EMoreData`. If this is the case, you will need to reissue the

`AwaitPush()` request to retrieve the remaining data. Note that once the complete message has been received, you can create a `CPushMessage` object using the headers and body.

`CWAPFullySpecCLPushService`, as you would expect, is the fully specified version of `CWAPBoundCLPushService`, to be used for listening to push messages from a single remote host.

Messaging

Series 60 features powerful capabilities for handling email, SMS and MMS messages. Messages may be created, sent, received, stored, manipulated and deleted. This feature is known as the messaging architecture, and this section shows you how you can make use of its features in your Series 60 applications. There are a number of useful things you may want to do with messages from within your own application, such as:

- Create and send a message.
- Perform some action when a new message is received.
- Manipulate an existing message, or group of messages: for example, searching through the body text of all received messages.

Series 60 includes three key APIs for performing these kinds of tasks:

- The **Client MTM** API.
- The **Send-As** API.
- The **CSendAppUi** class.

This section first provides an overview of the key concepts and terminology relating to the Series 60 messaging architecture and then goes on to discuss each of these APIs in detail.

Key Messaging Concepts

This subsection introduces some of the basic concepts of Series 60 messaging that underlie all three messaging APIs.

Messaging Server and Session

As mentioned earlier, the messaging architecture is built around a server, which manages all messaging resources on a phone. This is accessed from a client process through an instance of the session class `CMsvSession`. Any client process that uses messaging services must have at least one instance of this class. In almost all cases, one instance is all it needs, and this can usually be shared by reference between all classes that require access to the messaging server.

Messaging Entries

The data managed by the messaging server takes the form of a collection of **entries**. An entry can be one of four different types:

- A folder.
- A message.
- An attachment.
- A service.

An entry can have child entries and can itself be the child of another entry (its parent). In this way, a tree structure is built up which can be thought of as analogous to a file system with its folders, subfolders and files. At the root of the tree is the **root index entry** (to continue the file-system analogy, the root index entry plays a role like that of the root folder of a drive). This contains four standard folders: *Inbox*, *Outbox*, *Drafts* and *Sent Items*, which may in turn contain any number of messages and/or user-defined subfolders. Each subfolder itself may contain messages and/or further subfolders, and so on. A message may have attachments as child entries or, less typically, other messages.

The role of the fourth type of entry, the service entry, is less obvious. A service entry contains configuration data for an individual service, or account. Often you will find one service entry per MTM, but this is not necessarily so. For example, if a user has two MMS accounts set up on their phone, there will be two MMS service entries. Service entries are stored in the root index entry, alongside the standard folders.

MTMs: Message Type Modules

An MTM is a plug-in to the messaging architecture to provide support for a specific message type. It comprises a server-side DLL which interacts directly with the messaging server, and a suite of client-side DLLs which provide an API for client applications. Further details are provided below when the Client MTM API is discussed.

Generic Entry Handling/Unified Inbox

The messaging architecture is designed to allow all types of entry to be treated generically, as far as possible. Just as most filing systems allow you to, say, delete or rename a folder in exactly the same way as you would a file, the Series 60 messaging architecture allows you to use exactly the same APIs to perform most routine management tasks on any kind of entry. These tasks include deleting, moving, copying, navigating to parent and child entries, and finding details such as size and date. All of these tasks can be carried out without knowing (or caring) about the type of the entry.

The main classes used for this generic entry handling are `CMsvEntry`, `TMsvEntry` and `CMsvStore`.

Entry Storage

The Messaging Server is responsible for storing all types of entry and providing concurrent client processes with safe, shared access to them. The server MTM component interacts with the Messaging Server to handle sending and receiving of messages.

The Messaging Server can store information about each entry in three different locations: the messaging **store**, the messaging **index** and the file system. A basic understanding of the roles of these locations will help you to grasp other messaging concepts and APIs that you will see later in this chapter.

The Messaging Store

Each entry has a file store associated with it, used to persistently store its in-memory representation. The format of an entry's data in the Messaging Store depends on its type: note that there is no generic format for Messaging Store data. Message entries use it to store things like body text (where present) and headers (with the format dependent on the specific MTM). Service entries use it to store all their configuration information. Folder entries do not use the store (note they still have a store, but they leave it empty), and attachment entries may use it, based on the individual MTM implementation.

An entry's store is accessed using the `CMsvStore` class.

The Message Index

For every entry, regardless of its type, the Messaging Server maintains a *generic* set of summary information in the Messaging Index. The index is loaded into RAM when the Messaging Server starts, and it stays in memory until the server closes, so it provides a quick way to access information about an entry. Note that the index does not store *all* information about an entry — only *some generic* information. For example, it does not store the message body, but it does contain information about the size, date and type of the entry, its unique ID, and so on. The idea is that the index entry contains enough information about an entry to display a summary of it (for example, in an *Inbox* view) without opening a file store or needing to load an MTM. This enables summary views to display a list of messages quickly.

The index entry is represented by the `TMsvEntry` class.

The File System

Finally, each entry is assigned a specific folder in the file system, where it may choose to store further data. As with the store, this is optional and its usage is MTM-specific. Use of it tends to vary widely between MTM implementations, and the folder is typically only accessed directly by the Server MTM. As an application developer you will almost certainly never use it, but you will see APIs which refer to it (such as `CMsvEntry::GetFilePath()`), so it just helps to be aware of it. Otherwise, its use is beyond the scope of this book.

Key Messaging Classes and Data Types

This section presents a quick reference guide to some of the most commonly used messaging types.

CMsvSession

Represents a client-side session to the messaging server. Note that it is a C-class, not an R-class as you may have expected. Normally one instance per client thread is enough.

TMsvId

Simply a `typedef` of a `TInt32`. The messaging server assigns a unique `TMsvId` to each entry. In most cases this is performed dynamically, but there are a few fixed IDs—for example, for the standard folders (*Inbox*, *Outbox*, *Drafts* and *Sent*). Once you have an entry's ID you can use it to obtain any other information about the entry, including its index entry and store data.

TMsvEntry

Represents an individual index entry. As mentioned above, the index entry contains a restricted, generic set of summary information for an entry.

CMsvEntry

This is perhaps the most commonly misunderstood class in the messaging architecture. Perhaps because of its name (which looks misleadingly like `TMsvEntry`), it is often assumed that an instance of this class contains all the data belonging to an individual entry (such as its headers, body text, and such like). This is not the case, as you will see as soon as you look at its API. `CMsvEntry` is *not* an in-memory representation of an entry—likewise, calling `CMsvEntry::NewL()` does *not* create a new entry in the messaging server.

Instead, `CMsvEntry` is better thought of as an *entry handle* or *entry context*. `CMsvEntry` provides you with an interface onto a specific entry, through which you can obtain other objects which contain the entry's data. As such, a `CMsvEntry` can be reassigned to "point to" a different entry, simply by providing the new entry's ID. In fact, this is how a `CMsvEntry` object should be used—it is an expensive class to instantiate, so it is recommended that you "recycle" `CMsvEntry` objects wherever possible and avoid needlessly creating new ones.

CMsvStore

`CMsvStore` represents a store for an individual entry. Its API can be used to store and retrieve the body text of a message (provided it has one) and other data (such as headers).

CMsvEntrySelection

This is just an array of `TMsvIds`, and as such provides a common way for passing around information about groups of entries. For example, methods such as `ChildrenL()` in `CMsvEntry` (used to enumerate an entry's child entries) return a `CMsvEntrySelection` pointer.

CMsvOperation

Many long-running activities in messaging, such as sending all items from the *Outbox*, or checking for email on a POP3 server, require more than an active object to handle them. As well as being notified that the request has completed, applications frequently need to be able to get progress information while the request is still running (for example, in order to update a progress dialog of the form "`Downloading message x of y`"). As a result, most asynchronous functions in messaging, as well as taking a `TRequestStatus&` parameter identifying an active object which will handle their completion, also return a `CMsvOperation` object which can be used to obtain progress information before the request completes.

The Messaging APIs

The rest of this chapter covers the three messaging APIs: the Client MTM API, SendAs API and CSendAppUi API. Each is discussed in detail with the help of example application code, but first the three APIs are briefly introduced and the particular role performed by each of them is outlined.

Client MTM API

At the most basic level, the structure of the messaging architecture is similar to that seen in the previous chapter for the Serial Communications and Sockets subsystems. Most importantly, it shares with them the following key factors:

- A server provides shared, generic access to messaging resources.
- A client-side API is used to access the server from an application.
- Plug-ins are used to provide actual implementations of specific technologies.

A crucial difference between the messaging architecture and those covered earlier in this chapter is in the nature of the plug-ins. A plug-in to the messaging architecture is known as an **Message Type Module (MTM)**, and rather than a single DLL it comprises a suite of DLLs which work together to provide an implementation of a specific message type. As opposed to the other communications technologies in Series 60, where the plug-in is entirely server-side, an MTM consists of both server-side and client-side components. The reason is that messaging protocols vary so widely in their capabilities that it is impossible to write a "one shape fits all" client-side API to cover all of them completely. So the client-side MTM is used to implement and extend the basic messaging API for a given message type.

Fortunately, however, it is rare to use the specific client MTMs directly, and the messaging architecture is designed in such a way that client code can usually be written in a generic, extensible way. Each client-side MTM DLL implements a class derived from `CBaseMtm`, and many messaging tasks (including all the examples listed above) can be performed using just this base class interface. This chapter will show you how to write code that uses this polymorphic API to perform quite sophisticated manipulation of messaging resources in a generic way.

> **NOTE** The abbreviation MTM can be used with a number of slightly different meanings, which can cause some confusion. Fortunately, the context will usually clarify which meaning is intended.
>
> First, an MTM refers to the suite of DLLs that make up the plug-in as a whole ("the SMS MTM," "the POP3 MTM," and so on).
>
> Second, MTM is also used to refer to each DLL in the suite (for example, "the MMS server MTM," "the client MTM").
>
> Finally, MTM can be used to refer to (an instance of) the polymorphic class exported by an MTM DLL (for example, "once you have created an MTM registry, you are ready to construct an MTM object").

Send-As API

The Client MTM API is extremely powerful and flexible, but it is also quite complex. Many applications require only a limited subset of this functionality, most commonly just the ability to create and send a new message. The Send-As API provides a simple class (`CSendAs`) that offers precisely this capability. The obvious advantage over the Client MTM API is its simplicity, but this is also its disadvantage: Send-As is not as powerful or flexible as the Client MTM API, and it can only be used to address a very specific (but common) need.

CSendAppUi class

A shortcoming identified in the Send-As API is that it requires a message to be built up "programmatically"—the developer has to write code to set various attributes of a message. If, however, you want to allow the user to create message content *interactively*—for example, in an editor—then you have to write the editor yourself, or find a suitable one to reuse. This inevitably involves a lot of work that most developers would like to avoid.

Series 60 addresses this common complaint by providing the `CSendAppUi` class. This provides a number of facilities for developers to enable interactive message creation from their code, including APIs to access the standard message editors. Now, obviously, the pros and cons are reversed: `CSendAppUi` cannot be used to programmatically construct a message. So neither Send-As nor `CSendAppUi` is a complete solution, but between them the APIs address two distinct requirements of third-party messaging developers.

Using the Client MTM API

As mentioned earlier, the Client MTM API can be used generically up to a point, by making polymorphic use of the common client MTM base class `CBaseMtm`. However, if a client wishes to make particular use of a specific MTM, it will usually need to cast a `CBaseMtm` instance up into an instance of the actual MTM being used (for example, `CSmsClientMtm` for SMS). There is nothing wrong with doing this—it is the correct way to access a specific MTM's specialized API. You just need to be aware that once this is done, your code becomes nongeneric and will not be usable with other MTMs without (often substantial) modification.

This section documents the example application **MtmsExample**, which uses the SMS and MMS MTMs to create and send messages of those types. Shown below is the member data of the `CMtmsExampleAppUi` class which is used throughout the chapter:

```
CMsvOperation*       iOp;
CMsvEntrySelection*  iEntrySelection;

CMsvSession*         iSession;
CClientMtmRegistry*  iMtmReg;
CSmsClientMtm*       iSmsMtm;
CMmsClientMtm*       iMmsMtm;
TBool                iReady;

TMsvId               iSmsId;
```

The only type here that has not already been documented is `CClientMtmRegistry`—this is used for loading MTMs and obtaining handles to them.

Connecting a Session to the Messaging Server

Before you can do anything useful in this (or any other) messaging application, you will need to connect a session to the messaging server. In order to do this, you need to have a class which implements the mixin interface `MMsvSessionObserver`—an instance of this class will be passed by reference to the session, and will then be notified of events that occur in the server. In this case, the class in question is `CMtmsExampleEngine`:

```
class CMtmsExampleEngine : public CActive, public MMsvSessionObserver
    {
    ...
```

Chapter 10 Messaging

```
private: // from MMsvSessionObserver
    void HandleSessionEventL(TMsvSessionEvent aEvent, TAny*
aArg1, TAny* aArg2, TAny* aArg3);
    …
}
```

Notification of server events will be performed by means of a call to `HandleSessionEventL()`, passing parameters as described in the SDK documentation.

Connection to the server can be either synchronous or asynchronous. It is recommended that a UI process connects asynchronously, to prevent blocking the responsiveness of the UI with a potentially long synchronous call.

The following line connects the session asynchronously:

```
// Connect the session
iSession = CMsvSession::OpenAsyncL(*this);
```

The parameter (`*this`) is a reference to an `MMsvSessionObserver`—in this case, the `CMtmsExampleEngine` object. You are notified of successful connection by a call to this object's `HandleSessionEventL()` method. You must wait until this event is received before constructing any other messaging objects (for example, the client MTM registry, or a `CMsvEntry` object). In the **MtmsExample** example, this second-phase messaging construction is performed in the `CMtmsExampleEngine::CompleteConstructL()` function. The following code shows how this is called in response to the session connection notification.

```
void CMtmsExampleEngine::HandleSessionEventL(TMsvSessionEvent
aEvent, TAny* /*aArg1*/, TAny* /*aArg2*/, TAny* /*aArg3*/)
    {
    switch (aEvent)
        {
        // This event tells us that the session has been opened
        case EMsvServerReady:
            CompleteConstructL();
            break;
        …
        }
    }
```

If the session fails to connect, `HandleSessionEventL()` will be called with an event code of `EMsvServerFailedToStart`. In this case, the aArg1 argument will point to a standard `TInt` error code indicating the reason for failure.

As mentioned, you can also connect the session synchronously, although this is strongly discouraged in UI applications. Nevertheless, you may find this useful if you are writing something like a rough command-line test harness, where simplicity is more important than responsiveness. Here you would use code like this:

```
// Connect the session-using OpenSyncL()
// instead of OpenAsyncL()
iSession = CMsvSession::OpenSyncL(*this);

// Now continue immediately with other messaging code.
```

In this case, the session is ready to use when `OpenSyncL()` returns. `OpenSyncL()` will leave if any error occurs when trying to connect.

Constructing the MTMs

After successfully connecting your session to the messaging server, you are ready to construct the MTM objects. In order to do so, you first need to construct a client MTM registry. This is achieved as follows:

```
// Create a client MTM registry
iMtmReg = CClientMtmRegistry::NewL(*iSession);
```

Once the client MTM registry is constructed, you can use it to obtain handles onto client MTMs. To do this, you need to call `CClientMtmRegistry::NewMtmL()`, passing the type of MTM you require. The individual MTM UIDs are listed in Table 10–1. The header files that the UIDs are defined in can be found in the directory `\epoc32\include`, relative to the root of your SDK.

Table 10–1 MTM UIDs

MTM	UID Name	Defined In
SMS	KUidMsgTypeSMS	SMUT.H
MMS	KUidMsgTypeMultimedia	MMSCONST.H
POP3	KUidMsgTypePOP3	MIUTSET.H
IMAP4	KUidMsgTypeIMAP4	MIUTSET.H
SMTP	KUidMsgTypeSMTP	MIUTSET.H
Infrared	KUidMsgTypeIr	IRCMTM.H
Bluetooth	KUidMsgTypeBt	BTMSGTYPEUID.H

In each case, `NewMtmL()` returns a `CBaseMtm` pointer, which is then cast into the appropriate MTM class.

Chapter 10 Messaging

```
// Obtain MTMs from the MTM registry

iSmsMtm = static_cast<CSmsClientMtm*>(
iMtmReg->NewMtmL(KUidMsgTypeSMS));

iMmsMtm = static_cast<CMmsClientMtm*>(
iMtmReg->NewMtmL(KUidMsgTypeMultimedia));
```

This concludes the initialization phase: the client MTMs are now correctly constructed and ready for use.

Using the SMS MTM

This section shows you how to create and send an SMS message using the SMS MTM. Most of the methods of `CSmsClientMtm` used in this section are inherited from `CBaseMtm`, and as such this section comes close to describing generic MTM usage. In the one or two cases where SMS-specific APIs are used, these will be highlighted.

Using the SMS MTM in Series 60

Unfortunately, some system header files are missing from the public Series 60 SDKs, which can cause problems when you try to compile SMS MTM code. The problem occurs if you need to use the class `CSmsSettings` directly. The file that declares this class (`smutset.h`) is dependent on two files which are missing (`etelgprs.h` and `etelbgsm.h`). Fortunately, though, you do not have any direct dependency on the missing files (this means you do not need to use any objects or types they define), so you can remove the lines that include them and replace them with simple forward declarations.

Table 10–2 File Changes Required Before Using the SMS MTM

File	Replace...	With...

smutset.h	`#include <etelgprs.h>`	`class RGprs`
		`{`
		`public:`
		`enum TSmsBearer{};`
		`};`

smsuaddr.h	`#if !defined(__ETELBGSM_H__)`
	`#include <etelbgsm.h>`
	`#endif`
		`const TUint KGsmMax`
		`TelNumberSize = 100;`

gsmumsg.h	`#if !defined (__ETELBGSM_H__)`
	`#include <etelbgsm.h>`
	`#endif`
		`class RSmsStorage`
		`{`
		`public`
		`enum TStatus {};`
		`};`

The replacements you need to make are detailed in Table 10–2. All files referred to are found in the \epoc32\include folder relative to the root of your SDK.

Creating an Entry

The first step required to create a new message is to create a local index entry object. This is represented by the TMsvEntry class. The object you create here will eventually be copied into the messaging index, at which point you can discard your local copy. You use the index entry to define a number of key attributes of the entry you want to create, including:

- Its type (for example, a message, as opposed to a folder, service or attachment).

- Its MTM (SMS).

- The ID of the service entry it is associated with.

- Its creation date.

This is seen in the following code:

```
// Set attributes on index entry
TMsvEntry indexEntry;
indexEntry.SetInPreparation(ETrue);
indexEntry.iMtm = KUidMsgTypeSMS;
indexEntry.iType = KUidMsvMessageEntry;
indexEntry.iServiceId = iSmsMtm->ServiceId();
indexEntry.iDate.HomeTime();
```

The next step is to create an entry in the messaging server. This will cause the following to happen:

- The TMsvEntry object you have specified will be copied to the messaging index.

- It will be assigned a unique ID, which can subsequently be used to identify this entry.

- A messaging store will be assigned for this entry.

- A folder will be created in the file system for any files this entry wishes to store.

The entry is created using CMsvEntry::CreateL(), and so you need to create or obtain an instance of CMsvEntry. Remember that CMsvEntry is a context (or handle) on an entry, rather than the entry data itself.

In this case, as you already have an instance of the SMS MTM, you can use its CMsvEntry object to avoid creating your own. First you set its context to the *Drafts* folder (where you want your new message to be created), then you call CreateL() to create a new child entry based on the attributes you set in your local index entry:

Chapter 10 Messaging

```
// Create entry from this index entry
iSmsMtm->SwitchCurrentEntryL(KMsvDraftEntryId);
iSmsMtm->Entry().CreateL(indexEntry);
```

Note that `indexEntry` was passed by *reference* to `CreateL()` and (provided `CreateL()` did not leave) will now have been updated with the ID of the newly created message entry. The next step is to set the MTM's `CMsvEntry` context to this new entry so that you can go about modifying it:

```
// Set the MTM's active context to the new message
iSmsId = indexEntry.Id();
iSmsMtm->SwitchCurrentEntryL(iSmsId);
```

The first modification to make to the message is to add some body text. In this case, a constant string, defined in the source file, is added:

```
// Add body
CRichText& body = iSmsMtm->Body();
body.Reset();
body.InsertL(0, KSMSBody);
indexEntry.iDescription.Set(KSMSBody);
```

The Client MTM API treats the message body as a rich text object, although clearly in the case of SMS only plain text is supported. This is not a problem: just insert plain text into the `CRichText` object, without specifying any format layer information, and it will work fine.

> **NOTE** Note that the body text is being set twice: once in the entry and again in the index entry's `iDescription` field. Why?
>
> The reason is that in summary views (such as the *Drafts* or *Sent Items* views), the data that is displayed is from the index entry, not the full stored entry. As mentioned earlier, the index entry keeps a *copy* of some summary information about the entry. Unfortunately the index entry is not automatically updated by the MTM, so it is your responsibility to make sure the two are in sync.

The `TMsvEntry` class has two descriptor public data members, `iDescription` and `iDetails`. These are typically used for subject and sender/recipient information, respectively. In the case of SMS there is no subject field, so it is usual to store the body (or the first few characters of it) here.

The next step is to set a recipient for the message. This was passed in as a descriptor parameter (`aAddress`) obtained by a dialog at the UI.

```
// Add addressee
iSmsMtm->AddAddresseeL(aAddress);
indexEntry.iDetails.Set(aAddress);
```

That is all the data you need to set for this message. Again, note that the addressee value is copied into the index entry. However, as `indexEntry` is only a local variable, you now need to save it in order to commit your changes to the messaging index. This is performed through the `CMsvEntry` API:

```
// Update index entry
iSmsMtm->Entry().ChangeL(indexEntry);
```

Finally, all that remains is to save the full message data to the store:

```
// Update store entry
iSmsMtm->SaveMessageL();
```

Validating the Entry

Before you send the message you have created, you should validate it to check that the data you have added is compliant with the message type. This is achieved using the `CBaseMtm` function `ValidateMessage()`, which is implemented by each MTM to provide type-specific validation. The parameter sent to `ValidateMessage()` is a bit-set which indicates the parts of the message to check. The return value is an equivalent bit-set indicating which parts of the message were invalid. Here this has been encapsulated in the `CMtmsExampleEngine::ValidateSMS()` function:

```
TBool CMtmsExampleEngine::ValidateSMS()
    {
    TMsvPartList msgCheckParts =
        KMsvMessagePartBody |
        KMsvMessagePartRecipient |
        KMsvMessagePartOriginator |
        KMsvMessagePartDate;
    TMsvPartList msgFailParts =
iSmsMtm->ValidateMessage(msgCheckParts);
    return msgFailParts == KMsvMessagePartNone;
    }
```

Validation is not mandatory, but it can help you to catch errors early that could otherwise cause problems during sending, which is more difficult to debug.

Sending the Entry

Once the message has been composed and validated, you are almost ready to send it. There are just a couple of steps that need to be completed beforehand.

First, you need to set the SMS service center number in the message headers. (You could have performed this earlier, but it has been left until the last moment in case the user alters their service center settings between creating and sending the message. This way you are guaranteed to pick up the settings that are valid at send time.) The service center number is obtained from the SMS service entry and set in the message headers as follows:

Chapter 10 Messaging

```
// Set context to the SMS message
iSmsMtm->SwitchCurrentEntryL(iSmsId);

// Load the message
iSmsMtm->LoadMessageL();

// Set the SMS service center address
CSmsSettings& settings = iSmsMtm->ServiceSettings();
const TInt numSCAddresses = settings.NumSCAddresses();
if (numSCAddresses > 0)
    {
    CSmsNumber* serviceCentreNumber = NULL;

    // Get the default SC address, if valid,
    // Otherwise just get the first from the list.
    if ((settings.DefaultSC() >= 0) &&
        (settings.DefaultSC() < numSCAddresses))
        {
        serviceCentreNumber =
&(settings.SCAddress(settings.DefaultSC()));
        }
    else
        {
        serviceCentreNumber = &(settings.SCAddress(0));
        }

    iSmsMtm->
SmsHeader().SetServiceCenterAddressL(serviceCentreNumber->
Address());
    }
else
    {
    // Panic
    // There should never be a missing service number
    }

// Save the message
iSmsMtm->SaveMessageL();
```

You should be able to see what is going on here without too much trouble. The current SMS settings are obtained from the SMS MTM, and from this the number of service centers currently configured (typically just one). If there is a valid default defined, then this is used; otherwise the first service center in the list is chosen. Finally, the message is saved to commit the changes just made.

> **NOTE** This is the first (and last) use that is made of an API from `CSmsClientMtm`. If it were not for this segment, `iSmsMtm` could have been just a `CBaseMtm` pointer.

Chapter 10 Advanced Communication Technologies

The final step required before sending is to update the index entry: you need to set the "in preparation" flag to false and set an appropriate sending state:

```
// Update the index entry
TMsvEntry indexEntry = iSmsMtm->Entry().Entry();
indexEntry.SetInPreparation(EFalse);
indexEntry.SetSendingState(KMsvSendStateWaiting);
iSmsMtm->Entry().ChangeL(indexEntry);
```

Finally you are ready to send the message. This is achieved as follows, with the code explained below:

```
// Now send
Cancel(); // prepare iOp for use
iEntrySelection->Reset();
iEntrySelection->AppendL(iSmsId);
TBuf8<1> dummyParams;
iOp = iSmsMtm->
InvokeAsyncFunctionL(ESmsMtmCommandScheduleCopy,
*iEntrySelection, dummyParams, iStatus);
    SetActive();
```

Examining this code line by line:

- The code is within a member function of CMtmsExampleEngine, which is an active object. As it will later involve issuing an asynchronous request, the first step is to cancel the active object to ensure that it is not still waiting on a previously issued request.

- iEntySelection is a CMsvEntrySelection—in other words, an array of entry IDs. This is reset to remove any existing contents, and then the ID of the SMS message added as its only element.

- An empty descriptor is declared. This is required as a parameter to InvokeAsyncFunctionL(), but as you do not make any use of it, an empty one will do.

- Now the asynchronous function is called. CBaseMtm defines two generic functions intended for use by derived classes to add MTM-specific extensions: InvokeSyncFunctionL() and InvokeAsyncFunctionL(). Each MTM defines a list of IDs (known as **opcodes**) which can be passed in by the client and handled in the server MTM. The opcode defined by the SMS MTM for sending is ESmsMtmCommandScheduleCopy. Along with this you pass the ID array specifying the messages you want to send (in this case, just one), the "dummy" descriptor parameter, and this object's iStatus.

- Note that InvokeAsyncFunctionL() returns a CMsvOperation object by pointer. While you are waiting for your active object to complete (that is, before the RunL() is called), you can use this object to obtain progress information about the send operation.

Chapter 10 Messaging

- Finally, the active object is set active to ensure that the active scheduler will schedule it when its request completes.

Now all you need to do is check the value of your `iStatus` in your `RunL()` to ensure that sending succeeded. You should also call `iOp->FinalProgress()` to check the final progress information stored in the operation object.

Using the MMS MTM

Use of the MMS MTM in Series 60 is very similar to the code you have seen above to make use of the SMS MTM. This section will just highlight the differences between the two message types.

A Brief Introduction to MMS

A Multimedia Messaging Service (**MMS**) message can contain a number of multimedia parts, such as images, sounds and videos. These, along with text parts, are usually packaged up into a **presentation** to display them to the user. A presentation might consist of just a single image with a text caption below it, or at a more advance level it can contain a number of **slides**, each containing several multimedia objects, and timing information to control transition between those slides.

The presentation part of the message, which defines the layout and sequence of the slides, usually takes the form of a document written in an XML language called **SMIL** (Synchronized Multimedia Integration Language, pronounced "smile"). This is often referred to as the "SMIL part." When you use the MMS MTM in Series 60, you do not need to add your own SMIL part explicitly—if you do not, a simple one will be generated for you.

Creating the Message

The process of creating and sending an MMS message begins, as it did for SMS, with setting the MTM's context to the *Drafts* folder, as this is where new message entries should be created until editing is complete. After this, call `CreateMessageL()`, passing the ID of your MMS service entry, to create a new message. The service entry ID is easily obtained from the MTM:

```
// Create a new message using the default service,
// and store its ID for later reference
iMmsMtm->CreateMessageL(iMmsMtm->DefaultSettingsL());
iMmsId = iMmsMtm->Entry().EntryId();
```

After this, you just need to set the various parts of your message, then save it. In most cases this can be performed directly with an API from `CMmsClientMtm`:

```
// Add any phone number and email addressees
iMmsMtm->AddAddresseeL(aAddress);

// Add attachments (also known as "media objects")
AddMMSAttachmentsL();
```

Chapter 10 Advanced Communication Technologies

```
// Set subject
iMmsMtm->SetSubjectL(KMMSSubject);

// Finally, save the message to commit changes to store
iMmsMtm->SaveMessageL();
```

Adding Attachments

As you can see, a bespoke function has been written for this class to add the attachments (or media objects) to the MMS message. The MMS MTM adds some MMS-specific functions for adding attachments to a message, and use of the function `CBaseMtm::CreateAttachmentL()` is discouraged. The `AddMMSAttachmentsL()` function used in this example looks like this:

```
// System includes
#include <aknutils.h>      // CompleteWithAppPath()
…

// String constants
_LIT(KMMSText, "Look! Some penguins!");

// Filenames
_LIT(KMMSImageFilename, "image.jpg");
…

void CMtmsExampleEngine::AddMMSAttachmentsL()
    {
    TMsvId attachmentId = KMsvNullIndexEntryId;

    // Add a JPEG attachment
    TFileName attachmentName(KMMSImageFilename);
    User::LeaveIfError(CompleteWithAppPath(attachmentName));
    iMmsMtm->CreateAttachment2L(attachmentId, attachmentName);

    // Add the text attachment
    iMmsMtm->CreateTextAttachmentL(attachmentId, KMMSText,
KMMSTextFilename);
    }
```

Using `CreateAttachment2L()` ensures that the image will be referenced in the message's SMIL part, and included as a visible part of the presentation received by the addressee. Using `CreateAttachmentL()` will result in a binary attachment which is not referenced by the SMIL part.

Here a SMIL part has not been added explicitly. SMIL is a fairly complicated language, and it is easy to write a SMIL document which is incorrect, resulting in a corrupt message. However, if you do wish to add a SMIL part explicitly, it is added in the same way as an image, but make sure you then set it as the "root" part of the message:

Chapter 10 Messaging

```
// System includes
#include <aknutils.h>        // CompleteWithAppPath()
…

// String constants
…
_LIT(KMMSTextFilename,  "hello.txt");


// Filenames
_LIT(KMMSImageFilename,  "image.jpg");
_LIT(KMMSSmilFilename,  "simple.smil");
…

void CMtmsExampleEngine::AddMMSAttachmentsL()
    {
    TMsvId attachmentId = KMsvNullIndexEntryId;

    // Add a JPEG attachment
    TFileName attachmentName(KMMSImageFilename);
    User::LeaveIfError(CompleteWithAppPath(attachmentName));
    iMmsMtm->CreateAttachment2L(attachmentId, attachmentName);

    // Add the text attachment
    iMmsMtm->CreateTextAttachmentL(attachmentId, KMMSText,
KMMSTextFilename);

    // Add the SMIL part
    attachmentName = KMMSSmilFilename;
    User::LeaveIfError(CompleteWithAppPath(attachmentName));
    iMmsMtm->CreateAttachment2L(attachmentId, attachmentName);
    iMmsMtm->SetMessageRootL(attachmentId);
    }
```

Note the difference here in adding the text part: namely, that you pass a third parameter which specifies a filename. This is because the SMIL part references the other parts by filename, and since you are providing a static SMIL file (rather than getting the MTM to create one for you), you need to ensure that the filename of the text part matches what is in the SMIL.

For your interest, here is the SMIL part being added for this message:

```
<smil xmlns="http://www.w3.org/2000/SMIL20/CR/Language">
   <head>
      <layout>
         <root-layout width="160" height="140"/>
         <region id="Image" width="160" height="120" left="0" top="0"/>
         <region id = "Text" width="160" height="20" left="0" top="120"/>
      </layout>
   </head>
```

```
<body>
    <par dur="5s">
        <img src="image.jpg" region="Image"/>
        <text src="hello.txt" region="Text"/>
    </par>
</body>
</smil>
```

Validating the Entry

After saving the new message, you validate it in exactly the same way as was shown for SMS.

Sending the Entry

Sending an MMS looks slightly different from the way it did for an SMS. This is primarily because the MMS MTM implements sending by means of a `SendL()` command, rather than using the base class's `InvokeAsyncFunctionL()` as the SMS MTM did. This notwithstanding, the steps needed to send your MMS message are pretty similar to those used for SMS:

```
void CMtmsExampleEngine::SendMMSL()
    {
    iMmsMtm->SwitchCurrentEntryL(iMmsId);

    // Mark the message as complete
    TMsvEntry indexEntry = iMmsMtm->Entry().Entry();
    indexEntry.SetInPreparation(EFalse);
    indexEntry.SetVisible(ETrue);

    // Commit changes to index entry
    iMmsMtm->Entry().ChangeL(indexEntry);

    // Now send
    Cancel(); // prepare iOp for use
    iOp = iMmsMtm->SendL(iStatus);
    SetActive();
    }
```

Using the Send-As API

The Send-As API offers a far simpler interface than the Client MTM API that you saw in the previous section, but it is also far less powerful. It allows you to create a message, add simple details to it, such as recipients, body text and attachments, and save it to the *Drafts* folder. Ironically, given its name, Send-As does not have an API for sending the messages it creates! This is seen as one of its major limitations.

The Send-As API is provided through the class `CSendAs`. This owns a whole host of messaging objects (including instances of `CMsvSession`, `CMsvEntry` and `CBaseMtm`) and takes care of constructing and initializing them.

Chapter 10 Messaging

The code you see in this section is taken from the **SendAsExample** example application. Much of it is from the `CSendAsExampleEngine` class. The member data of this class is extremely basic: just a `CSendAs` object (`iSendAs`) and a reference to an observer (`MSendAsExampleEngineObserver& iObserver`) which has been implemented to pass information back up to the AppUI.

Creating a Send-As Object

The first step the engine needs to carry out is to construct its `CSendAs` object. This is a one-liner—just a call to `CSendAs::NewL()`. This requires only one parameter, a reference to an `MSendAsObserver` object. The engine implements this mixin, so it just passes a reference to itself.

It is interesting to note that once a `CSendAs` object has been used to create and save a message, most further calls to it will result in a leave. Therefore, if you wish to create a second message, the object has to be destroyed and a new one constructed. For this reason, destruction and construction of `iSendAs` has been encapsulated in a utility function called `ResetL()`:

```
void CSendAsExampleEngine::ResetL()
    {
    delete iSendAs;
    iSendAs = NULL;
    iSendAs = CSendAs::NewL(*this);
    ...
    }
```

This is the function called from `ConstructL()` to construct `iSendAs` when the application starts up.

Setting Capabilities

One line was omitted from the previous code snippet (indicated by the ... line): `ResetL()` also calls `iSendAs->AddMtmCapabilityL()`. `CSendAs` maintains an array of available MTMs, which at construction is initialized to list all MTMs installed on the device. If your code requires certain capabilities which may not be supported by all MTMs, this API is used to specify them. As the **SendAsExample** example goes on to send an image as a message attachment, this function is called with `KUidMtmQuerySupportAttachments` (defined in `mtmuids.h`). Here is the fully implemented `ResetL()` function:

```
void CSendAsExampleEngine::ResetL()
    {
    delete iSendAs;
    iSendAs = NULL;
    iSendAs = CSendAs::NewL(*this);
    iSendAs->AddMtmCapabilityL(KUidMtmQuerySupportAttachments);
    }
```

On a device with email, MMS, SMS, infrared and Bluetooth MTMs installed, this will filter out the SMS MTM.

> **TIP** The value `KUidMtmQuerySupportedBody` can be passed to `AddMtmCapabilityL()` to filter out MTMs which don't support body text (for example, MMS). However, beware: the Bluetooth and infrared MTMs will be left in the list, and yet both will leave if you attempt to save a message that has body text.

Choosing an MTM

At some stage, you will need to select the MTM you wish to use to create your Send-As message. A common way of doing this is to prompt the user to choose. `CSendAs` has a function, `AvailableMtms()`, which returns a descriptor array containing human-readable names for the MTMs in the list. This can be used to populate a menu or listbox from which the user can choose their preferred MTM.

In the **SendAsExample** example application, these strings are put into a pop-up menu for the user to choose from. `CSendAsExampleEngine` defines an inline `AvailableMtms()` method which just calls the same function on `iSendAs`. The AppUi then implements its pop-up menu with the following code:

```
void CSendAsExampleAppUi::DynInitMenuPaneL(TInt aResourceId,
CEikMenuPane* aMenuPane)
    {
    if (aResourceId == R_SENDASEXAMPLE_SEND_SUBMENU)
        {
        const CDesCArray& mtms = iEngine->AvailableMtms();
        const TInt count = mtms.Count();

        if (count == 0)
            {
            return;
            }

        aMenuPane->SetItemDimmed(ESendAsExampleCmdDummy, ETrue);
        TInt commandId = ESendAsExampleCmdMtmBase;
        CEikMenuPaneItem::SData data;
        data.iCascadeId = 0;
        data.iFlags = 0;

        for (TInt i = 0; i < count; i++)
            {
            data.iCommandId = ++commandId;
            data.iText = mtms[i];
            aMenuPane->AddMenuItemL(data);
            }

        iMaxMtmCmdId = commandId;
        }
    }
```

Chapter 10 Messaging

Note that the menu, by default, contains a single item associated with the command ID `ESendAsExampleCmdDummy`, which is ignored in `HandleCommandL()`. The text of this item is set in resource to "Unavailable." So, if no MTMs are found, the user will see "Unavailable" on the menu, with no command attached to it. If MTMs are present, the above code removes this menu item before adding the MTM strings.

Note also that the command IDs used for the MTMs are dynamically allocated, starting at a defined base ID. `HandleCommandL()` is implemented to recognize these dynamic IDs and act appropriately:

```
void CSendAsExampleAppUi::HandleCommandL(TInt aCommand)
    {
    switch (aCommand)
        {
        case EAknSoftkeyBack:
        case EEikCmdExit:
            {
            Exit();
            break;
            }

        // Handle dynamic command IDs allocated to MTM list
        default:
            if (aCommand > ESendAsExampleCmdMtmBase &&
                aCommand <= iMaxMtmCmdId)
                {
                const TInt index = (
aCommand - ESendAsExampleCmdMtmBase) - 1;

                // Get an address
                TBuf<ESendAsExampleMaxAddressLength> address;
                CAknTextQueryDialog* dlg = new (ELeave)
CAknTextQueryDialog(address, CAknQueryDialog::ENoTone);

                TInt ret = dlg->
ExecuteLD(R_SENDASEXAMPLE_ADDRESS_QUERY);

                if (ret == EAknSoftkeyOk)
                    {
                    iEngine->CreateMessageL(index, address);
                    }
                }
            break;
        }

    // Reset iMaxMtmCmdId
    iMaxMtmCmdId = ESendAsExampleCmdMtmBase;
    }
```

Creating and Saving the Message

Once the user has chosen an MTM, you can create a message, validate it and save it to the *Drafts* folder. All the code in this section is taken from `CSendAsExampleEngine::CreateMessageL()`.

First, you need to set the MTM you wish to use. You can do this using an index calculated from the MTM chosen by the user from the pop-up menu (see above):

```
void CSendAsExampleEngine::CreateMessageL(
TInt aMtmIndex, const TDesC& aAddress)
    {
    // Set the MTM
    iSendAs->SetMtmL(aMtmIndex);
```

You also need to create a message and set parameters on it. In this case a recipient and an attachment are set. Note that the API to `CSendAs` is similar to that of `CBaseMtm`.

```
    // Create the message
    iSendAs->CreateMessageL();
    iSendAs->AddRecipientL(aAddress);

    // Add an attachment
    TMsvId attachmentId;
    TFileName filename(KAttachmentFilename);
    User::LeaveIfError(CompleteWithAppPath(filename));
    iSendAs->CreateAttachmentL(attachmentId, filename);
```

Finally, as with the Client MTM API, you can validate a message before saving it. If validation fails, the message is abandoned and an error reported to the observer (the `AppUI`):

```
    // Validate the message
    TMsvPartList msgFailParts = iSendAs->ValidateMessage();

    if (msgFailParts == KMsvMessagePartNone)
        {
        iSendAs->SaveMessageL(iStatus);
        SetActive();
        }
    else
        {
        iObserver.HandleSaveMessageCompleteL(KErrCorrupt);
        iSendAs->AbandonMessage();
        ResetL();
        }
```

Using the CSendAppUi Class

The previous section showed you Send-As, an API that is present on all Symbian OS platforms for making simple use of the messaging architecture. However, it was noted that Send-As is not well suited to interactive use of messaging, where you want the user to edit the message as they would in the phone's **Messages** application.

To address this shortcoming, Series 60 introduces the CSendAppUi class to provide interactive messaging capabilities to applications. CSendAppUi is a class containing functions used to invoke messaging UI capabilities. Specifically, it allows you to do the following:

- Invoke a message editor for a specific message type, optionally passing in data such as body text, addressees and attachments, as shown in Figure 10–3.

- Add a "Send" item to your application menu, to invoke a cascading menu which dynamically lists available message types, optionally filtered by capability, as shown in Figure 10–4. Selecting a message type invokes the appropriate editor.

- Implement a context-sensitive menu listing available message types— again, dynamically generated and optionally filtered by capability, as shown in Figure 10–5. Selecting a message type invokes the appropriate editor.

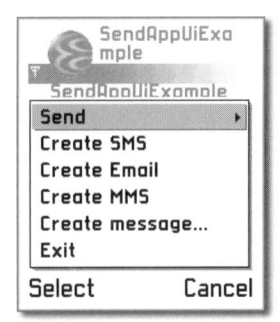

Figure 10–3 Main application menu of SendAppUiExample.

Figure 10–4 Cascading menu.

Figure 10–5 Context-sensitive menu invoked by the "Create Message…" menu item.

This section shows you how to use these features in your own application.

Note that it is not usual for an application to implement all the options available in this application. In particular (as you will see), the "Send" and "Create Message…" commands do more or less the same thing, and usually only the

"Send" command would be included on the application's main menu. Just be aware that this is an example application, and some of its functionality is intentionally duplicated to illustrate different ways of achieving the same thing.

> **WARNING**
>
> The name of the CSendAppUi class is misleading—it is not, in fact, an App UI! You might have expected that it would be a specialization of CAknAppUi, and that applications wishing to use its API should derive their own App UI class from it. Unfortunately, this is incorrect. In fact, CSendAppUi derives from CBase and has no special significance within the application architecture. (You would not get very far if you did try to derive your own App UI from it!) Common practice is for an application's App UI class to *own an instance of* CSendAppUi.

Getting Started

The code in this section is taken from the CSendAppUiExampleAppUi class in the **SendAppUiExample** example. Here is the relevant member data for that class:

```
private:    // Data
    CSendAppUi*             iSendAppUi;
    TSendingCapabilities    iSendAppUiCapabilities;
    CRichText*              iRichText;
```

iSendAppUi is the object you will use to provide the capabilities mentioned above. iSendAppUiCapabilities will be used throughout to specify a consistent set of capabilities which you wish to support in your user's messages. Both members are initialized in the CSendAppUiExampleAppUi::ConstructL() method:

```
    iSendAppUi = CSendAppUi::NewL(ECmdSendAppUiBaseCmdId);
    iSendAppUiCapabilities.iFlags =
TSendingCapabilities::ESupportsBodyText;
```

The value passed to CSendAppUi::NewL() is used as the starting value for the dynamic menu IDs which CSendAppUi creates—the dynamic values are assigned by incrementing this value. You should therefore ensure that you have no other command IDs with values higher than this, as they could clash with the dynamic IDs. The easiest way to do this is to define this value at the end of your command ID enum, as in SendAppUiExample.hrh:

```
// Command ids
enum
    {
    ECmdCreateSMS = 0x6000,
    ECmdCreateEmail,
    ECmdCreateMMS,
    ECmdCreateGeneralMessage,
    ECmdSendAppUiBaseCmdId
    };
```

Chapter 10 Messaging

`iSendAppUiCapabilities` is initialized to indicate that, for the purposes of this example, all messages should support body text.

Dynamic Menus

As well as the initialization performed at construction, you also need to put code in place to enable the dynamic menus that `CSendAppUi` will populate. Figure 10-6 shows which menu items in the application's main menu are specified in the resource file.

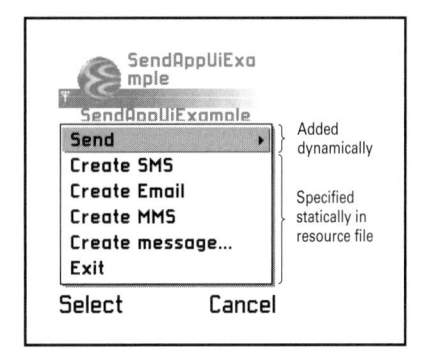

Figure 10-6 Dynamic and static menus in SendAppUiExample.

As you can see, just the **Send** menu item is added dynamically. This is performed in `CSendAppUiExampleAppUi::DynInitMenuPaneL()`:

```
void CSendAppUiExampleAppUi::DynInitMenuPaneL(
   TInt aMenuId, CEikMenuPane* aMenuPane)
   {
   // specify the sending capabilities that the message
   // types displayed must satisfy
   switch (aMenuId)
      {
      case R_SEND_EXAMPLE_MENU_PANE:
         iSendAppUi->DisplaySendMenuItemL(*aMenuPane, 0,
   iSendAppUiCapabilities);
         break;

      case R_SENDUI_MENU:
         iSendAppUi->DisplaySendCascadeMenuL(*aMenuPane,
   NULL);
         break;
      }
   }
```

`DynInitMenuPaneL()` is a standard App UI function used to dynamically alter a menu just before it appears. `aMenuId` gives the ID of the menu about to be shown. In this case, you are interested in modifying two menus:

- `R_SEND_EXAMPLE_MENU_PANE` identifies the application's main menu and is defined in this application's `.rsg` file. The subsequent call to `DisplaySendMenuItemL()` adds the "Send" item to it. The second parameter (0) indicates the position at which the item should appear on the menu. 0 represents the top of the menu, with subsequent menu items numbered sequentially (1, 2, 3…). So in this case, it is added as the top item.

- `R_SENDUI_MENU` identifies a dynamic, cascading menu which `CSendAppUi` creates for you and associates with the "Send" menu item. The call to `DisplaySendCascadeMenuL()` will populate the cascading menu with the list of available, filtered message types. The value of `R_SENDUI_MENU` is defined in the system header file `SendNorm.rsg`, so you will need to system include this in your own `.cpp` file.

Handling Commands

Now that you have implemented your dynamic menu code, you are ready to handle the commands that your application's menu will generate. As you would expect, this is performed in the framework function `HandleCommandL()`. In the case of the three "hard-coded" message types on the main menu, each is implemented by an appropriately named function in `CSendAppUiExampleAppUi`:

```
case ECmdCreateSMS:
   CreateSMSL();
   break;
case ECmdCreateEmail:
   CreateEmailL();
   break;
case ECmdCreateMMS:
   CreateMMSL();
   break;
```

These three functions are very similar, and each results in a call to the same function on `CSendAppUi`. `CreateEmailL()` is covered in detail later in this section.

The handler for the "Create message…" command looks very similar:

```
case ECmdCreateGeneralMessage:
   CreateGeneralMessageL();
   break;
```

`CreateGeneralMessageL()` will also be covered in detail later in this section — it assembles a body part, then calls a different API on `CSendAppUi` to show the menu listing available message types, as shown in Figure 10–5.

Finally, the command handler for the dynamic command IDs generated by `CSendAppUi` looks somewhat different:

```
default:
    // It should be a dynamic command ID
    // generated by SendAppUi.
    // If not, panic in debug.
    if (!HandleSendAppUiCommandL(aCommand))
        __ASSERT_DEBUG(EFalse, Panic(EBadCommandId));
    break;
```

Note that, unlike the previous handlers, this is not handled in response to a fixed command ID. Instead, any unhandled command ID is passed to the `HandleSendAppUiCommandL()` function, which returns `EFalse` if it is not one of the dynamic IDs generated by `CSendAppUi`, and otherwise handles it appropriately. This, too, is covered in detail later in this section.

Creating a Message

`CreateEmailL()`, `CreateSMSL()` and `CreateMMSL()` all follow the same basic sequence of steps:

- Construct some data representing parts of a message (for example, body text, addressees or attachments).

- Invoke `CSendAppUi` to create a new message, passing the data constructed in the previous step.

NOTE The message parts passed into `CSendAppUi` are optional—you do not need to pass any at all. You will remember that `CSendAppUi` is predominantly for interactive message creation: if you had to specify all the message parts in your code, it would be no great improvement over Send-As. However, this functionality exists because it is so commonly desired to prepopulate certain fields before letting the user take over.

The three functions differ only in the kind of data they preconstruct and in the message type ID passed to `CSendAppUi`. As they are so similar, only `CreateEmailL()` is detailed here.

The first step of this function is to create an array of addressees. As this is an example application, these are loaded from resource file:

```
// create an array for addresses and add two to it
CDesCArrayFlat* addressArray = new (ELeave)
CDesCArrayFlat(KDefaultGranularity);
CleanupStack::PushL(addressArray);

HBufC* string = StringLoader::LoadLC(R_EMAIL_ADDRESS1);
addressArray->AppendL(*string);
```

```
CleanupStack::PopAndDestroy(string);
string = NULL;

string = StringLoader::LoadLC(R_EMAIL_ADDRESS2);
addressArray->AppendL(*string);
CleanupStack::PopAndDestroy(string);
```

Once the descriptor array is populated with a couple of addresses, you are ready to get `CSendAppUi` to create a message. This is achieved using the function `CreateAndSendMessageL()`:

```
// Display the email editor with addresses
// already inserted
iSendAppUi->CreateAndSendMessageL(KUidMsgTypeSMTP, NULL,
NULL, KNullUid, addressArray, NULL, ETrue);

CleanupStack::PopAndDestroy(addressArray);
```

This switches into the standard Series 60 message editor for the specified message type (in this case, email). All the familiar functionality is available, such as pulling in addressees from the **Contacts** application and adding attachments from the phone's file system. Once the user has finished editing their message they have the same options as in the native **Messages** application: they can send the message, save it to *Drafts* or delete it.

A look at the declaration of `CSendAppUi::CreateAndSendMessageL()` (in `sendui.h`) reveals the purpose of its many parameters:

```
IMPORT_C virtual void CreateAndSendMessageL(
   const TUid aMtmUid,
   const CRichText* aBodyText = NULL,
   MDesC16Array* aAttachments = NULL,
   const TUid aBioTypeUid = KNullUid,
   MDesC16Array* aRealAddresses = NULL,
   MDesC16Array* aAliases = NULL,
   TBool aLaunchEmbedded = ETrue);
```

The only mandatory parameter is the MTM UID, to specify the type of message you wish to create. The subsequent parameters can all be used to specify optional message data, as you saw above with the addressee array. The final parameter (`aLaunchEmbedded`) specifies whether the editor should appear embedded in your application (that is, with your application's icon still present in the title bar), or exactly as it would when invoked from the **Messages** application (nonembedded). The most typical case is to specify `ETrue` here, but `CSendAppUiExampleAppUi::CreateSMSL()` passes `EFalse` so you can see what it looks like.

Note that `CSendAppUi` does **not** take ownership of any of the data passed to it — you are still responsible for deleting it.

CreateGeneralMessageL()

This function is called when you invoke the **Create Message...** menu item, and it results in a second menu listing the available message types, filtered by your criteria, as shown in Figure 10–5. The code is very similar to the `CreateEmailL()` code seen in the previous subsection. The main difference is in the call made to `CSendAppUi`. Here is the function in its entirety:

```
void CSendAppUiExampleAppUi::CreateGeneralMessageL()
    {
    __ASSERT_DEBUG(iRichText && iRichText->DocumentLength() == 0, Panic(ERichTextNotReady));

    // insert text into the message body
    HBufC* string = StringLoader::LoadLC(R_GENERAL_BODY_TEXT);
    iRichText->InsertL(0, *string);
    CleanupStack::PopAndDestroy(string);
    string = NULL;

    // set the title of the box which displays
    // the message types
    string =
StringLoader::LoadLC(R_GENERAL_MESSAGE_POPUP_TITLE);

    // display a list of the message types which
    // satisfy the specified
    // sending capabilities
    iSendAppUi->CreateAndSendMessagePopupQueryL(*string,
iSendAppUiCapabilities, iRichText, NULL, KNullUid, NULL,
NULL, NULL, ETrue);

    CleanupStack::PopAndDestroy(string);
    iRichText->Reset();
    }
```

There are only a few significant differences to note between this and the previous example:

- Instead of `CreateAndSendMessageL()`, the function calls `CreateAndSendMessagePopupQueryL()`. This adds the step of showing the pop-up list of message types before switching to the editor.

- Because you are not specifying the message type yourself, you pass a capabilities object to filter the list of message types you want to display. You will remember that this was initialized in `ConstructL()` to display only message types which support body text. (This is why MMS does not appear in the list—MMS does not technically support body text, although it may have a text attachment.)

- You also pass a descriptor as the first parameter to specify the title of the pop-up menu that appears.

Here is the declaration of `CreateAndSendMessagePopupQueryL()` from sendui.h:

```
IMPORT_C virtual void CreateAndSendMessagePopupQueryL(
   const TDesC& aTitleText,
   TSendingCapabilities aRequiredCapabilities,
   const CRichText* aBodyText = NULL,
   MDesC16Array* aAttachments = NULL,
   const TUid aBioTypeUid = KNullUid,
   MDesC16Array* aRealAddresses = NULL,
   MDesC16Array* aAliases = NULL,
   CArrayFix<TUid>* aMtmsToDim = NULL,
   TBool aLaunchEmbedded = ETrue);
```

Note that you can also optionally specify a list of MTMs which you wish to dim in the pop-up menu.

Once the user had chosen from the menu, the editor is displayed exactly as before.

HandleSendAppUiCommandL()

`HandleSendAppUiCommandL()` is the last of the functions called in `HandleCommandL()`, and it handles dynamic IDs generated by `CSendAppUi` for the items in the **Send** cascading menu, as shown in Figure 10-4. This function has a `TBool` return type, returning `EFalse` if the command ID passed to it is not recognized as a valid ID from the cascading menu. This check is made by calling `CSendAppUi::CommandIsValidL()`:

```
TBool CSendAppUiExampleAppUi::HandleSendAppUiCommandL(
   TInt aCommand)
   {
   // First check that this is a valid command
   if (!iSendAppUi->CommandIsValidL(aCommand,
iSendAppUiCapabilities))
       {
       return EFalse;
       }
```

After this, you can proceed to invoke the message editor in much the same way as before. You may choose to specify body text, as before. You should be careful to only set data which you have specified in your capabilities filter, otherwise you could try to add data to a message type that does not support it (such as trying to add attachments to an SMS).

```
   // insert text into the message body
   HBufC* bodyText = StringLoader::LoadLC(R_SMS_BODY_TEXT);
   iRichText->InsertL(0, *bodyText);
   CleanupStack::PopAndDestroy(bodyText);
```

```
    // display the editor associated with aCommand.
    // The displayed message will have some
    // text already inserted.
    iSendAppUi->CreateAndSendMessageL(aCommand, iRichText,
NULL, KNullUid, NULL, NULL, ETrue);

    iRichText->Reset();

    return ETrue;
```

As in `CreateEmailL()`, this calls `CSendAppUi::CreateAndSendMessageL()`, although here you are calling an overloaded version which takes a dynamic menu ID instead of an MTM. Since `CSendAppUi` constructed the dynamic menu, it is able to map the command ID onto an MTM itself. All other parameters to this overload of `CreateAndSendMessageL()` are identical to those seen earlier.

Watching for Incoming Messages

The previous sections have looked at various ways to create and send messages on a Series 60 device. This final section looks at the other side of the picture: how to receive notification when a new message arrives on the phone and obtain a handle on that message.

The code samples in this section are taken from the **MsgObserver** example application.

The APIs

The main classes involved in receiving message notification are `CMsvSession` and `MMsvSessionObserver`. In fact, these have already been introduced earlier in this chapter (see the subsection *Using the Client MTM API*). In that earlier subsection, you were more concerned with `CMsvSession` — here the observer API is of most interest.

`MMsvSessionObserver` is used to receive notification of events that occur in the messaging server. Some of the most commonly used of these events are:

- Successful completion of a call to `CMsvSession::OpenAsyncL()`.
- Creation of a new entry in the server.
- Changes to an existing entry in the server.

Notification of these events arrives through a call to `MMsvSessionObserver::HandleSessionEventL()`, which you need to implement in your own class. The whole API can be understood quite easily by looking at the declaration of `MMsvSessionObserver`, from `msvapi.h`:

```
class MMsvSessionObserver
    {
public:
    enum TMsvSessionEvent
        {
        EMsvEntriesCreated,
        EMsvEntriesChanged,
        EMsvEntriesDeleted,
        EMsvEntriesMoved,
        EMsvMtmGroupInstalled,
        EMsvMtmGroupDeInstalled,
        EMsvGeneralError,
        EMsvCloseSession,
        EMsvServerReady,
        EMsvServerFailedToStart,
        EMsvCorruptedIndexRebuilt,
        EMsvServerTerminated,
        EMsvMediaChanged,
        EMsvMediaUnavailable,
        EMsvMediaAvailable,
        EMsvMediaIncorrect,
        EMsvCorruptedIndexRebuilding
        };

public:
    virtual void HandleSessionEventL(TMsvSessionEvent aEvent,
TAny* aArg1, TAny* aArg2, TAny* aArg3)=0;
    };
```

Watching the Inbox

The **MsgObserver** example application watches for new messages in the *Inbox* and gets their body text when they arrive. Watching the *Inbox* is a common use of MMsvSessionObserver.

The first step required is to derive one of your classes from MMsvSessionObserver and open a session on the messaging server. In **MsgObserver**, this is performed by the AppUi class:

```
class CMsgObserverAppUi : public CAknAppUi,
    public MMsvSessionObserver
    {
    ...

private: //Data
    CMsgObserverContainer* iAppContainer;
    CMsvSession* iMsvSession;
    CMsvEntry* iMsvEntry;
    TMsvId iNewMessageId;
    };
```

The session is opened asynchronously in the `ConstructL()`, and construction of `iMsvEntry` is performed once this completes:

```
void CMsgObserverAppUi::ConstructL()
    {
    ...
    iMsvSession = CMsvSession::OpenAsyncL(*this);
    }

...

void CMsgObserverAppUi::HandleSessionEventL(TMsvSessionEvent
aEvent, TAny* aArg1, TAny* aArg2, TAny* /*aArg3*/)
    {
    switch (aEvent)
        {
        case EMsvServerReady:
            // Initialise iMsvEntry
            if (!iMsvEntry)
                {
                iMsvEntry = CMsvEntry::NewL(*iMsvSession,
KMsvGlobalInBoxIndexEntryId, TMsvSelectionOrdering());
                }
            break;

        ...
        }
    }
```

You are now ready to receive notifications from the messaging server. You just need to be aware of a feature that you may not have expected—when a new message arrives, you receive not one event but two. You receive first an `EMsvEntriesCreated` event and then an `EMsvEntriesChanged` event. This reflects the two distinct operations carried out by the messaging server on receipt of a message:

1. Create a new, empty entry of type "message." (Generates an `EMsvEntriesCreated` event.)

2. Use the storage locations associated with this entry (index entry, store and/or file folder) to store the message's data. (Generates an `EMsvEntriesChanged` event on completion.)

It is important that you understand the sequence of these two steps, because—as you can now see—it is no good just watching for `EMsvEntriesCreated` events. At the point when this event arrives, you cannot read any data from the entry, because it has not been stored yet. Likewise, you cannot just look out for `EMsvEntriesChanged` events in the *Inbox*, because you will have no way of distinguishing the arrival of a new message from a change to an existing message. So, instead, you need to do the following:

- Handle the `EMsvEntriesCreated` event and store the ID of the new entry. (The `aArg1` parameter points to an array of IDs, typically containing just one element.)

- Handle the `EMsvEntriesChanged` event and compare the ID of the changed entry (again, available via `aArg1`) with the ID you stored previously. If the two are equal, you now have a handle to the newly created and stored message.

This sequence of steps is shown in this abridged section of `CMsgObserverAppUi::HandleSessionEventL()`. Note that in both of these events `aArg2` identifies the ID of the entry's parent folder. You can use this to filter events to a specific folder—in this case, the *Inbox*.

```
case EMsvEntriesCreated:
   // Only look for changes in the Inbox
   if (*(static_cast<TMsvId*>(aArg2)) == KObservedFolderId)
      {
      CMsvEntrySelection* entries =
static_cast<CMsvEntrySelection*>(aArg1);
      iNewMessageId = entries->At(0);
      }
   break;

case EMsvEntriesChanged:
   // Only look for changes in the Inbox
   if (*(static_cast<TMsvId*>(aArg2)) == KObservedFolderId)
      {
      CMsvEntrySelection* entries =
static_cast<CMsvEntrySelection*>(aArg1);

      if (iNewMessageId == entries->At(0))
         {
         // It's the same message we received the
         // EMsvEntriesCreated event for
         ...
         }
      }
   break;
```

Finally, here is the code used in the **MsgObserver** example to handle the message. It uses the `CMsvStore::RestoreBodyTextL()` function to restore the message's body text to a rich text object, then passes it to the UI for display.

```
// Set entry context to the new message
iMsvEntry->SetEntryL(iNewMessageId);

// Open the store, read-only
CMsvStore* store = iMsvEntry->ReadStoreL();
CleanupStack::PushL(store);

// Get body text and send it to the container
```

```
if (store->HasBodyTextL())
    {
    CRichText* richText = CRichText::NewL(
        iEikonEnv->SystemParaFormatLayerL(),
        iEikonEnv->SystemCharFormatLayerL());

    CleanupStack::PushL(richText);
    store->RestoreBodyTextL(*richText);
    const TInt length = richText->DocumentLength();
    iAppContainer->SetTextL(richText->Read(0, length));
    CleanupStack::PopAndDestroy(richText);
    }
else
    {
    iAppContainer->SetTextL(KNoBodyText);
    }
CleanupStack::PopAndDestroy(store);
```

Telephony

Series 60 provides APIs that allow you to make use of its telephony features. The telephony services in Series 60 are provided by a server called **ETel**. ETel is very similar in design to the other servers described in the previous chapter, such as the Socket Server and the Comms Server:

- It provides a generic, high-level client API that abstracts clients from the differences in implementation detail between specific hardware types.
- Plug-in modules (here known as **TSY**s) are used server-side to add support for different hardware types, without affecting the client-side API.
- ETel provides safe management of multiple, concurrent client connections.

The ETel client-side API consists of a session class `RTelServer` plus subsession classes to implement three fundamental abstractions: **phones**, **lines** and **calls**.

A phone is represented by the subsession class `RPhone`. This represents a particular telephony device. Its API provides access to the status and capabilities of the device and allows the client to be notified if these change.

A line is represented by the subsession class `RLine`. A phone can have one or more lines. `RLine` presents a single line and can be used to obtain status and capability information about it (for example, its hook status). As with `RPhone`, it also provides the capability to be notified if changes occur to the line.

A call is represented by the subsession class `RCall`. A line may have zero or more active calls, and an `RCall` object represents one such call. The `RCall` API is used to perform tasks such as dialing a number, waiting for an incoming call, and hanging up a call. As with the other subsessions, it can also be used to notify a client of changes to a call.

Using the ETel API

The code samples in this section come from the **AnsPhone** example application, which demonstrates many common ETel features. The code is mostly taken from the following classes:

- `CAnsPhoneEngine`—main engine class.
- `CAnsPhoneCallMaker`—responsible for making calls.
- `CAnsPhoneCallWatcher`—responsible for receiving calls.
- `CAnsPhoneCallLog`—responsible for retrieving information about logged calls.

The **AnsPhone** example application uses these and other classes to perform the following tasks:

- Making a call.
- Receiving a call.
- Retrieving the last calling number.

These steps are described in detail in the following sections, with reference to the code where appropriate.

Getting Started

Before you can do any real work with ETel, you first need to set up your `RTelServer` session and a phone and line. This is performed in the `CAnsPhoneEngine` class, and this class's key member data is shown here:

```
RTelServer         iSession;
RLine              iLine;
RAnsPhonePhone     iPhone;
```

Note that `RAnsPhonePhone` is a thin class, publicly derived directly from `RPhone` and adding one or two additional capabilities.

`CAnsPhoneEngine` initializes the session and subsession objects in the function `CAnsPhoneEngine::TelStartL()`. First, it opens the `RTelServer` session:

```
_LIT(KTSY, "phonetsy");

...

void CAnsPhoneEngine::TelStartL()
    {

    ...
```

Chapter 10 Telephony

```
User::LeaveIfError(iSession.Connect());
// load the appropriate tsy
User::LeaveIfError(iSession.LoadPhoneModule(KTSY));
```

The function then lists the phones supported by the device and selects the first one from the resulting list:

```
// in order to get a handle on a line, you must get a
// handle on an RPhone object
TInt numberPhones = 0;
User::LeaveIfError(iSession.EnumeratePhones(numberPhones));

if (!numberPhones)
    {
    User::Leave(KErrNotFound);
    }

// use the 1st available phone
RTelServer::TPhoneInfo phoneInfo;
User::LeaveIfError(iSession.GetPhoneInfo(0, phoneInfo));
User::LeaveIfError(iPhone.Open(iSession, phoneInfo.iName));
```

Once the phone is opened, you need to enumerate the lines on that phone and find one that supports voice calls. Once you find a suitable line, you can open it:

```
// we must now find a line that will accept voice calls
TInt numberLines = 0;
User::LeaveIfError(iPhone.EnumerateLines(numberLines));
RPhone::TLineInfo lineInfo;
TBool foundLine = EFalse;

for (TInt a = 0; a < numberLines; a++)
    {
    User::LeaveIfError(iPhone.GetLineInfo(a, lineInfo));

    if (lineInfo.iLineCapsFlags & RLine::KCapsVoice)
        {
        foundLine = ETrue;
        break;
        }
    }

if (!foundLine)
    {
    User::Leave(KErrNotFound);
    }

User::LeaveIfError(iLine.Open(iPhone, lineInfo.iName));
}
```

Once successfully opened, this `RLine` handle is passed into classes like `CAnsPhoneCallMaker` and `CAnsPhoneCallWatcher` which use it to open an `RCall` object.

Making a Call

This step, as you may expect, is performed by the `CAnsPhoneCallMaker` class. This is an active object with the following key members:

```
private:
   // the state of this object governs what happens when the
   // StartL()/Stop()/RunL()/DoCancel() functions are called

   enum TState { ENoState, EDialling, EWatching };

private:
   MAnsPhoneCallMakerObserver&    iObserver;
   RLine&                         iLine;
   RCall                          iCall;
   RCall::TStatus                 iCallStatus;
   TState                         iState;
   TPtrC                          iNumber;
```

`iLine` is a reference to the `RLine` object which was opened in `CAnsPhoneEngine::TelStartL()`. `MAnsPhoneCallMakerObserver` is an observer class defined along with this class to allow `CAnsPhoneCallMaker` to pass events elsewhere in the application. Its detail is peripheral to this chapter, but you can follow it easily in the **AnsPhone** example code.

A number of asynchronous calls are involved in making a call with ETel. These are all handled by `CAnsPhoneCallMaker`, which means that its `RunL()` must be able to determine which asynchronous call has completed at any given time. In order to facilitate this, `CAnsPhoneCallMaker` uses the `iState` member to implement a simple state machine. Before each asynchronous call, the state is set appropriately, and it can then be checked at the start of the `RunL()` method.

Here is a high-level outline of the steps involved in making a call:

1. Open an `RCall` object.
2. Commence dialing on this `RCall` object. This is an asynchronous request.
3. When the call to commence dialing completes, check whether the caller has already hung up.
4. If not, issue a second asynchronous request to watch for the end of the call.
5. When this request completes, take appropriate steps to handle the end of the call.

These steps are covered in more detail below.

Steps 1 & 2: Open RCall and Dial

The process begins by opening the `RCall` object, then commencing dialing. This is performed by the `CAnsPhoneCallMaker::MakeCallL()` function, which looks like this:

```
void CAnsPhoneCallMaker::MakeCallL(const TDesC& aNumber)
    {
    iState = EDialling;
    iNumber.Set(aNumber);
    StartL();
    }
```

`StartL()` switches on `iState`. For the state `EDialling`, it opens its `RCall` object and commences dialing:

```
User::LeaveIfError(iCall.OpenNewCall(iLine));
iCall.Dial(iStatus, iNumber);
```

Finally, it sets itself active. Now it has to wait until its `RunL()` is invoked to see whether dialing was successful.

Step 3: Check If the Caller Has Hung Up

When `CAnsPhoneCallMaker::RunL()` is called, it first checks to see whether the caller has hung up before connecting:

```
// if the caller hangs up, then we will get
// KErrGeneral error here
if (iState == EDialling && iStatus.Int() == KErrGeneral)
    {
    Stop();
    iObserver.HandleCallHungUpL();
    return;
    }
```

Step 4: Watch for the End of the Call

Having ascertained that the asynchronous call to `RCall::Dial()` was successful, `CAnsPhoneCallMaker` now has to watch for the end of the call. It does this by advancing `iState` to `EWaiting`, then calling `StartL()`. This in turn calls the appropriate ETel API:

```
iCall.NotifyStatusChange(iStatus, iCallStatus);
...
SetActive();
```

Step 5: Handle the End of the Call

Now all that remains is to handle the end of the call. This is detected when `RunL()` is called with the following conditions:

- `iState` is equal to `EWatching`.
- `iCallStatus` is equal to `RCall::EStatusHangingUp` or `RCall::EStatusIdle`.

When this occurs, you need to cancel your request for status information on `iCall` and close the handle. This is performed in `CAnsPhoneCallMaker::Stop()`:

```
iCall.NotifyStatusChangeCancel();
iCall.Close();
```

Receiving a Call

Receiving a call is very similar to making a call: it, too, consists of a basic state machine built around a number of asynchronous requests to ETel. In the **AnsPhone** example, this state machine is encapsulated in the `CAnsPhoneCallWatcher` class. Here is its relevant member data:

```
private:
    // the state of this object governs what happens when
    // the StartL() / Stop() / RunL() / DoCancel()
    // functions are called.

    enum TState { EWaiting, EAnswering, EWatching };
private:
    MAnsPhoneCallWatcherObserver&   iObserver;
    RLine&                          iLine;
    RCall                           iCall;
    TName                           iCallName;
    RCall::TStatus                  iCallStatus;
    TState                          iState;
```

By now most of this should be familiar from `CAnsPhoneCallMaker`, above. The only new addition is `iCallName`, a descriptor which will be used to store the name that ETel assigns to an incoming call.

Here the state machine is obviously slightly different from that required for making a call. Here are its basic steps:

1. Wait on `iLine` for a call to come in. This is an asynchronous request.

2. When a call comes in, attempt to answer it. (Another asynchronous request.)

Chapter 10 Telephony

3. If answering completes successfully, wait for the end of the call. (A third asynchronous request.)

4. Finally, take appropriate steps to handle the end of the call.

These steps are covered in detail below.

Step 1: Wait for an Incoming Call

This is performed as follows in `CAnsPhoneCallWatcher::StartL()`:

```
// sets iCallName when it receives an incoming call
iLine.NotifyIncomingCall(iStatus, iCallName);
```

This is all that is required. Remember that `iLine` has already been opened by `CAnsPhoneEngine`.

Step 2: Attempt to Answer a Call

When `NotifyIncomingCall()` completes successfully, indicating that an incoming call has been received, two steps are required to handle it: open a call handle, and use it to answer the call. This is performed in `CAnsPhoneCallWatcher::StartL()`:

```
// a call has come in, so start watching the call
User::LeaveIfError(iCall.OpenExistingCall(iLine,
iCallName));
   iCall.AnswerIncomingCall(iStatus);
   ...
   SetActive();
```

Note that `iCallName`, which was earlier passed to `RLine::NotifyIncomingCall()`, has now been initialized by ETel and is used to identify the call you wish to answer.

Step 3: Wait for the Call to End

If the call to `RCall::AnswerIncomingCall()` completes successfully, then you just need to wait for the call to end. Again, this is performed in `CAnsPhoneCallWatcher::StartL()`, and in exactly the same way that `CAnsPhoneCallMaker` watches for the end of an outgoing call:

```
iCall.NotifyStatusChange(iStatus, iCallStatus);
   ...
   SetActive();
```

Step 4: Handle the End of the Call

At the end of a call, you should perform the same shutdown steps as in `CAnsPhoneCallMaker` and also return to the waiting state if you intend to listen for further incoming calls. This is carried out in `CAnsPhoneCallWatcher::RunL()`:

Chapter 10 Advanced Communication Technologies

```
User::LeaveIfError(iCall.GetStatus(iCallStatus));
...
if (iCallStatus == RCall::EStatusHangingUp)
    {
    ...
    iCall.Close();
    iState = EWaiting;
    }
StartL();
```

Setting `iState` to `EWaiting` and calling `StartL()` restarts the state machine.

Retrieving the Last Calling Number

This is the only part of this section not to use an ETel API, as ETel does not provide facilities for retrieving the last calling number. Instead, you need to use the **Log Engine** API. The **Log Engine** is responsible for storing details of various system events (including incoming and outgoing phone calls) in a database.

The class which encapsulates this functionality in the **AnsPhone** example is `CAnsPhoneCallLog`. This is an active object, with the following key data members:

```
MAnsPhoneCallLogObserver&    iObserver;
CLogClient*                  iSession;
CLogViewRecent*              iRecentView;
```

`MAnsPhoneCallLogObserver` is a locally defined observer interface to allow other objects in **AnsPhone** to receive information from `CAnsPhoneCallLog`.

`CLogClient` is a high-level session class to the **Log Engine**—it encapsulates an `R`-class session and adds further functionality.

`CLogViewRecent` is a view on the **Log Engine** database, not a view in the UI sense. This is used to retrieve data from the **Log Engine**.

Both of the **Log Engine** classes are active objects.

Getting the Number

Components in **AnsPhone** obtain the last calling number by calling `CAnsPhoneCallLog::GetNumberL()`. This asks `iRecentView` for a list of recent events as follows:

```
    if (iRecentView->IsActive())
       iRecentView->Cancel();

    if (iRecentView->SetRecentListL(KLogNullRecentList,
iStatus))
       SetActive();
```

`CLogViewRecent` potentially has access to a number of event lists—the value `KLogNullRecentList` passed as the first parameter indicates that the view is to include events from all lists.

You should also note that `SetRecentListL()` returns `TBool` (rather than the `void` return type you would normally expect from an asynchronous function). This is used to indicate whether the log contains any events—if the function is called before any call information has been logged (that is before a call has been made/received *and hung up*), it returns `EFalse` and *does not issue an asynchronous request*. So it is important that you only call the handling active object's `SetActive()` function if `SetRecentListL()` returns `ETrue`.

Once the `SetRecentListL()` request completes, obtaining the last calling number is very straightforward. This is how it is implemented in `CAnsPhoneCallLog::RunL()`:

```
const CLogEvent& event = iRecentView->Event();
iObserver.HandlePhoneNumberL(event.Number());
```

Summary

This chapter has covered the higher-level communications APIs provided by Series 60. These allow developers to access sophisticated communications capabilities without needing to make direct use of relatively low-level technologies such as sockets and serial communications.

The HTTP section introduced the HTTP Client API added in Series 60 2.x. This can be used to create an HTTP session, make a simple request/response transaction, and receive notification of events occurring during the processing of that transaction.

The WAP section discussed the WAP protocol implementation, WAP stack architecture and connectionless versus connection-oriented sessions. The different APIs provided by Series 60 1.x and Series 60 2.x were also examined.

As shown in the Messaging section, Series 60 provides a number of APIs for creating and sending messages. The Client MTM API is the most powerful and fully featured but is also rather complex. For developers who just need to create simple messages and store them in the *Drafts* folder, the Send-As API was introduced. Finally, the `CSendAppUi` class was introduced as a convenient way to incorporate creation and sending functionality into an application's UI.

To complete the chapter, the Telephony section examined some common Telephony APIs: initiating calls, receiving incoming calls, and obtaining the phone's last calling number.

chapter 11

Multimedia: Graphics and Audio

Using the advanced multimedia facilities of Series 60 to enable sound and graphics in your application

Chapter 11 Multimedia: Graphics and Audio

The multimedia architecture of Series 60 has been designed and optimized for mobile devices. While system resources may at times be constrained, the multimedia features available to developers are far from so, having a rich feature set more akin to a desktop computing environment. With relative ease the different components can be used for numerous tasks, ranging from drawing simple shape primitives to playing ring tones.

The media architecture has been updated for Series 60 2.x, and many related APIs have been deprecated. Although backward compatibility has been maintained where possible, the specific APIs for Series 60 2.x should generally be used when developing applications for Series 60 2.x devices. In this chapter, where APIs are platform specific, this will be clearly marked and both implementations will be discussed.

This chapter covers the following main topics:

- **Series 60 Graphics Architecture**—The three key components of the graphics architecture are the Window Server, Font and Bitmap Server and Multi Media Server. The role of each Server and the interaction between them is covered.

- **Basic Drawing**—All the essential APIs for any type of drawing application are included here, such as changing the brush type, the pen color and drawing lines. The screen coordinates and geometry are explained and the graphics context introduced.

- **Fonts and Text**—A number of fonts are available, and effects such as underlining can be applied.

- **Shapes**—Regular shapes including rectangles, ellipses and polygons can all be drawn easily with the Series 60 APIs. Arcs and pies can also be drawn.

- **Bitmaps**—APIs are provided for the loading and manipulation of bitmaps for display in your application. The technique of bitmap masking, which is important for good bitmap presentation, is also explained here.

- **Animation**—This section describes the animation architecture, and techniques that help to produce smooth animations, such as using off-screen bitmaps, and double buffering. The Direct Screen Access Framework, which is particularly useful for fast animation games programming, is also introduced.

- **Image Manipulation**—Images can be easily rotated and scaled, and APIs are available to convert between image types.
- **Audio**—This section concentrates on adding sound to your application. Both recording sound and playing sound are covered.

The following example applications accompany this chapter in order to illustrate the functionality discussed:

- **BasicDrawing**—A simple application that draws different types of colored line across the screen.
- **FontsAndText**—An application that renders text and manages fonts.
- **Shapes**—Draws bar charts, pie charts, and line charts to represent a particular data set, illustrating how to draw many common shapes such as rectangles and circles.
- **Bitmap**—An application to show you how to use bitmaps.
- **Animation**—Demonstrates a basic animation sequence.
- **ClientAnimation**—Shows you how to perform double-buffered animation.
- **Skiing**—A simple skiing game that demonstrates the use of the Direct Screen Access Framework.
- **ImageManip**—An application that allows an image to be rotated, scaled, and converted into a different image format.
- **MultiMediaF**—A Series 60 2.x-specific version of **ImageManip** that uses the Multi Media Framework.
- **AnsPhone**—An application to record and play back telephone messages.
- **AudioPlayer**—An application that can play simple tones or streamed audio data.

It is desirable that you have a basic comprehension of computer graphics before reading this chapter. From a Series 60 standpoint, a basic appreciation of the Series 60 Application Framework, as covered in Chapter 4, is also desirable—not necessarily for understanding the concepts presented here, but more for being aware of the presence of certain classes within the example applications.

The implementations of some APIs are asynchronous in nature, and so it is recommended that you read the **Using Asynchronous Services with Active Objects** section of Chapter 3 before trying to use such APIs. This will also help you understand why certain functions work the way they do.

Overview of Series 60 Graphics Architecture

Graphics programming in Series 60 focuses primarily on interactions with the three servers: the **Window Server**, the **Font and Bitmap Server** and the **Multi Media Server**. Each plays a vital role in providing an all-encompassing set of APIs for performing virtually any kind of graphics operation.

The Window Server gives client applications access to the essential input and output resources of the device keypad and screen, respectively. The Font and Bitmap Server, as its name suggests, manages fonts and bitmaps in memory, so that they can be used in the most efficient manner possible. Finally, the Multi Media Server is employed for image conversion and manipulation.

While communication with these various components does involve a Client/Server relationship, many of the provided classes abstract away the complexities of this, and the Client/Server interactions are effectively hidden.

Window Server

The Window Server enables applications to interact with a device's screen and its input mechanisms, such as the keypad. Consequently, any application with a user interface will need to use the Window Server. Client applications will invoke asynchronous requests, via an interface to the Window Server, which decides on the application that should have access to the device screen, and receive user input.

This interface is a "session" with the Window Server and is represented by the RWsSession class. Figure 11-1 shows how the Window Server controls device hardware through these sessions.

The aspects of RWsSession that concern you are: windows, events, and the client-side buffer.

- Windows represent an area on the screen that can be drawn to. Applications will redraw using the RWindow class, which represents a handle to a standard window, derived from the abstract base class RDrawableWindow. The various types of window are extensively illustrated in the SDK documentation.

- There are three types of Window Server event that applications can be made aware of: a priority key press, such as the power button, triggers a TWsPriorityKeyEvent; a standard key press triggers a TWsEvent; and a TWsRedrawEvent occurs when a redraw is necessary.

- The Client/Server nature of the Window Server means that, in order for it to carry out drawing operations, a context switch is necessary from your application. There can be severe overhead in this, so a buffer of drawing events is maintained client-side—in other words, in the application. This is dispatched to the Window Server in one go, thereby minimizing the number of context switches. Flushing of this buffer takes place for a variety of

Chapter 11 Overview of Series 60 Graphics Architecture

reasons, but generally it will happen when the buffer is full or you have explicitly called `RWsSession::Flush()`.

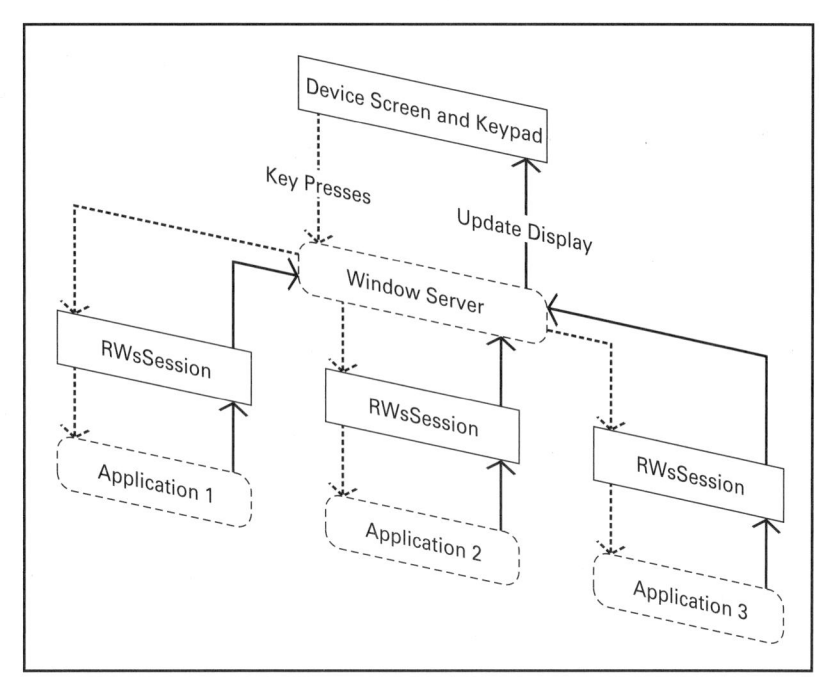

Figure 11-1 The flow of control for applications through sessions with the Window Server.

TIP The presence of the client-side buffer means that when debugging, stepping through a drawing command will not result in the emulator display being updated. To overcome this, the debugging hotkey sequence **Ctrl+Alt+Shift+F** can be used to enable autoflushing of the client-side buffer. It should be noted that this will be detrimental to performance if permanently enabled, however it is a useful debugging aid. A full list of emulator keyboard shortcuts is available in Appendix A.

In truth, discussing `RWsSession` is analyzing matters at a very "low level," and any application that employs the Series 60 application framework will have access to a variety of classes that hide such explicit Window Server interaction. These classes allow for relatively straightforward drawing to, and interrogation of, the screen, and they are conceptually similar to the mechanisms employed on other platforms. For example, classes derived from `CGraphicsContext` are analogous to the `Graphics` class in Java. Occasionally, however, you will be required to pass `RWsSession` objects around your code—so that these classes can have access to the Window Server—and so an appreciation of `RWsSession` is useful.

Font and Bitmap Server

The Font and Bitmap Server exists to allow fonts and bitmaps to be used in a manner that is both memory and speed efficient. As with the Window Server, applications are clients of the Font and Bitmap Server, and this allows fonts and bitmaps to be shared across the system, thereby reducing memory overhead.

Fonts and bitmaps are loaded into memory once, and a reference counter is used to determine how many clients require access to each resource. When the reference counter is zero, it indicates the resource is no longer needed and can be removed from memory. Parallels with the Window Server are again evident, with the `RFbsSession` class performing a function similar to that of `RWsSession`, in communications with the Font and Bitmap Server. However, in practice, an `RFbsSession` object is rarely seen, as the more readily usable classes of the Font and Bitmap Server, such as `CFbsBitmap` (which represents a bitmap), encapsulate it.

The Font and Bitmap Server plays another vital role in providing off-screen memory for bitmap manipulation, and in implementing double buffering, which is essential for flicker-free redrawing—for example, while animating. By using the `CFbsBitmapDevice`, `CFbsBitGc` and `CFbsBitmap` classes together, you can perform off-screen graphics operations. This is demonstrated in the **Bitmaps** section of this chapter.

The Window Server and the Font and Bitmap Server

Viewing the Window Server and the Font and Bitmap Server might lead you to conclude that they exist as distinct entities. While it is true that their roles differ, ultimately it is the coming together of the two that allows you to perform serious graphics operations on a Series 60 device. If graphics are prepared off screen using the Font and Bitmap Server classes, the Window Server classes will be needed in order to actually draw them to the screen. Indeed, the construction of an `RFbsSession` takes place when an `RWsSession` is created. Consequently, all Window Server clients are also Font and Bitmap Server clients.

Although the Window Server provides access to the device screen, this is not the same as having direct access to the screen. This is achieved through specialized classes and is discussed, with an example, in the **Direct Screen Access** section of this chapter.

Multi Media Server

While the Window Server and Font and Bitmap Server are somewhat entwined, the Multi Media Server is more independent, playing a key role in the management of images by providing the tools for image conversion and manipulation. Using the specialist classes of the Multi Media Server, images can be converted to and from a variety of industry standard file formats. The image formats supported are shown in Table 11–1.

Chapter 11 Overview of Series 60 Graphics Architecture

Table 11-1 Image Formats Supported by the Multi Media Server

Image Format	
BMP (Bitmap)	TIFF (Tagged Image File Format)
GIF (Graphics Interchange Format)	WBMP (Wireless Bitmap)
MBM (Multi Bitmap)	ICO (Icon)
PNG (Portable Network Graphics)	WMF (Windows Meta File)
SMS OTA (Over the Air)	

At the same time, classes are provided to perform rotation and scaling of images. Ultimately, images loaded and manipulated by the Multi Media Server will need to be stored in memory to be of practical use for display and interrogation purposes. This is achieved by using the bitmap classes provided, so once again there is a degree of crossover between the key graphics Servers.

The libraries to which the various components mentioned so far belong are listed in Table 11-2.

Table 11-2 Series 60 Graphics Libraries*

Library (.lib)	Role	Key Elements
GDI	Basic framework from which all graphics derive.	Fonts, geometry, color, graphics devices and contexts.
BITGDI	Basic framework from which all bitmapped graphics derive.	Bitmapped graphics devices and graphics contexts.
FBSCLI	Font and Bitmap Server client-side.	Bitmaps, fonts, Font and Bitmap Server sessions.
WS32	Window Server client-side.	Window types, Window Server sessions, Window Server bitmaps, Window Server graphics contexts and devices.
MEDIACLIENTIMAGE	Multi Media Server client-side.	Image conversion, rotation and scaling.

*Audio libraries are covered in Table 11-13.

Basic Drawing

This section looks at the classes that form the foundations for performing basic drawing. You should be familiar with the application framework, as shown in Chapter 4, and also with the use of controls, as shown in Chapter 5, since most of the examples will perform drawing within a `CCoeControl`-derived class. This hides away most of the complexities of explicitly having to create a Window Server client, and it provides a window to draw to.

You can create your own base for drawing without `CCoeControl`, but this is more complex and is unnecessary for illustrating the basic concepts. The function calls for drawing shapes, bitmaps and fonts are exactly the same, whether or not a `CCoeControl` is used, as drawing always takes place to a window using a graphics context.

The following code snippets are taken from the **BasicDrawing** application's `Draw()` function, and this simply draws some horizontal lines on the screen with differing colors and widths, as shown in Figure 11–2.

Initially a handle on the system graphics context is obtained. A graphics context performs various drawing functions, and later in the function it will be used to draw lines between two points on the screen using the `DrawLine()` method:

Figure 11–2 Basic drawing screen.

```
void CBasicDrawingContainer::Draw(const TRect& /*aRect*/)
const
    {
    // Get graphics context and clear it
    CWindowGc& gc = SystemGc();
    gc.Clear();
    ...
```

The `TPoint` class is used to represent coordinate points on the screen, while `TSize` represents a graphical object's width and height:

```
...
// Set up points and increment value for drawing
TPoint point1(KPoint1Begin, KPoint1End);
TPoint point2(KPoint2Begin, KPoint2End);

// Set up pen sizes for drawing
TSize penSize(KPenSize, KPenSize);
...
```

In order to make the drawn lines visible, the graphics context's pen is set to an appropriate color. The `TRgb` class represents color, and colors are defined using the `AKN_LAF_COLOR()` macro (see the subsection *Color and Display Modes* later in this chapter).

Chapter 11 Basic Drawing

The pen is used to determine the physical appearance of the line to be drawn. A pen has a style property, and this sets whether the line is solid, dotted or dashed. A line is then drawn on the screen from the location specified by the *x* and *y* coordinates belonging to `point1` to the location specified by `point2`:

```
…
//Set up drawing colors by using the AKN_LAF_COLOR() macro
TRgb colorBlue = AKN_LAF_COLOR(KColorBlue);
TRgb colorRed = AKN_LAF_COLOR(KColorRed);
TRgb colorGreen = AKN_LAF_COLOR(KColorGreen);
TRgb colorYellow = AKN_LAF_COLOR(KColorYellow);

// Set pen to be blue and solid. Then draw line
gc.SetPenColor(colorBlue);
// make pen solid
gc.SetPenStyle(CGraphicsContext::ESolidPen);
gc.DrawLine(point1, point2);
…
```

The function continues with similar operations being performed. In the first block the pen color is changed from blue to red, and the pen style is updated. You can also see that the pen size is altered using the `TSize` object, `penSize`. `TSize` is a generic class embodying size and is used in conjunction with various objects that are drawn to the screen, such as shapes, bitmaps and fonts.

`TPoint` and `TSize` have their mathematical operators overloaded; this allows a developer to quickly apply transformations, such as increasing the pen size and incrementing the point coordinates:

```
…
// Draw Red Line, increase pen size and set pen to be
dotted
  gc.SetPenColor(colorRed);
  // make pen dotted
  gc.SetPenStyle(CGraphicsContext::EDottedPen);
  // set pen size to (2,2)
  gc.SetPenSize(penSize);
  // increment starting point of line
  point1 += KIncrementPoint;
  // increment ending point of line
  point2 += KIncrementPoint;
  gc.DrawLine(point1, point2);

  // Draw Green Line
  gc.SetPenColor(colorGreen);
  // make pen dashed
  gc.SetPenStyle(CGraphicsContext::EDashedPen);
  // increase the size of the pen
  penSize += KIncrementPenSize;
  // set pen size to increment value
  gc.SetPenSize(penSize);
  // increment starting point of line
```

```
        point1 += KIncrementPoint;
        // increment ending point of line
        point2 += KIncrementPoint;
        gc.DrawLine(point1, point2);
        ...
    }
```

This `Draw()` function covers many of the fundamental concepts that form the basis for drawing in a Series 60 application—in particular, screen geometry, graphics devices and contexts, color and pens.

Screen Coordinates and Geometry

The coordinate system of the screen of a Series 60 device has the standard origin (0, 0) at the top left-hand corner of the screen. Moving rightward and downward from the origin increases the *x* and *y* coordinates, respectively. Measurements are in pixels, and the minimum Series 60 screen size is 176 by 208 pixels. Graphics tend to be drawn in a rectangular fashion, regardless of their true shape, and the geometry classes available reinforce this. The classes of note are `TPoint`, `TSize` and `TRect`. Their members are public for ease of use.

- `TPoint`—A two-dimensional *x*, *y* Cartesian coordinate position represented using two `TInt` members: `iX` and `iY`. Both members can be reset by passing the new coordinate values into the `SetXY()` function.

- `TSize`—Contains two public `TInt` members, `iWidth` and `iHeight`, to represent a two-dimensional size value. The `SetSize()` function is equivalent to `SetXY()` of `TPoint` and changes the size to the parameters passed in.

- `TRect`—Represents a rectangular area derived from two `TPoint` members. The top left corner of the rectangle is stored in the `iTl` member, the bottom right corner in the `iBr` member. A somewhat confusing aspect is that `iBr` lies (1, 1) pixels outside the rectangle, such that `iBr.iX—iTl.iX` is the width of the rectangle. `TRect` has many useful functions—for example, `Center()` returns the center point of the rectangle, and `Intersects()` determines whether this `TRect` overlaps another.

Throughout the examples in this chapter `TPoint`, `TSize` and `TRect` will be employed extensively. They are essential for drawing graphics to the screen with the desired positions and size. Figure 11–3 shows the relation between these classes.

Graphics Devices and Graphics Contexts

A graphics device represents what is being drawn to—for example, the screen, a bitmap in memory, or even a printer. Attributes of graphics devices include the size of the device area and the number of colors supported by that device.

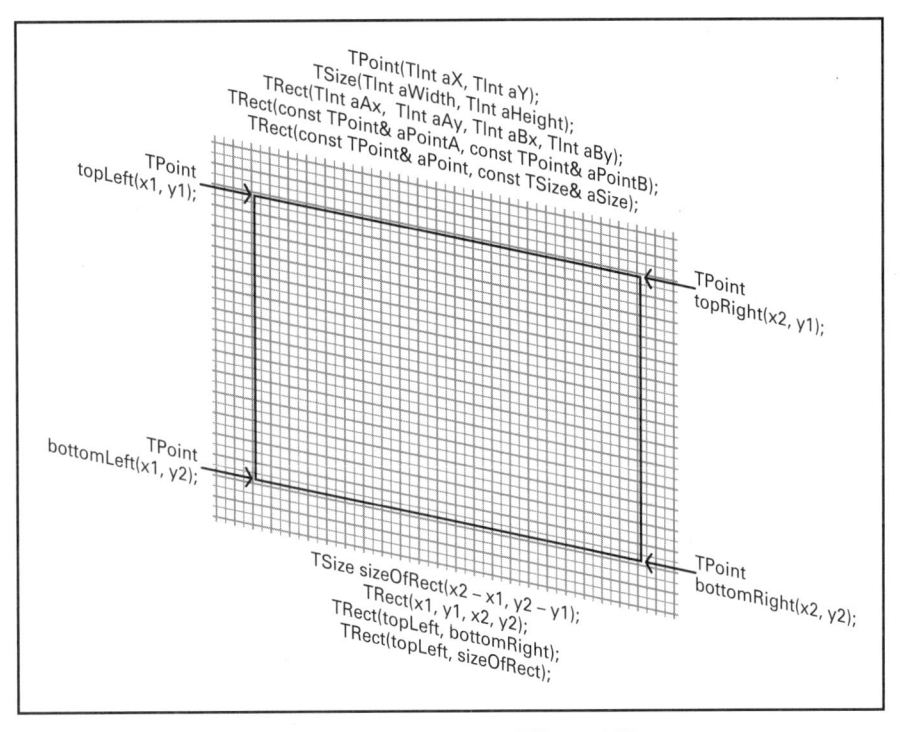

Figure 11–3 Relation of the geometry classes TPoint, TSize and TRect.

In order to be able to perform drawing operations they must have an associated graphics context and consequently a `CreateGraphicsContext()` function. The abstract class `CGraphicsDevice` provides an interface for all graphics devices, while `CBitmapDevice` is an interface for bitmapped graphics devices derived from `CGraphicsDevice`.

`CGraphicsDevice` provides functions for obtaining a device's current display mode, its size and the number of font typefaces associated with it.

`CBitmapDevice` allows you to obtain information about particular pixels and device scan-lines. `CWsScreenDevice` and `CFbsBitmapDevice` are concrete graphics device implementations representing the device screen via the Window Server and an in-memory bitmap, respectively. Their associated graphics contexts are `CWindowGc` and `CFbsBitGc`. By drawing inside a `CCoeControl` class, the **BasicDrawing** example does not require an explicit reference to a particular graphics device.

Figure 11–4 shows some of the various types of graphics devices and how they are architecturally related.

Table 11–3 provides an overview of the classes shown in Figure 11–4.

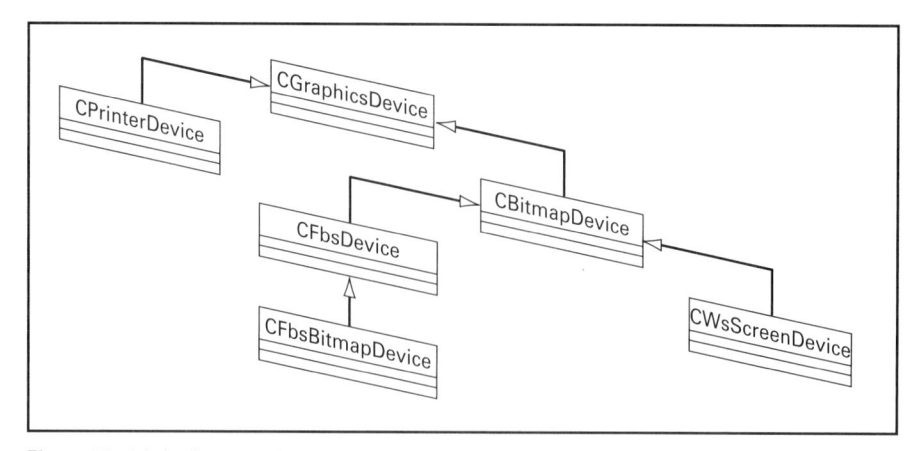

Figure 11-4 Inheritance relationship of a selection of graphics device types.

Table 11-3 Overview of Graphics Device Classes

Graphics Device	Description
CGraphicsDevice	Abstract base class for graphics devices.
CBitmapDevice	Abstract base class for bitmapped graphics devices.
CFbsDevice	Abstract base class for devices that use the Font and Bitmap Server for drawing fonts and bitmaps.
CPrinterDevice	Abstract base class for devices that make use of printing.
CWsScreenDevice	Concrete representation of the device screen in software employing the Window Server.
CFbsBitmapDevice	Concrete implementation using a bitmap maintained by the Font and Bitmap Server as a graphics device.

> **NOTE** Graphics devices and other key graphics classes often have methods that refer to **Twips**. Twips are a device-independent method for measurement, with a single twip equal to 1/1440th of an inch. This contrasts with pixels, whose size is device dependent.

Graphics contexts are used to perform the physical drawing to a graphics device. Consequently, they contain numerous functions for drawing shapes, text and bitmaps. As it would be cumbersome to have to provide attributes — for example, the color of a shape or the font of some text — every time a drawing function is called, they are contained within settings variables of the graphics context. This is illustrated in the **BasicDrawing** example, where the pen color and style are set before a line is drawn, and any subsequent drawing operations will use these properties unless they are explicitly changed:

```
// Set pen to be blue and solid. Then draw line
gc.SetPenColor(colorBlue);
// make pen solid
gc.SetPenStyle(CGraphicsContext::ESolidPen);
gc.DrawLine(point1, point2);
```

The variety of graphics context functions and types will become evident as the chapter unfolds and the example applications illustrate bitmap, text, shape and animation drawing.

Color and Display Modes

Colors are represented by the `TRgb` class, which is a 32-bit value employing 8 bits each for Red, Green and Blue, with 8 bits spare. As Series 60 is highly customizable for different licensees, the color scheme of the user interface can be changed. In order to ensure consistency between colors and the palette in use, it is necessary to employ the `AKN_LAF_COLOR()` macro to define colors. This can be found in the `aknutils.h` header file, which should be included by your application code.

Table 11–4 lists some common integer values and the corresponding color they will produce when passed into the `AKN_LAF_COLOR()` macro. (A full list of values can be found in the Series 60 Style Guide, in the SDK documentation.)

An example of how to use the `AKN_LAF_COLOR` macro is shown here:

```
const TInt KColorYellow = 5;
TRgb colorYellow = AKN_LAF_COLOR(KColorYellow);

// Set pen to be yellow
gc.SetPenColor(colorYellow);
```

TIP If the color scheme is changed by the user, then applications are informed of the change with the value `KEikMessageColorSchemeChange` passed to `CCoeControl::HandleResourceChange()`. Any derived control class may override this method to take appropriate action. If you have stored a `TRgb` value as member data, it should be reset whenever the palette changes.

The range of colors available for use is subject to two key factors: the number of colors the device can physically display, and the current display mode. Display modes are the color depths available and are represented by the `TDisplayMode` enumeration.

Table 11–4 Color Palette Values for the AKN_LAF_COLOR() Macro

Integer Value	Corresponding Color	Integer Value	Corresponding Color
0	White	137	Dark Brown
5	Yellow	141	Dark Purple
9	Light Yellow	146	Light Green 2
13	Light Purple	159	Green 2
17	Dark Yellow	172	Grey 12
20	Light Red	176	Dark Violet
23	Orange	185	Green
35	Red	210	Blue
43	Grey 3	215	Black
51	Light Brown	216	Grey 1
76	Light Green	217	Grey 2
84	Light Violet	218	Grey 4
86	Grey 6	219	Grey 5
92	Light Purple	220	Grey 7
95	Brown	221	Grey 8
105	Purple	222	Grey 10
107	Dark Red	223	Grey 11
120	Light Blue	224	Grey 13
129	Grey 9	225	Grey 14
131	Dirty Dark Yellow		

Chapter 11 Basic Drawing

Obviously the number of colors in a display mode cannot exceed the amount a device is physically able to display. Series 60 devices support a minimum 12-bit display of 4096 colors expressed by the `EColor4K` display mode. This uses 4 bits for Red, Green and Blue with 4 bits spare to make 16-bits (2 bytes) for ease of use—it is faster to move bytes than to perform bit operations. Some current Series 60 devices support a 16-bit display, expressed by `EColor64K`.

Devices have a "preferred display mode" to which graphics will be converted for display. As time will be spent performing any conversion, there can be a performance bottleneck if your graphics are not in the "preferred display mode." For static graphics this is not so much of a problem, but for animations and games you may need to take account of this fact—multiple versions of bitmaps can be created, with only the applicable ones installed, or loaded at runtime.

Pens and Brushes

Pens provide attributes of the outline, and brushes attributes of the filled regions, of drawable items, such as shapes. Both have color and style properties, the styles being represented by the `TPenStyle` and `TBrushStyle` enumerations, respectively. An example of a pen style would be `EDashedPen` that will draw a dashed outline, while `EVerticalHatchBrush` will fill a region with a vertical line pattern. For a full listing of the styles available look in the `\epoc32\include\gdi.h` header file in the root directory of your SDK.

The View from a Window and the Relationship with CCoeControl

From the **BasicDrawing** application it would appear that there are no references to windows, yet drawing takes place in a window. How can this be? By employing the application framework, and in particular using a class derived from `CCoeControl`, the window-related aspects of drawing are hidden.

Indeed, every `CCoeControl` has an association to a window, either by owning it or by lodging in the window of another control. `CCoeControl` has two functions with respect to this, `CreateWindowL()` and `SetContainerWindowL()`, one of which should always be called at construction time. It is possible to gain a handle on a `CCoeControl`'s window by calling its `Window()` function.

The Window Server will redraw areas of a window, should they be marked as invalid. This can occur because all, or part, of a control has been temporarily overlapped by another, and the overlapping control has been dismissed. The region of the window that was hidden by the overlapping control will now be marked as invalid, and the control will have to redraw that area of the window that it occupies.

To "force" a redraw of the control, the `DrawNow()` or `DrawDeferred()` function of `CCoeControl` can be called, or areas of the control's window can be marked as invalid—this is achieved by obtaining a handle to the window and calling its `Invalidate()` function. Invalidating an area will cause a redraw initiated by the Window Server to inform the control's window to begin redrawing. This latter mechanism is used in the **Animation** example.

> **WARNING** When redrawing without using the `DrawNow()` function, care must be taken not to draw outside the boundary of the control. This is easily achieved through clipping, which is implemented through the `SetClippingRect()` function of a graphics context. This should be provided with a `TRect` parameter of suitable dimensions.

Another advantage of employing a `CCoeControl` to handle drawing is that a variety of useful functions are provided "for free" that enable you to perform a wide variety of tasks—ranging from querying screen devices to employing resource files for handling localized descriptor strings. The control environment member variable of `CCoeControl`, `iCoeEnv`, is particularly valuable in this respect. Table 11–5 lists some of the functions this provides. For a list of all of the functions available, look in the file `\epoc32\include\coemain.h` in the root directory of your SDK, or in the SDK documentation for `CCoeEnv`.

Table 11–5 Useful Control Environment Functions

CCoeEnv Function	Description
`AppUi()`	Obtains a handle to the application AppUi, as type `CCoeAppUiBase`.
`FsSession()`	Obtains a handle to a File Server session.
`WsSession()`	Obtains a handle to the application's Window Server session.
`NormalFont()`	Obtains a handle to the Normal font of the system.
`ScreenDevice()`	Obtains a handle to the screen device.
`ReadResource()`	Reads a descriptor resource into a descriptor variable.
`AllocReadResourceL()`	Reads general resource into a descriptor for later conversion into true class type—for example, a User Interface component declared in a resource file.
`SimulateKeyEventL()`	Simulates key presses and can aid debugging.

Drawing without the application framework is possible, but it would be much more complex, as you would have to write your own Window Server client to provide the functionality offered by `CCoeControl`. One advantage would be less overhead when communicating with the Window Server, but a full illustration is beyond the scope of this chapter. However, examples are provided by the SDK. (In Series 60 2.x they can be found in the directory `\Examples\graphics\ws` in the root directory of your SDK—in earlier versions of the SDK they reside in `\Epoc32Ex\graphics\ws`.) As touched upon earlier, drawing will still take place via a graphics context, so all of the drawing concepts covered here are still valid.

Fonts and Text

Series 60 provides a wealth of functions for managing fonts and rendering text. Consequently, it is relatively straightforward to draw some simple text, and the following function, taken from the **FontsAndText** example, will perform such a task:

```
void CFontsAndTextBasicContainer::Draw(const TRect& aRect) const
    {
    // Obtain graphics context and clear it
    CWindowGc& gc = SystemGc();
    gc.Clear();

    // Set position of text relative to
    // the rectangle of the control
    TPoint textPoint(aRect.Width() / KHorizOffset,
aRect.Height() / KVertOffset);

    // Use resource architecture to obtain localizable string
    TBuf<KTextMaxLength> text;
    StringLoader::Load(text, R_FONTSANDTEXT_LOCALIZATION_TEXT);

    // Gain handle to Normal font of the system
    const CFont* normalFont = iEikonEnv->NormalFont();
    gc.UseFont(normalFont);

    gc.DrawText(text, textPoint); //Draw text
    gc.DiscardFont(); // Discard font
    }
```

Initially a handle to the system graphics context is acquired. It is cleared, and the point on the screen where the bottom left-hand corner of the text is to be placed is created. You then decide which font to use—in this case the "Normal" font of the system. This represents an extremely simple way of getting access to a font, and a variety of similar system fonts will be discussed later in this section. Next, the font to be used by the graphics context must be explicitly set. The most important step follows with the call to the graphics context to draw the text. The first parameter is the text literal, and the second is the position of the text. Finally, the system is informed that the font in use is no longer required.

570   Chapter 11   Multimedia: Graphics and Audio

WARNING   If a font type has not been set, the system will panic. However, this will not become evident until some time after the `DrawText()` function appears to have been called. This is because of the delay between function calls being made in code and the client-side buffer being dispatched to the Window Server.

The display produced by the `CFontsAndTextBasicContainer::Draw()` function is shown in Figure 11–5.

Text and Font Measurements

In order to draw text with precision on the screen, you must have an appreciation of the character metrics and their roles and relationships with each other. These are highlighted in Figure 11–6 and Table 11–6.

Figure 11–5 Simple text.

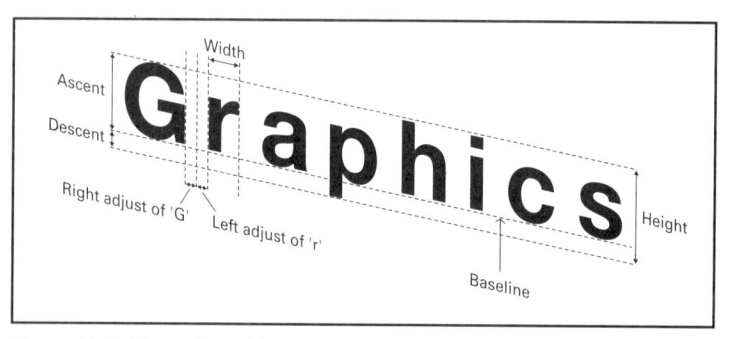

Figure 11–6 Illustration of font metrics and their relationship to a font.

This knowledge has practical applications, as there is an overloaded version of the `CGraphicsContext::DrawText()` function illustrated in the **Metrics** view of the **FontsAndText** example (Figure 11–7) that requires to be passed into it the text to be drawn, a rectangular area, the baseline of the text and a horizontal justification value.

Text can be drawn exactly in the center of a rectangle by calculating the baseline to be half the height of the rectangle (in this case it happens to be the rectangle of the screen, but potentially it could be any rectangular area), plus half the ascent of the font. The baseline is then passed into the `DrawText()` function.

Figure 11–7 Metrics.

```
   TInt baseline = (aRect.Height() / KBaselineDivider) +
(titleFont>AscentInPixels() / KBaselineDivider);
   ...
   gc.DrawText(text, aRect, baseline,
CGraphicsContext::ECenter);
```

Table 11–6 Definition of Font Metrics

Metric	Description
Baseline	Horizontal line on which text sits.
Ascent	Distance above the baseline the font ascends.
Descent	Distance below the baseline the font descends.
Height	Distance between the ascent and the descent.
Width	Horizontal distance occupied by a character.
Left Adjust	Spacing distance before the character.
Right Adjust	Spacing distance after the character.
Move	The sum of the width, left adjust and right adjust.

Key Font Classes and Functions

Numerous classes are provided for obtaining and using fonts. Interaction with fonts extends beyond just using them for drawing, and you need to communicate with the Font and Bitmap Server to derive information about the fonts available and ensure the fonts are used in as an efficient manner as possible. The key classes are listed in Table 11–7.

Table 11–7 Key Font Classes

Class	Description
`MGraphicsDeviceMap`	Assists in mapping from real-world size perspective—for example, from twips to (device-based) pixels.
`CTypefaceStore`	Provides an abstract interface to the store of font typefaces. `CFbsTypefaceStore` is a concrete implementation of this and represents the typeface store of the Font and Bitmap Server.
`TTypefaceSupport`	Provides essential information relating to a particular typeface, such as its name, its maximum and minimum heights, and whether it is scalable.
`TFontSpec`	Specification of the font in terms of style, typeface and height.
`CFont`	Provides an abstract interface to a font and has functions for obtaining baselines, ascents and text measuring.

Using the Key Font Classes to Enumerate Through All Available Fonts

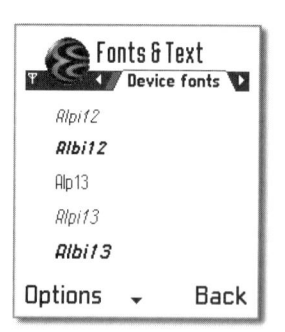

All of these classes are employed in the **Device Fonts** view of the **FontsAndText** example application, where the name of each font available is printed in that font, as shown in Figure 11-8.

In this example, the number of fonts present on the device is calculated, providing the counter limit for a loop to increment through the various fonts.

Figure 11-8 Device fonts.

```
// Obtain the number of typefaces available
// on this device.
iNumTypefaces = iCoeEnv->ScreenDevice()->NumTypefaces();
…
TBuf<KMaxFontName> fontName;
CFont* fontToUse;
TRgb colorBlack = AKN_LAF_COLOR(KColorBlack);

for (TInt i = iCurrentScrollNum; i < iNumTypefaces; i++)
    {
    // Get the i-th font on the device.
    iCoeEnv->ScreenDevice()->
TypefaceSupport(*iTypefaceSupport, i);
    // Get the font name.
    fontName = iTypefaceSupport->iTypeface.iName.Des();

    // Create font specification.
    TFontSpec fontSpec(fontName, KFontSpecSize);
    iDeviceMap->GetNearestFontInTwips(fontToUse,fontSpec);

    // Increment baseline to be 1.5 x height of font.
    textPoint.iY += (fontToUse->HeightInPixels() *
KBaseLineIncrementer);

    // Use font with graphics context.
    gc.UseFont(fontToUse);
    gc.SetPenColor(colorBlack);
    gc.DrawText(fontName, textPoint);
    gc.DiscardFont();
    iDeviceMap->ReleaseFont(fontToUse);
    }
```

Discarding and releasing the font are analogous. However, discarding is performed at the graphics context level, while releasing is performed by the screen device, completing the process of informing the Font and Bitmap Server

Chapter 11 Fonts and Text

that the application no longer requires the font. Forgetting to discard and release a font before an application exits will result in a panic. This panic is similar to a memory leak—though in this case the term "resource leak" expresses the error more appropriately.

Another issue facing developers is that different Series 60 licensees can potentially provide different fonts as part of their customization of the platform. Consequently, fonts must be used in the most generic way possible—via the application environment variable `iEikonEnv`. The following functions from the class `CEikonEnv`, found in `\epoc32\include\eikenv.h` in the root directory of your SDK, give access to device-independent fonts:

```
IMPORT_C const CFont* AnnotationFont() const;
IMPORT_C const CFont* TitleFont() const;
IMPORT_C const CFont* LegendFont() const;
IMPORT_C const CFont* SymbolFont() const;
IMPORT_C const CFont* DenseFont() const;
```

As mentioned earlier `CCoeEnv` provides the "Normal" font function (`NormalFont()`), and as `CEikonEnv` is derived from `CCoeEnv`, access is provided to this also. These functions are used in the **FontsAndText** example in the **Effects** view, as shown in the following code taken from `CFontsAndTextEffectsContainer::Draw()`:

```
...
// Draw black vertical text in title font.
const CFont* font = iEikonEnv->TitleFont();
TPoint textPoint(aRect.Width() / KLeftTextHorizAlign,
aRect.Height() / KLeftTextVertAlign);
gc.UseFont(font);
gc.SetPenColor(KRgbBlack);
gc.DrawTextVertical(text, textPoint, EFalse);
gc.DiscardFont();
...
```

Text Effects

A number of additional functions can be applied to change the appearance of text, ranging from rendering text with different colors, to changing the orientation of the text. Many of these are illustrated in the **Effects** view of the **FontsAndText** example application, as shown in Figure 11–9, and are graphics context functions.

Drawing text in different colors is simple to achieve—the pen color of the graphics context is the color of the text to be drawn, and so code similar to the following should suffice in order to draw green text:

Figure 11–9 Text effects.

```
gc.SetPenColor(colorGreen);
```

Additionally, if underlined text and strikethrough are desired, the following functions will turn these effects either on or off:

```
gc.SetUnderlineStyle(EUnderlineOn);
gc.SetStrikethroughStyle(EStrikethroughOn);
gc.SetUnderlineStyle(EUnderlineOff);
gc.SetStrikethroughStyle(EStrikethroughOff);
```

Text can also be drawn with a vertical orientation, in one of two ways: downward or upward. The `DrawTextVertical()` function performs this operation—its third parameter is a Boolean value, which if set to false will draw the text downward, and if set to true will draw it upward:

```
// Draw text vertically downwards
gc.DrawTextVertical(KHelloSeries60(), textPoint, EFalse);

// Draw text vertically upwards
gc.DrawTextVertical(KHelloSeries60(), textPoint, ETrue);
```

Shapes

Drawing shapes is extremely straightforward, as a graphics context provides a series of functions that draw particular shapes. Examples include `DrawRect()` and `DrawEllipse()` for drawing rectangular shapes and ellipses, respectively.

While the role of the function is quite obvious from its name, it is often the parameters required by these functions that can determine the appearance of the desired shape. Whether or not the shape is filled or outlined depends on the current pen and brush settings. These are determined by the `TPenStyle` and `TBrushStyle` enumerations.

In order to draw an outline shape, the brush style would be set to null and the pen style to a value other than null. Conversely, in order to draw a filled shape without an outline, you would set the pen style to null and the brush style to any value but null. Recalling that pens and brushes also have a color property, it is possible to have separate colors for the outline and filled regions of shapes.

```
// filled shape no outline
gc.SetPenStyle(CGraphicsContext::ENullPen);
gc.SetBrushStyle(CGraphicsContext::ESolidBrush);
// draw

// outline shape no fill
gc.SetPenStyle(CGraphicsContext::ESolidPen);
gc.SetBrushStyle(CGraphicsContext::ENullBrush);
// draw
```

Chapter 11 Shapes

```
// outline shape with fill
gc.SetPenStyle(CGraphicsContext::ESolidPen);
gc.SetBrushStyle(CGraphicsContext::ESolidBrush);
// draw
```

The **Shapes** example converts a data set, provided by the user, into simple graphical charts and covers drawing many of these shapes.

Rectangles

Drawing rectangles is achieved through calling the `DrawRect()` function of a graphics context. This takes a single `TRect` parameter, which is the area the drawn rectangle will occupy.

```
gc.DrawRect(rectOfCurrentBar);
```

Rounded rectangles can also be drawn—this function takes a second parameter, of type `TSize`, for the elliptical shape that generates the roundedness of the edges, as demonstrated in Figure 11–10.

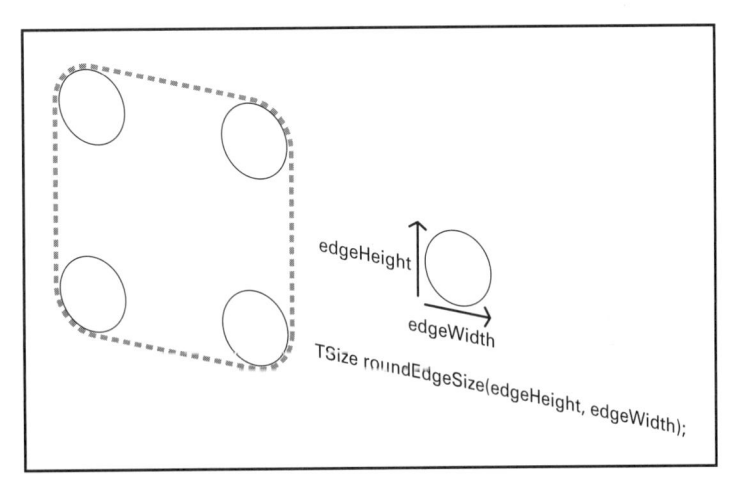

Figure 11–10 Role played by the TSize parameter in creating a rounded rectangle using the DrawRoundRect() function.

```
gc.DrawRoundRect(rect, roundEdgeSize);
```

In the **Shapes** example, the `DrawRect()` function is used to draw the bars of the bar chart view.

Ellipses

Ellipses are drawn using the `DrawEllipse()` method and require a single parameter of type `TRect`. This is the rectangular area within which the ellipse will be centered.

```
gc.DrawEllipse(ellipseRect);
```

Arcs and Pies

Arcs and pie slices are drawn in fundamentally the same way. The first parameter is the rectangular area within which resides the elliptical region, of which the arc or pie is a fraction. The other two parameters are two points through which radials can be drawn from the center of the ellipse. The arc or pie is drawn between the points where these radials intercept the edges of the ellipse. The pie area includes the radials to the intersection; the arc produces a chord when filled.

```
gc.DrawPie(KPieRect, pieStartPoint, pieEndPoint);
```

This is illustrated in Figure 11–11.

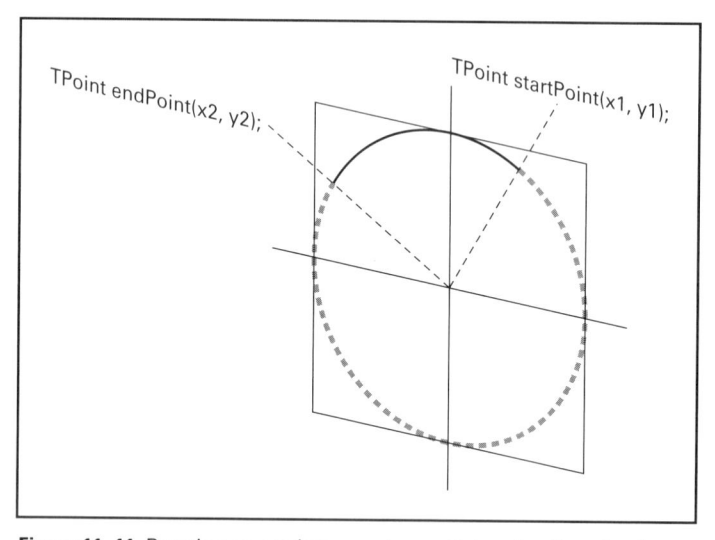

Figure 11-11 Drawing an arc between two radials using DrawArc().

`DrawPie()` would result in filling the area bounded by the radials and the arc.

NOTE When drawing pies, arcs, or any other shapes, you must remember that the Cartesian coordinate system differs from the coordinate system of a computer screen, or a Series 60 device—the origin is at the top left-hand corner, and the *y* axis increases downward.

Polygons

There are two separate functions for drawing polygons: `DrawPolyLine()` and `DrawPolygon()`. `DrawPolyLine()` will draw an outline shape, while `DrawPolygon()` will be filled. Polygons comprise of an array of points, and these are provided to either function in order to render the desired shape.

Chapter 11 Shapes

```
// Create Array of points (1 is added to
// allow for origin point)
   CArrayFix<TPoint>* polyLineArray = new (ELeave)
CArrayFixFlat<TPoint>(numElements + 1);

   CleanupStack::PushL(polyLineArray);
   polyLineArray->AppendL(origin);

   // Iterate value model. Calculate points based on
   // values and append to points array.
   for (TInt i = 0; i < numElements; i++)
       {
       // Calculated height of line based on input data
       heightOfCurrentLine = (iShapesModel.ElementAt(i) *
aAxisRect.Height()) / maxValue;
       xOfCurrentLine = origin.iX + ((i + 1) *
widthOfLineSpacer);
       yOfCurrentLine = origin.iY - heightOfCurrentLine;
       polyLineArray->AppendL(TPoint(xOfCurrentLine,
yOfCurrentLine));             }

   gc.DrawPolyLine(polyLineArray);
```

The code snippet above is taken from the **Shapes** application and is used to generate a line graph based on the data provided by the application's data model.

A special parameter that can be passed to the `DrawPolygon()` function determines how the shape will be filled if lines of the polygon should cross over each other. The `TFillRule` enumeration has two values, `EAlternate` and `EWinding`, with `EAlternate` filling only odd-numbered bounded regions, as illustrated in Figure 11–12.

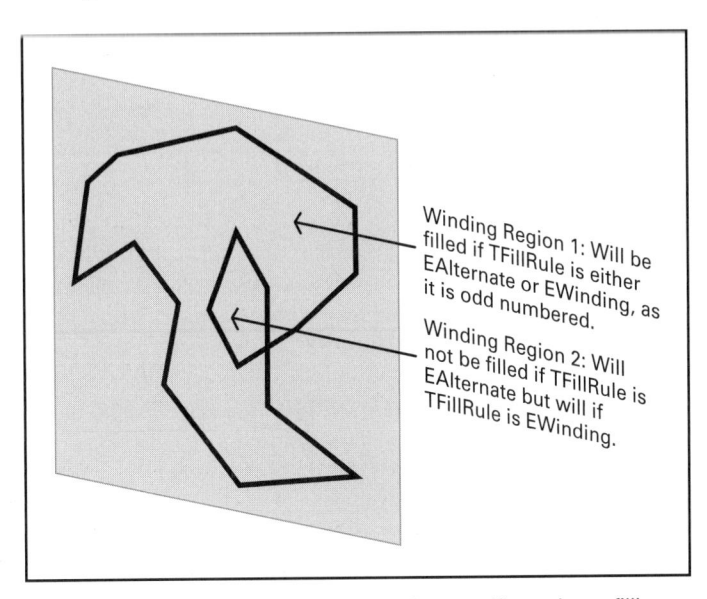

Figure 11–12 How the TFillRule enumeration specifies polygon filling.

Bitmaps

Bitmapped graphics have a variety of applications, ranging from displaying corporate logos to rendering the characters of a game. Consequently, Series 60, through graphics contexts and specific bitmap classes, provides numerous APIs for manipulating and drawing bitmaps. Both the Window Server and Font and Bitmap Server can play vital roles, and as a result there are two types of bitmap class. A `CFbsBitmap` is a bitmap managed by the Font and Bitmap Server, while a `CWsBitmap` is an extension for more efficient use by the Window Server. Functions with a `CWsBitmap` parameter tend to be faster than their `CFbsBitmap` equivalents.

Due to the memory constraints of Series 60 devices, and the storage demands of bitmapped graphics, bitmap (`.bmp`) files should be converted into Symbian OS multibitmap (`.mbm`) file format. There is also scope for using GIF and JPEG file types, although these must first be converted to one of the bitmap types for display—this is discussed in the **Image Manipulation** section of this chapter.

Generating Bitmaps for Application Use

A PC utility is provided by the SDK called **bmconv**, which, when supplied with a list of `.bmp` (Windows bitmap) files, will generate a single `.mbm` (Symbian OS multibitmap) file, containing all of the bitmap images. More information on **bmconv** can be found in Chapter 2.

While this can be used directly, it is far simpler to employ the `START BITMAP...END` command within the application's `.mmp` file to automatically run this tool for you.

If desired, all of the application's bitmaps can be contained within a single `.mbm` file, although it may be more manageable to group related graphics into separate `.mbm` files. For example, in a game, `levels.mbm` might contain all the images for the various levels, while `player.mbm` may contain the bitmaps for the central character. It would be rare and inadvisable to have an `.mbm` file for each `.bmp` file.

Handles to individual bitmaps within an `.mbm` are obtained through an enumeration. This enumeration is contained within a header file with an `.mbg` extension, which is generated as part of the conversion process. The `.mbg` file and enumeration will have the same name as the `.mbm` file—for example, `levels.mbm` will have a corresponding `levels.mbg` file, and the enumeration name will be `TMbmLevels`. The enumeration value for a specific bitmap is then of the form `EMbmLevelsMybitmap`, where `Mybitmap` is the name of the `.bmp` file of interest.

The process is shown in Figure 11–13, with an outline of the commands within `START BITMAP` provided in Table 11–8.

The pregenerated `.bmp` files are referenced within the scope of the `START BITMAP` command via the `SOURCE` command. The `SOURCEPATH` value is the

Chapter 11 Bitmaps

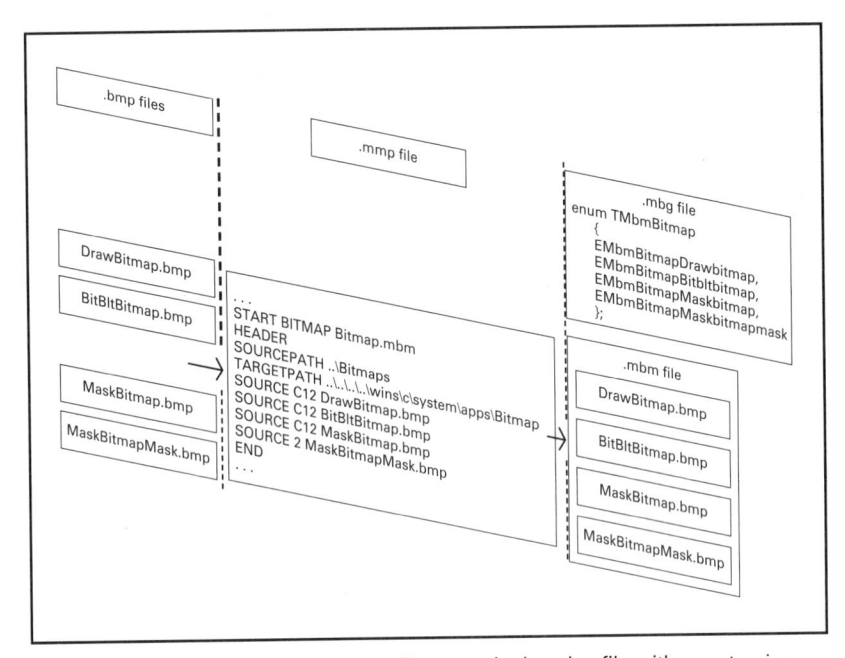

Figure 11-13 Converting multiple .bmp files to a single .mbm file with a partnering .mbg file.

Table 11-8 Overview of the START BITMAP Commands

Command	Purpose
START BITMAP	For each .mbm required, a separate START BITMAP is required.
[TARGETFILE target-file]	Name of the .mbm file to be produced.
[TARGETPATH targetpath]	If not specified, the .mbm file will be built in the same directory as the application.
[HEADER]	This command is needed to ensure the .mbg file is generated in the \epoc32\include directory, in the root directory of your SDK.
[SOURCEPATH sourcepath]	Directory of the .bmp files that follow in subsequent SOURCE statements.
SOURCE color-depth source-bitmap-list	The bitmaps that follow will all be of the color depth specified. The color depth must be of the form [c]depth, where the (optional) "c" indicates a color image and depth represents the color depth. Multiple SOURCEPATH and SOURCE statements can be defined—source bitmaps specified with each SOURCE statement should exist in the directory denoted by the SOURCEPATH statement above it.
END	This command concludes the .mbm creation.

location where the `.bmp` files are being stored. There can be multiple SOURCEPATH statements; however, the `.bmp` files in subsequent SOURCE statement must be contained within that path.

Loading and Drawing Bitmaps

With an `.mbm` file in place, bitmaps can be loaded and drawn—the path to the bitmap is declared within the application code and is used by the `CFbsBitmap::Load()` function:

```
_LIT(KDrawBitmapPath,
"\\system\\apps\\Bitmap\\Bitmap.mbm");
...
iBitmap = new (ELeave) CFbsBitmap();
User::LeaveIfError(iBitmap->Load(KDrawBitmapPath,
EMbmBitmapDrawbitmap));
...
```

In order to draw the bitmaps, a graphics context is required. Graphics contexts that derive from `CBitmapContext` are used for drawing bitmaps and provide two named functions, `DrawBitmap()` and `BitBlt()`. These two functions each have a variety of overloaded variations to suit differing needs and are illustrated in the SDK documentation. Clipping is performed automatically, so this need not concern you.

`DrawBitmap()` is the slower of the two, as it always performs bitmap scaling as part of its operation, whereas `BitBlt()` performs a block transfer of the bitmap. The **Bitmap** example application demonstrates both approaches, with `DrawBitmap()` being called by the `CDrawBitmapContainer::Draw()` function and `BitBlt()` by `CBitmapBitBltContainer::Draw()`.

```
...
gc.DrawBitmap(TRect(topLeft, bitmapSize), iBitmap);
...
gc.BitBlt(topLeft, iBitmap);
...
```

Bitmap Masking

Though drawing of bitmaps always occurs within rectangular areas, the bitmaps themselves can vary greatly in shape. Often it is useful that portions of the rectangular area are actually transparent to give the bitmap the appearance of its true shape without drawing over portions of a background. This is achieved through a technique known as masking.

Masking requires a second image that is used to determine which areas of the visible bitmap will actually be displayed. Masking bitmaps generally have a color depth of two (representing image/no image), and the `BitBltMasked()` function is passed both the masking image and displayed image as parameters.

`BitBltMasked()` is called by `CBitmapBitBltMaskContainer::Draw()` in the **Bitmap** example:

```
...
    gc.BitBltMasked(topLeft, iBitmap, TRect(TPoint(0, 0),
TSize(KSizeWidth, KSizeHeight)), iMask, ETrue);
...
```

Figure 11-14 shows how a bitmap can be masked.

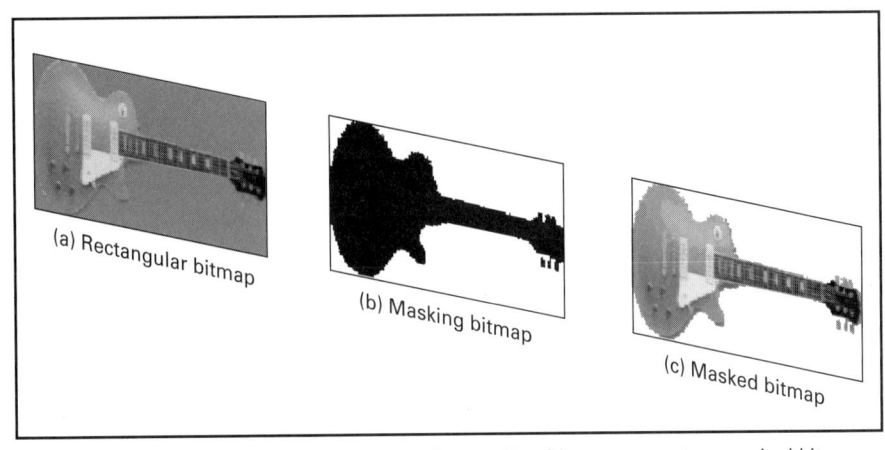

Figure 11-14 Using a rectangular bitmap and a masking bitmap to create a masked bitmap.

TIP When developing controls that require images, such as information dialogs, it is advisable always to use bitmap masking. This is because Series 60 can be greatly customized by Licensees, and the background colors of a particular control may vary from device to device.

Bitmap Functions

The `CFbsBitmap` class has numerous useful functions for obtaining information about a bitmap's attributes, as listed in Table 11-9.

Table 11-9 Useful CFbsBitmap Methods

Function	Role
DataAddress()	Returns the address of the first (top left) pixel.
DisplayMode()	The display mode of the bitmap.
ExternalizeL()	Externalizes the bitmap to a stream.
GetPixel()	Returns the RGB value (TRgb) of a given pixel.
InternalizeL()	Internalizes the bitmap from a stream.
Save()	Saves the bitmap to a named file.

Animation

Programming animation on Series 60 can be seen from two perspectives: using an animation DLL framework, or using client-side application code. Dedicated animation classes are provided as part of the platform and can be used to create a DLL to perform the animation inside a server. However, this approach is not suitable in all cases. Performing animation "client-side" (in application code) is at times preferable.

When all of the images that comprise the animation have been predetermined, then it is relatively simple to employ the DLL animation architecture. The `CAknBitmapAnimation` provides support for this. A bitmap animation DLL is provided as part of Series 60, and access to it is wrapped up inside `CAknBitmapAnimation`.

If the image to be displayed can dynamically change during program execution, then a client-side approach is best. An example of using an Animation Server might be the repeated animation of a corporate logo, while a client code example could be displaying the collision of objects in a game. Client-side animation uses double buffering to ensure that the animation does not suffer from flicker during redraws, and it requires an understanding of Active Objects.

The Animation Architecture

Employing the animation architecture directly can be quite complex, requiring the implementation of a specific DLL for each desired animation. The main advantage of this approach is that the animation is able to run within the thread of the Window Server. As this is a high-priority thread, performance benefits are guaranteed. Client-side approaches must use Active Objects, which will not enjoy the priority advantages afforded by the thread of the Window Server.

The provided animation classes use a Client/Server mechanism in order to animate, with the client and server being wrapped up inside the DLL. The client in this case should not be confused with the alternative "client-side" approach.

Animation clients should derive from the `RAnim` class and are used to pass the Animation Server relevant information, such as the number of frames per second. Deriving from `RAnimDLL` provides the DLL interface. On the server side, `CAnim`-derived classes must be implemented to perform the animation, with `CAnimDLL` being used for the DLL interface.

Implementing your own animation DLL is beyond the scope of this chapter, however the SDK documentation provides an in-depth explanation of the key animation classes and how to use them. An example can also be found in the directory `\Series60Ex\Animation` in the root of your SDK.

In truth, there is no real need to implement a DLL for animating bitmaps, as the `CAknBitmapAnimation` class (as used in the **Animation** example) provides an interface to an animation DLL—you need only provide it with some images and attribute data.

Chapter 11 Animation

The simplest way to use the `CAknBitmapAnimation` class is to define its relevant parts inside a resource file and then initialize the class with this data. Three essential resources are needed to construct an animation: the bitmap animation data (`BMPANIM_DATA`), an array of bitmap frames (`ARRAY`), and the individual bitmap frames themselves (`BMPANIM_FRAME`). Table 11–10 lists the elements of the `STRUCT` definitions from the `Avkon.rh` file. `ARRAY` is ignored, as it is generic and not specific to animation as such. Resource files are covered in more detail in Chapter 5.

Table 11–10 Animation Resource Components

STRUCT	Element Type	Element Name	Role
BMP_ANIMDATA	WORD	`frameinterval`	Delay between frames.
	WORD	`playmode`	Three playmodes—see `\epoc32\include\avkon.hrh` in the root directory of your SDK: Play—play once. Cycle—repeat from the first frame to the last. Bounce—play the first frame to the last and then back down to the first, repeatedly.
	BYTE	`flash`	Causes the animation to flash, or not, while playing.
	LTEXT	`bmpfile`	The `.mbm` file containing the animation frame.
	LLINK	`frames`	Reference to the `ARRAY` resource of `BMPANIM_FRAME` definitions.
	LLINK	`backgroundframe`	Reference to the background frame definitions.
BMPANIM_FRAME	WORD	`time`	Time that frame is displayed for.
	WORD	`posx`	*x* coordinate of top left of frame.
	WORD	`posy`	*y* coordinate of top left of frame.
	WORD	`bmpid`	ID (from the `.mbg` file) of the desired bitmap from the `.mbm` file for this frame.
	WORD	`maskid`	ID (from the `.mbg` file) of the desired masking bitmap from `.mbm` file for this frame.

Chapter 11 Multimedia: Graphics and Audio

The following resource structures show part of the concrete implementation from the **Animation** example. However, it should be noted that all elements do not have to be assigned values for the resource to be valid:

```
RESOURCE BMPANIM_DATA r_animation_example_data
    {
    frameinterval = 50;
    playmode = EAknBitmapAnimationPlayModeBounce;
    flash = 0;
    bmpfile = ANIMATION_BMPFILE_NAME;
    frames = r_animation_example_array ;
    }

RESOURCE ARRAY r_animation_example_array
    {
    items =
        {
        BMPANIM_FRAME
            {
            time = 200;
            bmpid = EMbmAnimationDisplayimageframe1;
            },
        ...
        BMPANIM_FRAME
            {
            time = 1000;
            bmpid = EMbmAnimationDisplaytextframe;
            }
        };
    }
```

The animation control can be fully constructed via the resource and started as shown here:

```
// instantiate CAknBitmapAnimation
iAnimation = CAknBitmapAnimation::NewL();

TResourceReader reader;
iCoeEnv->CreateResourceReaderLC(reader,
R_ANIMATION_EXAMPLE_DATA); // map resource to resource reader

// provide animation control with animation data
iAnimation->ConstructFromResourceL(reader);
// animation control needs a window
iAnimation->SetContainerWindowL(*this);
iAnimation->StartAnimationL(); // start animation
CleanupStack::PopAndDestroy(); // resource reader created
```

Initially the CAknBitmapAnimation object is created. However, it does not contain any data relating to the animation itself—as the animation has been defined in a resource file, you need to read in the resource, and this is achieved by creating a TResourceReader object pointing to the animation resource.

Chapter 11 Animation

The `CAknBitmapAnimation::ConstructFromResourceL()` can then be used to provide the animation data to the animation control. A window to draw to is required by the animation control, and this is assigned through the `SetContainerWindowL()` function in the **Animation** example application.

Should you choose to write your own `CAknBitmapAnimation`-derived class, then there is also the option of the animation control being "window-owning" by calling the `CreateWindowL()` function. The animation can be started and stopped by using the `StartAnimationL()` and `CancelAnimation()` methods, respectively.

Practical use of `CAknBitmapAnimation` can be found throughout Series 60, as it is employed by numerous user interface controls, such as `CAknNoteControl` and `CAknQueryControl`.

Off-Screen Bitmaps and Double Buffering

Often, when programming animations, you do not have a set of predefined bitmaps, and you have to construct each animation frame in code. Before you can display the new frame on the screen, you need to clear the previous one. If you are using the system graphics context, clearing and then constructing each frame, then there is a high probability that the resulting animation will flicker. This is a common problem on all computing platforms, because screens are continually refreshed, and a refresh may take place while the frame has been cleared, or is currently being constructed.

In order to overcome this, the technique of **double buffering** is employed—creating a bitmap within memory and using it to "build up" the next animation frame off screen. Then, when it is fully "prepared," it is quickly drawn to the screen in one go. The three key areas of graphics devices, graphics contexts and bitmaps are interwoven in order to implement double buffering on Series 60.

Recalling earlier discussions relating to graphics devices, the "device" being drawn to here is an in-memory bitmap. An ideal choice for this class is the `CFbsBitmapDevice`, which enables a bitmap managed by the Font and Bitmap Server to act as a graphics device. Consequently, this requires an associated bitmap to be provided. The bitmap device also has to provide client code with its graphics context using the `CreateContext()` function. Note that the `CDoubleBufferedArea` class is not part of Series 60 and has been written for illustration purposes as part of this text.

```
void CDoubleBufferedArea::ConstructL(TSize aSize,
TDisplayMode aDisplayMode)
    {
    // Constructing off-screen play area:
    // create in memory/offscreen bitmap
    iAreaBitmap = new (ELeave) CFbsBitmap();

    // set size and display mode of bitmap
    iAreaBitmap->Create(aSize, aDisplayMode);
```

```
// create bitmap device
iAreaBitmapDevice = CFbsBitmapDevice::NewL(iAreaBitmap);

// create graphics context for drawing to bitmap}
iAreaBitmapDevice->CreateContext(iAreaBitmapContext);
}
```

Now, using the graphics context of the bitmap device, it is possible to perform all of the graphics operations illustrated throughout this chapter, such as rendering shapes, text and bitmaps. However, rather than being drawn to the screen, they are now being drawn to the bitmap in memory:

```
    iDoubleBufferedArea->
GetDoubleBufferedAreaContext().Clear();
    TRgb colorBlack = AKN_LAF_COLOR(KColorBlack);
    iDoubleBufferedArea->
GetDoubleBufferedAreaContext().SetBrushColor(colorBlack);
    iDoubleBufferedArea->
GetDoubleBufferedAreaContext().DrawRect(Rect());
    ...
```

As a final step, you blit the off-screen bitmap using the screen's graphics context:

```
    ...
    gc.BitBlt(TPoint(0,0), &(iDoubleBufferedArea->
GetDoubleBufferedAreaBitmap()));
    ...
```

There is now no danger of drawing to the screen while the next frame is still being "prepared," as this has all taken place in memory.

The Client-Side Approach to Animation

If you want to perform double-buffered animation within application code, a dedicated drawing function and timing mechanism are essential. In the **ClientAnimation** example, as with all of the other examples, drawing takes place within a `CCoeControl`-derived class.

Potentially, `CCoeControl::DrawNow()` could be repeatedly called via a timing mechanism. This would in turn call `CCoeControl::Draw()`, within which the animation code would be implemented. However, the motivation for providing a separate dedicated drawing function is that if you choose not to use a `CCoeControl` in your own implementation, the example is still relevant.

```
void CClientAnimationContainer::DrawFrame()
    {
    Window().Invalidate(Rect());
    ActivateGc();
    Window().BeginRedraw(Rect());
    DrawToDoubleBufferedArea();
    CWindowGc& gc = SystemGc();
```

Chapter 11 Animation

```
    gc.BitBlt(TPoint(0, 0), &(iDoubleBufferedArea->
GetDoubleBufferedAreaBitmap()));
    Window().EndRedraw();
    DeactivateGc();
    }
```

When using your own draw function instead of the `CCoeControl::Draw()` function, some extra housekeeping operations are required. Recalling that the Window Server redraws invalid regions of windows, the area to be drawn to must be invalidated before redrawing can begin. Following this, the system graphics context has to be activated so that an object exists, with which drawing can actually take place.

Before drawing commences, the Window Server is informed that an attempt is being made to draw to a portion of the screen by calling the `RWindow::BeginRedraw()` function. This has the effect of setting the Window Server to clip drawing to the area passed into `BeginRedraw()`. Drawing can now take place as normal, remembering that in this example drawing takes place to the off-screen bitmap. The off-screen bitmap is then drawn to the screen. Before deactivating the graphics context, `RWindow::EndRedraw()` is called to inform the Window Server it can clean up objects it created for the redraw process.

TIP If `ActivateGc()` has been called and is called again, without a previous call to `DeActivateGc()`, a panic will occur.

In order to perform animation you need to repeatedly call the redraw function, and for this a timer is necessary. The most suitable and easy to use timer, for this task, is the `CPeriodic` timer:

```
    // function definition from CPeriodic class in e32base.h
    IMPORT_C void Start(TTimeIntervalMicroSeconds32 aDelay,
TTimeIntervalMicroSeconds32 anInterval, TCallBack aCallBack);
```

Upon starting the timer, a callback function and two time intervals are passed in. The first time interval is the delay from the calling of the `CPeriodic::Start()` function to the calling of the callback function for the first time:

```
    // function use in ClientAnimation example application
    iTimer->Start(0, KAnimFrameTime,
TCallBack(CClientAnimationContainer::AnimationCallback,
this));
```

The timer will then repeatedly invoke the callback function with a delay of the second interval specified, until it is stopped — calling the `CPeriodic::Cancel()` function stops the timer. By making the callback call your redraw function, animation occurs:

```
// Callback function used to call DrawFrame()
TInt CClientAnimationContainer::AnimationCallback(TAny* aPtr)
    {
    ((CClientAnimationContainer*)aPtr)->DrawFrame();
    // returning 1 ensures the callback function
    // is repeatedly called
    return 1;
    }
```

Note that there are numerous ways to implement the timing mechanism, not just by using `CPeriodic`. An alternative approach might be to derive from `CTimer` and in the overridden `RunL()` function perform the necessary drawing. This alternative approach is used in the subsequent Direct Screen Access example game, **Skiing**.

`CPeriodic` is itself derived from `CTimer`. More traditional approaches, such as using a while loop with a time checking and waiting mechanism, are greatly discouraged. The Series 60 View Server has been designed to detect and terminate unresponsive applications. Animating via a while loop, even for a relatively short period of time, will result in the View Server considering your application to be unresponsive, causing it to panic and close your application. By using the system timing classes outlined, which are Active Objects, you avoid this problem.

Direct Screen Access

Using the drawing mechanisms outlined so far, the influence of the Window Server is evident at every step. There is good reason for this, as Series 60 devices must be aware of many external events outside the realm of the currently active application. For example, users need to be informed of incoming messages or telephone calls. Consequently, there is a need for extra processing in management of the device screen, and this can result in a performance overhead.

For standard applications, where redrawing is infrequent, this is not such a concern. However, for graphically intensive applications, such as games, things are much more problematic.

The Direct Screen Access (**DSA**) classes are designed to work in conjunction with the Window Server for external event management, while providing improved performance when accessing the screen. All of the methods already covered for drawing images, text and shapes can be used with Direct Screen Access. In fact, from a programmer's point of view, the only thing that changes is the graphics context object.

Direct Screen Access architecture is examined in this section. The example application illustrating Direct Screen Access is a simple game, **Skiing**.

Architectural Overview

The Direct Screen Access architecture comprises four key classes, each playing a vital role in the drawing and screen management process. Their roles are defined in Table 11–11.

Table 11–11 Direct Screen Access Classes

Class	Role
RDirectScreenAccess	Makes asynchronous requests to directly use the device screen.
CDirectScreenAccess	Active Object which reissues requests and manages an abort on completion.
MAbortDirectScreenAccess	Mechanism to stop the Active Object, so that the Window Server can process an external event.
MDirectScreenAccess	Mechanism to restart the Active Object when external event processing is complete.

The two areas to be addressed in order to understand Direct Screen Access are illustrated in the following two diagrams. In Figure 11–15, a request is issued to the Window Server to operate in Direct Screen Access mode, which if successful will provide an area on the screen that can be drawn to.

In Figure 11–16, an event takes place that demands that the Window Server take control of a portion of the screen—for example, displaying a menu. This will

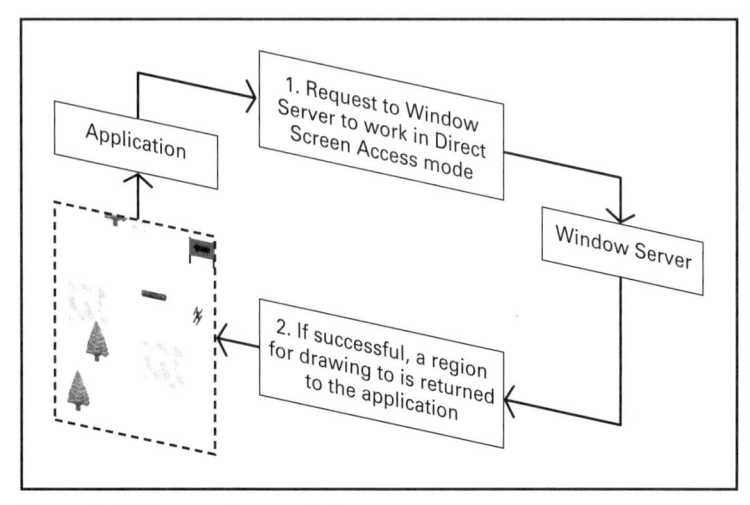

Figure 11–15 Requesting use of direct screen access.

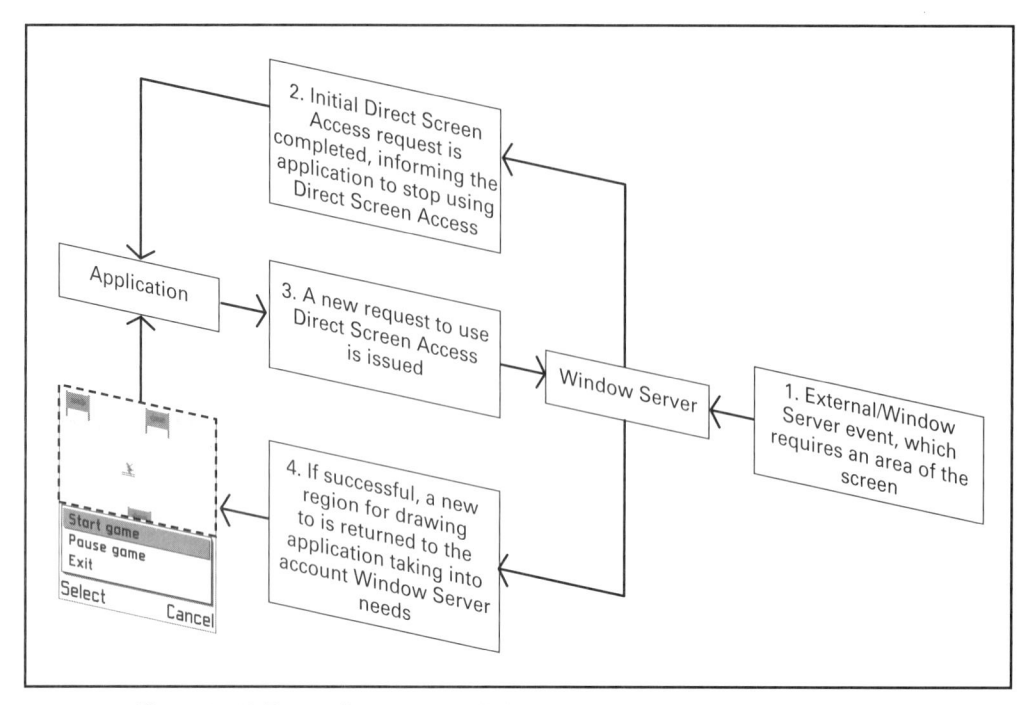

Figure 11–16 Responding to external/window server events when using direct screen access.

result in Direct Screen Access being stopped. Then a new request is made; however, this time the region available for drawing will differ, to account for the Window Server requirements (in this case, displaying a menu).

Key Classes for Direct Screen Access

Table 11–11 provided a brief introduction to the DSA classes. A more thorough examination of each of the key classes is provided here, and their role is further explained with support from code snippets taken from the **Skiing** application.

RDirectScreenAccess

The `RDirectScreenAccess` class represents the client interface to the Direct Screen Access Server. Using the `RDirectScreenAccess::Request()` function, a request is made to the Window Server to operate in Direct Screen Access mode. If the request is granted, the region of the screen available to be directly drawn to is returned. It is important to understand that this region may change if menus or dialogs are displayed—for example, system notifications, such as "message received" notes.

The `Request()` function is asynchronous and must be passed a `TRequestStatus` object—this request will complete only if Direct Screen Access must be aborted. `RDirectScreenAccess` also has a `Cancel()` method which tells the Window Server that Direct Screen Access mode is no longer required.

Chapter 11 Direct Screen Access

In the example **Skiing** game, exposure to `RDirectScreenAccess` is minimal as the necessary interactions are encapsulated by the use of the `CDirectScreenAccess` class. If you were to take the step of writing your own wrapper class then the key interaction issues would be:

- Handling the completion of `RDirectScreenAccess::Request()` (abortion of DSA) via a Window Server session.
- Managing the graphics device and context being drawn to.
- Keeping track of the region of the screen being drawn to, as this can change if menus or dialogs are displayed.

In most cases, it is far simpler to merely use the provided `CDirectScreenAccess` class, but advanced developers need to be aware of the above issues relating to the use of `RDirectScreenAccess`.

CDirectScreenAccess

`CDirectScreenAccess` hides the complexity of managing `RDirectScreenAccess` by creating a high-priority Active Object "wrapper" around it. Following a successful request to use Direct Screen Access, you are free to draw using the graphics context provided by the `CDirectScreenAccess::Gc()` function.

If Direct Screen Access has to be aborted, the `TRequestStatus` that was earlier passed into `RDirectScreenAccess::Request()` completes, and this will result in the `CDirectScreenAccess::RunL()` function being called. The application is informed of completion via a `MDirectScreenAccess` observer object that is passed in to `CDirectScreenAccess` at construction (see "MAbortDirectScreenAccess and MDirectScreenAccess" later in this section).

`CDirectScreenAccess` is a high-priority Active Object, so the `RunL()` method will be called quickly.

TIP As applications need to be aware of external events, you may find it useful to write an "Adapter" or "Wrapper" class around `CDirectScreenAccess` in order to make handling these events more manageable. In the **Skiing** game, the `CDSAWrapper` class performs this task.

In the example **Skiing** game `CDirectScreenAccess` is constructed in the following way:

```
...
    iDSA = CDirectScreenAccess::NewL(iWs, iScreenDevice,
iWindow, aObserver);
...
```

The parameters passed into `CDirectScreenAccess::NewL()` all play a vital role. The Window Server session of your application, `iWs`, is used as the channel for interaction with the Window Server. The screen device, `iScreenDevice`, is used to provide the framework with a pointer to the graphics device in use by the Window Server. The window parameter, `iWindow`, represents a handle to the window that will be drawn to. The final parameter, `aObserver`, is the class being used to manage the aborting and restarting of Direct Screen Access.

As `CDirectScreenAccess` is an Active Object, it needs to be active in order to handle the completed requests of `RDirectScreenAccess`. This is achieved by simply calling the `CDirectScreenAccess::StartL()` function, which, in turn, issues `RDirectScreenAccess::Request()` and stores the region available to draw to:

```
...
if (iDSA && !iDSA->IsActive())
    {
    iDSA->StartL();
    }
...
```

In order to perform drawing, a handle to the graphics context being used by the Direct Screen Access framework is needed. Once all drawing is complete, it is necessary to update the screen by calling the `Update()` function of the graphics device:

```
...
// obtain handle to DSA graphics context
CFbsBitGc* gc = iDSA->Gc();
...
// call some drawing functions on gc
...
// Update DSA graphics device
iDSA->ScreenDevice()->Update();
...
```

MAbortDirectScreenAccess and MDirectScreenAccess

`MAbortDirectScreenAccess` and `MDirectScreenAccess` are the mixin interface classes that specify the functions used to stop and then restart Direct Screen Access. Each adds a single function definition—`AbortNow()` in the case of `MAbortDirectScreenAccess`, and `Restart()` in the case of `MDirectScreenAccess`.

Both of these functions have a single parameter of type `RDirectScreenAccess::TTerminationReasons`, which provides you with the reason why Direct Screen Access had to be stopped. Note that different approaches may be required when handling an abort or restart, depending on the value of this parameter.

Chapter 11 Direct Screen Access

The `MDirectScreenAccess` observer reference passed into the constructor for `CDirectScreenAccess` allows your application to be informed when Direct Screen Access must be stopped (by calling `AbortNow()`) or can be restarted (by calling `Restart()`). Note that `MDirectScreenAccess` is derived from `MAbortDirectScreenAccess` and so actually provides both of these function definitions. Your concrete observer class must derive from `MDirectScreenAccess` and implement these functions—in the **Skiing** game, the `CSkiGameLoop` class performs this task.

As a rule of thumb, your observer class should at least have a reference to the `CDirectScreenAccess` class. It is highly likely that when `AbortNow()` is called, `CDirectScreenAccess::Cancel()` will be called. Likewise, when `MDirectScreenAccess::Restart()` is invoked, a resulting call to `CDirectScreenAccess::StartL()` will probably take place.

Implementation Considerations

In order to make applications respond properly to external events, it is essential that the `CAknAppUi::HandleForegroundEventL(TBool aForeground)` function be overridden. If `aForeground` is `ETrue`, the application has come into focus and Direct Screen Access should be started; if it is `EFalse`, then focus has been lost and Direct Screen Access should be stopped. The implementation from the **Skiing** game example is given here:

```
void CSkiingAppUi::HandleForegroundEventL(TBool aForeground)
    {
    // This is necessary so that the application behaves
    // itself just in case there's another application
    // that is using Direct Screen Access
    if (aForeground)
        {
        if (!iGamePaused)
            {
            iAppContainer->StartDSA();
            }
        SetKeyBlockMode(ENoKeyBlock);
        }
    else
        {
        iAppContainer->StopDSA();
        SetKeyBlockMode(EDefaultBlockMode);
        }
    }
```

An issue of concern when employing the Direct Screen Access classes is that only one application can instantiate them at any given point in time—when stopping Direct Screen Access due to `HandleForegroundEventL()` or similar, you should delete your DSA class instances so that another application can potentially use Direct Screen Access.

TIP From within an application's AppUi class, calling `CAknAppUi::SetKeyBlockMode(ENoKeyBlock)` enables applications to be responsive to users pressing more than one key simultaneously. This is essential for games, when implementing functionality such as running and jumping at the same time.

In order to ensure smooth drawing, you should use double buffering before drawing to the graphics context provided by Direct Screen Access. Choosing an appropriate redraw rate is also important, and you should be aware that this will be hardware dependent—do not exceed the physical refresh rate of the device.

Image Manipulation

Image manipulations are performed using the Series 60 multimedia architecture. Dedicated APIs are provided for image rotation, image scaling and converting to and from a variety of industry-standard file types, such as JPEG and GIF. Between Series 60 1.x and 2.x the architecture was revised and improved upon, although the key functionality remains similar.

Under Series 60 1.x the architecture is referred to as the Multi Media Server, while for Series 60 2.x it is the Multi Media Framework. Code written for Series 60 1.x is still compatible with Series 60 2.x, but you cannot take advantage of any enhancements, such as multithreading and the ability to write new conversion plug-ins. As most operations take place asynchronously, when using the multimedia APIs, an understanding of Active Objects is imperative.

From an implementation perspective, the key difference between the alternative Series 60 versions is handling the asynchronous nature of the image manipulations. Under Series 60 1.x an observer mechanism, based on the `MMdaImageUtilObserver` class, is used. By deriving from this class and implementing its virtual functions, applications are informed of the completion of a manipulation and can respond appropriately.

When developing for Series 60 2.x, a `CActive`-derived class should be created and the `TRequestStatus` member `iStatus` passed into the function performing the manipulation. This will result in the `RunL()` of the `CActive`-derived class being called when the manipulation is finished.

The example code shown in this section is taken from two example applications: **ImageManip** for Series 60 1.x code and **MultiMediaF** for Series 60 2.x code.

Image Conversion

Conversion can take place to and from images stored in files or descriptors. Following a decode operation, if an image is to be displayed onto the device screen, a bitmap object of type `CFbsBitmap` must be created.

Chapter 11 Image Manipulation

Conversely, if the image is already displayed on screen—for example, within a camera application—and needs to be saved as a JPEG, its data will currently be contained within a `CFbsBitmap` object. Consequently, the `CFbsBitmap` class plays an important role in the conversion process, acting as a bridging point between both encoding and decoding.

Decoding

The principles for decoding an image are the same, regardless of the Series 60 version being developed for. First, a channel to the data store, such as a GIF file, is opened, and this provides essential image attributes such as the image size.

When this is complete, a `CFbsBitmap` object can be instantiated with respect to the attributes obtained; however, it is still lacking the image data that will be displayed. The conversion operation takes place, resulting in the image data being stored inside the bitmap object, ready for use by application code.

Series 60 1.x

`CMdaImageFileToBitmapUtility` and `CMdaImageDescToBitmapUtility` are used to decode from files and descriptors, respectively. Both derive from `CMdaImageUtility` and provide two asynchronous functions, `OpenL()` and `ConvertL()`, which will be used in the decoding process. At the time of its creation, the decoding class is passed a reference to an `MMdaImageUtilObserver` object to inform applications that the asynchronous functions have completed.

```
void CImageManipAdapter::DecodeOpenL()
    {
    iFileToBitmap->OpenL(KPathOfGifFile);
    }
```

`OpenL()` will invoke `MMdaImageUtilObserver::MiuoOpenComplete()`:

```
void CImageManipAdapter::MiuoOpenComplete(TInt aError)
    {
    if (aError == KErrNone)
        {
        switch (iManipulationState)
            {
            ...
            case EDecode:
                TRAPD(err, DecodeConvertL());
                break;
            ...
            }
        }
    }
```

If the manipulation state is set to `EDecode`, then `DecodeConvertL()` is called. From this function, a call to `CMdaImageUtility::FrameInfo()` is made, and this provides the vital image data to construct an appropriate bitmap:

```
void CImageManipAdapter::DecodeConvertL()
  {
  TFrameInfo frmInfo;
  iFileToBitmap->FrameInfo(KGifIndex, frmInfo);
  iImage->Create(frmInfo.iOverallSizeInPixels,
iDeviceDisplayMode);
  iFileToBitmap->ConvertL(*iImage, KGifIndex);
  }
```

`ConvertL()` will call `MMdaImageUtilObserver::MiuoConvertComplete()`:

```
void CImageManipAdapter::MiuoConvertComplete(TInt aError)
  {
  if (aError == KErrNone)
     {
     switch (iManipulationState)
        {
        case EDecode:
        case EEncode:
        case ERotating:
        case EScaling:
           iManipulationState = EDoNothing;
           iMultimediaController.RedrawView();
           break;
        default:
           break;
        }
     }
  }
```

Series 60 2.x

Unlike Series 60 1.x, there is a single decoding class in Series 60 2.x, `CImageDecoder`; however, it has two synchronous functions to perform the "opening" step. `CImageDecoder::FileNewL()` is used for files and `CImageDecoder::DataNewL()` for descriptors.

The Multi Media Framework has an extensible plug-in architecture for image **codecs** (coder/decoders), and both functions are overloaded so that the correct codec can be selected to decode the image. These overloaded functions have the option to perform codec selection in one of four ways: automatically, via a MIME type, using an image type UID, or using a plug-in UID. A plug-in UID is passed into the "opening" function if two plug-ins are available that can decode the same image type, but a preference exists as to which one should perform the decoding.

To discuss all of the approaches in detail is beyond the scope of this chapter, but further information can be found in the SDK documentation, where each approach is extensively illustrated. In most cases, automatic detection will suffice, as many industry-standard file formats are already provided for.

Chapter 11 Image Manipulation

```
void CMultiMediaFAdapter::DecodeOpenAndConvertL()
  {
  delete iDecoder;
  iDecoder = 0;
  iDecoder = CImageDecoder::FileNewL(iFs, KPathOfGifFile);
  ...
```

Once a handle to the image store has been acquired, the bitmap object has to be set up correctly using data obtained from a call to the `CImageDecoder::FrameInfo()` function. This is analogous to `CMdaImageUtility::FrameInfo()` from Series 60 1.x. `CImageDecoder` also has a number of other frame interrogation functions, such as `FrameData()`, which returns image quality and copyright information.

```
  ...
  TFrameInfo frmInfo = iDecoder->FrameInfo(KGifIndex);
  TRect rectOfImage = frmInfo.iFrameCoordsInPixels;

  delete iImage;
  iImage = 0;
  iImage = new (ELeave) CFbsBitmap();
  iImage->Create(rectOfImage.Size(), iDeviceDisplayMode);
  ...
```

As `CImageDecoder::FileNewL()` and `CImageDecoder::DataNewL()` are synchronous, there is no need for specific management of the completion of either operation. The conversion step, however, is asynchronous and is performed by `CImageDecoder::Convert()`.

`Convert()` takes a `TRequestStatus` object as a parameter, and uses this to inform the calling class that the conversion has completed. By ensuring the calling class is an Active Object (derived from `CActive`), its `TRequestStatus` member, `iStatus`, can be passed into the `Convert()` method, and when the conversion is finished the `RunL()` method of the calling class will be invoked. It is worth remembering that `RunL()` will be called only if the Active Object is "active" — in other words, its `iActive` member is `ETrue`. This requires `SetActive()` to be called immediately after `CImageDecoder::Convert()`:

```
  ...
  // Decode to the bitmap asynchronously.
  // RunL will be called when decoding is completed.
  iDecoder->Convert(&iStatus, *iImage, KGifIndex);
  SetActive();
  }
```

Encoding

Encoding, like decoding, is a three-stage process. The data store, be it a file or descriptor, must be created as the destination for the encoding operation. An image data object is then instantiated with the correct data settings relating to the file type. Finally, the conversion takes place with respect to the data settings passed into the conversion function.

Series 60 1.x

`CMdaImageBitmapToFileUtility` is used to encode to files, while `CMdaImageBitmapToDescUtility` encodes to descriptors. Both have a `CreateL()` function in order to construct the data store.

```
void CImageManipAdapter::EncodeOpenL()
    {
    // set up the jpg saving information that is required
    iJpgFormat->iSettings.iQualityFactor = KJpgQuality;

    iJpgFormat->iSettings.iSampleScheme =
TMdaJpgSettings::TColorSampling(TMdaJpgSettings::EColor420);

    TMdaPackage* codec = NULL;
    iBitmapToFile->CreateL(KPathOfSaveJpgFile, iJpgFormat,
codec, NULL);
    }
```

`MMdaImageUtilObserver::MiuoCreateComplete()` is used to notify applications that creation has completed.

Like their decoding equivalents, `CMdaImageBitmapToFileUtility` and `CMdaImageBitmapToDescUtility` derive from `CMdaImageUtility` and have an asynchronous `ConvertL()` method. `ConvertL()` can be passed image format data—for example, the compression rate if the image is to be a JPEG, and codec information. The image format is contained within the `TMdaClipFormat` class, while `TMdaPackage` encapsulates the codec.

```
void CImageManipAdapter::MiuoCreateComplete(TInt aError)
    {
    if (aError == KErrNone)
        {
        switch (iManipulationState)
            {
            case EEncode:
                TRAPD(err, EncodeConvertL());
                break;
            default:
                break;
            }
        }
    }
```

Once the conversion is complete, the observer function `MiuoConvertComplete()` is called:

Chapter 11 Image Manipulation

```
void CImageManipAdapter::MiuoConvertComplete(TInt aError)
   {
   if (aError == KErrNone)
      {
      switch (iManipulationState)
         {
         case EDecode:
         case EEncode:
         case ERotating:
         case EScaling:
            iManipulationState = EDoNothing;
            iMultimediaController.RedrawView();
            break;
         default:
            break;
         }
      }
   }
```

Series 60 2.x

`CImageEncoder` is employed to encode bitmaps and has two overloaded functions for creating the data store that will receive the converted image. Like `CImageDecoder`, `DataNewL()` is used if a descriptor is the destination, while `FileNewL()` is used if a file is.

Both functions are overloaded to give the same flexibility previously seen when decoding—this allows encoding to take place via a supplied MIME type, image type UID, or encoder UID. The creation of the data store is a synchronous operation, so explicit completion handling is unnecessary:

```
void CMultiMediaFAdapter::EncodeOpenAndConvertL()
   {
   // Create a new file ready for encoding.
   delete iEncoder;
   iEncoder = 0;

   iEncoder = CImageEncoder::FileNewL(iFs,
KPathOfSaveJpgFile(), KJpgMime(), CImageEncoder::EOptionNone);
   ...
```

Before converting through the `CImageEncoder::Convert()` method, it is possible to set certain attributes relating to the image frame data using the `CImageFrameData` class. An example of this might be setting the number of colors in the image palette of a JPEG. Once the desired elements have been set up, the `CImageFrameData` can be passed in to `CImageEncoder::Convert()`.

```
   ...
   // Set up the jpg saving information that is required.
   TJpegImageData* jpgData = new (ELeave) TJpegImageData();
   // jpgData will be deleted by the Multi Media Framework
   CleanupStack::PushL(jpgData);
```

```
    jpgData->iQualityFactor = KJpgQuality;
    jpgData->iSampleScheme = TJpegImageData::EColor420;

    // Create the new image data.
    delete iJpgImageData;
    iJpgImageData = 0;
    iJpgImageData = CFrameImageData::NewL();

    // Set the saving information.
    // Passes ownership if successful.
    User::LeaveIfError(iJpgImageData->
AppendImageData(jpgData));
    // jpgData now owned by iJpgImageData.
    CleanupStack::Pop(jpgData);

    // Begin conversion process
    // RunL will be called when encoding is completed.
    iEncoder->Convert(&iStatus, *iImage, iJpgImageData);
    SetActive();
}
```

`CImageEncoder::Convert()` is also called within an Active Object class—as with `CImageDecoder::Convert()`, the `RunL()` method of the Active Object is called when the asynchronous processing completes:

```
void CMultiMediaFAdapter::RunL()
    {
    switch (iManipulationState)
        {
        ...
        case EEncode:
            {
            iManipulationState = EDoNothing;
            iMultimediaController.RedrawView();
            break;
            }
        ...
        }
    }
```

A key enhancement of the Series 60 2.x Multi Media Framework is the ability for each encoding and decoding operation to take place within its own thread, as previously only an Active Object-based implementation existed.

The encoding or decoding object is notified of the desire to use threads by passing in the enumeration value `CImageDecoder::EOptionAlwaysThread` to the `DataNewL()` or `FileNewL()` function. Plug-ins themselves can be designed to run in their own thread, and the flag will be ignored should a plug-in be selected where this is the case. Because the resources of a Series 60 device are limited, it is not always best to use multithreading, as there is an overhead in switching between threads. However, the nonpreemptive nature of

Active Objects means that there is a potential loss of responsiveness while an Active Object executes. It is also worth noting that each new image to be encoded or decoded requires a new `CImageEncoder` or `CImageDecoder` object to be created.

Image Rotation

The image rotation classes are very similar on both versions of Series 60. Specific rotation angles cannot be user defined—however, the `TRotationAngle` enumeration provides a selection of valid values.

Series 60 1.x

The `CMdaBitmapRotator` class performs image rotation through its asynchronous `RotateL()` function. `CMdaBitmapRotator::RotateL()` is overloaded to allow for two types of rotation: one to a separate target bitmap object, and the other to the original bitmap object itself.

```
void RotateL(MMdaImageUtilObserver& aObserver, CFbsBitmap&
aSrcBitmap, CFbsBitmap& aTgtBitmap, TRotationAngle aAngle);

void RotateL(MMdaImageUtilObserver& aObserver, CFbsBitmap&
aSrcBitmap, TRotationAngle aAngle);
```

`TRotationAngle` allows images to be rotated clockwise in 90-, 180- and 270-degree increments.

```
    iRotator->RotateL(*this, *iImage,
CMdaBitmapRotator::ERotation90DegreesClockwise);
    iRotator->RotateL(*this, *iImage,
CMdaBitmapRotator::ERotation180DegreesClockwise);
    iRotator->RotateL(*this, *iImage,
CMdaBitmapRotator::ERotation270DegreesClockwise);
```

Application code is informed of the completion of a rotation through the `MMdaImageUtilObserver` mechanism with a resulting call to `MMdaImageUtilObserver::MiuoConvertComplete()`.

Series 60 2.x

`CBitmapRotator` carries out image rotation via the asynchronous `Rotate()` function. Once more a `TRequestStatus` object is used to inform the calling class that the operation is complete. As with `CMdaBitmapRotator::RotateL()`, there are two versions of the `CBitmapRotator::Rotate()` function available, depending on whether or not the resultant rotation is placed into a separate bitmap object.

```
    void Rotate(TRequestStatus* aRequestStatus, CFbsBitmap&
aSrcBitmap, CFbsBitmap& aTgtBitmap, TRotationAngle aAngle);

    void Rotate(TRequestStatus* aRequestStatus, CFbsBitmap&
aBitmap, TRotationAngle aAngle);
```

Chapter 11 Multimedia: Graphics and Audio

The `TRotationAngle` enumeration has been enhanced to also include `EMirrorHorizontalAxis` and `EMirrorVerticalAxis`, which, as their names suggest, will mirror the image along the horizontal and vertical axis, respectively.

```
    iRotator->Rotate(&iStatus, *iImage,
CBitmapRotator::ERotation90DegreesClockwise);
    iRotator->Rotate(&iStatus, *iImage,
CBitmapRotator::ERotation180DegreesClockwise);
    iRotator->Rotate(&iStatus, *iImage,
CBitmapRotator::ERotation270DegreesClockwise);
    iRotator->Rotate(&iStatus, *iImage,
CBitmapRotator::EMirrorHorizontalAxis);
    iRotator->Rotate(&iStatus, *iImage,
CBitmapRotator::EMirrorVerticalAxis);
```

Image Scaling

Scaling an image is programmatically similar to rotation.

Series 60 1.x

`CMdaBitmapScaler` executes scaling using the asynchronous `ScaleL()` function. Again, `MMdaImageUtilObserver::MiuoConvertComplete()` is called on completion. Like rotation, scaling can result in a new bitmap object, or merely be performed on an existing bitmap.

```
void CImageManipAdapter::ScaleImageL()
   {
   if (iScaleState == EScaleDown)
      {
      iScaler->ScaleL(*this, *iImage, TSize(KScaledWidthOne,
KScaledHeightOne));
      iScaleState = EScaleUp;
      }
   else
      {
      iScaler->ScaleL(*this, *iImage, TSize(KScaledWidthTwo,
KScaledHeightTwo));
      iScaleState = EScaleDown;
      }
   }
```

Series 60 2.x

The `CBitmapScaler` class is used to scale bitmaps by calling its `Scale()` method. `Scale()` is overloaded to allow for scaling to a separate `CFbsBitmap` object. As `Scale()` is asynchronous, it should be called by an Active Object, which provides the `TRequestStatus` parameter, as required.

```
void CMultiMediaFAdapter::ScaleImage()
   {
   if (iScaleState == EScaleDown)
```

Chapter 11 Image Manipulation

```
      {
      iScaler->Scale(&iStatus, *iImage,
TSize(KScaledWidthOne, KScaledHeightOne));
      iScaleState = EScaleUp;
      }
   else
      {
      iScaler->Scale(&iStatus, *iImage,
TSize(KScaledWidthTwo, KScaledHeightTwo));
      iScaleState = EScaleDown;
      }
   SetActive();
   }
```

As has been discussed, there are subtle differences between the multimedia architectures of Series 60 1.x and Series 60 2.x. There is a facility to use multithreading on Series 60 2.x, while handling the asynchronous nature of the manipulation APIs is performed by an observer on Series 60 1.x, and a user-defined Active Object on Series 60 2.x. Table 11–12 summarizes the main differences between platform versions.

Table 11–12 Overview of Differences between Series 60 1.x and 2.x

Operation	Series 60 1.x	Series 60 2.x
Execution	Active Objects only	Threads or Active Objects
Asynchronous Function Handling	`MMdaImageUtilObserver` mechanism	`CActive::RunL()` approach
Decoding	`CMdaImageDescToBitmapUtility` `CMdaImageFileToBitmapUtility`	`CImageDecoder`
Encoding	`CMdaImageBitmapToDescUtility` `CMdaImageBitmapToFileUtility`	`CImageEncoder`
Rotation	`CMdaBitmapRotator`	`CBitmapRotator`
Scaling	`CMdaBitmapScaler`	`CBitmapScaler`
Import Library	`MediaClientImage.lib`	`ImageConversion.lib` (encode/decode) `BitmapTransforms.lib` (rotation/scaling)
Include Header	`MediaClientImage.h`	`ImageConversion.h` (encode/decode) `BitmapTransforms.h` (rotation/scaling)

TIP
You should use an image-manipulation "Adapter" class to contain all of the platform-dependent manipulation code needed—application code then only has to interact with this adapter. For Series 60 1.x your adaptor should derive from `MMdaImageUtilObserver`, while on Series 60 2.x it should derive from `CActive`. The adapter class will probably need to be implemented as some form of state machine to keep track of the current operation—an enumeration value is often used to represent the current state.

Audio

The multimedia architecture is used to perform audio operations, such as recording and playing. A variety of different audio formats—for example, `.wav` and `.midi`—are supported for use within applications.

To handle the various formats a plug-in architecture is used to match the file format to an appropriate codec (coder/decoder). Besides this, it is also possible to explicitly select a codec, which is necessary when dealing with raw audio data.

Utility classes are available for carrying out the essential audio tasks, and provision is made for recording, streaming, converting between audio formats, tone playing and audio file playing. From a technical standpoint, an observer mechanism is used, with each utility having a dedicated observer class and callbacks. This allows applications to respond to the asynchronous nature of interacting with the audio utility classes.

Consequently, in a similar vein to the approach used in managing image manipulation, it can be useful to write an "Adapter" class, deriving from the observer, which implements the callbacks. The libraries of interest when developing applications that use sounds are listed in Table 11–13.

Table 11–13 Series 60 Audio Libraries

Library (.lib)	Role	Key Elements
`MediaClientAudio.lib`	Playing and recording of sounds. Conversion between audio formats.	Audio sample editor, audio sample player and audio tone player.
`MediaClientAudioStream.lib`	Allows multimedia clients to stream audio data.	Audio streaming.

Unlike image manipulation, the changes between Series 60 1.x and Series 2.x, have been more "behind-the-scenes," so differences will be highlighted as and when they appear. The various aspects of Series 60 sound are illustrated across two example applications, **AnsPhone** and **Audio**.

Recording

Recording is made possible through the `CMdaAudioRecorderUtility` class. In addition to recording in different audio formats, this class also offers a playback facility. While this can save resources, by removing the need to instantiate a separate player object, use of one of the player objects discussed later in this section, such as `CMdaAudioPlayerUtility`, is recommended for dedicated playback.

In terms of physical storage, recording can take place to a file, a descriptor and in Series 60 2.x, a URL. Recording is essentially a three-stage process of gaining a handle to a data store, setting the desired configurations and then recording.

Accessing the data store and physically recording are asynchronous operations, and the `MMdaObjectStateChangeObserver::MoscoStateChangeEvent()` function informs applications of the completion of each operation. This architecture requires the class calling the asynchronous functions to derive from `MMdaObjectStateChangeObserver` and implement `MoscoStateChangeEvent()`.

The **AnsPhone** example application has a class `CAnsPhoneEngine`, which derives from `MMdaObjectStateChangeObserver` and implements `MoscoStateChangeEvent()`:

```
class CAnsPhoneEngine
: public CBase,
  public MMdaObjectStateChangeObserver,
  public MAnsPhoneTimerObserver,
  public MAnsPhoneCallWatcherObserver,
  public MAnsPhoneCallMakerObserver,
  public MAnsPhoneCallLogObserver
  {
  ...
public:  // MMdaObjectStateChangeObserver
   virtual void MoscoStateChangeEvent(CBase* aObject, TInt
aPreviousState, TInt aCurrentState, TInt aErrorCode);

   ...
   CMdaAudioRecorderUtility*    iSound;
   ...
   // Recording message settings
   TMdaFileClipLocation    iMessageLocation;
   TMdaAudioDataSettings   iMessageSettings;
   TMdaWavClipFormat       iMessageFormat;
   TMdaPcmWavCodec         iMessageCodec;
   ...
   };
```

Chapter 11 Multimedia: Graphics and Audio

`CMdaAudioRecorderUtility` provides an `OpenL()` function that opens a channel to the data store that should be used as the destination for the recording or, if the class is being employed for playback, the audio data that will be played. As the correct codec for handling raw audio data cannot be selected automatically, an explicit reference to a codec may be necessary.

In addition to `OpenL()`, functions are also available for specific destination types. Series 60 1.x offers `OpenFileL()` and `OpenDesL()` for files and descriptors, respectively, while Series 60 2.x also has `OpenUrlL()`.

In the **AnsPhone** application, `CMdaAudioRecorderUtility::OpenL()` is used in the `RecordMessageL()` method:

```
iSound = CMdaAudioRecorderUtility::NewL(*this);
...
iSound->OpenL(&iMessageLocation, &iMessageFormat,
&iMessageCodec, &iMessageSettings);
...
```

Once the open operation has completed, the function `MoscoStateChangeEvent()` (from the `MMdaObjectStateChangeObserver` mixin) will be called.

Before recording, a variety of configuration settings are available, including: volume, sample rate and balance. These settings are extensively covered in the SDK documentation. The **AnsPhone** application sets the gain, volume and recording position before recording.

Once the required configuration is set, `CMdaAudioRecorderUtility::RecordL()` can be invoked, remembering that `MoscoStateChangeEvent()` will be called once more when recording has finished:

```
void CAnsPhoneEngine::MoscoStateChangeEventL(CBase*
/*aObject*/, TInt /*aPreviousState*/, TInt aCurrentState,
TInt aErrorCode)
    {
    ...
    switch (iState)
        {
        case ERecordInit:
            // Record from the telephony line
            // and set to max gain
            if (iIsLocal)
                {
                iSound->SetAudioDeviceMode(
CMdaAudioRecorderUtility::ELocal);
                iSound->SetGain(iSound->MaxGain());
                }
            else
                {
                iSound->SetAudioDeviceMode(
```

Chapter 11 Audio

```
CMdaAudioRecorderUtility::ETelephonyNonMixed);
        TInt maxGain = iSound->MaxGain();
        iSound->SetGain(maxGain / 2);
        }

        // Delete current audio sample from beginning of file
        iSound->SetPosition(TTimeIntervalMicroSeconds(0));
        iSound->CropL();

        // start recording
        iSound->RecordL();
        iState = ERecord;
        break;
        ...
    }
```

Note that the above function is actually called from `MoscoStateChangeEvent()` within a trap harness, as it may leave.

Another important consideration is the `TDeviceMode` enumeration, which determines how the audio device handles playback and recording. This enumeration is defined in the file `\epoc32\include\MdaAudioSampleEditor.h` in the root directory of your SDK. Table 11–14 examines the options available.

Table 11–14 Audio Device Modes

Enumeration Value	Effect
`EDefault = 0`	If a call is currently taking place, then the line and the device microphone will be recorded, and playback will be made down the line. If there is no call, then the device microphone will be recorded, and playback will be through the device speaker.
`ETelephonyOrLocal = EDefault`	As with `EDefault`.
`ETelephonyMixed = 1`	If a call is in progress, then the audio source is the device microphone, mixed with the telephony line. Playback will be both through the device speaker and down the phone line. If there is no call, then recording is not attempted to the device microphone, and playback will not take place to the device speaker.

Chapter 11 Multimedia: Graphics and Audio

Table 11-14 continued

Enumeration Value	Effect
`ETelephonyNonMixed = 2`	If there is a call in progress, then only the telephone line is used for recording and playback—it is not mixed with the device microphone or speaker.
`ELocal = 3`	The telephone line is ignored. The device microphone will be recorded, and the device speaker is used for playback.

> **WARNING** There is a known issue with Series 60 1.x and some early devices—the default mode will always be used by `CMdaAudioRecorderUtility`, despite any values passed into `SetAudioDeviceMode()`. An up-to-date list of known Series 60 issues can be found at **www.forum.nokia.com**.

Although primarily used for recording, as highlighted earlier, `CMdaAudioRecorderUtility` can be used for playback purposes. Opening the data source and configuring the playback options are the same as for recording; however, the `CMdaAudioRecorderUtility::PlayL()` function is used to play the audio data.

```
void CAnsPhoneEngine::MoscoStateChangeEventL(CBase*
/*aObject*/, TInt /*aPreviousState*/, TInt aCurrentState,
TInt aErrorCode)
    {
    ...
    switch (iState)
        {
        ...
        case EPlayInit:
            {
            // Play through the device speaker
            // and set to max volume
            if (iIsLocal)
                {
                iSound->SetAudioDeviceMode(
CMdaAudioRecorderUtility::ELocal);
                }
            else
                {
```

Chapter 11 Audio

```
            iSound->SetAudioDeviceMode(
CMdaAudioRecorderUtility::ETelephonyOrLocal);
            }

            iSound->SetVolume(iSound->MaxVolume());

            // Set the playback position to the start
            // of the file
            iSound->SetPosition(TTimeIntervalMicroSeconds(0));
            iSound->PlayL();
            ...
    }
```

Tones

The `CMdaAudioToneUtility` class plays audio tones. Tones can be sine waves, or Dual Tone Multi Frequency (**DTMF**) telephony signals (tone-dialing "touch-tones"). As with recording, an observer is used for notifications and callbacks—in this case `MMdaAudioToneObserver`.

For notification purposes, `MMdaAudioToneObserver` provides two pure virtual methods: `MatoPrepareComplete()`, to inform the client that the tone player is configured properly to play, and `MatoPlayComplete()`, which is called once the tone has finished playing. Both of these functions accept a `TInt` parameter to indicate if any errors have occurred. If `KErrNone` is passed into either function, then the action was successful; otherwise the error value must be interpreted, and appropriate action taken.

Following the construction of `CMdaAudioToneUtility`, a tone can be generated by the player through one of the variants of the `CMdaAudioToneUtility::PrepareToPlayXXX()` function. These variants cover playing DTMFs and tone sequences from files and descriptors. The `PrepareToPlayTone()` function will play a single tone based on the frequency and duration parameters provided to it.

```
void CAudioPlayerEngine::PlayToneL()
    {
    iState = EPlaying;
    iPlayerTone = CMdaAudioToneUtility::NewL(*this);
    iPlayerTone->PrepareToPlayTone(KToneFrequency,
TTimeIntervalMicroSeconds(KToneDuration));
    }
```

Once `MatoPrepareComplete()` has been called, the player can be configured in many ways—for example, setting the volume, and in Series 60 2.x, the balance. A fuller description of the configuration options can be found in the SDK documentation.

Chapter 11 Multimedia: Graphics and Audio

`CMdaAudioToneUtility::Play()` can now be called, and `MatoPlayComplete()` will inform the application that playing has completed. Playing can be aborted before completion by calling `CMdaAudioToneUtility::CancelPlay()`.

```
// implementation of MMdaAudioToneObserver virtual function
void CAudioPlayerEngine::MatoPrepareComplete(TInt aError)
    {
    if (aError)
        {
        Stop();
        }
    else
        {
        iPlayerTone->SetVolume(iPlayerTone->MaxVolume() / KToneVolumeDenominator);
        iPlayerTone->Play();
        }
    }
```

Audio Data

The playing of audio data, such as .wav and .midi, is made possible by the `CMdaAudioPlayerUtility` class and its associated observer, `MMdaAudioPlayerCallback`. This observer has two functions to inform the client application of the current status of `CMdaAudioPlayerUtility`: `MapcInitComplete()` indicates that the data source has been opened successfully, while `MapcPlayComplete()` signals that playing has concluded. As with the callback functions used in tone playing, both accept a `TInt` parameter to indicate if any errors have taken place.

While the differences between Series 60 1.x and Series 60 2.x are minimal in the other audio utility classes, they are much more noticeable with `CMdaAudioPlayerUtility`, as its feature set has greatly increased.

Under Series 60 1.x, `CMdaAudioPlayerUtility` has three instantiation functions for reading audio data from files, descriptors and read-only descriptors, named `NewFilePlayerL()`, `NewDesPlayerL()`, and `NewDesPlayerReadOnlyL()`, respectively. These functions are asynchronous, and on completion will invoke `MapcInitComplete()`.

```
void CAudioPlayerEngine::PlayWavL()
    {
    TFileName wavFile(KToPlayFileWav);
    User::LeaveIfError(CompleteWithAppPath(wavFile));
    iState = EPlaying;
    iPlayerFile =
CMdaAudioPlayerUtility::NewFilePlayerL(wavFile, *this);
    }
```

Chapter 11 Audio

In addition to the functions above, when targeting Series 60 2.x there is also `CMdaAudioPlayerUtility::NewL()`, which does not require a specific data source—however, on completion, it too will call `MapcInitComplete()`. Series 60 2.x-specific "open" and "open and play" functionality allows a single player instance to open multiple data sources at different times. Note that in addition to standard files and descriptors, URLs can also be opened for playing:

```
OpenUrlL(const TDesC& aUrl, const TDesC8& aMimeType = KNullDesC8);

OpenAndPlayUrlL(const TDesC& aUrl, const TDesC8& aMimeType = KNullDesC8, TInt aPriority = EMdaPriorityNormal, TMdaPriorityPreference aPref = EMdaPriorityPreferenceTimeAndQuality);
```

Using Series 60 1.x, it is possible to achieve a similar effect to the "open and play" facility of Series 60 2.x by calling `CMdaAudioUtility::Play()` from within the implementation of `MapcInitComplete()`:

```
void CAudioPlayerEngine::MapcInitComplete(TInt aError, const TTimeIntervalMicroSeconds& /*aDuration*/)
    {
    if (aError == KErrNone)
        {
        iPlayerFile->SetVolume(iPlayerFile->MaxVolume() / KToneVolumeDenominator);
        iPlayerFile->Play();
        }
    else
        {
        Stop();
        }
    }
```

Remember, though, that this is optional, and an application might have an explicit "play" feature, which could be invoked at any time, not just immediately after opening the data source.

An audio clip can be stopped at any time by calling `CMdaAudioPlayerUtility::Stop()`. Additionally on Series 60 2.x `CMdaAudioPlayerUtility::Close()` should be invoked, as this will allow new audio data to be played with the same `CMdaAudioPlayerUtility` object.

> **TIP** In Series 60 1.x, MIDI files are not supported for playback in the emulator and will cause a panic. They will, however, play on most Series 60 1.x-based hardware. This is illustrated in the example application, **Audio**.

Chapter 11 Multimedia: Graphics and Audio

The player can be configured like the other audio utilities—for example, setting the volume and balance. Other functions associated with playing, such as getting the duration of a clip and setting the play position within a clip, can be particularly useful when implementing a player control that displays the length of the clip, or has a graphical slider which increments during playback of the audio track. Full details of the APIs available in each version of the platform can be found in the relevant SDK documentation.

Streaming

Audio streams allow applications to play audio without having the entire contents of the audio data. Audio data is accessed and buffered incrementally, and every attempt is made to ensure smooth and continuous playback. To perform this task, the CMdaAudioOutputStream and MMdaAudioOutputStreamCallback classes are used. These two classes provide the mechanisms to manage the flow of buffered audio data within the multimedia architecture, and to pass it on to the lower-level sound device.

Data must be presented to the player in descriptor format, and the only supported audio format that the streamed player can support is 16-bit pulse-code modulation (**PCM**). Usually PCM data contains a 44-byte header that contains information about the audio data payload. Since the streaming audio player accepts only 16-bit PCM, it does not require this header information.

The desired properties of the audio device can be set up in different ways through the CMdaAudioOutputStream class. The SetAudioPropertiesL() method can be used to set the sample rate and number of channels used. Additional attributes, such as the volume, can be controlled by passing a TMdaAudioDataSettings object into the Open() method—calling Open() prepares the player object for streaming:

```
void CAudioPlayerEngine::PlayStreamL()
  {
  // open the file and load it into the buffers
  RFs fs;
  CleanupClosePushL(fs);                           // PUSH
  User::LeaveIfError(fs.Connect());

  RFile file;
  CleanupClosePushL(file);                         // PUSH

  TFileName streamFile(KToPlayFileStream);
  User::LeaveIfError(CompleteWithAppPath(streamFile));
  User::LeaveIfError(file.Open(fs, streamFile, EFileRead |
    EFileShareReadersOnly));

  TInt fileSize = 0;
  file.Size(fileSize);
  iStreamData = new (ELeave) TUint8[fileSize];
```

Chapter 11 Audio

```
    iStreamBuffer = new (ELeave) TPtr8(iStreamData, fileSize,
fileSize);
    file.Read(*iStreamBuffer);

    CleanupStack::PopAndDestroy(2);               // file & fs

    iState = EPlaying;
    iPlayerStream = CMdaAudioOutputStream::NewL(*this);
    iPlayerStream->Open(&iStreamSettings);
    }
```

`MMdaAudioOutputStreamCallback` is used to handle the asynchronous nature of buffering data for streaming audio. It provides three virtual functions to respond to opening the player, copying a buffer of data to the player, and the termination of playing:

- `MaoscOpenComplete()` notifies the client that the stream has been opened following a call to `CMdaAudioOutputStream::Open()` and indicates that the audio output stream is ready for use:

```
void CAudioPlayerEngine::MaoscOpenComplete(TInt aError)
    {
    if (aError == KErrNone)
        {
        TRAPD(err, iPlayerStream->WriteL(*iStreamBuffer));
        }
    else
        {
        Stop();
        }
    }
```

- `MaoscBufferCopied()` is called after data has been placed in the buffer by `CMdaAudioOutputStream::WriteL()`. This informs the client application that the audio data was written successfully. The copied buffer can now be destroyed, freeing up memory, or more data can be written by calling `WriteL()` again:

```
void CAudioPlayerEngine::MaoscBufferCopied(TInt aError, const
TDesC8& /*aBuffer*/)
    {
    if (aError == KErrNone)
        {
        TRAPD(err, iPlayerStream->WriteL(*iStreamBuffer))
        iPlayerStream->SetVolume(iPlayerStream->MaxVolume() /
KToneVolumeDenominator);
        }
    }
```

- `MaoscPlayComplete()` informs the client that streaming has been terminated:

```
void CAudioPlayerEngine::MaoscPlayComplete(TInt /*aError*/)
    {
    iState = EStopped;
    delete iPlayerStream;
    iPlayerStream = NULL;
    delete iStreamBuffer;
    iStreamBuffer = NULL;
    delete [] iStreamData;
    iStreamData = NULL;

    TRAPD(err, iObserver.HandlePlayingStoppedL());
    }
```

> **WARNING** Be aware that, if the buffer size is too small, it could cause the audio to sound as if it were stuttering. This is because the audio device is consuming the buffer faster than the new buffers are being copied.

Summary

This chapter has covered a lot of ground, owing to the wealth of graphical and multimedia APIs provided by Series 60. Sound and graphics can add much user value to your application, so mastering these techniques is important if you want to create desirable applications.

This chapter introduced the fundamental drawing APIs that you will soon become familiar with, such as those for drawing shapes and managing fonts. More advanced graphics techniques, including double buffering and image scaling and rotation, were introduced as you progressed through the chapter. Techniques for producing smooth animations were explained and the Direct Screen Access framework discussed. While animation may not always be appropriate, it can sometimes enhance your application greatly. Direct Screen Access should probably be used exclusively for games.

The addition of sound can also benefit the usability of your application—but take care not to overdo things. This chapter has covered the main APIs for recording and playing sounds and provided example applications to demonstrate their use.

Chapter 11 Summary

It is important to remember that the media architecture has been updated for Series 60 2.x. This means that when you need to consider carefully your target platform(s) before creating your application. Many of the media-related APIs have been deprecated in Series 60 2.x, and alternative APIs introduced. This chapter has pointed out the main changes to the media APIs in Series 60 2.x and should therefore help you to make an informed choice about which to use.

Having read this chapter, you now have strong foundations with which to create visually compelling applications, potentially enhanced further by the wide variety of audio utilities available.

chapter 12

Using Application Views, Engines and Key System APIs

Covering how to invoke published standard application views, the use of a number of application engines and several useful system functions from within applications

A number of the standard Series 60 applications, such as Contacts and Calendar, publish selected views for use by other applications—for example, to allow listing and selection of a contact from within another application.

Also, the engines for some standard applications have been implemented as servers—the Calendar engine is implemented in this way. An engine written as a server enables data sharing between multiple applications that need access to the data, including your own, if required.

Many other useful public APIs are available for use by developers—some are Symbian OS APIs and others are Series 60 specific. The number and variety of these APIs is increasing with every revision of Series 60.

This chapter describes and illustrates some of the most important functionality available at the time of writing. The topics covered in this chapter are:

- **Using Standard Application Views**—Many of the key applications in Series 60 devices publish one or more views for use by other applications. Examples of how to use such views are given for Phonebook, Calendar, Photo Album and Messaging. Additional information is given on invoking other applications that do not allow view switching and using library controls that provide access to key functionality, such as the Speed Dial dialog.

- **Application Engines**—Several applications also share their data with other applications by providing a server-based engine. This section provides information on how you can use the engines from the Logger application, the Camera application, Phonebook, Calendar and Photo Album applications for your own needs.

- **Accessing System Capabilities**—Occasionally some applications may need to access hardware settings, or to sense their current state, in order to provide the level of functionality required. This section demonstrates how to use the Hardware Abstraction Layer (HAL) APIs to get, and sometimes set, various hardware settings. Use of the System Agent APIs to receive notification about system and hardware state changes is also shown.

A number of example projects are provided to illustrate the majority of the APIs described in this chapter. Some of the simpler APIs are just described, but for the more complex techniques, complete example projects are provided.

Chapter 12 Using Standard Application Views   619

Table 12–1 summarizes the example applications and the techniques they demonstrate.

Table 12–1 Techniques and Example Applications

Technique	Details	Example Application
View Switching	Calendar, Phonebook, Camera, Photo Album, Profiles, Messages	**ViewManager**
View Switching	Phonebook	**CallSummary**
View Switching	Media Gallery application (replaced Photo Album/Images application in Series 60 2.x)	**FilmReel2**
Nonswitchable Applications	Invokes the WAP browser	**ViewManager**
Invoke Exported UI Controls	Invokes the Speed Dial Dialog	**CallSummary**
Application Engines	Log engine	**CallSummary**
Application Engines	Phonebook	**CallSummary**
Application Engines	Speed Dial	**CallSummary**
Application Engines	Calendar (Agenda) engine	**AlarmOrganiser**
Application Engines	Camera engine	**FilmReel**
System APIs	System Agent	**SystemAgent**
System APIs	Hardware Abstraction Layer	**HalView**

All the project files, source files and deployment information associated with the examples in this book are available online, as noted in the Preface.

Using Standard Application Views

The Avkon View-Switching Architecture is the Series 60 Application Framework that enables view switching across applications. As described in Chapter 4, for an application to use the view architecture, the main user interface class, (`AppUi`), must be derived from `CAknViewAppUi`, and all views must be derived from `CAknView`.

The design of view-switching applications is described in detail in Chapter 4. Here you can see how to use the views that are made available by some of the key applications in Series 60 devices. The code presented in this section is taken from the **ViewManager** example project.

To activate an external application's view you use `CCoeAppUi::ActivateViewL()`, which will have been inherited from `CAknViewAppUi`. The function `ActivateViewL()` has two overloaded forms:

```
void ActivateViewL(const TVwsViewId& aViewId);

void ActivateViewL(const TVwsViewId& aViewId,
TUid aCustomMessageId, const TDesC8& aCustomMessage);
```

The `TVwsViewId` class is constructed using the `UID` of the application to switch to, and the `UID` of the view required within the application. If an application wishes to allow other applications to switch to any of its views, it should export the required `UID`s in a header file.

The second form of `ActivateViewL()` has two further parameters allowing custom-defined messages to be passed to the view being activated. The messages are passed to the view's `DoActivateL()` method. In this way, a view can take an action as it is being activated—for example, a specific dialog pane could be displayed.

Phonebook View Switching

The Phonebook application (aka Contacts) has three main views: contacts view, groups view and a focused contact view. Note that to successfully switch to the group view or focused contact view from your application, it may be necessary for the Phonebook application to already be running in the background. To switch to the contacts view, you can use:

```
const TUid KPhoneBookUid = { 0x101f4cce }; //from PbkUID.h
...
const TUid KPhoneBookContactViewUid = { 1 };
ActivateViewL(TVwsViewId(KPhoneBookUid,
KPhoneBookContactsViewUid));
```

To switch to the groups view, you can use:

```
const TUid KPhoneBookGroupViewUid = { 2 };
ActivateViewL(TVwsViewId(KPhoneBookUid,
KPhoneBookGroupsViewUid));
```

The focused contact view displays a specified contact. In order to switch to this view you need to use `ActivateViewL()` with three parameters. The phonebook API provides a helper class called `CPbkViewState` for use with this version of `ActivateViewL()`. To switch to the focused contact view, first construct a `CPbkViewState` object, and then set the ID and index of the contact and the field you want to focus on:

Chapter 12 Using Standard Application Views

```
CPbkViewState* pbkViewParam = CPbkViewState::NewLC();
// First contact
pbkViewParam->SetFocusedContactId(0);
pbkViewParam->SetFocusedFieldIndex(3);
```

To create a descriptor of the required format use `CPbkViewState::PackLC()` and pass it to `ActivateViewL()`.

```
HBufC8* paramBuf = pbkViewParam->PackLC();
const TUid KPhoneBookUid = { 0x101f4cce }; //from PbkUID.h
const TUid KPhoneBookFocusedViewUid = { 4 };
const TVwsViewId viewId(KPhoneBookUid,
KPhoneBookFocusedViewUid);

// Activate the view
ActivateViewL(viewId, CPbkViewState::Uid(), *paramBuf);
```

Not all application APIs provide a helper class to assist in the creation of a message to pass into `ActivateViewL()`.

To create a descriptor to pass to an application without using a helper class, you need to know the format the descriptor needs to take. For the Phonebook application, this information can be found in `CPbkViewState.h`.

One way to write binary data to a descriptor is by using `RDesWriteStream`, and once created you can pass the descriptor to `ActivateViewL()` in the usual way.

```
TBuf8<16> param;
RDesWriteStream stream(param);
stream.PushL();
stream.WriteInt8L(1);   // opcode EFocusedContactId
// Contact ID of last contact
stream.WriteInt32L(numberOfContacts - 1);
stream.WriteInt8L(4);   // opcode EFocusedFieldIndex
stream.WriteInt32L(3);  // field index 3
stream.CommitL();
CleanupStack::PopAndDestroy();   // stream
```

Using `RDesWriteStream` rather than a helper class, although less straightforward, has the advantage of there being no dependency on `CPbkViewState.h` and `PbkView.lib` at compile time or `PbkView.dll` at runtime.

Calendar View Switching

The Calendar (aka Agenda) application does not take any messages to allow you to specify a particular date to switch to. However, using `ActivateViewL()` with only one parameter, you can switch to the month, week, or day view.

If the Calendar application is not already running in the background when you switch views, the view shown by Calendar will probably default to the day view for the current date. Even if you specify the month or week view, it will probably be displayed only briefly, before the day view is then displayed. This behavior is not as might be expected, but it has been observed on Series 60 1.x and 2.x devices and emulators.

If Calendar is currently running, and you switch to one of its views, the correct view will be shown, and the date displayed will be the last date that was selected by the user.

```
const TUid KCalendarUid = { 0x10005901 };
// 0x01 Month view
// 0x02 Week view
// 0x03 Day view
const TUidKCalendarMonthViewUid = { 0x01 };
ActivateViewL(TVwsViewId(KCalendarUid,
KCalendarMonthViewUid));
```

While the Calendar application does not take any messages, the entries are accessible via the agenda model; see the subsection entitled, *Calendar Engine Access* later in this chapter.

Camera View Switching

To switch to a Camera view use:

```
const TUid KCameraUid = { 0x1000593F };
// 0x01    standby mode
// 0x02    viewfinder mode
// 0x04    Name base, and Quality settings
const TUid KViewFinderModeUid = { 0x02 };
ActivateViewL(TVwsViewId(KCameraUid, KViewFinderModeUid));
```

Like the Calendar application, the Camera application does not take any messages.

Note that on an emulator there is no camera hardware. Instead, special provisions have been made to allow applications to use the camera APIs while running on an emulator—see the *Camera APIs* subsection later in this chapter.

Photo Album View Switching

The Photo Album (Series 60 1.x) application also does not take any messages. It offers three views that can be switched to:

```
const TUid KPhotoAlbumUid = { 0x101F4CD1 };
// 0x01    Imagelist view—the default view
// 0x03    Messaging Picture Grid view
// 0x04    Messaging Picture view
const TUid photoAlbumViewUid = { 0x01 };
ActivateViewL(TVwsViewId(KPhotoAlbumUid,
photoAlbumViewUid));
```

Chapter 12 Using Standard Application Views 623

In Series 60 2.x you need to switch to the main view of the Media Gallery application, as this has replaced the Photo Album/Images application. This is illustrated in the **FilmReel2** application described in *Camera API Changes in Series 60 2.x* later in this chapter.

Profiles View Switching

The Profiles application offers two switchable views: the main view and a settings view.

```
const TUid KProfileUid = { 0x100058F8 };
// 0x01 main view - the default view
// 0x02 settings view
const TUid profileViewUid = { iViewIndex + 1 };
ActivateViewL(TVwsViewId(KProfileUid, profileViewUid));
```

Messaging View Switching

You can switch to three messaging views. Neither the main view, {0x01}, nor the delivery report view, {0x03}, take a message ID or message, whereas the folder view takes a message ID but not a message. Message IDs for the folder view are defined in `msvstd.hrh` and allow selection of the folder to view.

```
   const TUid KFolderViewUid = { 2 };
   const TUid KMsvGlobalInBoxIndexEntryUid =
{ KmsvGlobalInBoxIndexEntryIdValue };
   TBuf8<255> customMessage;
   ActivateViewL(TVwsViewId(KMessagingUid, KFolderViewUid),
KMsvGlobalInBoxIndexEntryUid, customMessage);
```

It is not possible to switch to a view in order to write a new message; If this functionality is required, you can use `CSendAppUi`, as described in Chapter 10.

Nonswitchable Applications

Occasionally, you may require additional functionality that is not available through view switching. For example, launching a specific WAP or Web page by view switching to the browser application may not be possible on a particular Series 60 device. An alternative, when view switching does not provide the required functionality, is to use the application architecture framework. This can be used to launch an application with a specified document, or, if the application is already running, to send a message to load one.

To find out if an application is running, you can create a list of all running applications using `TApaTaskList` and then search the `TApaTaskList` object for the application of interest.

```
const TInt KWmlBrowserUid = 0x10008D39;
TUid id(TUid::Uid(KWmlBrowserUid));
TApaTaskList taskList(CEikonEnv::Static()->WsSession());
TApaTask task = taskList.FindApp(id);
if (task.Exists())
    {
    ...
    }
```

It is possible to send a message to a running application using `TApaTask::SendMessage(TUid aUid, const TDesC8& aParams)`. The parameter `aUid` should be `KUidApaMessageSwitchOpenFileValue` to open the file passed in `aParams`, or `KUidApaMessageSwitchCreateFileValue` to create a new file. The message sent is handled by `CEikAppUi::ProcessMessageL(TUid aUid, const TDesC8& aParams)`. In the browser application, `ProcessMessageL()` has been implemented to ignore any value passed to `aUid` and launch the link passed by `aParams`.

```
if (task.Exists())
    {
    HBufC8* param8 = HBufC8::NewLC(link->Length());
    param8->Des().Append(*link);
    // UID is not used
    task.SendMessage(TUid::Uid(0), *param8);
    CleanupStack::PopAndDestroy();
    }
```

If the application is not running, then start it using `RApaLsSession`. This creates a session with the systems application server, which maintains a cached list of all installed applications on the device. Once connected to the application server, you can launch the application with a specified document using `RApaLsSession::StartDocument(const TDesC& aFileName, TUid aAppUid, TThreadId& aId)`, where `aFileName` is the document the application is to be launched with and `aAppUid` is the UID of the application to launch. On return, `aId` will contain the ID of thread created to launch the application. The browser application uses the descriptor passed to `aFileName` as the link to be launched in the browser window.

```
RApaLsSession appArcSession;
// Connect to AppArc server
User::LeaveIfError(appArcSession.Connect());
TThreadId id;
appArcSession.StartDocument(*link,
TUid::Uid(KWmlBrowserUid), id);
appArcSession.Close();
```

Application Engines

Some of the key applications in Series 60, such as Contacts and Calendar, have server-based engines that provide APIs to allow other applications to access their data. In the majority of cases such applications will also provide switchable views to provide direct functionality to other applications. If a view suitable for your needs is not available or you simply want to access some of the data held, then you can use the engine API to get access to the data you require.

Log Engine

The log engine records events such as telephone calls, faxes, SMS messages and emails in a database and is normally accessed by the log viewer application. The log engine API provides other applications with access to the log engine. Entries in the log engine can be read, filtered, added and deleted. It is also possible for new event types to be added.

Reading Log Entries

The log engine is implemented as a server. In order to read entries, you need to make a connection to the engine using CLogClient, which allows views of events in the log engine to be created. Creation of a view is required to filter events in the log, to navigate through the list of events and to access an event at a particular position in the log.

The API provides three classes to generate views on the log engine events. Most useful are CLogViewEvent, which generates a filtered view of events, and CLogViewRecent, which gives a view containing recent events in the event log.

To use CLogViewEvent, a filter must be created using CLogFilter to specify which events should appear in the view. Each property of the filter has a getter and setter function—the most important filter property being the event type. You can set a filter's event type using CLogFilter::SetEventType(TUid aType), where aType is the UID associated with the event type of interest. Commonly used UIDs include KLogCallEventTypeUid for voice calls and KLogShortMessageEventTypeUid for SMSs. To find a filter's event type use CLogFilter::EventType().

A filter can contain a contact, a number, a remote party, the duration, a duration type and a direction. A string such as "incoming" or "outgoing" is used to express the direction of an event, and this can be found by passing one of the resource IDs: R_LOG_DIR_IN, R_LOG_DIR_OUT, R_LOG_DIR_IN_ALT, R_LOG_DIR_OUT_ALT, R_LOG_DIR_FETCHED or R_LOG_DIR_MISSED to CLogClient::GetString().

```
TBuf<64> directionStr;
iLogClient.GetString(directionStr, R_LOG_DIR_IN);
ilogEvent->SetDirection(directionStr);
```

Once the filter has been created, it can be employed to generate a view of the events in the log engine using `CLogViewEvent::SetFilterL(const CLogFilter& aFilter, TRequestStatus& aStatus)`. Note that this function is asynchronous, as creating the view can take some time. If there are no events in the view, the function returns `EFalse`, if there are events, the asynchronous request is issued and the function returns `ETrue`.

```
TBool eventsInView = iLogView->SetFilterL(*iLogFilter,
iStatus);
...
if (eventsInView)
    {
    // if there are events in the filtered view
    // the asynchronous request has been issued
    SetActive();
    }
else
    {
    ...
    }
```

`CLogViewEvent` is derived from `CLogView`, which provides the following functions to navigate the events in a view: `FirstL(TRequestStatus& aStatus)`, `LastL(TRequestStatus& aStatus)`, `NextL(TRequestStatus& aStatus)` and `PreviousL(TRequestStatus& aStatus)`.

```
if (iLogView->NextL(iStatus))
    {
    // if there is another entry, issue the request
    // to move to the next entry.
    SetActive();
    }
```

The class `CLogEvent` encapsulates the details of an event in the log engine. Each event is given a unique ID when it is added, and it can be accessed using `CLogEvent::Id()`. Each `CLogEvent` attribute has a getter and setter function.

```
CLogEvent* logEvent = CLogEvent::NewL();
logEvent->SetEventType(KLogCallEventTypeUid);
logEvent->SetDurationType(KLogDurationValid);
...
logEvent->SetDirection(directionStr);
...
logEvent->SetDuration(duration);
...
logEvent->SetNumber(number);
...
```

```
    logEvent->SetContact(contactId);
    ...
    logEvent->SetTime(time);

void CCallArray::AddEntryL(const CLogEvent& aEvent)
    {
    ...
    TTimeIntervalSeconds timeInSeconds = aEvent.Duration();
    ...
    entryParameters.iDuration = timeInSeconds.Int();
    entryParameters.iContactId = aEvent.Contact();
    const TDesC& number = aEvent.Number();
    ...
    }
```

It is possible to add your own event type(s) to the log engine. A new event type is defined with `CLogEventType`, then registered with the log engine using `CLogClient::AddEventType()`. In addition to the standard information stored for an event, `CLogEvent` provides the means to store and link your own data. Each event can have a link associated with it using `CLogEvent::SetLink()` and can be accessed using `CLogEvent::Link()`. The link can be used to relate the event to an entity in another application, such as the message ID for emails. An event can also have data associated with it; the data is stored as a heap descriptor, allocated by `CLogEvent::SetDataL()` and accessed using `CLogEvent::Data()`.

To add an event to the log engine, create a `CLogEvent` object and add it using `CLogClient::AddEvent(CLogEvent& aEvent, TRequestStatus& aStatus)`. Ensure the `CLogEvent` object remains in existence and valid until the request completes.

```
    CLogEvent* logEvent = CLogEvent::NewL();
    logEvent->SetEventType(KLogCallEventTypeUid);
    logEvent->SetDurationType(KLogDurationValid);
    ...
    iEventArray.Append(logEvent);
    ...
    // Issue a request to add the next event to the log
    iLogClient.AddEvent(*(iEventArray[iEntryAdded]), iStatus);
    SetActive();
```

To delete an event from the log engine, use `CLogClient::DeleteEvent(TLogId aId, TRequestStatus& aStatus)`.

Camera APIs

Series 60 1.x and 2.x provides support for devices with a built-in camera, with access to the camera hardware being controlled by the Camera Server. Only one client may use the camera hardware at any given time, so if you try to connect to the Camera Server when it is already in use, an error value is returned. The Series 60 1.x Camera Server is fairly simple, it supports taking pictures of two fixed types: either 24-bit 640 by 480 pixel images or 12-bit 160 by 120 pixel images—Series 60 2.x expands on the options available. The simplicity of the API has not proven an obstacle to the development of a large range of applications that use the Camera Server, such as video recording. The accompanying example applications, **FilmReel** and **FilmReel2** for Series 60 2.x, take a sequence of low-resolution pictures, which are aggregated in a larger image that is saved to the Photo Album.

Using Camera APIs with an Emulator

To make the Camera Server functional on a Series 60 1.x emulator, you need to copy two `.mbm` files into the root of the emulator's `c:` drive. A `.zip` archive containing suitable files is available from **http://www.forum.nokia.com**, or as part of the online materials for this book, as described in the Preface.

Two image files (`valo_vga.mbm` and `valo_qqvga.mbm`) are provided to emulate high-resolution and low-resolution images from the camera. Copy both files to the root of the emulator's `c:` drive—this is located in one of the following directories in the root of your SDK, as appropriate to your IDE and SDK configuration:

```
For  \Epoc32\wins\c\         —for Visual C++
For  \Epoc32\winsb\c\        —for C++Builder
For  \Epoc32\winscw\c\       —for CodeWarrior
```

Instead of real images, a static image from one of the `.mbm` files will be returned to the application using the Camera Server, when running on an emulator. Note that for the example application, **FilmReel**, to work correctly on the emulator, it is essential that the image files are in the correct location.

For Series 60 2.x SDKs, the Camera Server captures an image of a small box, which slowly moves across the screen, as can be seen in the standard Camera application's viewfinder. The two images for Series 60 1.x SDKs are not required when using the APIs on a Series 60 2.x SDK. Instead, the camera API captures a gray image with the targeting square in the position at the time of image capture. The image is then available to the Media Gallery application.

Camera Server

Before you can take a picture with the camera, you need to make a connection to the Camera Server and then turn the camera on. The code used below is taken from the file `CameraManager.cpp`.

The client-side handle to the Camera Server is of type `RCameraServ`, and connecting to it is performed by simply calling `Connect()`:

```
User::LeaveIfError(iCameraServer.Connect());
```

> **TIP** Note that this code will leave if you cannot access the camera hardware. This might be due to another application already using it, or even to the absence of camera hardware on the device.

You then need to turn the camera on. The Camera Server provides an asynchronous function, `TurnCameraOn()`, as turning the camera on can take quite some time. Usually a request to turn on the camera would be encapsulated within an active object:

```
iCameraServer.TurnCameraOn(iStatus);
SetActive();
```

When the camera is no longer required, it can be turned off using `TurnCameraOff()` and the connection to the Camera Server terminated using `Close()`:

```
iCameraServer.TurnCameraOff();
iCameraServer.Close();
```

Note that the camera will also be turned off automatically after a few minutes of inactivity. There will be no notification of this behavior, so if the camera has been turned off and you try to take a picture, then an error is returned in the `TRequestStatus` of the calling `GetImage()` function, as described next. You must turn the camera back on before trying to retake the picture.

Taking a Picture

The main point of using a camera is to take pictures; this is achieved using the asynchronous function `GetImage()`:

```
void GetImage(TRequestStatus& iStatus, CFbsBitmap& aBitmap);
```

In order to use `GetImage()`, you need to pass in references to a `TRequestStatus` and a bitmap object. Do not worry about setting the correct width, height, or bit depth for the bitmap object, as the Camera Server does this automatically. Usually this request to take a picture would be encapsulated within an active object:

```
iCameraServer.GetImage(iStatus, aBitmap);
SetActive();
```

When the request completes, the value of `iStatus` will indicate if a picture was successfully taken.

Settings

Two camera settings can be changed: the image quality and the lighting conditions. Starting with the latter, the Camera Server has two lighting states: normal and night. The night setting should be used when the overall lighting is dim or dark, but it can also sometimes give better results in bright conditions. To change the lighting settings use the logically named `SetLightingConditions()`, which takes a parameter of either `RCameraServ::ELightingNormal` or `RCameraServ::ELightingNight`.

```
iCameraServer.SetLightingConditions(
RCameraServ::ELightingNormal);
```

The other setting you can change is the image quality. The camera API supports two image-quality settings: high and low. To change the image quality use `SetImageQuality()` with a value of either `RCameraServ::EQualityHigh` or `RCameraServ::EQualityLow`. High-quality images are 640 by 480 pixels with 16 million colors, and low-quality images are 160 by 120 pixels with 4096 colors.

```
iCameraServer.SetImageQuality(RCameraServ::EQualityHigh);
```

A high-quality image is 32 times the size of a low-quality image, and this might be an issue, especially if numerous images are being stored.

TIP The speed with which pictures are taken depends on the image quality and the lighting conditions, with low-quality images being significantly quicker to take than high-quality images. To implement a viewfinder, create a timed loop to take and display pictures using the low-quality image mode. This will give less image flicker than the high-quality image mode.

Camera API Changes in Series 60 2.x

The Camera APIs have changed between Series 60 1.x and 2.x. To illustrate the use of the Series 60 2.x Camera APIs, an example application called **FilmReel2** is provided. The main differences between this and the Series 60 1.x example application, **FilmReel**, are in the `CFilmReel2Container` class. **FilmReel2** uses the Series 60 2.x `CCamera` class instead of `CCameraManager`—the `CActive`-derived class that encapsulates the Series 60 1.x Camera API, `RCameraServ`, as used in the **FilmReel** application.

`CCamera` provides an interface that an application can use to interact, and to acquire images from the camera. In **FilmReel**, an Active Object-derived class was used to handle the asynchronous events. However, Series 60 2.x provides a mixin class `MCameraObserver`, and this is registered with `CCamera` to notify the application of various key events. You must provide implementations of the pure virtual methods declared by `MCameraObserver`. An appropriate method will be called when one of the specified events occurs—for example, when the camera is ready, when an image has been taken, or when an error has occurred.

To capture an image, you need to have created an instance of `CCamera` — this is achieved by calling the static `NewL()` function:

```
iCamera = CCamera::NewL(aObserver, 0);
```

Once you have created an instance of the camera, the camera device must be reserved. To achieve this, call the `Reserve()` method:

```
iCamera->Reserve();
```

The `Reserve()` method performs an asynchronous function call, and once the camera has been reserved, it will call the `MCameraObserver::ReserveComplete()` method. If the camera device was reserved successfully, the error code returned is `KErrNone`.

After successfully reserving the camera, the next step is to power it on, by calling the `PowerOn()` method. The `PowerOn()` method performs an asynchronous action, and once it has completed it calls the `MCameraObserver::PowerOnComplete()` callback method.

```
iCamera->PowerOn();
```

After the camera has been powered on successfully, and before an image can be captured, the image format and size must be specified. The formats supported are available from `TCameraInfo::iImageFrameFormatsSupported`, and the sizes supported are available from `CCamera::EnumerateCaptureSizes()`. To set the format and size the client calls `CCamera::PrepareImageCaptureL()`, specifying the image format and size. This must be called at least once before taking an image.

```
iCamera->PrepareImageCaptureL(
CCamera::EFormatFbsBitmapColor4K, 1);
```

Once the camera has been set up, the client application can call `CCamera::CaptureImage()` to capture an image. If an image has been captured, it will call `MCameraObserver::ImageReady()`. This will return the image if successful. Note, the image format and size vary, depending on the camera hardware used.

Once the application has finished using the camera, it should be released, and this is achieved by calling `CCamera::Release()`.

The `CCamera` interface also offers the following configuration features:

- Flash mode.
- Zoom settings for both optical and digital.
- Brightness.
- Contrast.
- Exposure.

The availability of these features will depend on the hardware capabilities on the device.

Note also that **FilmReel2** switches to the main view of the Media Gallery application, as this has replaced Photo Album/Images in Series 60 2.x.

At the time of writing, when building for target, it was necessary to build for thumb, since no ECam.lib was available for armi. Hence the .pkg file supplied with the example project (to build the .sis installation package file) specifies that the target build location for the executable files is thumb. Also, the Platform UID in the .pkg file is set to 2.0 to ensure the application is not installed on a device based on an earlier version of Series 60.

Phonebook Engine

The phonebook engine (aka Contacts) API allows access to its database of contacts and provides classes that enable the Phonebook application's standard dialogs to be used, from within your own application.

The phonebook API is a Series 60 implementation of the Symbian OS Contacts engine API. Much of the Contacts engine API is duplicated by the phonebook API, and it should be used only when the phonebook API is not sufficient.

Creating and Accessing Phonebook Items

The phonebook engine is implemented by CPbkContactEngine. If a default database exists, CPbkContactEngine::NewL() connects to the database; otherwise it creates it. An overloaded version of this function exists for opening a database other than the default.

```
iPbkContactEngine = CPbkContactEngine::NewL();
```

Each entry in the phonebook database is made up of a number of fields that can be added or removed from an entry. Each field stores a single value that can be text, binary, or a date and time. CPbkFieldInfo objects contain information about the field types, and TPbkContactItemField objects contain the actual field data in an entry.

The fields supported by the phonebook database can be found using CPbkContactEngine::FieldsInfo(), which returns an array of CPbkFieldInfo objects.

```
  const CPbkFieldsInfo& fieldsInfo = iPbkContactEngine->
FieldsInfo();
  TInt numberOfFields = fieldsInfo.Count();

  for (TInt i = 0; i < numberOfFields; i++)
     {
     CPbkFieldInfo* fieldInfo = fieldsInfo[i];
     ...
     }
```

Chapter 12 Application Engines

`CPbkFieldInfo` provides the means to discover the attributes of a field, such as whether it contains a telephone number or text.

```
CPbkFieldInfo* fieldInfo = fieldsInfo[i];
if (fieldInfo->IsPhoneNumberField())
    ...
```

Each entry in the phonebook database is given a unique ID (`TContactItemId`), used when specifying entries. To access the fields of any contact entry you need to open it for reading or writing. To open for reading use `CPbkContactEngine::ReadContactL()`, which takes a `TContactItemId` and returns a `CPbkContactItem*` for reading.

To modify a contact entry, open the entry for writing, make the changes and then commit them to the phonebook database. To open for writing use `CPbkContactEngine::OpenContactL()`, which takes a `TContactItemId` and returns a `CPbkContactItem*` for editing. Once the modifications have been made, commit them using `CPbkContactEngine::CommitContactL()`, taking the `CPbkContactItem` to be committed.

The class `CPbkContactItem` represents an individual entry in the phonebook database. It provides access and search functions for the array of `TPbkContactItemField` that it owns, as well as utility functions to obtain the email, telephone, SMS and MMS details for an entry. The title of an entry is found using `CPbkContactItem::GetContactTitleL()`. The title field mapping varies according to which fields contain data. For example, if no first name is defined but a last name is, the last name is returned as the title. Similarly, if no first or last name is defined, the business name is returned as the title.

```
// open contact for reading
CPbkContactItem* contact = iPbkContactEngine->ReadContactLC(aContactId);

// get the contacts name
HBufC* name = contact->GetContactTitleL();
```

You can search for a particular field in a contact (`TPbkContactItemField`) by field ID (`TPbkFieldId`) or by field definition (`CPbkFieldInfo`). To search by ID use `CPbkContactItem::FindField()`, which returns a `TPbkContactItemField` based upon the field ID passed as a parameter. The list of field Ids can be found in `PbkFields.hrh`.

```
TPbkContactItemField* phoneNumber = contactItem->FindField(EPbkFieldIdPhoneNumberGeneral);
```

Other useful functions include `CPbkContactItem::FindNextFieldWithText()` and `CPbkContactItem::FindNextFieldWithPhoneNumber()`, which find the next entry containing text and the next phone number, respectively.

A `TPbkContactItemField` field can be one of four types, the two important ones being text and date/time. There is no numeric field data type—telephone numbers and so on, are stored in text fields. As well as the data, the field owns a content type describing the type of data stored, a label describing the field for the user, and attribute flags.

The content type can be discovered with `TPbkContactItemField::StorageType()`. If the data is text, there are several ways of accessing it. `TPbkContactItemField::Text()` returns a `TPtr` to the text stored in a field, while `TPbkContactItemField::GetTextL()` fills the descriptor passed in with the text stored in the field. To add or edit existing text use `TPbkContactItemField::TextStorage()`, which returns a pointer to the field's text storage (a `CContactTextField` object). Using `CContactTextField::SetTextL()`, you can then change the text data stored by the field.

```
// find the phonenumber field of the contact, and add
// aCallInfo's telephone number
TPbkContactItemField* phoneNumber = contactItem->
FindField(EPbkFieldIdPhoneNumberGeneral);

phoneNumber->TextStorage()->SetTextL(
aCallInfo.iTelephoneNumber);
```

Similarly, if the data stored in the field is a date/time, `TPbkContactItem::Time()` returns the data as a `TTime`. If the data is to be edited, `TPbkContactItem::DateTimeStorage()` returns a pointer to the field's date storage (a `CContactDateField` object). To set the time for this field use `CContactDateField::SetTime()`, which takes a `TTime` parameter.

To add a new field to an entry, create a `CPbkFieldInfo` to describe the field you wish to add, then use `CPbkContactItem::AddFieldL()` to add the field to the entry. Alternatively use `CPbkContactItem::AddOrReturnUnusedFieldL()`, which finds an unused field or creates a new one based upon the `CPbkFieldInfo` parameter passed to the function.

To add a new entry to the phonebook database, create a `CPbkContactItem`. The simplest way to do this is by using `CPbkContactEngine::CreateEmptyContactL()`, which returns a `CPbkContactItem` with default fields.

```
CPbkContactItem* contactItem = iPbkContactEngine->
CreateEmptyContactL();
```

Alternatively you could use the `CPbkContactItem NewL()` or `NewLC()` methods, which construct a wrapper around, and take ownership of, a `CContactItem` that has been constructed using the contact model API.

```
static CPbkContactItem* NewL(CContactItem* aItem, const
CPbkFieldsInfo& aFieldsInfo, MPbkContactNameFormat&
aNameFormat);
```

Chapter 12 Application Engines

A reference to the phonebook engine's `MPbkContactNameFormat` can be obtained using `CPbkContactEngine::ContactNameFormat()`.

Once the `CPbkContactItem` has been constructed, add it to the phonebook database using `CPbkContactEngine::AddNewContactL()`.

```
TContactItemId contactId =
iPbkContactEngine>AddNewContactL(*contactItem);
```

To make a copy of an entry, `CPbkContactEngine::DuplicateContactL()` takes the ID of the entry to be copied and returns the ID of the new entry.

To delete a single entry use `CPbkContactEngine::DeleteContactL()`, which takes the ID of the entry to be deleted. If more than one entry is to be deleted, use `CPbkContactEngine::DeleteContactsOnBackgroundL()`, which takes an array of IDs in the form of a `CContactIdArray`. This is an asynchronous function, so, in most cases, it will be preferable to other nonasynchronous functions for deleting multiple entries.

Receiving Notification

It is possible to ask the phonebook engine to notify an application when changes are made to the phonebook database. This can be important when dealing with long-running applications, where changes can be made to the database when the application is in the background. `CPbkContactChangeNotifier` manages the task of registering an observer with the phonebook engine. When a `CPbkContactChangeNotifier` object is constructed, it registers the observer as an observer of the relevant phonebook engine, and on destruction removes the observer from the phonebook engine's list of observers.
`CPbkContactChangeNotifier::NewL()` can be used to register an observer with the phonebook engine, taking as parameters the engine to be observed and the observer to be called when the engine changes.

Alternatively, `CPbkContactEngine::CreateContactChangeNotifierL()` can be used, taking as a parameter the observer to be called and returning a `CPbkContactChangeNotifier`.

In order for a class to be used as a phonebook engine observer, it needs to inherit from `MPbkContactDbObserver`, which contains two functions: `DatabaseEventHandledL()` and `HandleDatabaseEventL()`, both of which are passed a `TContactDbObserverEvent`. An observer of the engine is notified of any change event twice; `HandleDatabaseEventL()` is called first by `CPbkContactEngine` for all of its observers, then `CPbkContactEngine` calls `DatabaseEventHandledL()` for each of the observers to ensure that they have all been notified of the change event.

The default implementation of `DataBaseEventHandled()` does nothing; otherwise the observer class will receive two notifications of every phonebook engine event. The `TContactDbObserverEvent` passed to `DatabaseEventHandledL()` and `HandleDatabaseEventL()` contains the type of event that has occurred and the ID of the phonebook database entry affected. The list of event types can be found in cntdbobs.h. Commonly reported events types include: EContactDbObserverEventContactChanged, EContactDbObserverEventContactDeleted, EContactDbObserverEventContactAdded and EContactDbObserverEventSpeedDialsChanged.

CPbkContactEngine functions such as CPbkContactEngine::AddNewContactL() take a second parameter, TBool aImmediateNotify, which by default is false. If true, the observers are notified immediately of the event (for example, a contact has been added); however, the observers will be informed of the event twice.

Phonebook Searching

To find a particular entry in the database, the phonebook engine API provides the means to iterate though all the entries or search directly. To iterate through the phonebook database use CPbkContactIter.

An iterator can be constructed using CPbkContactIter::NewL(), CPbkContactIter::NewLC() or CPbkContactEngine::CreateContactIteratorLC(). To iterate through the entries use CPbkContactIter::FirstL() to move to the first entry and CPbkContactIter::NextL() to move to the next. Both functions return the ID of the entry the iterator has been advanced to, or KNullContactId if none are found. A pointer to the current phonebook database entry is obtained using CPbkContactIter::CurrentL(), or a copy of the current entry is obtained using CPbkContactIter::GetCurrentL().

When searching the phonebook database for a telephone number, use CPbkContactEngine::MatchPhoneNumberL() rather than iterating through all of the entries. CPbkContactEngine::MatchPhoneNumberL() searches for the telephone number passed in as a descriptor (the second parameter passed is the number of characters to use in the search), returning an array of IDs of matching entries in the form of a CContactIdArray.

To search for text other than a telephone number, you can search synchronously or asynchronously. To search synchronously, CPbkContactEngine::FindLC() scans the phonebook database for the text and returns a CContactIdArray containing the IDs of matching entries. It is possible to pass a CPbkFieldIdArray of field IDs to be searched, but in most cases FindLC() searches other fields as well as those specified, so the matched entries need to be checked. As searching for text can take some time, especially if there are many entries in the phonebook, searching asynchronously is generally a better choice.

To search the phonebook asynchronously use CPbkIdleFinder. To obtain a CPbkIdleFinder call CPbkContactEngine::FindAsyncL(), the default implementation takes a descriptor parameter with the text to be searched for.

`FindAsyncL()` creates a `CPbkFindlerIdle` object that searches asynchronously for the text in the phonebook database. To check if `CPbkIdleFinder` has finished searching, `MIdleFindObserver`, the third parameter of `CPbkContactEngine::FindAsyncL()`, allows the user of `CPbkIdlefinder` to monitor the search progress. `MIdleFindObserver::IdleFindCallBack()` is called for every 16 items searched, and also at the end of the search. This callback greatly assists in implementing a progress dialog or search status update. Discovering if `CPbkIdlefinder` has finished is achieved by calling `CPbkIdleFinder::IsComplete()`. When the search is complete, the array of IDs of matching entries is obtained using `CPbkIdleFinder::TakeContactIds()`.

Speed Dialing and Common Dialogs

The phonebook engine manages the mapping of speed dial numbers to telephone numbers within the phonebook database. `CPbkContactEngine` provides four functions to manage speed dial mappings.

```
    void SetFieldAsSpeedDialL(CPbkContactItem& aItem, TInt
aFieldIndex, TInt aSpeedDialPosition)   // Sets the requested
field of the contact as a speed dial number.

    TContactItemId GetSpeedDialFieldL(TInt aSpeedDialPosition,
TDes& aPhoneNumber) const // Gets the number mapped to the
requested speed dial location.

    void RemoveSpeedDialFieldL(TContactItemId aContactId, TInt
aSpeedDialPosition) // Removes the speed dial mapping from a
contact.

    TBool IsSpeedDialAssigned(const CPbkContactItem& aItem,
TInt aFieldIndex) const    // Returns if an item's field is
assigned to a speed dial mapping.
```

`CSpdiaControl` is a utility class that provides services to both the Speed Dial and Phonebook applications for getting and setting a speed dial number configuration. Use it in your applications to provide a dialog for assigning a speed dial position to a telephone number.

```
    // Launch the speed dial dialog to add a speed dial link
    // to the selected contact
    CSpdiaControl* speedDialControl =
CSpdiaControl::NewL(*iPbkContactEngine);

    return speedDialControl->ExecuteLD(aId,
(*iFieldInfoArray)[0]-> FieldId());
```

As well as the speed dial selection dialog, the phonebook API offers other common dialogs accessible to external applications, such as `CPbkPhoneNumberSelect` to select a phone number, `CPbkContactEditorDlg` to edit a contact, and `CPbkDataSaveAppUi` to implement "Save to phonebook" menu functionality.

```
    void CPhoneBookEngine::DynInitMenuPaneL(TInt aResourceId,
CEikMenuPane* aMenuPane)
        {
        iPbkDataSaveAppUi->DynInitMenuPaneL(aResourceId,
aMenuPane);
        if (aResourceId == R_CALLVIEW_MENU_PANE)
            {
            // Adds Phonebook data save menu items
            // to the menu pane
            iPbkDataSaveAppUi->AddMenuItemsL(aMenuPane,
ECallSummaryAddContactDB);
            }
        }

    TBool CPhoneBookEngine::HandlePhoneBookCommandL(
TInt aCommand, const TDesC& aNumber)
        {
        return iPbkDataSaveAppUi->HandleCommandL(aCommand,
*iFieldInfoArray, aNumber);
        }
```

Compact Business Cards and vCards

With the `CBCardEngine` class, the phonebook API supports the transmission of vCard and compact business cards to and from the phonebook database, by infrared, SMS and so on.

A `CBCardEngine` object is created using `CBCardEngine::NewL(CPbkContactEngine* aEngine)`, where `aEngine` is the open phonebook engine object for import or export. The `CBCardEngine` uses `RReadStream` and `RWriteStream` for the transfer of vCard data to and from the phonebook database. This allows the streams to be stored in a file or a stream buffer hosted by a descriptor.

To read a vCard record from a stream into the phonebook database, pass a reference to an open `CPbkContactItem` and the stream to read the vCard from to the function
`CBCardEngine::ImportBusinessCardL(`
`CPbkContactItem& aDestItem, RReadStream& aSourceStream)`. It is up to you to commit any changes to the phonebook database.

To write a vCard record into a stream, use
`CBCardEngine::ExportBusinessCardL(`
`RWriteStream& aDestStream, CPbkContactItem& aSourceItem)`, where `aSourceItem` is the contact entry you wish to export to the stream.

Calendar Engine Access

The Calendar (aka Agenda) engine/model API provides access to the time-management data used by the Calendar and To-do applications. The model allows entries to be added, edited and deleted. You can also use it to access details associated with entries, such as alarm and synchronization information.

The API is very large, and only a brief glimpse of its power and versatility can be given here; however, its use is extensively covered in the SDK documentation—see the Symbian API Guide, Application Engines and API Reference sections.

Agenda Models and Concepts

There are four types of agenda entry, each with different properties:

- **Events** (`CAgnEvent`)—are "all-day" entries that may have a display time, but do not have an actual start time. Events may have multiple day durations using the optional end date information.

- **Anniversaries** (`CAgnAnniv`)—are events that can occur only once a year, such as birthdays. Information about the base year can be stored.

- **Appointments** (`CAgnAppt`)—have a start time and date, and an end time and date.

- **To-dos** (`CAgnToDo`)—represent a task to be carried out and may have a display time and date, and either a due or crossed-out date to indicate when the task is due by or when it was performed. A to-do also contains details about the to-do list it belongs to and its priority.

The models for accessing the agenda data are arranged in a three-layer hierarchy:

- **Entry model** (`CAgnEntryModel`)—is the base model. An entry object represents each entry, with all repetitions stored in the one object.

- **Indexed model** (`CAgnIndexedModel`)—extends the entry model, indexing the data to allow filtering before entries are read.

- **Instance model** (`CAgnModel`)—extends the indexed model and is the most suitable for applications with a user interface, as separate objects are used to represent each instance of a repeating object. This will be the focus of the discussion.

Agenda Server

Before creating an instance of one of the model classes, you need to create an instance of the Agenda Server client, `RAgendaServ`, using `RAgendaServ::NewL()`. Then connect it to the server by calling `RAgendaServ::Connect()`. This will start the server if it is not already running.

WARNING Series 60 applications should always access the Agenda model in Client mode. "Normal" mode is obsolete and does not allow concurrent access to agenda files. Beware of using methods specifically overloaded for "normal" mode.

Once an instance of a model class has been created using its `NewL()` function, call `CAgnEntryModel::SetServer()` to associate the Agenda Server with the model. Most calls to the Agenda Server are made indirectly via a model class. When `SetServer()` is called, the model's mode is automatically set to client, but this can also be explicitly set using `CAgnEntryModel::SetMode()`.

```
iAgendaServer = RAgendaServ::NewL();
iAgendaServer->Connect();
iModel = CAgnModel::NewL(this);
iModel->SetServer(iAgendaServer);
iModel->SetMode(CAgnEntryModel::EClient);
```

Having created an instance of the model class and set the associated server, it is necessary to open the file containing the agenda data. This is achieved using the model class's `OpenL()` function, which takes the name of the file containing the agenda data with the default times to display for the different types of entry. The `CAgnIndexedModel` and `CAgnModel` versions of `OpenL()` are further overloaded to take a `MAgnProgressCallBack` observer, which reports the progress of the load.

```
iModel->OpenL(*agendaFile, KDefaultTimeForEvents,
KDefaultTimeForAnnivs, KDefaultTimeForDayNote, &iObserver,
EFalse);
```

Notification of Changes and Progress Monitoring

Some functions of the agenda model can block the model while a long-running task completes. When the model is created, it is possible to specify an observer derived from `MAgnModelStateCallBack`, which provides a function `StateCallBack()`, and called with `EBlocked` when the model enters a blocked state, and `EOk` when it leaves the blocked state.

```
    class CAlarmOrganiserEngineBase: public CBase, public
MAgnModelStateCallBack
        {
        ...
    private: // from MAgnModelStateCallBack
        void StateCallBack(CAgnEntryModel::TState aState);
        ...
        };

    void CAlarmOrganiserEngineBase::ConstructL()
        {
```

Chapter 12   Application Engines

```
    …
    iModel = CAgnModel::NewL(this);
    …
}
```

The state of the model can be found at other times by using `CAgnEntryModel::State()`, which returns the current state of the model.

To provide information on the progress of agenda operations that may take an extended time to complete—for example, merging and tidying agenda files—the file operation functions are overloaded to take an observer derived from `MAgnProgressCallBack` supplying two functions: `Progress()` and `Completed()`. These functions are called by the model to keep the user informed of the current state of the operation. `Progress()` supplies the percentage of the operation that has completed, `Completed()` when the operation has fully completed, rather than `Progress()` being called at 100 percent. The frequency with which `Progress()` is called is set by the callback frequency parameter when calling the function that issues the long-running request.

```
    iModel->OpenL(*agendaFile, KDefaultTimeForEvents,
KDefaultTimeForAnnivs, KDefaultTimeForDayNote, &iObserver,
EFalse, EOpenCallBackHigh);
```

Alternatively, you could wait for the file to be loaded by using the function `RAgendaServ::WaitUntilLoaded()`; however, this is a synchronous operation and will freeze the UI, making the machine unresponsive, so it is not recommended.

Reading Agenda Entries

Reading entries depends on whether the instance or the entry model is being used. To get entries from the instance model, you can use one of the `CAgnModel::PopulateXXXInstanceListL()` functions, which return a list of IDs matching the criteria given in the arguments to the function. This can include a filter.

```
    void PopulateDayInstanceListL(CAgnDayList<TAgnInstanceId>*
aList, const TAgnFilter& aFilter, const TTime& aTodaysDate)
const;

    void PopulateDayDateTimeInstanceListL(
CAgnDayDateTimeInstanceList* aList, const TAgnFilter&
aFilter, const TTime& aTodaysDate) const;

    void PopulateMonthInstanceListL(CAgnMonthInstanceList*
aList,const TAgnFilter& aFilter,const TTime& aTodaysDate)
const;

    void PopulateTodoInstanceListL(CAgnTodoInstanceList*
aList,const TTime& aTodaysDate) const;
```

```
    void PopulateSymbolInstanceListL(CAgnSymbolList*
aList,const TAgnFilter& aFilter, const TTime& aTodaysDate)
const;
```

The `CAgnXXXList` classes, such as `CAgnDayDateTimeInstanceList`, are wrapper classes around an array of IDs. Having obtained a list of IDs of interest, you can get the instance from its ID using `CAgnModel::FetchInstanceLC()`, which returns a pointer to a `CAgnEntry`.

```
    CAgnEntry* entry =
iModel->FetchInstanceLC((*dayList)[ii - 1]);
```

When the instance is retrieved, its type is unknown. The base class for the agenda entry classes, `CAgnEntry`, provides the means to discover the type of an entry enabling you to cast it to the appropriate class. Calling `CAgnEntry::Type()`, will return a value of `EAnniv`, `EAppt`, `EEvent` or `ETodo`. You can then cast the instance to the correct class using `CAgnEntry::CastToAnniv()`, `CAgnEntry::CastToAppt()`, `CAgnEntry::CastToEvent()` or `CAgnEntry::CastToToDo()`, respectively.

```
    CAgnEntry* entry = iArray[aIndexOfAlarm];
    switch (entry->Type())
       {
       case CAgnEntry::EAnniv:
       case CAgnEntry::EAppt:
       case CAgnEntry::EEvent:
          {
          ...
          }
       case CAgnEntry::ETodo:
          {
          CAgnTodo* todo = entry->CastToTodo();
          ...
          }
       ...
       }
```

Entries in the agenda model can also contain text that is stored in rich text format. The text associated with an entry can be accessed using `CAgnEntry::RichTextL()`, which returns a pointer to a rich text object.

```
    TBuf<256> text;
    entry->RichTextL()->Extract(text, 0, entry->RichTextL()->
DocumentLength());
```

As well as rich text, an entry can contain other information such as repeat details, category information, location, meeting attendees and so on. Each entry data item can be accessed through a range of specifically named methods provided by `CAgnEntry`—see the SDK documentation for further details.

Searching Instances and Filtering

`TAgnFilter` can be passed to the "populate list" functions to filter the IDs returned. There are also more specialized filters derived from `TAgnFilter`, such as `TAgnsrvTidyFilter`, used when tidying an agenda file—see the SDK documentation for further details.

After creating a `TAgnFilter`, you need to set the details of the filter. For example, to create a filter that filters out the instances which do not have an alarm set, use `TAgnFilter::SetIncludeAlarmedOnly()`.

```
TAgnFilter filter;
filter.SetIncludeAlarmedOnly(iAlarmed);
```

Filters can also include or exclude: to-dos, anniversaries and events, timed or untimed appointments, repeating entries, and crossed-out entries. It is possible to extract information about a filter; for example, `TAgnFilter::AreAlarmedOnlyIncluded()` returns whether the filter is restricted to alarmed entries or not.

As well as being able to populate lists with instances meeting certain criteria, it is possible to search for instances, or instances with a particular property. The agenda model API offers several functions to help with this. One option is to iterate though the entries using an iterator created by `RAgendaServ::CreateDateIterator()` or `RAgendaServ::CreateEntryIterator()`. `CAgnModel` also offers an implementation with `CAgnModel::NextDayWithInstance()` and `CAgnModel::PreviousDayWithInstance()`, which find the next or previous day that has an instance in the agenda model. This can then be used within one of the list-populating functions to find the relevant instances.

```
TTime day;
...
day = iModel->NextDayWithInstance(day, filter, day);
```

To search for instances matching a search string use `CAgnModel::FindNextInstanceL()` or `CAgnModel::FindPreviousInstanceL()`. Once a match is made, a search should be made of the rest of the instances for that day, and a `CAgnDayList` is returned containing any matches.

Adding, Editing and Deleting Entries

To add a new entry to the model, create it with `CAgnAppt::NewL()`, `CAgnEvent::NewL()`, `CAgnAnniv::NewL()`, `CAgnToDo::NewL()`, or the corresponding `NewLC()` function. To make the entry repeating, create an instance of `CAgnRptDef`, defining the required repeating pattern, and pass it to `CAgnEntry::SetRptDefL()` which sets the repeat details of the entry. Once you have constructed and set the contents of the entry, then add the entry to the model using `CAgnModel::AddEntryL()`, which takes the entry and returns an entry ID. If the entry is a to-do, a second parameter must be passed indicating the position in the to-do list the entry will occupy.

`CAgnEntryModel` and `CAgnIndexedModel` are concerned primarily with entries in the agenda file, while the `CAgnModel` is concerned with instances rather than entries. Therefore, if you are using `CAgnModel`, use `UpdateInstanceL()`, `DeleteInstanceL()` and `FetchInstanceL()`, rather than `UpdateEntryL()`, `DeleteEntryL()` and `FetchEntryL()`. Always use `CAgnModel::AddEntryL()` to add a new entry to an agenda model, as it does not have an instance date at the point that it is added.

If the entry is nonrepeating, then there are no added complications of dealing with an instance rather than with an entry. After getting the entry to edit using `CAgnModel::FetchInstanceLC()`, update the entry in the model using `CAgnModel::UpdateInstanceL()`, which returns the ID of the entry.

```
  CAgnEntry* entry =
iModel->FetchInstanceLC((*dayList)[ii - 1]);
  ...
  iModel->UpdateInstanceL(entry);
```

If the entry is repeating, then as well as passing the entry, you need to pass an enumeration parameter, `TAgnWhichInstances`, which can be `ECurrentInstance`, `EAllInstances`, `ECurrentAndFutureInstances`, or `ECurrentAndPastInstances` and describes whether you are changing all, or a subset, of the instances.

`CAgnModel::UpdateInstancesL()` defaults to `EAllInstances`, where the changes are made to all instances of the entries. Otherwise, as you are in effect trying to split the entry by defining instances that are distinct in properties, a new entry is created. Choosing `ECurrentInstance` results in the creation of a new nonrepeating entry, while choosing either `ECurrentAndFutureInstances` or `ECurrentAndPastInstances` results in a new repeating entry.

To delete an instance use `CAgnModel::DeleteInstanceL()`, which takes either a pointer to the entry, or the ID of the instance. If the instance is not repeating, then it is deleted. If the entry is repeating, a `TAgnWhichInstances` enumeration parameter is passed, which defaults to `EAllInstances`. This deletes the entry and all instances. If `ECurrentInstance` is used, the current instance is deleted, but before this can happen, the current instance needs to be made a repeat exception. A repeat exception is a date that would normally be included in an entry's repeat date, but is excluded; for example, an entry is repeated every Monday, except dates which are public holidays. Then the public holidays are exceptions. A repeat exception is created using `CAgnRptDef::AddExceptionL()`, which takes a `TAgnException`, which is a wrapper class around the date which is to be the exception.

Calendar Alarms

A Calendar (aka Agenda) event can have an associated alarm, which is handled by the alarm server. In order for alarms set in the Calendar (Agenda) file to be handled by the alarm server, they must be added to the alarm queue.

To create an association with the alarm server, create a `CAgnAlarm` object, which represents the interface between the agenda model and the alarm server. A connection to the alarm server is made during construction of the `CAgnAlarm` object. Once created, register the `CAgnAlarm` object with the model using `CAgnEntryModel::RegisterAlarm()`. Once registered, whenever an alarmed entry is added to the file, the model queues the next few outstanding alarms with the alarm server:

```
// Create a CAgnAlarm and associated it with the model
iAgnAlarm = CAgnAlarm::NewL(iModel);
iModel->RegisterAlarm(iAgnAlarm);
```

Before closing the agenda file that contains the current model, you need to call `CAgnAlarm::OrphanAlarm()`. This allows the alarms to be processed after the session is closed:

```
// Orphan the alarm, before we close the file
iAgnAlarm->OrphanAlarm();
delete iAgnAlarm;
```

To set an alarm for an entry use `CAgnBasicEntry::SetAlarm()`, setting the number of days before the start date that the warning alarm is to be sounded, and the time the alarm is to be sounded. The time the alarm is to sound is expressed as the number of minutes after midnight.

```
TTime time = entry->InstanceStartDate();
TInt hours = time.DateTime().Hour();
TInt minutes = time.DateTime().Minute();
entry->SetAlarm(0, minutes + 60 * hours);
```

For to-do entries, the alarm is set relative to the due date. For the alarm to be relative to the start date, call `CAgnToDo::SetAlarmFromStartDate()` after setting the alarm:

```
CAgnTodo* todo = entry->CastToTodo();
todo->SetAlarm(0, 0);
todo->SetAlarmFromStartDate();
...
iModel->UpdateInstanceL(entry);
```

To remove an alarm from an entry, call `CAgnEntry::ClearAlarm()`:

```
entry->ClearAlarm();
iModel->UpdateInstanceL(entry);
```

If you need to inquire about the details of any alarm associated with an agenda entry, there are methods to return this data. The method names are fairly self-explanatory. To find the number of days prior to the date of the agenda entry, use `CAgnBasicEntry::AlarmDaysWarning()`, followed by `CAgnBasicEntry::AlarmTime()` to find the time of the alarm. If you are only enquiring as to whether the entry has an alarm, use `CAgnBasicEntry::HasAlarm()`.

```
if (aEntry->HasAlarm())
    {
    ...
    }
```

Photo Album Engine

The photo album engine provides an API whose purpose is to assist in the management of image files. As well as providing utility functions, it provides a variety of common dialogs and controls that can be used in your own applications.

Note that in Series 60 2.x the Photo Album (aka Images) application has been replaced by the Media Gallery application. As a result, some of the photo album classes and methods are deprecated for use in Series 60 2.x applications. Search for "Deprecated List" in the Series 60 2.x SDK documentation for specific details.

Accessing Photo Album Folders

The majority of the photo album API is concerned with the management of images within the photo album folders, the settings for which can be found via the static `TPAlbSettings` functions. These return the settings for the various image folders.

```
IMPORT_C static HBufC* RootImageFolderLC();
...
IMPORT_C static TInt SetDefaultImageFolderL(TDesC& aPath);
```

Two classes enable transfer of images between an application and the image folders: `CPAlbImageUtil` moves images to the image folders and provides the image-saving API for the photo album. `CPAlbImageFactory` retrieves bitmaps from the image folders.

To place an image in the image folders, `CPAlbImageUtil::MoveImageL()` will move the image, and `CPAlbImageUtil::CopyImageL()` will copy the image, leaving the original in place. Both functions take a `TBool` parameter, `aReplace`, which will replace an existing file of the same name in the image folders if true.

Chapter 12 Application Engines

If `aReplace` is false, then `CPAlbImgaUtil` will generate a unique name for the image file based upon its original name.

To generate a thumbnail of the image being moved or copied, you can use an overloaded version of `CPAlbImageUtil::MoveImageL()` or `CPAlbImageUtil::CopyImageL()` that takes a reference to a `CFbsBitmap`, which will be filled with a thumbnail of the image being transferred.

```
iImageUtil = CPAlbImageUtil NewL();
iImageUtil->MoveImageL(aSourceFile, ETrue, aThumbBitmap);
```

The API also enables folder management tasks such as deleting and renaming image files using `CPAlbImageUtil::DeleteImageL()` and `CPAlbImageUtil::RenameImageL()`. Both take a `CPAlbImageData` object that encapsulates image information, providing member functions to access the name of the file containing the image, the path to the file and inquiring whether a thumbnail image exists.

After moving or copying an image file, you can get a pointer to the `CPAlbImageData` describing the image using `CPAlbImageUtil::ImageData()`. A pointer to the `CPAlbImageData` object is used to synchronize an image file and its thumbnail representation using `CPAlbImageUtil::SynchronizeL()`. The synchronization process completes by generating a thumbnail for the image file if one does not already exist, and if the thumbnail exists but not the image file, then the thumbnail is deleted.

`CPAlbImageFactory` allows retrieval of both the original image and its thumbnail from the image folders. As the creation of a bitmap from the image stored in the image folders can take some time, `CPAlbImageFactory` provides asynchronous functions to generate the bitmaps, with the assistance of the mixin class `MPAlbImageFactoryObserver`. Its single function, `MPTfoCreateComplete(CPAlbImageFactory* aObj, TInt aError, CFbsBitmap* aBitmap)`, is called on completion of an asynchronous request for a bitmap with `aObj`, a pointer to the `CPAlbImageFactory` which made the request, and `aBitmap`, the requested bitmap. The ownership of the returned bitmap is passed to the observing class. To get a bitmap from an image in the image folders use `CPAlbImageFactory::GetImageAsync()`, which takes either the full path filename of the image, or a `CPAlbImageData`:

```
void CImageObserverClass::GetImage(TFileName aFilename)
    {
    iImageFactory = CPAlbImageFactory::NewL(this);
    iImageFactory->GetImageAsync(aFileName);
    }

void CImageObserverClass::MPTfoCreateComplete(
CPAlbImageFactory* aObj, TInt aError, CFbsBitmap* aBitmap)
    {
    if (aError == KErrNone)
        {
```

```
            iBitmap = aBitmap;
        ...
        }
    ...
    }
```

To fetch a thumbnail use a `CPAlbImageData` object to specify the thumbnail required and call `CPAlbImageFactory::GetThumbnailAsync(CPAlbImageData)`. If the thumbnail does not exist, it is created and stored in the thumbnail directory. You can also generate thumbnails with the synchronous functions `CPAlbImageFactory::CreateThumbFromBitmapL()` and `CPAlbImageFactory::CreateThumbnailFromBitmapL()`.

Using Photo Album UI Components

In Series 60 1.x the Photo Album application provides several UI components that might be useful in your applications, such as selection dialogs and an image viewer control. However, for Series 60 2.x applications see the note earlier under *Photo Album Engine* about deprecated classes and methods.

The selection dialogs allow applications to present the user with a familiar interface for the selection of images. There are two selection dialogs, `CPAlbImageFetchPopupList` and `CPAlbPictureFetchPopupList`, for selecting images and pictures, respectively. Pictures essentially differ from images in their folder location and the filename extension used. For a picture file to be compatible with photo album it must have a filename extension of .ota, and the target folder is always the "Pictures" folder inside photo album.

To present the user with an image selection dialog, construct the `CPAlbImageFetchPopupList` dialog, passing a pointer to a `CPAlbImageData` object, which, when the dialog returns, will contain the details of the image selected:

```
CPAlbImageData* data = CPAlbImageData::NewL();
CleanupStack::PushL(data);
TBool result(CPAlbPictureFetchPopupList::RunL(data));
...
CleanupStack::PopAndDestroy(data);
```

To display an image in your own application, you normally need to load the image, convert it into a `CFbsBitmap`, then draw the bitmap onto the screen, as described in Chapter 11. The photo album API, however, provides `CPAlbImageViewerBasic` that hides these steps.

To use a `CPAlbImageViewerBasic` control to display an image, create the control and then call `CPAlbImageViewerBasic::LoadImageL()`, which takes the name of the image file to load and the color depth of the image as parameters. As well as displaying the image, `CPAlbImageViewerBasic` can be used to rotate, pan and zoom the image:

```
#include <AknUtils.h>    // Needed for CompleteWithAppPath()

// Note no drive letter given
_LIT(KFileName, "\Images\mini.jpg");
    ...
    CPAlbImageViewerBasic* iViewer =
CPAlbImageViewerBasic::NewL(this, Rect());
    TFilename filename = KFileName;
    // filename = c:\Images\mini.jpg
    CompleteWithAppPath(filenamc);
    iViewer->LoadImageL(filename, EColor4K);
```

Note the use of `CompleteWithAppPath()` here to complete the full filename — the drive letter and any missing path is provided from the applications installation location. Take care here, `CompleteWithAppPath()` may return an error code if there is a problem, but the filename will not be changed if an error occurred.

Accessing System Capabilities

When developing for Series 60, there are APIs available to query the system capabilities, share state data with other applications and register for notification of changes to system states. These APIs are implemented in two main components: the Hardware Abstraction Layer (HAL) and System Agent (SA) components.

The example applications **HalView** and **SystemAgent** described in this section show further use of these two components.

Hardware Abstraction Layer

HAL provides you with a very simple-to-use API, which can query attributes of the device. The API itself is very small, with only three functions, all provided by the class `HAL`:

```
static TInt Get(TAttribute anAttribute, TInt& aValue);
static TInt Set(TAttribute anAttribute, TInt& aValue);
static TInt GetAll(TInt& aNumEntries, SEntry*& aData);
```

The class `HAL` (`hal.h`) derives from `HALData` (`hal_data.h`), which contains the definition of the enumeration `TAttribute`. This simply defines 70+ attributes that can be queried, covering such information as device manufacturer, battery status, available memory and so on. You will need to include the `hal.h` header file in your code and add `hal.lib` library into your project (`.mmp`) file to get it to link correctly.

Note that the simplicity of the API restricts its use to that of state information, as the value returned is purely an integer value. Note also that the degree to which each device supports all of these fields depends on the manufacturer.

Two of the API methods allow you to retrieve information from HAL. Either an array containing all of the entries, or a single specific entry, can be retrieved. In this example, all of the states are read in one call, plus the amount of RAM in the device is read directly:

```
TInt numHalEntries;
HAL::SEntry* halEntries;
HAL::GetAll(numHalEntries, halEntries); // All entries

TInt ram;
HAL::Get(HAL::EMemoryRam, ram); // Get one specific entry
```

Some of the fields are recognized as being dynamic; in other words, their value may change each time they are queried. This can be determined by the status of the `iProperties` attribute of an `SEntry` structure returned from `HAL::GetAll()`. If this equals `HAL::EEntryDynamic` (0x2), then the field is dynamic; however, this does not immediately mean it is settable.

To find out if a field is settable, an attempt must be made to set it through the `HAL::Set()` method. If a field is not settable, then the return value from the call will be one of the standard system error codes, typically `KErrNotSupported`. For example:

```
TInt err = HAL::Set(HAL::ECaseSwitchDisplayOn, 1);
```

TIP The attribute `HALData::EMachineUid` can be used to identify the phone model that your application is installed on at runtime—see Table 12-2. This makes it possible to customize the behavior of the application to use any licensee-specific APIs that may be available. Getting this ID is illustrated next.

A call to the `HAL::Get()` function will return a unique code identifying the Series 60 product at runtime.

```
#include <hal.h> // also link to hal.lib
...
TInt mUid = 0;
HAL::Get(HALData::EMachineUid, mUid);
```

Be aware, however, that the machine UID is no guarantee of the software version on the device—you need to use `SysUtil::GetSWVersion(TDes &aValue)` to get the software version. See the `SysUtil` class in the Series 60 SDK documentation for more details of this and many other useful methods.

Chapter 12 Accessing System Capabilities

Table 12-2 Machine IDs

Product	Machine UID
Nokia 7650	0X101F4FC3
Nokia 3650	0X101F466A
Nokia N-Gage	0X101F8C19
Nokia 6600	0X101FB3DD
Siemens SX1	0X101F9071
Sendo X	0x101FA031
Series 60 1.2 emulator	0X10005F62

Phone IMEI Number

Another class related to HAL is PlpVariant, which is most commonly used to access the IMEI number of a phone. The IMEI, or *International Mobile Equipment Identity*, number is a unique 15-digit code assigned to each GSM (*Global System for Mobile*) mobile device. This number can be retrieved manually from a Series 60 device by entering the sequence "*#06#" into the Phone application. It is also printed on the compliance plate, which is usually located under the battery. Programmatically, this number can be accessed by the following code, which returns a string containing the IMEI number:

```
// A typedef for a TBuf
TPlpVariantMachineId machineId;
PlpVariant::GetMachineIdL(machineId);
```

Note that this will work only on target—the emulator will not return a valid value. Also, do not forget to #include <plpvariant.h> in your code and link against plpvariant.lib.

System Agent

System Agent is a Symbian OS component designed to allow a generic framework of notifications and system event state monitoring. System Agent is a Symbian OS Server, providing clients with access to the information through client-side handles.

Chapter 12 Using Application Views, Engines and Key System APIs

The two main classes involved are `RSystemAgent` and `RSAVarChangeNotify`. `RSystemAgent` is used to get data and also to register interest in a change to a value. `RSAVarChangeNotify` is used to tell System Agent that a state has changed. Once System Agent is notified of a change, it pushes this information out to all parties who have registered an interest in that state.

All System Agent information is stored via a `UID`. The System Agent values that are supported by default are stored in `sacls.h`. The list includes `KUidPhonePwr`, `KUidSIMStatus`, `KUidNetworkStatus`, `KUidNetworkStrength`, `KUidChargerStatus`, `KUidBatteryStrength`, `KUidCurrentCall`, `KUidDataPort`, `KUidInboxStatus`, `KUidOutboxStatus`, `KUidClock`, `KUidAlarm` and `KUidIrdaStatus`.

The following example code shows how System Agent could be used inside an active object:

```
RSystemAgent sysAgent;
TSysAgentEvent sysAgentEvent;
sysAgent.Connect();
sysAgentEvent.SetRequestStatus(iStatus);
// aUid is the UID we want to monitor.
sysAgentEvent.SetUid(aUid);
sysAgent.NotifyOnEvent(sysAgentEvent);
```

The example shows the connection to the System Agent server, and the call to `NotifyOnEvent` tells System Agent that you wish to be notified when the value of the item associated to `aUid` changes. The application will be notified through the completion of the request associated with the `TRequestStatus`, which is set in the `TSysAgentEvent`. This will result in the `RunL` on the active object to be invoked. The **SystemAgent** example that accompanies this chapter monitors several values, one of which is identified through the UID `KUidInboxStatus` from the include file `sacls.h`. This enables the application to monitor when new unread messages are available in the Messaging Inbox.

More advanced monitoring of system data can be achieved through the use of `TSysAgentCondition`, which allows for notification when one of the conditions `ESysAgentEquals`, `ESysAgentNotEquals`, `ESysAgentGreaterThan` or `ESysAgentLessThan` is met. In this case, `RSystemAgent::NotifyOnCondition()` is used.

It is also possible to add your own data to System Agent by passing a unique UID to the `RSAVarChangeNotify::NotifySaVarChangeL()` method:

```
const TUid KUidSystemAgentStateVariable = { 0x101F6119 };
...
iSAVarChangeNotify.NotifySaVarChangeL(
KUidSystemAgentStateVariable, iSystemAgentState);
```

This will notify all other System Agent clients who have shown an interest in monitoring this data. When there is no longer a need for System Agent to store this value, it can be released via the following call:

```
iSAVarChangeNotify.ReleaseVariable(
KUidSystemAgentStateVariable);
```

Vibration API Support

Introduced in Series 60 2.x, this public interface allows you to control the device vibration feature. Vibration might be used in an application such as a game to signal a collision or an explosion, or for giving tactile feedback to other game events, such as resonance when a ball hits an object such as a racquet or a bat.

To use this API, include the `vibractrl.h` header in the source code and the `vibractrl.lib` in the `.mmp` file. The factory functions to construct an instance of the control object are:

```
CVibraControl* NewL()
CVibraControl* NewL(MVibraControlObserver *aCallback)
CVibraControl* NewLC(MVibraControlObserver *aCallback)
```

Some key methods are:

```
virtual TInt StartVibra(TUint16 aDuration) = 0
virtual TInt StopVibra(void) = 0
virtual TVibraModeState VibraSettings(void) const = 0
```

`StartVibra()` initiates the device vibration feedback, where `aDuration` is the interval in milliseconds. A value of 0 specifies that the vibration should continue indefinitely. Vibration can be stopped before the specified duration has elapsed with a call to `StopVibra()`.

The `StartVibra()` method does not block, but returns immediately, so that the vibration happens simultaneously as the application continues to run. If `StartVibra()` is called again, before the first vibration completes, then the first vibration is interrupted and the second vibrations starts immediately—the periods of vibration are not cumulative.

TIP The vibration settings in the user profile must be active for the vibration to be activated via this API. Also, specific Series 60 devices may have implementation-defined or hardware-imposed limits to the duration of the vibration feature (to conserve power). In such circumstances any vibration will cut off at that limit, even if the duration parameter is greater than the limit.

Use `VibraSettings()` to retrieve the current vibration settings from the current user profile. Then, if vibration is not active but is needed by the application—typically a game—the user could be informed and offered the option to enable the vibration feature for the duration of the game. This method returns `TVibraModeState` with the possible states of `EVibraModeON`, `EVibraModeOFF` or, if an error occurs, `EVibraModeUnknown`.

The mixin interface `MVibraControlObserver` class declares:

```
    virtual void VibraModeStatus(CVibraControl::TVibraModeState
aStatus) = 0
    virtual void VibraRequestStatus(
CVibraControl::TVibraRequestStatus aStatus) = 0
```

`VibraModeStatus()` is called when the vibration setting in the user profile is changed, so your application can be aware of changes the user makes. `VibraRequestStatus()` is called when the device vibration feature is requested.

The argument `aStatus` is a `TVibraRequestStatus` and indicates the current `VibraControl` request status. In addition to the return value supplied by the `StartVibra()` method (Symbian OS standard error codes), more detailed status information is returned by the callback method `MVibraControlObserver::VibraRequestStatus()`. Possible values for `aStatus` are `EVibraRequestOK`, `EVibraRequestFail`, `EVibraRequestNotAllowed`, `EVibraRequestStopped`, `VibraRequestUnableToStop` and `EVibraRequestUnknown`.

Summary

The decision whether to integrate directly with an application engine, switch views to an external application, or both (for example, edit an agenda entry and then switch to the Calendar application to view the entry) can be a difficult one. If you want to edit or change data, then using the application engines might be the correct decision, while if you only want to display the information, switching views to an external application might be wiser. By employing external views and using the application engines, with much-reduced effort and resources you will be able to present the user with a familiar, clear interface and effectively share data with other applications in a safe manner.

Many publicly available application engine APIs are contained within the SDK documentation. Details of other APIs can sometimes be found by searching the SDK header files in the `\epoc32\include` directory in the root of your SDK. Examples presented here include the log engine, agenda model, phonebook engine and many others. As the platform develops over time, you can expect much more functionality to become available—the examples presented in this chapter represent the general principles of using such Symbian and Series 60 APIs. Hopefully this experience will prepare you for using new APIs as they emerge.

chapter 13

Testing and Debugging

Finding errors and preparing your application for release

Chapter 13 Testing and Debugging

Testing and debugging are important aspects of developing an application, and they should be performed at every stage of the software development life cycle. This chapter gives you an overview of how to test and debug a Series 60 application.

Without going into specific detail about particular products, the tools and techniques available are highlighted, and hints and tips provided.

This chapter covers the following main topics:

- **Quality Assurance**—Software development is still a relatively new discipline. However, a lot of collective wisdom has been gathered on the subject, and these observations and experiences have generated standards and guidelines for the development process. Following these standards will ensure a better product and help reduce the tedious and potentially costly debugging and maintenance phases of software development. This section will identify and illustrate good coding practices to help avoid errors in the first place.

- **Testing**—Although Quality Assurance can go a long way toward producing robust, high-quality applications, confidence that an application has met the required quality standards can be secured only after testing. Formal testing involves a distinct shift in the software life cycle—a handover of the product from developers to an objective testing body that will assess it. However, informal testing should be performed by developers themselves as the code evolves. This section details the various types of testing tools (and some specific products) that are available. Also provided is an overview of existing formal testing methods and procedures, to enable you to support formal testing efforts.

- **Debugging**—It is often stated that debugging is as much an art as a science. This is because the origins of unexpected behavior in an application are often difficult to track down, and the process may require lateral and creative thinking. This section focuses on the features commonly offered by IDE debuggers, and how they can be used to aid in the debugging process.

This chapter provides information that should help you to produce an application suitable for commercial release. The three sections, **Quality Assurance**, **Testing**, and **Debugging**, can be read independently, but they should all be considered as part of an overall strategy to help you create a bug-free application. The **Quality Assurance** section will help you to avoid errors in the first place, **Testing** provides information on the tools and methodologies available to ascertain the existence of errors, and **Debugging** shows how you can locate and remove errors.

An example application, **Testing**, is provided to accompany this chapter. It illustrates how to write a test harness, and in particular how you can test the response of your application to Out-Of-Memory conditions. Details of how to download the full buildable source can be found in the Preface.

Quality Assurance

Creating high-quality software is inherently difficult. Fortunately, as the software development discipline has matured, effective proactive strategies have been identified in order to streamline and stabilize the development process. This collective wisdom is called Quality Assurance. It is a process-oriented methodology—the philosophy is that improving the software development process automatically improves the end product. Quality Assurance is multifaceted; this section focuses on those aspects that are immediately relevant to Series 60 code generation.

While the later **Testing** and **Debugging** sections are of primary importance, once you have written your application, the advice and tips given in the **Quality Assurance** section should be heeded before you begin development. This should help you to prevent problems.

Coding Standards

As discussed in Chapter 3, Series 60 offers well-defined coding standards. Compliance is essential, because it offers numerous benefits—for example, the standards encompass memory management techniques (in other words, proper use of the Cleanup Stack, two-phase construction and so on) to ensure that applications make responsible use of inherently limited memory.

The standards also provide class and function naming conventions, and these greatly improve the readability of code. Naming conventions convey important information about the purpose and behavior of classes and functions, without requiring full inspection of code logic. This information can aid in the debugging and maintenance of code, especially when these tasks are not being carried out by the original developer. For these reasons, coding standards are an important part of Quality Assurance.

TIP Formal code reviews by peers or mentors are another good way to locate common programming errors and enforce coding standards.

Coding standards are covered in more detail in Chapter 3.

Common Coding Mistakes

Quality Assurance seeks, in part, to avoid past mistakes. In this spirit, common coding mistakes by Symbian OS developers are listed below, so that you are less likely to repeat them:

- After becoming accustomed to the automatic zeroing of all member data for CBase-derived classes, developers often forget to initialize member data for non-CBase-derived classes. Automatic zero-initialization is CBase-specific behavior, so T-classes and all standard types should initialize their data to suitable values on construction.

- Developers who write a C-class and then forget to derive it from CBase will leave themselves vulnerable to memory leaks. This is because objects pushed onto the Cleanup Stack that are not derived from CBase are pushed as TAny*. No destructor will be called when these objects are explicitly removed from the stack, either via PopAndDestroy() or when the stack implicitly performs cleanup upon leaving. (Note that it is fine for pointers to some objects, such as T-class objects, to be pushed onto the Cleanup Stack as TAny*, because they do not have destructors.) For this reason, remember to derive all C-classes from CBase.

- Dangling pointers are easily created by forgetting to set pointers to NULL when explicitly deleting their corresponding objects. It is important to do this even if the next line of code allocates a new value, because the allocation itself may leave. One exception to this rule is in destructors, as member data pointers will fall out of scope as their class instance is destroyed.

- Constructors and destructors should not call leaving functions. Doing this compromises the failsafe memory management measures offered by two-phase construction. Therefore, never call in a C++ constructor or destructor a function that can leave.

- Virtual functions should not appear in constructors and destructors. The intention of a virtual function is to provide transparent access to its most derived implementation. This polymorphism feature is not available in constructors and destructors because the derived class will not exist at this time. In other words, the constructor of base class A does not know that it is actually creating a derived class B, and therefore a "virtual" function call within the constructor will only call class A's implementation.

- Explicitly deleting an item that has already been placed on the Cleanup Stack is a mistake. By placing an item on the Cleanup Stack, you are relinquishing control of that item. By explicitly deleting the item, the Cleanup Stack is left with a dangling pointer to invalid memory—its subsequent attempt to delete this memory will cause an access violation.

- Developers often forget to adhere to function naming standards—in particular, make sure any functions that can leave have a trailing "L".

(Symbian has developed a simple tool called **Leavescan** to check this for you—this tool will be covered in the **Testing** section.)

- Do not export inline functions, as this causes problems for dependent components at link time. Inline functions should be declared `inline` and an implementation should be provided in a separate `.inl` file.

- Some of the application objects in a UI application—for example, the `CAknApplication`-derived object—are created *before* the Cleanup Stack. If a leave occurs before the Cleanup Stack is created, then the application will crash.

- Developers often forget that their applications may eventually be ported to another language, and restrict descriptor sizes. Make sure that any descriptors that hold user-visible text have space to expand when localized, and that any controls that use such text can resize to fit.

- Do not ignore compiler warnings. Although your application may still run, warnings often illustrate instances of bad coding practices.

Defensive Programming

Defensive programming is the practice of anticipating the possibility of application failure—often creating a system to test for errors, and receiving notification when they occur. Defensive programming makes your code easier to maintain, while helping to avoid coding errors. In addition to general coding rules (such as always initializing local variables before use), a number of macros are provided that you should use to actively check the state of your code. Typically, this simply means checking, at critical points within the code, that the values of variables (or any resources whose corresponding values are susceptible to change) fall within expected boundaries.

Assertions

Assertions are designed to catch **programming** errors. They allow you to perform logic tests upon code and to specify the consequences of failure. (Failures not caused by programming errors, such as bad data, or out-of-memory conditions, should be handled by Leave exceptions.) Assertions, as shown below, are facilitated by a distinct coding mechanism. It is important to understand that this new mechanism does not offer new functionality as such—you could perform the same tests and specify the same consequences through normal programming techniques. Nevertheless, it is good practice to use assertions, because they clearly define the developer's intentions and visibly differentiate these critical "logic checks" from normal program logic.

There are two types of assertion:

- `__ASSERT_ALWAYS` catches programming errors in all builds.

- `__ASSERT_DEBUG` catches programming errors for debug builds only.

They each have two parameters: the test to perform (a statement, clause or function that returns a Boolean value) and the code to call if that test fails—it is common to Panic the application, providing distinct panic numbers for each assertion made, so it is easy to locate the origin.

Assertions are most often used in debug builds—here their main role is to speed up development by facilitating early detection of errors. But, as shown by the existence of `__ASSERT_ALWAYS`, they are sometimes used in release builds as well. In release builds, assertions generally have a very specific role: to validate parameters that are passed into public APIs. They are used in this way in order to clearly define the range of parameters with which the API was designed to operate, and to offer a firm refusal should the inputs fall outside of clearly documented boundaries.

Side Effects of Assertions
Sometimes, inconsistent behavior arises between release and debug builds. If this situation occurs, there may be an assertion side effect in your testing code.

When resolving such discrepancies, you should first look out for any code in an `#ifdef _DEBUG` preprocessor directive. Such code will be included only in the debug version, and therefore it is an obvious place to look when locating differences. For exactly the same reason, another likely source of behavioral differences is `__ASSERT_DEBUG` statements. Like conditional code specified by an `#ifdef _DEBUG` condition, debug assertions run only in the debug version. One common mistake that creeps into assertions is accidentally using an assignment operator in place of an equality operator in the assertion test:

```
__ASSERT_DEBUG((aValue = 5),
User::Panic(KMyPanicDescriptor));
```

The above code demonstrates the error. Instead of comparing `aValue` to 5, the above code *sets* it to 5. What is even more confusing is that the assignment operator returns "true," and so the assertion will always succeed.

Class Invariants

In brief, an invariant is a definition of an object's state that remains true throughout the object's lifetime—typically, an object specifies allowable values and/or ranges for some or all of its member variables. You test an object's integrity by checking its state against its invariance definition at the beginning and end of every member function in a nontrivial class.

Note that **invariants are tested only in debug builds**, but it is important that the testing function is defined for all builds to avoid any binary compatibility issues. For efficiency, all test code can be compiled out of release builds using preprocessor directives, as it will never actually be called.

The macro `__DECLARE_TEST` should be inserted as the *last item* in the class declaration. (This is because it switches back to `public` access as part of the macro definition—adding it as the last item avoids the possibility of changing the accessibility of other methods or attributes.)

Chapter 13 Quality Assurance

This macro declares the class method void __DbgTestInvariant() const, and this method should be defined to check the object's state. If an illegal state is found, then it should call User::Invariant(), which will cause a Panic. Panics raised in this way are more difficult to track down than the numerically distinct panics raised by assertions. One way to proceed, however, is to put a breakpoint in the __DbgTestInvariant() function itself, just prior to panicking. This makes it possible to trace the stack to the function that first violated the object's invariance.

When testing invariance, if a base class defines a __DbgTestInvariant() function, then that should be called before any further checking. Note that a __DbgTestInvariant() function should not make any calls to other class functions that test invariance, which could potentially result in an infinite recursive loop.

The macro __TEST_INVARIANT is used to call __DbgTestInvariant() in debug builds. This should be called at the start of each function, and also at the end of each non-const function. Note that a function may have multiple returns (in other words, multiple endings), in which case invariance should be tested at every return point. Static functions cannot test invariance!

As a simple example, consider the code below—it can be assumed that __DECLARE_TEST is the last item in the declaration of class TEgInvariant:

```
void TEgInvariant::DoSomeThing()
    {
    // Calls __DbgTestInvariant() in debug builds.
    __TEST_INVARIANT;
    // Do something with iData
    // ...
    // Calls __DbgTestInvariant() in debug builds.
    __TEST_INVARIANT;
    }
```

The __TEST_INVARIANT macro invokes a call to __DbgTestInvariant(), and this function tests that the value of iData remains within its upper bound:

```
void TEgInvariant::__DbgTestInvariant() const
    {
#if defined(_DEBUG)
    if (iData > 100) // Test iData is valid.
        {
        User::Invariant();
        }
#endif
    }
```

NOTE There is a difference between conceptual and bitwise const-ness. The bitwise representation of an object may not change in a const function, but an object that is owned by the class through a pointer may still be changed. If this is the case, invariance should also be tested at the end of the function.

Heap Testing

Series 60 provides macros to test heap allocation. Macros are also provided to simulate out-of-memory (OOM) conditions in order to test cleanup code, and how an application responds if system memory is exhausted. These macros are defined only for debug builds and so can be left in deliverable code. In fact, the application framework for GUI applications provides heap checking as standard, and this will detect any memory leaks when the application is closed.

An example of how to test for OOM is given in the **Out-Of-Memory Testing** section, later in this chapter.

TIP Memory leaks often occur when applications exit, as a result of incomplete cleanup code. Therefore it is important to test code to destruction. You can accomplish this by using the **Back** or **Exit** command to terminate your application, rather than just closing the emulator.

It is most common to test the User heap, as this is where dynamic memory is generally allocated for the current thread. However, it is also possible to test the Kernel heap, or a specific heap (for example when using multiple threads, with multiple heaps), and details can be found in the SDK documentation under *Memory Allocation*. The main heap macros are listed in Table 13–1.

WARNING If you use the C standard library, then you have to call `CloseSTDLib()` from `\epoc32\include\libc\sys\reent.h` before the terminating `__UHEAP_MARKEND` macro of the application is called and after the C standard library's DLL is no longer needed. The reason is that the C standard library stores reentrant information in the Thread Local Storage (TLS), which is not automatically cleaned up.

Heap testing is covered further in "Resource Failure Methods" later in this chapter.

Testing

Testing is the process of finding discrepancies between specified and actual behavior. These discrepancies are caused by errors in the source code. Subsequently determining and correcting their cause is known as debugging—this will be covered in the **Debugging** section.

Software failures can also occur when an application hits environmental constraints, such as running out of memory. Although these eventualities are often neglected by application specifications, it is still very important that applications gracefully handle them. Therefore, it is important that testing

Table 13-1 User Heap Macros
All macros refer to the current thread's heap and are defined only for debug builds.

Macro Name	Description
__UHEAP_MARK	Marks the start of heap checking. Must be matched by a corresponding call to __UHEAP_MARKEND or __UHEAP_MARKENDC. If previous calls to __UHEAP_MARK heap are still open, a new nested level is created.
__UHEAP_MARKEND	Marks the end of heap checking from an earlier call to __UHEAP_MARK. If there is no matching call, then a USER 51 panic is raised. All memory allocated at the current nest level must have been deleted or it will panic with the address of the first orphaned heap cell.
__UHEAP_MARKENDC(aValue)	Marks the end of heap checking from an earlier call to __UHEAP_MARK. If there is no matching call, then a USER 51 panic is raised. It expects aValue heap cells to still be allocated at the current nest level or it will panic with the address of the first orphaned heap cell.
__UHEAP_CHECK(aValue)	Checks that the number of cells allocated at the current nested level of the heap is the same as aValue. If not, it panics with the line number and source file this statement exists at.
__UHEAP_CHECKALL(aValue)	Checks that the total number of cells allocated on the heap is the same as aValue. If not, it panics with the line number and source file this statement exists at.
__UHEAP_FAILNEXT(aAfter)	Simulates heap allocation failure after every aAfter call to perform a heap allocation, for example, calls to new.
__UHEAP_SETFAIL(aNature, aFrequency)	Simulates heap allocation failure depending on the values of the supplied parameters. aNature: the nature of the failure that is imitated*. aFrequency: the frequency of failure.
__UHEAP_RESET	Cancels simulated heap allocation failure.

For example, RHeap::ERandom *or* RHeap::ETrueRandom—*see the definition of* TAllocFail *in the documentation for* RHeap *in the SDK documentation.*

procedures incorporate testing, such as **out-of-memory (OOM)** testing, to ensure that applications operate successfully across a range of environmental situations.

This section considers some basic testing strategies and offers information about how you can implement them. The tools that are available to aid testing will be examined, along with some useful techniques for testing Series 60 applications. It is beyond the scope of this section to discuss software testing techniques in general, but many good sources exist on the subject. The software testing FAQ at **http://www.faqs.org/faqs/software-eng/testing-faq/index.html** should be a good starting point for information on this subject.

Strategies for Testing

Exhaustive testing takes a lot of time and effort to perform, but putting some thought into the methodologies covered below can help to lower the overhead involved.

Test Teams and Procedures

The testing setup for a project needs to be decided before that project commences. The first order of business is to organize a testing team for your project. It is important that any programmer involved in writing code for an application, or a component of an application, should not be included in the testing team responsible for that particular application or component. The reason is that their familiarity with the code compromises the whole process— as they know the code intimately, they may be inclined to test based on how the application *does* work rather than how it *should* work. Significant benefits derive from the objective viewpoint that independent testers bring. Apart from ensuring your application's integrity, testers often give essential feedback concerning the usability of the UI.

Second, you need a test plan or test specification. This will specify the methods, inputs and expected results for every feature to be tested. It should not only include the testing of the correct use of the application, but should also consider how the application responds to incorrect input, unexpected events and so on. The test plan can be formulated prior to, or in parallel with, development. However, you should be careful not to write a test plan that simply describes how the application has been implemented—it must describe how you want the application to work.

Unit, Integration and System Testing

Software systems often comprise a number of different components. The typical testing approach is to begin by testing these smaller, more fundamental constituents in isolation. This means it is possible to begin testing before a complete application is written. Once confidence in the individual components is secured, they can be assembled and tested as cohesive subsystems. Because the internal integrity of each smaller component has already been ensured, problems that arise in subsystem tests are likely to be grounded in integration difficulties. By progressing in this manner, from small individual components to holistic testing in multiple increments, the software can be comprehensively tested in an efficient manner. This is the rationale behind *unit, integration and system testing*.

As the above overview suggests, a unit is an individual software component, and unit testing will test its behavior in a stand-alone environment. Sometimes this will involve the creation of test code that is not part of the release implementation, since usually it is other internal components that invoke its behavior. (Series 60 allows for the specification of such test code as a test project in the `bld.inf` file—see the entry for `prj_mmpfiles` in the SDK documentation for further details.)

Integration testing ensures that combined components behave correctly together. In general, all of the components will have been separately unit tested, and so the focus of such testing is on communication between components. This level of integration testing is sometimes referred to as "*integration testing in the small*."

System testing will discover defects that are properties of the entire system (or at least a deliverable product). It is a high level of integration test, but it also encompasses testing that the application interacts correctly with its environment (sometimes known as *compatibility testing* or "*integration testing in the large*"). Testing on hardware is important when performing this sort of end-to-end testing, as there can be subtle differences between hardware and emulator testing (see *Differences of Testing on Target versus Emulator* later in this chapter).

Functional and Structural Testing

While unit, integration and system testing addresses the scope and progression of the testing effort, *functional* (or behavioral) and *structural* testing are two different methodologies used to generate the tests themselves, regardless of scope.

Functional testing checks that the behavior of your component corresponds to its specification. It is generally "*black box*" testing. There is little or no need to understand the internal workings. The values of parameters passed into this "black box" are either random, or based on attributes of the specification.

Structural (or "*white box*") testing, however, uses specific internal knowledge of the component to guide selection of the test data, in order to test structural integrity. Developers may be able to provide significant guidance to testing plans formulated under this methodology, since they will be intimately aware of potential weaknesses.

Boundary-value analysis is a type of structural testing that uses test values that are on the edge (inside and outside) of allowable value ranges. This sort of testing can be much more useful than just testing random values.

Both of these methodologies are important in *designing* component tests, but their differences become transparent in the implementation of the test plan.

Performance, Stress and Recovery Testing

Stress testing involves subjecting a system to a load that exceeds expected operational ranges. This testing is particularly important for applications designed to run on devices with very limited resources. Essentially, the system is pushed to breaking point to ensure that an application fails cleanly, without losing data or orphaning resources.

Performance testing is similar to stress testing, but with a reduced load that is more representative of normal use. Contrary to stress testing, the aim of performance testing is to benchmark an application's responsiveness under normal loadings. For example, consider an application that must receive multiple SMS messages or handle multiple key presses. Performance testing would determine how many messages or key presses the application could handle within a specified period. This benchmark would then be used to measure the performance impact of subsequent code changes.

TIP The emulator keyboard shortcut **Ctrl+Alt+Shift+Z** sends the keys A through J in fast sequence to the application, to test its ability to handle rapidly repeated keys. A full list of emulator shortcut keys can be found in Appendix A.

With Series 60 development, a common revelation of stress testing is that the application UI's responsiveness may suffer from long-running background tasks. Such tasks can prevent applications from handling user input in a timely manner, but this can typically be remedied by breaking the offending task into multiple subtasks using Active Objects.

Performance tests should also detect problems caused by occurrence of events which are external to the application under test—for example, how the application reacts to an incoming voice call, SMS, system notification, or alarm. A game application, for instance, should pause when moved to the background by a system event. This sort of behavior is important in reducing processor load as well as in the usability aspects. A guide to enabling good behavior in your application is given in Chapter 4.

It is important to realize that Series 60 devices will typically be running several applications at once, so performance tests should consider this—high demands on system resources can degrade the performance of other applications.

WARNING In cases of low memory, Series 60 may automatically close down background applications to free resources. This may bring its own problems to your test plan!

Recovery testing involves creating test cases for extraordinary situations, such as a dropped connection during data transfer, or a sudden loss of device power. The focus of recovery testing is to ensure that the application handles these situations as responsibly and predictably as possible. Specifically, the tests should aim to verify that there is no data corruption, and where it makes sense, the application should try to revert to the state it was in before the test. For example, it should try to reestablish a lost connection and continue data transfer.

Acceptance Testing

If an application has been written for a particular customer, or group of customers, then *acceptance testing* may be appropriate. This type of testing involves the end user and is carried out to ensure that the application meets the specifications of the customer. Your application may work perfectly, but if it does not match the requirements of the target users it will fail acceptance testing. Note that if you are writing an application for a particular client, you should be mindful of any criteria they have specified throughout the development process.

As the type of testing necessary will depend upon the nature of the application and the customer, no further information is given here.

Tools and Techniques for Testing

A number of tools are available that you can use to help test your Series 60 application. These tools can be broadly split into static and dynamic categories, and further split into manual and automatic sections.

Static tools test software without executing it. In essence they are code review tools, which some might argue are technically quality-assurance tools, not testing tools, but we will treat them alike here. An example of an automatic static tool is the compiler, which will automatically scan source code and find any coding errors. An example of a manual static tool is a source code comparison tool, where the user can track changes between versions of files. Static test tools should typically be used before dynamic tools, and any errors or warnings addressed.

Conversely, dynamic test tools test software by executing it. An example of an automatic dynamic tool is a code coverage tool, which can be useful in tracking down redundant code. An example of a manual dynamic tool is a debugger, where the user can step through code statements to check an application's behavior.

All testing should be automated wherever possible. This not only eliminates many of the tedious tasks associated with manual testing, but also removes the vulnerability to human error.

Suggested Tools

The following is a list of testing and quality assurance tools that should prove useful in testing your Series 60 applications. Some of these tools are free, while others are commercial and will require licensing. This is not, by any means, an authoritative list, and a quick search of the Internet will doubtlessly reveal many more. However, those listed below generally provide specific support for Symbian OS and Series 60 development.

Static Tools

Code comparison tools are static manual testing tools. They can be useful during testing, as they provide a way to determine the code changes that have occurred between different versions of the software and can therefore help you to pinpoint where errors have been introduced. Many code comparison tools are available, from simple text output **diff** tools to fully featured GUI tools such as **Windiff** and the comparison tools found with most configuration control software. One particularly useful commercial tool is **Beyond Compare** from http://www.scootersoftware.com. It allows comparison between two versions of a directory hierarchy—enabling you to easily examine any differences in the directory structure, as well as any changes to a particular file.

Table 13–2 provides details of some automatic static testing tools that you may find helpful throughout the development process. The list starts with the compilers and IDEs that support Series 60 development, and moves on to tools that can detect bad coding practice and deviation from the coding standards.

Dynamic Tools

The most frequently used manual dynamic tools are the debuggers from the IDEs listed in Table 13–2. These allow you to step through code and examine variable values, memory addresses and so on, as the application is running. Further information on debugging Series 60 applications is available in the **Debugging** section of this chapter.

Automatic dynamic testing tools basically fall into two main categories: code coverage and scripted testing.

Code coverage tools can be used in conjunction with other testing tools, providing a method of examining which parts of your application have been covered by your test cases. This is structural testing at its most comprehensive, and it allows you to add further test cases as required for completeness. Code coverage analysis also allows you to trace any redundant code, which is a useful quality-assurance tool.

Table 13–3 provides details of some of the code coverage tools available for Symbian OS. Note that some tools can be tricky to set up for Symbian OS applications, so you should consult the documentation of your chosen tool for guidance.

TIP Profiling is the process of generating a statistical analysis of a program—showing, for example, the percentage of program execution time used by each function—and can be very useful in finding inefficiencies in code. Many of the code coverage tools and IDEs available provide support for profiling on the emulator, but profiling can also be achieved on target through the use of `RDebug`. Details are available from FAQ-0426 on the Symbian Knowledgebase—see **http://www3.symbian.com/faq.nsf/**.

Table 13–2 Automatic Static Tools Useful for Testing

Tool	Description	Further Details
Borland C++ Builder 6 Mobile Edition / Borland C++ BuilderX	IDE compatible with Series 60.	Further details are available from **http://www.borland.com/mobile**.
Metrowerks CodeWarrior Development Studio for Symbian OS v2.5	IDE compatible with Series 60.	Different editions of the CodeWarrior Development Tools are available. More information is available from **http://www.metrowerks.com**.
Microsoft Visual C++ 6.0 and .NET	IDE compatible with Series 60.	Further details are available from **http://msdn.microsoft.com/visualc/**.
GNU C++ compiler, gcc	The cross-compiler used for target builds.	The GNU compiler is provided with the SDK, but source code and further information can be found at **http://www.symbian.com/developer/downloads/tools.html**, if required. Note that **gcc** may come up with different warnings than the compiler in your chosen development IDE. You should aim for zero warnings on all compilers.
PC-Lint	Detects potential C++ problems and bad coding practices.	Available from Gimpel Software (**http://www.gimpel.com**), PC-Lint provides better error information than most compilers, and it can be customized to suppress particular warnings. You should aim to use this on all code at least once. Information regarding the configuration of PC-Lint for Symbian OS (and hence Series 60) development can be found in FAQ 0449 on the Symbian Knowledgebase—see **http://www3.symbian.com/faq.nsf/**.
Leavescan	Checks that code is leave-safe.	The name of any function that may potentially leave must have a trailing L (or LC)—in accordance with Symbian OS coding standards (see Chapter 3). This tool is available free of charge from Symbian. Further information can be found in FAQ 0291 on the Symbian Knowledgebase—see **http://www3.symbian.com/faq.nsf/**.
EpocCheck	Checks Symbian OS coding conventions.	A Perl script that checks function naming conventions, that member data is not pushed onto the Cleanup Stack, and that there are no IMPORT_C/EXPORT_C mismatches (a DLL API specification—see the SDK documentation for details). This tool is available free of charge from Symbian. Further information can be found in FAQ 0347 on the Symbian Knowledgebase—see **http://www3.symbian.com/faq.nsf/**.

Table 13–3 Code Coverage Tools

Tool	Details	Further Information
BullseyeCoverage	Code coverage tool, formerly known as C-Cover.	Further details are available from http://www.bullseye.com/.
LDRA Testbed	Code coverage tool, also provides statistical analysis.	Further details are available from http://www.ldra.co.uk/.
Metrowerks CodeTEST	Code coverage tool, also provides performance and memory analysis, and software execution trace.	Further details are available from http://www.metrowerks.com.
Testwell CTC++	Provides code coverage for Symbian OS emulators, through integration with Microsoft Visual C++.	Further details are available from http://www.testwell.fi/ctcdesc.html.

Scripted testing involves creating test harnesses to test code at the unit level (possibly creating a batch script to run several tests in sequence) or by using tools that automatically drive your application engine or user interface. Such tools may be designed to run from test scripts written purely as a list of text instructions, or they may encompass capture and replay methods to record GUI test input.

Some scripted testing tools are listed in Table 13–4.

Console Applications

Often with large applications, or when porting between different UI platforms, it makes sense to split the application engine into a separate DLL. This is discussed further in Chapter 4. Splitting the engine into its own DLL allows you to write console-based test harnesses to test the engine functionality. This is covered in more detail in the *Test Harnesses* subsection later in this chapter.

Resource Failure Methods

Series 60 provides methods to assist in performance testing, as mentioned in "Heap Testing" earlier in this chapter—methods such as __UHEAP_FAILNEXT. There is also a GUI interface to these methods on the debug emulator, which can be accessed using the keyboard shortcut **Ctrl+Alt+Shift+P**, as shown in Figure 13–1.

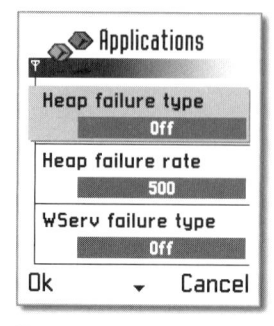

Figure 13–1 Resource failure testing tool.

This tool allows you to dynamically set heap, Window Server and file access failures to occur in your application, in order to test how your application copes with such events. Your application

Chapter 13   Quality Assurance

Table 13-4 Scripted Testing Tools

Tool	Description	Further Information
Nokia Testing Suite	A free automated testing tool that allows you to emulate user activities on a Nokia Series 60 device or a Series 60 emulator. As well as user tests, test scripts are provided for performance testing applications—these tests must pass in order to gain Nokia approval for your application.	This tool uses test scripts specifying a sequence of key presses and special commands, and involves connecting an application on a device with an application on a PC via infrared or Bluetooth. Further information on this tool is available from the Application Testing section of Forum Nokia—see http://www.forum.nokia.com.
SymbianOsUnit	A generic C++ unit testing framework for Symbian OS applications. Automated unit testing can be accomplished on both emulator and target.	This tool is provided as an open-source project by Penrillian (http://www.penrillian.com) and is available under the GNU Lesser General Public License (LGPL). Further information can be found at http://www.symbian.com/developer/downloads/tools.html.
Mobile Innovation TRY	Executes text-based test scripts on a device (or emulator) that emulate user input. Test output can be validated by text or screenshot comparison.	Further details are available from http://www.mobileinnovation.co.uk/.
TestQuest Pro	Emulates the actions of a manual tester to facilitating testing on an emulator or device.	Further details are available from http://www.testquest.com/.
Digia Quality Kit	A suite of automated testing tools.	Further details are available from http://www.digia.com/.

should not leak memory as a result of resource failure conditions, and ideally it should be able to recover from such conditions and continue operating. At the very least the user should be warned, and no data should be lost. Note that the settings provided by this tool will affect all applications running on the emulator.

The keyboard shortcut **Ctrl+Alt+Shift+Q** will turn heap failure mode off, but Window Server and file access errors have to be turned off using the dialog itself.

There are other debug keyboard shortcuts defined for displaying resources used, but their usefulness on a Series 60 emulator is limited, as they use the (debug emulator only) `CEikonEnv::InfoMsg()` method to display information, and this is often truncated by the width of the emulator screen. However, a complete list of emulator keyboard shortcuts is provided in Appendix A.

Debug Output

There are several ways to enable debug output in Series 60 applications, but many of these are poorly documented. Besides user methods, such as outputting text to screen using dialogs, notes, and the emulator-only `CEikonEnv::InfoMsg()`, there are two basic methods: file logging and serial output.

Serial Output

The undocumented class `RDebug` is defined in `e32svr.h` and resides in `euser.dll`. It offers various debugging APIs, such as profiling (as mentioned in "Dynamic Tools" earlier in this chapter), but of particular importance here is the `Print()` function.

`RDebug::Print()` is a `printf()`-style variable argument function, which takes a descriptor and an optional list of parameters. The descriptor may contain just text or include formatting information for the following arguments—as defined in *Format string syntax* in the SDK documentation. Some basic formatting characters are given in Table 13–5.

Table 13–5 Basic Format Characters

Format Character	Parameter Type	Description
%d	TInt	Decimal
%u	TUint	Decimal
%x	TUint	Hexadecimal
%b	TUint	Binary
%e	TReal	Exponential form
%f	TReal	Fixed form
%g	TReal	Either fixed or exponential form, depending on which can display the greatest number of significant figures
%s	TText*	Zero-terminated C-style string in either narrow or Unicode build
%S	TDesc*	Descriptor
%%		An actual percentage ('%') character

Note the upper-case "S" for descriptors.

Chapter 13 Quality Assurance

On the emulator, `RDebug` will print data to the IDE output window. On target, it will send data over the serial port (up to 80 characters sent to `COM1`, writing directly to the UART). As serial ports are generally not part of the hardware on Series 60 devices, this still may not fulfill your logging needs! If you have a device or a reference board with an RS-232 serial port, then it is important to ensure that the port is not used for anything else during logging. A HyperTerminal can be used to connect to the device, but be aware that the required communications settings are device specific. Similarly, it may be possible to connect over IR, but again any details are device specific.

Despite its name, `RDebug` code will exist in all builds, so it is important that any logging code is removed from release builds. A macro is probably the simplest way to ensure this:

```
// Note that _DEBUG is automatically defined
// for debug builds only.
#ifdef _DEBUG
#define TRACE(a) RDebug::Print a
#else
#define TRACE(a)
#endif
```

Then debug build-only trace code can be added as:

```
// Note the double parentheses, which allow the
// macro to have a variable argument list.
_LIT(KTraceOutput, "Value of iArray[%d] is %d");
TRACE((KTraceOutput, j, iArray[j]));
```

File Logging

There are various APIs for file logging on Series 60. You can use the `RFile` API directly, specifying your own log file format, you can use `CLogFile` from the Series 60 examples (`\Series60Ex\HelperFunctions`) as used in the `TestFrame` example (`Series60Ex\TestFrame`), or you can use `RFileLogger`.

`RFileLogger` is a simple class that can write descriptor text, `printf()`-style formatted text, and hexadecimal dumps to a file with optional time and date information. The class is defined in `flogger.h` and implemented in the library `flogger.dll`. (Note that the `.mmp` file statement `debuglibrary` can be used to specify `.lib` files for debug-only builds.)

The following code can be used to open the logging session, typically in the `ConstructL()` of the class you are logging:

```
// Connect to server.
User::LeaveIfError(iFileLogger.Connect());

// Open log file and leave if there is an error.
iFileLogger.CreateLog(KLogDirectory, KLogFile,
EFileLoggingModeOverwrite);
```

```
User::LeaveIfError(iFileLogger.LastError());

// Set timing format.
TBool useDate = ETrue;
TBool useTime = ETrue;
iFileLogger.SetDateAndTime(useDate, useTime);
```

If you are logging from multiple files, then the session can be opened in one class and passed by reference to other classes as needed.

iFileLogger is a member instance of RFileLogger, it would be sensible to make it mutable so that you can access the non-const Write() functions within constant methods that require logging.

KLogDirectory is the name of an *existing* subdirectory of c:\Logs. If the directory does not exist, then no errors will occur in creating, writing to or closing the log, but **no actual output will be written**. The function RFileLogger::LogValid() can be used to test that the log can be written to, or RFs can be used to create the directory if necessary—see Chapter 3 or the SDK documentation for details of creating a directory.

KLogFile is the name of the file to log to, and the last parameter specifies the writing mode, either EFileLoggingModeOverwrite or EFileLoggingModeAppend. Note that the log will always be created in the c:\Logs hierarchy—it is not possible to log to removable media.

SetDateAndTime() specifies whether the date and/or time will be output in each line of the log.

Use of the three basic logging functions is demonstrated in the code snippet below. Write() is used for simple descriptors, WriteFormat() is a variable-argument function for formatted descriptors (as described in Table 13–5), and HexDump() is a hexadecimal dump:

```
_LIT(KText, "Plain descriptor");
iFileLogger.Write(KText);

_LIT(KTraceOutput, "Value of iArray[%d] is %d");
iFileLogger.WriteFormat(KTraceOutput, j, iArray[j]));

_LIT8(KMemory, "hello this is a memory dump");
iFileLogger.HexDump(NULL, NULL,
KMemory().Ptr(),KMemory().Size());
```

Be aware that static versions of these functions are also available, but these are much less efficient, because they make a new connection to the file server for each line of logging. They should be used only if the required logging is very infrequent.

The above example would produce output of:

Chapter 13 Quality Assurance

```
14/08/2003           2:36:50        Plain descriptor
14/08/2003           2:36:50        Value of iArray[4] is 10
14/08/2003           2:36:50           0000 : 68 65 6c 6c 6f 20 74
68 69 73 20 69 73 20 61 20  hello this is a
14/08/2003           2:36:50           0010 : 6d 65 6d 6f 72 79 20
64 75 6d 70                      memory dump
```

To close the logging session, use the following code (typically in the destructor of the class you are logging):

```
iFileLogger.CloseLog(); // Close the log file.
iFileLogger.Close();    // Close the server connection.
```

Again, you may wish to use compiler directives to conditionally compile the logging code only for debug builds, or just make use of the fact that nothing will be written if the specific logging directory does not exist. The latter approach requires that you refrain from writing code to generate the logging directory, and that you remember to include the required libraries in all builds.

WARNING Note that logging code will cause loss of speed and performance and will also increase the size of the binary.

To read the log file on the emulator, just look in \Epoc32\Wins\c\Logs in the root installation directory of your Series 60 SDK. To get the log file off a target device, use a file browser that supports sending (such as **FExplorer** from http://www.gosymbian.com), or a connectivity solution that lets you transfer specific files, as may be supplied with your Series 60 device. A file browser can also be used to add/remove the logging directory to enable/disable logging, and may also provide functionality to read small log files on the device.

One further point to note is that disk space can be severely limited on target devices. SysUtil::FFSSpaceBelowCriticalLevelL() can be used to check that there is enough space on disk to write to, but this may be overkill when writing a debug log. However, you should note that your log could end due to running out of space, rather than an error at that point in the code!

Differences of Testing on Target vs. Emulator

Your application may run fine on the emulator, but not on target, or vice versa, and this may be due to inherent differences between the two platforms. Although the emulator provides a full target environment, the fact that there *are* differences means that it is important to test your applications on target devices, in case any problems may occur. As well as obvious differences, such as availability of specific hardware, there are some more subtle differences, and the main ones are listed below.

The Thread/Process Model

On target, each application runs in its own process. On the emulator, however, each application runs in its own thread (WINS meaning WINdows, Single process). This potentially has an impact on the memory model, as threads share writable memory, whereas processes do not. This means that bad pointers may corrupt another application's (or even the system kernel's) memory on the emulator. Also, it is possible to design applications that share memory on the emulator without using specific shared memory APIs. Such applications will not work on a target device.

TIP Random pointers are likely to result in a `Kern-Exec 3` panic. Further information can be found in the SDK documentation.

Similarly, process-relative handles can be used only within the process that created them. For example, when passing a handle across a Client/Server boundary, you must use `RHandlebase::Duplicate()` to convert it into a valid handle in the other process. This limit will not be imposed on the emulator.

Hardware Limits

An application must function within certain bounds determined by the target hardware. When running applications on the emulator the same boundaries may not apply; for example, you would not need to be concerned about whether you have enough disk space. In this subsection the various constraints that you need to be consider are highlighted.

Heap and Stack Sizes

Target hardware supports different (specifiable) stack and heap sizes for applications, whereas the emulator uses default values that may differ from those on hardware. The default heap size is likely to be much bigger on the emulator than on target! Note that it may be possible to set the emulator stack size used in your IDE. See Chapter 2 for more information on setting stack and heap sizes.

If your application exceeds the available stack size, a `__chkstk` error will occur when linking during a target build. You should look to split up functions causing the problem or try increasing the stack size. You should also be aware that recursive functions can still blow the stack at runtime—this will lead to a `Kern-Exec 3` panic.

Machine Word Alignment

Series 60 devices use 32-bit ARM chips with RISC architecture for cost and power efficiency. This means that all memory words must be aligned to 32-bit machine word boundaries.

WARNING Dereferencing a pointer whose address is not a multiple of 4 will result in an access violation.

32-bit (or larger) struct and class members will be 32-bit aligned by the compiler, with appropriate padding, so `sizeof` may return a larger value on target than on the emulator. For example, the following structure would have a size of 6 bytes on the emulator and 8 bytes on target:

```
struct SInfo
    {
    TText8 iText;     // Stored at offset 0, size 1 byte
    TText8 iText2;    // Stored at offset 1, size 1 byte
    TInt32 iInteger;  // Stored at offset 4, size 4 bytes
    };
```

`iText` will lie on a 32-bit boundary, `iText2` will be stored in the same 32-bit word as `iText` and `iInteger` will be aligned to the next available 32-bit word.

C-style arrays will also be aligned, but are rarely used in Symbian OS. However, code such as the following would generate an access violation the second time through the loop on target, as `p` would not be a 32-bit multiple, despite this code's being valid on the emulator:

```
TText8 array[200];
for (TInt i = 0; i <= 196; i++)
    {
    TInt* p = (TInt*)array[i]; // Needs a cast.
    // Four bytes from the array makes one integer.
    TInt n = *p;
    }
```

Any code that casts one type of packed structure to another might need to be implemented using `Mem::Copy()` —for example:

```
TText8 array[200];
for (TInt i = 0; i < 196; i++)
    {
    // Really a TAny*, so no cast needed!
    TAny* p = array[i];
    TInt n;
    Mem::Copy(&n, p, sizeof(TInt));   // Copy byte by byte.
    }
```

Note that in code without casts, the compiler will ensure machine word alignment, so no special coding is required.

Out-Of-Disk Errors

The emulated disk space available is constrained only by the available space on your PC's disk. It is important that applications do not assume that sufficient disk space will be available for every write that they make, as it would be on the emulator. Disk space on a device can be quickly used up.

Writable Static Data

The emulator will allow non-const static or global data in applications or DLLs — whereas the **Petran** tool will give an error at the link stage, when building for target. **Petran** converts windows (Portable Executable format) executables to ARM format, and although it is possible to "eliminate" such errors by specifying the -allowdlldata flag, executing code containing writable static data will cause an immediate panic on target devices! To implement writable static data you should use Thread Local Storage (TLS) instead — see the SDK documentation for more details.

> **TIP** Note that "const" C++ objects are not constant until their constructor is called, so these also count as writable data.

For further information about **Petran** and how to locate writable static data in your application, consult FAQ-0329 on the Symbian Knowledgebase — see **http://www3.symbian.com/faq.nsf**.

Timing Differences

Most code will generally run faster on the emulator than on a target device because the processor speeds are usually much higher on PCs.

Furthermore, the standard clock tick interval is different: the interval is 1/10 second on the emulator and 1/64 second on target. Any RTimer::After() requests will be rounded up to the corresponding resolution, and this can lead to subtle timing differences. These may not be a problem for general timeout purposes, but they can affect applications needing higher timing resolutions, such as for animations.

> **NOTE** Although perhaps obvious, it is important to understand that there is a minimum wait of one tick, so the minimum wait time is 6.4 times larger on the emulator than on target.

Directory Differences

There are some differences between the directories used on the emulator and a target device. First, note that there are actually two emulators: the debug and the release emulator. Each has its own ROM (z:) drive, but they share all writable drives.

Generally, when developing, you will be using the debug emulator, and this uses a path such as \epoc32\release\wins\udeb relative to the root of your SDK. Within this directory will be the emulated z: drive, the emulator itself (epoc.exe) and any DLLs required by the emulator. The first main difference is that DLLs on the emulator typically exist *outside* of the emulated drive system. In other words, the emulator DLLs are stored on your PC at a level higher than the emulated `z:` drive, so they cannot be seen by the emulator's file system. On target such DLLs will be stored in `\system\libs` on the relevant drive.

The shared writable emulator drives—for example, `c:`—exist in a directory such as `\epoc32\wins` relative to the root of your SDK.

NOTE The actual paths used depend on the IDE you have installed. For example, `wins` may be replaced by `winscw` or `winsb`. If you are using the release emulator rather than the debug emulator, then replace `udeb` with `urel` in any paths that specify it.

The second main difference is that the emulator build tools build applications on the `z:` drive, as the tools were originally designed for building the standard applications that are supplied on your device's ROM drive. Third-party applications will be installed on one of the writable drives on the device—typically the Flash Filing System (FFS), or `c:` drive; but also possibly on a removable media drive, such as `e:` (if available). Your code should not make any assumptions about the drive that it is installed on. `CompleteWithAppPath()` is one way to find out the drive your application is installed to at runtime—see the SDK documentation for further details.

GUI applications run as polymorphic DLLs on both the emulator and target devices. However, Symbian OS executables run differently on each. On target, executables are typically stored in `\system\programs` on the applicable device drive. Emulated executables run as a single Windows executable, which includes the emulator code. Hence these are typically stored in the same directory as the emulator—for example, `\epoc32\release\wins\udeb`—which is above the emulated file system.

Server programs run as an executable on target, but as a DLL on the emulator. The macro `__WINS__` can be used to conditionally compile any code that is relevant only to emulator implementations, such as `E32Dll()`.

Test Harnesses

As mentioned earlier (in the "Console Applications"), it often makes sense to split the application engine and GUI, and this allows you to write console-based unit tests. Not only are these much simpler than UI-based test applications, they can also be portable across different UI platforms. Furthermore, they allow for automatic testing, where a batch of executables are run, and their output (in text format) is compared to a standard set of results using a **diff**-style tool.

The class `Console` found in `e32base.h` can be used to create a `CConsoleBase`-based console application (see `e32cons.h`). Although this provides a `Printf()` method for drawing text to the screen, there is a better way to create console-based test executables—the poorly documented `RTest` class, and this is covered in some detail below.

Also of note, the `TestFrame` example, which comes with the SDK, demonstrates a Series 60 GUI-based test harness. This allows you to test that an application panics, leaves and handles invariants correctly by locally overriding the standard `User` calls, although its use requires manual intervention. Further details of this are given in the SDK documentation—it will not be covered in any depth here.

RTest

The console-based test class `RTest` is defined in `e32test.h`. It provides logging to console screen and file, but it cannot be used for testing GUIs, only application engines. In addition, there can be problems if you are trying to write tests for communications, because connection dialogs cannot be displayed in a console application.

`RTest` is constructed with the title of the test, which is used to identify it in output messages. A call to `Title()` will output the text passed into the constructor, plus the operating system build version.

```
_LIT(KTestName, "My test");

RTest test(KTestName);

// Note that, as with all console applications, you
// will have to create your own Cleanup Stack.
CleanupClosePushL(test);

test.Title();
```

Tests are automatically numbered—and the numbering can be nested to multiple levels. The `Start()` method opens a new nested level and sets the subtest number to `1`. A call to `Next()` will increment the number at the current level, and a call to `End()` will close the nested level. Nested subnumbers will appear separated by a dot (".")—for example, `001.02.01`.

Each `Start()` must have a matching `End()`—if there is no closing `End()` call, then the "…test completed…" notification will not be displayed; if there are too many calls to `End()`, then a panic is generated and `"End without matching Start()"` is displayed.

Tests are carried out using `operator()`—the expression in the parentheses is evaluated, and if false, then an error message will be printed and the test will panic. For example, code such as:

```
_LIT(KTestA, "Test A");
_LIT(KTestB, "Test B");

test.Start(KTestA); // 001
test(ETrue);        // Used to verify how far you have got

test.Next(KTestB);  // 002
test(aBool);
test.End();
```

will produce output similar to the following if `aBool` is true:

```
RTEST TITLE: My test 1.02(320)
Epoc/32 1.02(320)
RTEST: Level   001 Next test - Test A
RTEST: Level   002 Next test - Test B
RTEST: SUCCESS : My test test completed O.K.
```

or similar to this if `aBool` is false:

```
RTEST TITLE: My test 1.02(320)
Epoc/32 1.02(320)
RTEST: Level   001 Next test - Test A
RTEST: Level   002 Next test - Test B
RTEST: Level   002 : FAIL : My test failed check 1 at line
Number: 58
RTEST: Checkpoint-fail
```

> **TIP** Naming your `RTest` instance "test" has an interesting side effect: a macro defined in `e32test.h` expands `test(x)` to `test(x, __LINE__)`. Hence all `operator()` calls are automatically expanded to include the line number of the file they reside in. (The constructor is overloaded to ignore this line number.) Note, however, that this can sometimes run against coding standards—in which case simply add the __LINE__ preprocessor directive into your code by hand.

All output from the executable is printed to the console window. On the emulator, it will also be printed to your IDE's debug window (if applicable) and appended to the file `Epocwind.out` in your temporary directory (as set by the PC system environment variable `TEMP`—for example, `c:\Temp`). Note that this file is always appended to, so when running batches of tests, it is important to delete any existing file first. All tests will then be logged to the same file.

On target, as with the `RDebug` class (see "Serial Output" earlier in this chapter), any output will also be sent out across the serial port (if available), where it can be logged to file via a connected HyperTerminal.

Typical tests would be to check expected return values from functions, or to validate the internal state of some data. If a test is designed to leave, then it can be useful to add a `test(EFalse);` line after the line that is expected to leave. If the code reaches this point, then it will panic.

Other methods supported by the `RTest` class include:

- `Printf()`, which outputs user-formatted text. Note that new lines are not automatically generated.

- Two `Panic()` overloads, which call `User::Panic()` as expected.

684   Chapter 13   Testing and Debugging

- `Getch()`, which gets keyboard input and can be useful for pausing the console in manual tests or when running on target without connection to a HyperTerminal.

- `SetLogged()`, which can be used to turn output on or off, and `Logged()`, which can be used to determine whether or not logging is turned on.

The console is closed using the `Close()` method. If this method is left out, the test log will not show any errors, but the executable will panic on exit.

```
CleanupStack::Pop();    // test
test.Close();
```

Out-Of-Memory Testing

The test harness code shown so far is not actually taken from a real example. It is designed purely to show how the API works. However, the **Testing** example shows how a real out-of-memory (OOM) test harness can be created.

The code shown below is not the complete source, but it should be enough to demonstrate the principles. This brings together what you have learned so far about `RTest` and the heap checking and failure macros.

The basic mechanism for OOM testing is using `__UHEAP_SETFAIL` to create heap allocation errors. Using a loop, you can comprehensively test how a given code segment responds to every possible heap allocation failure, increasing the failure interval on each iteration until the code finally completes without failure—if the code segment makes X memory allocations, then the loop must repeat X times. This sort of testing proves that your code will not orphan resources or otherwise corrupt data, due to OOM conditions.

```
RTest test(KTestTitle);
…

TInt error = KErrNoMemory;
TInt failAt = 0;

while (error != KErrNone)
   {
   failAt++;// Increase the failure interval.

// Set the failure interval and start nested heap checking.
   __UHEAP_SETFAIL(RHeap::EDeterministic, failAt);
   __UHEAP_MARK;

// Run the test code in a trap harness.
   TRAP(error, DoTestL());
   …
// End nested heap checking and reset (turn off)
// heap failure.
   __UHEAP_MARKEND;
```

```
__UHEAP_RESET;
...
// Test that we have no unexpected errors.
// This will Panic and break out of the loop on failure.
test((error == KErrNoMemory) || (error == KErrNone));
}
```

Within each iteration of the loop, the failure interval is increased. By setting the type to `EDeterministic`, you know that the allocation will fail at the `failAt` interval specified. A nested heap check is created using the `__UHEAP_MARK` and `__UHEAP_MARKEND` macros, and the code to be tested is called from within a trap harness.

If the code being tested leaves, then we can test that the error returned is either: `KErrNoMemory` (as the allocation has failed) or `KErrNone` (as the failure interval is now greater than the number of allocations in the test code, hence the test has passed).

If the code being tested leaks any memory, then the heap check should pick it up. Remember, though, that the code being tested may be designed to reserve resources—particularly in the case of testing class methods. For example, class members may be allocated that would be deleted in the class's destructor, not if the method leaves. In such cases the `__UHEAP_MARKENDC()` macro could be used to handle the nondeleted resources, or the resources should be deleted in the test harness before the `__UHEAP_MARKEND` macro is called.

A further complication occurs if your test code uses a server session that caches memory. For example, if you create a `CFbsBitmap` object, then this will use an `RFbsSession`, and that will cache filename data on the heap. This memory is not freed immediately after the `CFbsBitmap` is deleted, so the heap check will fail. One possible solution might be to allocate and delete the `CFbsBitmap` before the start of the heap check, in order to preinflate the server's memory cache. The server will then use the existing cache for the next `CFbsBitmap` object you create, and you will not experience an allocation/deallocation mismatch inside your test.

The test loop will exit when either the test is passed (`error == KErrNone`) or `RTest` or the heap checking macros panic. With the code shown above, there is a slight danger of the test getting into an infinite loop if the test code still runs out of memory with an arbitrarily large value of `failAt`. In this case, you may do better to use a `for` loop with an arbitrarily large end value, and `break` if `error` equals `KErrNone`.

Any leave codes can be checked by looking at the value of `error` in a debugger. Note, however, that other side effects of OOM conditions may cause panics or access violations elsewhere in the code. If the test does not pass, then you will need to examine the code being tested in a debugger to establish the exact cause of failure.

Chapter 13 Testing and Debugging

Also note that Cleanup Stack code may also cause heap allocation failure, so bear this in mind if counting the allocations to find the erroneousness line. (You will generally find that it is quicker to step through in a debugger!)

Although it is possible to run such a test harness on target (and there may be occasions where code will react differently on target to OOM conditions), the current Series 60 SDKs do not include target debugging libraries. In other words, you cannot build for `armi udeb`. The heap checking and failure tools will not work in release builds, so no OOM testing will actually occur, although the test will probably still appear to pass.

Running Test Executables

Series 60 provides no default mechanism to run executables on target. Usually this type of application is limited to system servers, so to run an executable on target, you need to:

- Package your executable up in a `.sis` file (see Chapter 2 for details) and install it on your device. Executables sent directly to a device will be blocked from running by the system security measures.

- Locate an application that can launch executables, such as the file browser **FExplorer** from **http://www.gosymbian.com**, or the **ExeLauncher** utility provided with this book—this can be downloaded with the example applications as shown in the Preface.

The **ExeLauncher** utility allows you to run any executables in the directory `c:\EMCC\Exes`, so you must make sure that your `.sis` file installs them there. Also, all executables must have the suffix `.exe`.

Then, to run an executable, simply browse the required file using one of the tools mentioned above and launch it using the selection key or relevant menu item.

Debugging

Debugging is the process of locating, isolating and fixing coding errors. An error may be syntactical, logical or simply an unwanted side effect of the code, so debugging can be an art form as much as an applied science.

As with testing, it is easiest to break the debugging process down into the smallest constituent parts and address each error as it appears. By definition, testing identifies only that there is an error—but the distinction between testing and debugging can be blurred. Hence, some of the techniques already covered in this chapter could be equally identified as debugging.

While debugging techniques can be used to fix the problems highlighted by formal testing, they will also prove useful during the normal development process. As you build and run your code on a day-to-day basis, you will doubtless encounter errors, and the techniques you learn in this section will help you to locate and fix them.

Debugging an Application on the Emulator

The main tool you will use for debugging is likely to be the debugger integrated with your IDE. Different IDEs (and sometimes different editions of the same IDE) provide different debugging features, so it is worth researching both current and forecasted support before choosing one that is right for you.

Traditionally, most Symbian OS development involved using Microsoft Visual C++. Now, for Series 60, the development options have expanded to include Metrowerks CodeWarrior and Borland C++ Builder—these tools provide native support for Series 60, and new features are being added all the time.

Due to the differences in IDEs, only general debugging tips will be given in this section. For more specific details you should refer to your IDE help facilities or consult the documents available on Forum Nokia—see **http://forum.nokia.com**.

General IDE Overview

The debugging facilities of each available IDE vary, but the windows and dialogs available are generally very similar (albeit sometimes with different names). For example:

- **Call stack**—This is a list of the functions that have been called, and it can usually be accessed only when program execution is halted—for example, a breakpoint has been reached. The most recently called function is at the top, and you can trace where each function was called from.

- **Memory window**—This shows the contents of memory and can be used to look at the contents of data outside of the current function and also to examine when memory changes.

- **Watch window**—This will show the values of user-selected variables or expressions. Sometimes it is possible to change the values and see how the code reacts. Some windows will show all variables with relevant local scope.

- **Breakpoints, threads and modules**—These will show lists of the currently set breakpoints, currently running threads, and currently attached libraries. Breakpoints sometimes allow you to stop the execution at a point in the program based on a set of criteria.

You should familiarize yourself with these terms, as they will be referred to throughout the rest of this section.

Chapter 13 Testing and Debugging

Debugging Tips

The following section deals with some common debugging issues and provides some helpful hints to aid you in debugging your Series 60 application.

Viewing the Contents of Descriptors

Not all IDEs support viewing the contents of descriptors. However, with the simple method described here, the contents of *any* descriptor can be easily viewed:

- Make sure that any IDE options to show Unicode strings are selected.

- Stop the debugger at the required line by using a breakpoint, or by stepping through the code.

- Enter the name of the descriptor into the watch window and expand it to establish the value of its `iType` attribute.

- Apply one of the casts shown in Table 13–6 to the required descriptor in the watch window, depending on the value of `iType` and whether it is a Unicode or narrow (8-bit) descriptor. Note that in the table the name of the descriptor is assumed to be `aDes`.

Table 13–6 Casts for Viewing Descriptor Values in an Expression Window

Character Set	Value of `iType`	Cast
Unicode	0	`(TText16*)(&aDes) + 2`
	1	`(TPtrC16*)&aDes`
	2	`(TPtr16*)&aDes`
	3	`(TText16*)(&aDes) + 4`
	4	`(TText16*)(*((int*)&aDes + 2)) + 2`
Narrow	0	`(char*)(&aDes) + 4`
	1	`(TPtrC8*)&aDes`
	2	`(TPtr8*)&aDes`
	3	`(TText8*)(&aDes) + 8`
	4	`(TText8*)(*((int*)&aDes + 2)) + 4`

Note that in Series 60, all descriptors are Unicode unless their class name ends in an "8".

For example, Figure 13–2 shows a `TDesC` being decoded to view its contents in Visual C++.

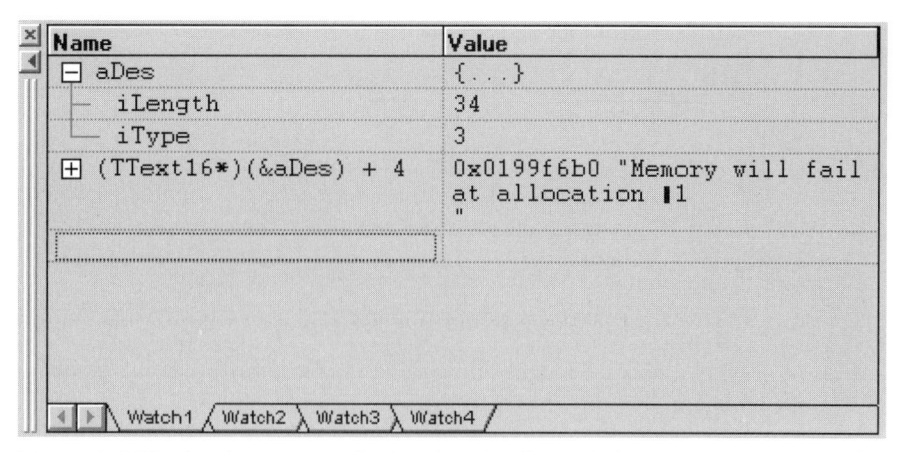

Name	Value
⊟ aDes	{...}
iLength	34
iType	3
⊞ (TText16*)(&aDes) + 4	0x0199f6b0 "Memory will fail at allocation ❙1 "

Figure 13-2 Viewing the contents of a descriptor in Microsoft Visual C++.

TIP When using a class such as `CEditableText`, any descriptors returned from its member functions may contain, along with readable text, some unrecognized characters. These are typically text-formatting characters such as `EParagraphDelimiter`.

Debugging Macros

There are a few macros which are useful to know when it comes to Series 60 debugging. Some of these are well known (such as `_DEBUG`, which is defined only for debug builds and can be used for conditional compilation—this is used with the `TRACE` macro defined earlier in "Serial Output"), and some have been covered earlier in this chapter (such as assertions, invariants and heap checking macros).

However, there is one important macro not yet covered: `__DEBUGGER()`, defined in `e32def.h`. This macro can be used to programmatically stop execution before a panic (since, after a panic occurs, all contextual debugging information is lost), and as such is a very powerful tool for quickly tracking down programming errors.

This macro applies only to debug emulator builds, and it will cause the emulator to stop the debugger at the current line if "just-in-time" (**JIT**) debugging is enabled—this means that if the debugging emulator is running outside of the debugger, then it will attempt to open the currently defined JIT debugger and attach it to the running emulator process.

Just-in-time debugging is enabled by default, but it can be turned off for the emulator in the `epoc.ini` file (see the SDK documentation for details). It must also be enabled in the IDE.

"Debugging" Resource Files

There is no real way to debug resource files—they define code resources that do not actually get executed. However, problems in resource files will produce one of two immediately visible symptoms:

- The resource file will not build.
- The resource file complies, but does not do what is intended.

The first issue is sometimes the more difficult one to solve. If the compiler does not correctly identify syntax errors within the file, then the solution is generally to simplify the resource file, and gradually build it back up until you can identify the problem.

If the resource file compiles successfully but causes problems at runtime, then the file should be checked for logical errors. The SDK documentation and the information in Chapters 5 through 8 detail many of the resource structures used.

Incorrect display of user-visible text may occur because the text defined in resource is overwritten in the code, or because the required localization files do not exist (in which case the name of the resource will be displayed instead). Check that the correct localized file is included and that the correct language is specified. Also check that the correct resource name is used in both the resource file and the localization file.

Note also that antinesting macros are not used in resource files, so check that files are not included multiple times.

Removing Screen Flicker
Screen flicker is generally caused by `CCoeControl`-derived classes clearing their rectangle before drawing to it. Also, excessive use of `DrawNow()`, rather than the buffered `DrawDeferred()`, can make matters worse (this is covered in more detail in Chapter 11).

In order to find the source of the problem, try using the debug emulator and arrange your screen so that you can see both the emulator and your IDE. Run the application causing the flicker, and use the keyboard shortcut **Ctrl+Alt+Shift+F** to enable automatic Window Server flushing—this will let you see the effect of each draw command, as drawing will no longer be buffered. (This will also slow down the application and make flicker from other sources more pronounced!)

Place a breakpoint in the suspect `Draw()` method(s) and use the keyboard shortcut **Ctrl+Alt+Shift+R** to cause a full-screen redraw—the application will stop at your breakpoint. By stepping through the code, you will be able to observe the effect of each draw command on the emulator. Details of all the available keyboard shortcuts are available in Appendix A.

Emulator Lockup
Sometimes the emulator will freeze after a given time—particularly if the debugger has been stopped for a long time at a breakpoint. A simple solution is to press **F11** twice. This generates a case open/close event, which wakes up

Chapter 13 Debugging   691

the emulator. It should be noted that this keyboard shortcut is part of Symbian OS and has been inherited from earlier Psion machines that made use of it. This event is not used within Series 60, so it should cause no unwanted side effects.

Finding Memory Leaks

Memory leaks occur when allocated memory is orphaned. There are two basic types of memory leak: *Static* leaks are repeatable under the same conditions each time. They are caused by mismatched allocation and deallocation, and so always occur in the same place. *Dynamic* leaks, however, are nonrepeatable—typically being caused by an error or race condition. These are trickier to find, as they will not occur on every run, or will appear to occur in different places.

Either type of leak will cause an error on the debug emulator when the application closes, as the heap checking macros that automatically surround the framework of every Series 60 UI application will panic. Make sure that the line `JustInTime 0` is not present in the file `\Epoc32\Data\Epoc.ini` in the root directory of your SDK, or is changed to read `JustInTime 1`. This ensures that the debugger will be halted if your application panics.

> **TIP** Some versions of the SDK display only a "Program closed" error message if an application panics. Creating an empty file called `ErrRd` in `\Epoc32\Wins\C\System\Bootdata` will ensure that the traditional panic code dialog is displayed.

The following description is valid for locating memory leaks in Microsoft Visual C++. Other IDEs may require the use of other methods or may provide further tools for tracking memory leaks—see the documentation of your chosen IDE for further help.

Note that debug-build emulator code that may be available with other Symbian OS platforms is not necessarily provided by Series 60, so methods documented elsewhere may not work with a Series 60 SDK.

This method is useful mainly for locating static memory leaks:

- First run the application that contains your memory leak (by pressing **F5**)— it will halt the debugger when the panic occurs, but as the necessary source and debug-build object code is not supplied with the SDK, no useful information can be gleaned here.

- Continue the debugger (press **F5**) until the panic dialog appears—the address of the heap cell that leaked will be printed in the debugging output window (and also in the panic dialog if the `ErrRd` tip above is followed). For example, the text might read `Panic ALLOC: 10052684`. *Make a note of this value.*

- At this point it is important to stop debugging and restart the emulator. Every time the application needs to be run, the emulator must be restarted, or the memory leak will occur at a different memory address.

- Set a breakpoint (press **F9**) in the `Application` class's `AppDllUid()` method. This allows you to stop the application before it allocates any memory.

- Run the application again, and when it stops at the breakpoint, open the breakpoints dialog (**Ctrl+B**) and temporarily disable this breakpoint by deselecting its tick box. Then, using the **Data** tab of the dialog, set a new breakpoint at the leaking address by typing `0x`, followed by the address previously noted, into the expression textbox. Using the previous example, this would be: `0x10052684`. Also set the number of elements to watch to `4` (the size of a pointer in bytes).

- Continue execution (press **F5**)—you should get a dialog containing text such as *Break when '0x10052684' (length:4) changes* each time the memory at the watch point changes, and the debugger will stop. The call stack window can be used to trace the line of user code that caused the allocation (where applicable). Note that this memory location may be allocated and deallocated several times. *It is the last allocation that causes the leak.* Also, note that the memory may be changed outside an allocation—an existing object may simply be updated. However, the memory leak will generally link to a call to `NewL()`, `NewLC()`, `new (ELeave)` or similar in your code.

Once the erroneous line is located, the source code must be examined to determine the cause of the error—typically you will not have deleted some allocated memory.

Further contextual information may possibly be gleaned from a memory address by casting it to a `CBase*` in a watch window (in case it is a `CBase`-derived object), or casting it according to the rules for type 0 Unicode or narrow descriptors in Table 13–6 (`HBufC*` and `HBufC8*`). Also, examining the contents of that address in a memory window may provide further clues.

As can be seen, this is not a trivial exercise, so it is important to follow coding guidelines and test regularly (small changes narrow the field of possibilities)! All memory allocated must be deleted once, and once only, and any objects deleted outside of a destructor should immediately have their pointers set to `NULL`, to avoid dangling pointers.

Debugging an Application on Target

The subsection *Differences of Testing on Target vs. Emulator* earlier in this chapter explained that the behavior of an application when run on the emulator may differ from its behavior when run on target hardware. If you encounter errors that occur only when your application is run on the device, then debugging on target may be necessary.

The GNU debugger (**gdb**) can be used for on-target debugging of Series 60 devices, and in this subsection we will show you how to set up and run **gdb** from a Windows command prompt.

The Borland C++ Builder IDE also provides support for on-target debugging using **gdb**, and the setup for this is the same as described here. For further information on using on-target debugging from within Borland C++ Builder you should consult its documentation.

> **NOTE** Metrowerks CodeWarrior also supports on-target debugging in its Professional and OEM editions, but this does not use **gdb**. See its IDE documentation for further details on any support provided.

To be able to carry out on-device debugging you need a copy of the GNU debugger, **gdb**, on your PC, and a copy of the GNU debugger stub, **gdbstub**, on your device. Required versions are provided with the Series 60 SDK. You also require a serial link for communication between PC and device.

Setting up the Serial Link

Obviously it is important that the serial link between the PC and the Series 60 device is configured correctly. In this discussion, an infrared connection is used—it is also possible to configure a connection over Bluetooth or (where available) a serial cable.

To ensure that communication between **gdb** and **gdbstub** is possible, you need to make certain that no other application is using the required COM port on your PC—make sure that you close any connectivity applications or port monitors. Note also that, due to some limitations in **gdb**, it is necessary that the COM port used is configured to COM1, COM2, COM3 or COM4.

As most infrared devices generally add their own communication stack on top of the COM stack, you might need to install a virtual COM driver to correct this. One suitable application is **IrCOMM2k**, which can be found at **http://www.ircomm2k.de/** or **http://sourceforge.net/projects/ircomm2k**.

Installing and Configuring Gdbstub

Some Series 60 devices may have **gdbstub** preinstalled. If not, then you need to install the file \epoc32\release\armi\urel\gdbstub.sis from the root of your SDK to your Series 60 device. This will create the executable gdbstub.exe in the \System\Programs directory of your device, and the libraries gdbseng.dll and gdbseal.dll in the \System\Libs directory.

Chapter 13 Testing and Debugging

Depending on the version of gdbstub.sis used, a configuration file (gdbstub.ini) may also be created in the directory c:\gdbstub on your device. If not, then you will have to create this file yourself. The configuration file settings for infrared communication are shown below. If you wish to use Bluetooth, or some other serial link, then you will need to amend the settings as shown in the *GDB stub configuration file syntax* section of the SDK documentation.

```
[COMMSERV]
PDD=EUART%d
LDD=ECOMM
CSY=IRCOMM
PORT=0
RATE=115200
```

Once the necessary configuration file has been created, **gdbstub** is ready to be executed. Note that you will need a file manager application (such as **FExplorer** from **http://www.gosymbian.com**) installed on the device to enable you to locate and execute **gdbstub**.

Creating a GDB Initialization File

To initialize the **gdb** environment on your PC, you should create an initialization file, gdb.ini, in your application's group directory. An example of the format of this file is shown below along with an explanation of the purpose of each line:

```
symbol-file
//c/symbian/6.1/series60/epoc32/release/armi/udeb/sample.sym
epoc-exec-file c:\system\apps\sample\sample.app
target epoc com1
break NewApplication
source //c/symbian/6.1/series60/epoc32/gcc/share/epoc-des.ini
```

- symbol-file—specifies the symbol file used to provide the debugging information for your example. This will be in the \epoc32\release\armi\udeb\ directory relative to the root of your SDK, so the path defined here will depend on where your SDK is installed. Note also that the file path is specified using forward slashes ("/"), not backslashes ("\"). In this case the example application is called sample.app, so the symbol file is correspondingly called sample.sym.

- epoc-exec-file—specifies where the actual application will be located on the device. Note that this example determines that the application is installed to the device's c: drive.

- target—specifies which COM port on your PC to use (in this case, which port the IR transmitter is attached to).

- `break`—sets a breakpoint on the function `NewApplication()`.
- `source`—loads the **gdb** script to support Unicode descriptor debugging. Again, the actual value of this depends on where your SDK is installed, and it may actually be in a separate `Shared` directory outside of the root of your SDK. As before, the path is specified using forward slashes.

Building the Application

Note that you need to build a debug target (`armi udeb`) version of your application, rather than the usual `armi urel` release build. Most Series 60 SDKs do not actually come with debug libraries, so it is possible that the build process will fail due to linker errors! If this is the case, you should copy everything from your SDK's target release directory (`\epoc32\release\armi\urel`) to your SDK's target debug directory (`\epoc32\release\armi\udeb`) and then rebuild.

Installing the Application on a Device

Once your application has built successfully, you will need to create a new debugging `.sis` file. Do not forget to specify the `udeb` directories in your `.pkg` file, so as to package the correct (debugging) version of the binaries. This `.sis` file should be installed onto your Series 60 device in the usual way—further details can be found in Chapter 2.

Debugging on a Device

The first step is to start up **gdbstub** on the Series 60 device. Remember that it is located at `\system\programs\gdbstub.exe`—you will need a file manager application to navigate to it and execute it, as previously noted. When it is running, you should see the infrared icon flashing—this means that the **gdbstub** has been started.

To make the connection, you then need to start **gdb** on your PC, making sure that the Series 60 device and the PC have a line-of-sight connection. Open up a Windows command prompt and navigate to your application's `group` directory (where you placed the `gdb.ini` file) and type:

```
gdb -nw
```

Gdb should start up and display some information, including the breakpoints that have been set:

```
GNU gdb 4.17-psion-98r2
Copyright 1998 Free Software Foundation, Inc.
GDB is free software, covered by the GNU General Public
License, and you are
welcome to change it and/or distribute copies of it under
certain conditions.
```

```
Type "show copying" to see the conditions.
There is absolutely no warranty for GDB.  Type "show
warranty" for details.
This version of GDB has been modified by Symbian Ltd. to add
EPOC support.
Type "show epoc-version" to see the EPOC-specific version
number.
This GDB was configured as "—host=i686-pc-cygwin32 —
target=arm-epoc-pe".
Breakpoint 1 at 0x10001012: file
..\\..\\..\\..\\..\\SYMBIAN\\6.1\\SERIES60_
1.2\\SERIES60EX\\HELLOWORLD\\SRC\\Helloworld.cpp, line 15.
(gdb)
```

If you do not see the Symbian modification notice, then you may be using the wrong version of **gdb**—check that your PC PATH environment is set up to correctly find the version of **gdb** in your Series 60 SDK.

If all is well, then the infrared connection will have been established and the infrared icon on the phone will be permanently on. To start debugging you should type run at the command prompt. Further information about the available debugging commands can be found under *How to use GDB* in the SDK documentation, and also at **http://www.refcards.com/about/gdb.html**.

Closing a Session

When you have finished debugging, you need to close the session. Type q or quit at the Windows command prompt to stop the debugging session and exit **gdb**. Note, however, that it is not possible to bring **gdbstub** to the foreground on a device, as it is an .exe, so to close the session on the device, you will need to manually reset the phone (switch it off and back on again).

Summary

This chapter has rounded off the book by covering many of the tools and methodologies that are necessary to produce high-quality code.

It has introduced you to the broad discipline of Quality Assurance, whose guiding principle is that *to produce good software, you must focus primarily upon the development process*. From a development perspective, this means that you should strive to consistently apply Series 60 coding standards in all of the Series 60 projects you are involved in. Although learning these standards takes some effort, you will be rewarded in the long run, as your project's debugging and maintenance phases will be significantly shortened.

Chapter 13 Summary

This chapter has also highlighted the vital role of testing in ensuring software quality, giving you an insight into the various methodologies and techniques that can be used, as well as examining the specific static and dynamic testing tools available for Series 60.

Finally, an overview of debugging was given, covering both emulator and on-target debugging. The general features of all debuggers were introduced, and specific hints and tips were provided to enable you to quickly locate and correct errors in your Series 60 code.

Appendix

Emulator Shortcut Keys

The Series 60 emulator is the primary development and debugging tool for Series 60 Symbian OS application development. It provides PC keyboard combinations for accessing special testing and logging functionality and also for mimicking the hardware keys of an actual device. Most of these shortcuts are specifically for the debug emulator, but some of the hardware emulation ones will work with the release emulator also. Note, too, that some of the shortcuts defined are inherited from Symbian OS and are not really relevant for Series 60.

The available keyboard shortcuts are as listed in Tables A–1 through A–5.

Appendix Emulator Shortcut Keys

Table A–1 Drawing

Shortcut Key	Description
Ctrl + Alt + Shift + **R**	Redraws the whole screen, to test an application's redraw functionality.
Ctrl + Alt + Shift + **F**	Enables Window Server auto-flush for all applications using the current control environment. This can be used to slow down drawing for finding flicker problems.
Ctrl + Alt + Shift + **G**	Disables Window Server auto-flush for all applications using the current control environment. This is the default setting.
Ctrl + Alt + Shift + **M**	Displays a "Move me!" dialog to test partial redraw, which can be moved using the navigation keys. Note that multiple instances can be nested, but there is no benefit in this.
Ctrl + Alt + Shift + **Enter**	Displays a "Move me!" dialog, as above.

Table A–2 Window Server*

Shortcut Key	Description
Ctrl + Alt + Shift + **E**	Enables logging of all Window Server messages (if logging is set up).
Ctrl + Alt + Shift + **D**	Disables Window Server logging, if active.
Ctrl + Alt + Shift + **W**	Dumps the full window tree from the Window Server to the log (if logging is set up). If disabled, logging will be temporarily enabled to achieve this.
Ctrl + Alt + Shift + **K**	Kills the foreground application.
Ctrl + Alt + Shift + **X**	Shuts down the Window Server.
Ctrl + Alt + Shift + **H**	Dumps the list of cells allocated on the Window Server's heap to the log (if logging is set up).
Ctrl + Alt + Shift + **U**	Cycles the display through its possible sizes. Note that a display of size 176 by 208 pixels is the only currently available option in Series 60. (This shortcut is inherited from Symbian OS.)
Ctrl + Alt + Shift + **O**	Cycles the screen orientation through its possible rotations. For Series 60, these are 180 degrees apart. Note that for legibility it flips the emulator rather than the screen.

Appendix Emulator Shortcut Keys

Table A–2 continued

Shortcut Key	Description
Ctrl + Alt + Shift + **T**	Displays the task switcher—a list of currently running applications, which can be switched to. This is similar to a long press of the **Applications** key, which should be used in preference, as it is guaranteed to produce up-to-date results. (This shortcut is inherited from Symbian OS.)
Ctrl + Alt + Shift + **I**	Dumps the control tree to the IDE debug output window for the next window clicked on.
Ctrl + Alt + Shift + **J**	Draws colored borders around all controls.
Ctrl + Alt + Shift + **L**	Shuts all dialogs using `AknDialogShutter`.

* For further information on Window Server logging, see "How to set up window server logging" in the Symbian OS API Guide section of the SDK documentation. Note, however, that there is a known defect with Window Server logging in Series 60 1.x—see Forum Nokia (**http://www.forum.nokia.com**) for further details of known Series 60 defects.

Table A–3 Resource Allocation

Shortcut Key	Description
Ctrl + Alt + Shift + **A**	Displays the number of heap cells allocated by the current application. Note that this value cannot be clearly seen on a Series 60 emulator screen! (This shortcut is inherited from Symbian OS.)
Ctrl + Alt + Shift + **B**	Displays the number of File Server resources in use by the current application. Note that this value cannot be clearly seen on a Series 60 emulator screen! (This shortcut is inherited from Symbian OS.)
Ctrl + Alt + Shift + **C**	Displays the number of Window Server resources in use by the current application. Note that this value cannot be clearly seen on a Series 60 emulator screen! (This shortcut is inherited from Symbian OS.)
Ctrl + Alt + Shift + **P**	Displays the dialog for the resource failure tool. The application heap can be set to fail at a random or deterministic rate, Window Server operations can be set to fail at a random or deterministic rate, and File access can be set to fail at a deterministic rate. See Chapter 13 for more information on Heap Testing.
Ctrl + Alt + Shift + **Q**	Turns off heap failure mode. Note that the resource failure tool dialog will not be updated if it is open.

Appendix Emulator Shortcut Keys

Table A–4 Hardware Emulation*

Shortcut Key	Description
` (The key to the left of 1)	Equivalent to the Selection key (center of the navigation keys).
Alt + **1**	Equivalent to the left soft key.
Alt + **2**	Equivalent to the right soft key.
Cursor keys	Equivalent to the navigation keys.
0–9, ***** and **#**	Equivalent to the ITU-T numeric keypad.
Alt + **0**	Equivalent to the power-off key. Note that in most emulators the power key has no effect.
Alt + **3**	Equivalent to grip open, to emulate a device with a hideaway keypad opening.
Alt + **4**	Equivalent to grip closed, to emulate a device with a hideaway keypad closing.
Alt + **5**	Equivalent to the side key.
Alt + **S**	Equivalent to the **Send** call key.
Alt + **E**	Equivalent to the **End** call key. Note that this will switch to the phone application on target only.
Home	Equivalent to the **Applications** key.
Ctrl + **H** or **Backspace**	Equivalent to the **Clear** key.
Shift	Equivalent to the **Edit** key.

* Note that the mappings of some of these keys may be changed in the `\Epoc32\Data\Epoc.ini` file.

Appendix Emulator Shortcut Keys

Table A-5 Miscellaneous

Shortcut Key	Description
Ctrl + Alt + Shift + **V**	Toggles verbose information messages. Text in calls to `CEikonEnv::VerboseInfoMsg` will be displayed.
Ctrl + Alt + Shift + **N**	Toggles the Active Scheduler shaker.
Ctrl + Alt + Shift + **Y**	Mounts a file system as the `x:` drive for testing removable media support. Will actually fail with a system error! (This shortcut is inherited from Symbian OS.)
Ctrl + Alt + Shift + **Z**	Sends the keys *A* though *J* in fast sequence to the application, to test its ability to handle fast repeated keys.
(Numeric keypad) **+**	(The "+" key on the numeric keypad)—Toggles the Front End Processor (FEP) status display. This shows the status and current mode of the FEP.
Alt + **F**	Toggles the FEP state on or off.
Esc	Cancels any shown dialogs or menus.
F1	Shows the current options menu. (Note that this will occur even if no **Options** soft key is currently specified!)
F5	Simulates changing removable media cards—not used in Series 60. (This shortcut is inherited from Symbian OS.)
F9	Power off. (Turns off the emulator.)
F10	Simulates an emergency shutdown (in other words, what would occur if battery levels dropped below the minimum level). No user operations are allowed. **F9** will power it back up.
F11	Simulates a device's case being opened or closed. Not used in Series 60—see grip open/close events. (This shortcut is inherited from Symbian OS.)*

* *Note, however that pressing **F11** twice (to "open and close the case") can be used to wake up a frozen emulator. (When waiting at an application breakpoint for a long time, the emulator can go into sleep mode.)*

Glossary

Active Object—A class responsible for issuing requests to asynchronous service providers and handling those requests on completion. Must be derived from `CActive`.

Active Scheduler—A class responsible for scheduling events to the Active Objects in an event-handling program. Class must be derived from `CActiveScheduler`.

AIF (Application Information File)—File which contains the caption, icon, capabilities and MIME priority support information for an application. Has the file extension `.aif`.

AIF Builder—A GUI application for generating an application information file, including its bitmaps. The SDK application runs on a development machine (not the device or Emulator).

API (Application Programming Interface)—The visible public behavior a system object or component exposes to other objects or components.

Apparc (Application Architecture)—Provides part of the basic framework for Symbian OS applications along with CONE.

Application Framework—Handles application start-up and accessing the application data (its document).

Application Launcher—The default view of a Series 60 device, displaying the applications available for selection in grid or list view modes.

AppWizard—Tool that can be integrated with IDEs, enabling you to quickly produce skeleton GUI applications.

Glossary

ARM Processor—A processor running 32-bit (or 16-bit) embedded RISC from ARM.

ARM4—32-bit instruction set and binary interface for ARM-based processors. If an application is compiled for ARM4, it can only call functions compiled for ARM4 or ARMI. ARM4 code runs faster than THUMB code but takes up more ROM space.

ARMI—32-bit instruction set and binary interface with interworking for ARM-based processors. If an application is compiled for ARMI, it can call functions compiled for ARM4, THUMB or ARMI. ARMI code runs faster than THUMB code but takes up more ROM space. This is the suggested default build format for compatibility.

ASCII (American Standard Code for Information Interchange)—A standard for the code numbers used by computers to represent all the upper- and lower-case Latin letters, numbers, punctuation, and so on. There are 128 standard ASCII codes, each of which can be represented by a 7-digit binary number.

Avkon—Series 60 standard UI library and application framework—built on Symbian Uikon technology.

BC (Binary Compatibility)—A library is binary compatible, if a program linked dynamically to a former version of the library continues running with newer versions of the library without the need to recompile.

Bearer—A telephone network used to carry a call.

Black-box testing—Testing carried out without knowledge of the component under test.

Blit—Block transfer copy of pixel data from memory to a graphics device.

Bluetooth (BT)—An open standard for wireless transmission of voice and data between mobile devices.

Bmconv—Tool for converting bitmaps between Windows and Symbian OS formats.

BMP (bitmap file)—Provides the pixel patterns used by pictures, icons and masks, sprites and brush styles for filling areas of the display.

Boolean—An expression or variable that can have only a true or false value.

CA (Certificate Authority)—A trusted third-party organization or company that issues digital certificates used to create digital signatures and public-private key pairs.

Certificate Generator—Used to create a certificate request file, which is sent to a certification authority.

Cipher suite—A mechanism for data encryption used by secure sockets.

Glossary

Cleanup Stack—The stack of references to partially constructed items maintained by `CleanupStack::PushL()` and `CleanupStack::Pop()`, which will be cleaned up, should a leave occur.

Client—A program which requests services from another program.

Clipping region—The region to which graphics primitives are clipped.

Codec (coder/decoder)—Mechanism for converting data into a different format—and back again.

Coding standards—Standards defined by Symbian or others that dictate good coding practice.

COM—A serial communication port which supports the RS-232 standard of communication.

CommDB (communications database)—Provides systemwide storage for communications-related settings.

Comms module (CSY)—Serial communications module. Provides an implementation of a serial port to the Serial Communications Server.

Compact Business Cards—Business card in an old, Nokia-defined format for the Nokia 9000 smart phone.

Competence Center—Series 60—Nokia's system for awarding a limited number of companies with an audited high level of competence in Series 60 software engineering.

Competence Center—Symbian—Symbian's system for awarding a limited number of companies with an audited high level of competence in Symbian OS software engineering.

Compound control—A control that contains one or more simple or compound controls.

CONE (Control Environment)—Provides an active-object interface to the Window Server's asynchronous services, and provides framework for controls and app UI.

Console applications—Applications that do not have a GUI interface.

Container control—A compound control.

Context-sensitive menu—Provides users with a particular menu dependent upon the state of the application when the menu is requested.

Control—A rectangular area of the screen that may respond to user input events.

Crop—To eliminate unwanted portions of a picture when it is output.

CSV (comma-separated value)—A file format whereby a comma separates each distinct data entry.

CSY (serial communications module)—Serial communications module. Provides an implementation of a serial port to the Serial Communications Server.

Datagram—A packet of information used in a connectionless network service that is routed to its destination using an address included in the datagram's header.

Daytime—An Internet protocol for determining the time at a remote computer's location—given in human-readable form.

Descriptor—A class derived from `TDesC`, which describes an area of memory used as either a string or some binary data.

Dialog—A control, normally invoked by the selection of a menu command, allowing interaction between the user and the program.

Dial-Up Networking (DUN)—A way to connect to a network by dialing in over phone lines to a modem.

Direct Screen Access (DSA)—A way of drawing to the screen without using the window server—avoids Client/Server communication and is therefore much faster.

DLL (Dynamic Link Library)—Dynamic Link Libraries are loaded in response to an explicit API call made by an executing program.

DNL (Dynamic Navigational Link)—Applications that use the view architecture can allow other applications to send messages specifying a view to display, and possibly relevant accompanying data, called a Dynamic Navigational Link (DNL) message.

DNS (Domain Name Server)—A database of Internet names and addresses that translates the names to the official Internet Protocol numbers and vice versa.

Document embedding—Storing one document inside another so there is an association between them—for example, storing a picture inside a text document.

Double buffering—A technique to smooth animation by building an image in an off-screen buffer prior to display.

DTMF (Dual Tone Multi-Frequency)—A method used by the telephone system to communicate the keys pressed when dialing. Pressing a key on a phone's keypad generates two simultaneous tones, one for the row and one for the column. These are decoded by the exchange to determine which key was pressed.

ECom—A framework that facilitates the use of plug-in modules.

Editor—A UI control for data entry.

Glossary

EIKON—EIKON consists of a programming framework together with a set of concrete controls and standard dialogs. EIKON was replaced in Symbian OS v6.0 by UIKON and platform-specific UI libraries.

Emulator—An implementation of the Symbian platform hosted on PCs running Microsoft Windows. The Emulator is the primary development environment for the Symbian platform.

Engine—The UI independent portion of an application, concerned with data manipulation and other fundamental operations independently of how these are eventually represented to the user.

EPOC—The original name for Symbian OS. It defines an Operating System designed specifically for mobile, ROM-based computing devices.

ESOCK—The Symbian OS sockets framework. It provides an abstract sockets interface, for which extensions to support particular sockets protocols, such as TCP/IP, can be written.

ETel—The Symbian OS telephony framework. It provides an abstract telephony interface, for which extensions to support particular telephony protocols or devices can be written.

Event—A loose term used to describe the cause of the completion of a request to an event source.

Event source—An asynchronous service provider that causes requests to complete when some event—usually not directly solicited—occurs.

Exception—A program condition that causes it to leave.

Externalization—The process of writing an object's data to a stream.

Factory function—A function that returns a pointer of its own type—an object construction mechanism.

FFS (Flash Filing System)—A persistent file system that uses flash RAM for storage. Typically the `c:` drive.

FEP (Front End Processor)—A framework enabling the input of characters— for example, T9, handwriting recognition or voice input.

FIFO (first-in, first-out)—A queued type of buffer.

File—A collection of data in persistent storage, accessible via the File Server.

File Server—The server thread that mediates all file system operations. All application programs that use files are clients of this thread.

Flush—To empty a buffer, sending its contents to the next stage in processing.

Fold—The removal of differences between characters that are deemed unimportant for the purposes of inexact or case-insensitive matching. As well as ignoring differences of case, folding ignores any accent on a character.

Form — UI component to present related data entry fields.

Framework — A component that allows its functionality to be extended by writing plug-in modules — framework extensions. The extension developer writes classes that derive from interfaces defined by the framework. The framework loads the required extensions during runtime.

FTP (File Transfer Protocol) — Protocol used on the Internet for exchanging files.

GCC — GNU C++ compiler.

GDI (Graphics Device Interface) — The Symbian OS component related to graphics manipulation, graphics contexts, and bitmaps.

GET — HTTP request method used to ask for a specific document.

GIF (Graphics Interchange Format) — A common format for image files.

GNU — An organization devoted to the creation and support of open-source software.

GPRS (General Packet Radio Service) — A GSM data transmission technique that does not set up a continuous channel from a portable terminal for the transmission and reception of data, but transmits and receives data in packets.

Granularity — The number of elements by which the array capacity of an array is increased.

Grid — A UI component that provides functionality for row and column layout.

GSM (Global System for Mobile communication) — A digital mobile telephone system that uses a variation of time-division multiple access. It is the most widely used of the three digital wireless telephone technologies (TDMA, GSM and CDMA).

GT (Generic Technology) — Symbian OS generic platform used as the basis of a UI platform such as Series 60.

GUI (Graphical User Interface) — Systems of windows, dialogs and other controls that the user interacts with.

HAL (Hardware Abstraction Layer) — Used to provide a generic interface to the hardware and "hide" hardware-specific functions.

Handle — A way of identifying an object that is owned or managed by another thread or process.

Heap — An area of memory used for dynamic memory allocation.

HTML (HyperText Markup Language) — Coding language used to create hypertext documents for use on the Internet.

Glossary

HTTP (HyperText Transfer Protocol)—The underlying protocol used by the Internet. HTTP defines how messages are formatted and transmitted, and what action Web servers and browsers should take in response to various commands.

HyperTerminal—A Windows program that you can use to connect a host computer to your hardware device, using serial communications devices on the host computer and your hardware device connected via a null modem cable.

IAS (IrDA Information Access Service)—Layer of the IrDA protocol stack that is responsible for service discovery.

ICO—Image format for an icon.

IDE (integrated development environment)—A system for supporting the process of writing software. Such a system may include a syntax checker, graphical tools for program entry, and integrated support for compiling, running and debugging the program—relating compilation errors back to the source.

IMAP4 (Internet Message Access Protocol, v.4)—An open standard for remotely accessing Internet email stored on a server.

IMEI (International Mobile Station Equipment Identification)—A phone serial number. To displayed a phone's IMEI number, enter the sequence *#060#.

Installation File Generator—The `makesis.exe` tool that is provided with the SDK to create an installation file.

Interface (ECom)—A class that defines the services offered by the plug-in DLLs.

Internalize—The process of reading data from a stream and assigning that data to an object, possibly constructing a new object from the data.

Internationalization—Making an application available to a international market by supporting different languages, and so on.

Internet access point (IAP)—A series of settings that can be used to access the Internet, such as username, password, ISP details and so on.

Internet Assigned Numbers Authority (IANA)—Central registry for various Internet protocol parameters, such as port, protocol and enterprise numbers, and options, codes and types.

Internet Control Message Protocol (ICMP)—The part of the IP protocol that handles error and control messages.

Inter-Process Communication (IPC)—Communication across thread and process boundaries. Used by servers in Symbian OS.

IP (Internet Protocol)—The network layer protocol for the TCP/IP protocol suite.

IP Address—A globally unique number assigned to each device on the Internet, used to make a connection to that specific machine.

IR (infrared)—Electromagnetic waves in the frequency range just below visible light—corresponding to radiated heat.

IrCOMM—IrDA protocol that provides COM (serial) port emulation for legacy COM applications, printing and modem devices.

IrDA (Infra Red Data Association)—The body that sets the standards used in infrared communications.

IrLAN—IrDA protocol used to support IR wireless access to local area networks.

IrLAP—IrDA protocol that provides a point-to-point link.

IrLMP—IrDA protocol that provides support for multiple sessions over the single point-to-point link.

IrMUX—IrDA protocol that provides unreliable datagram service.

IrOBEX (Infrared Object Exchange)—Protocol used to exchange vCards and vCals.

IrTranP—IrDA protocol that can be used for transferring pictures between digital cameras and Symbian OS phones.

ISP (Internet Service Provider)—A company that provides access to the Internet, usually for a fee.

Java—A high-level programming language developed by Sun Microsystems.

Java MIDP (Mobile Information Device Profile)—A set of Java APIs that are specialized for use in mobile phones.

JPEG (Joint Photographic Experts Group)—A graphic image file or an image compression algorithm.

Just-in-time (JIT)—Just-in-time debugging—the debugger attaches to a process that is at the point of dying.

Kernel—The core of the Operating System. It manages memory, loads processes and libraries, and schedules threads for execution. Symbian OS's Kernel is called `ekern.exe`.

L2CAP (Bluetooth Logical Link Control and Adaptation Protocol)—Controls how multiple users of the link are multiplexed together, handles packet segmentation and reassembly, and conveys quality-of-service information.

Leave—Symbian OS exception handling. To leave is to invoke the function `User::Leave()`. This causes a return to the current trap harness and is equivalent to `throw` in C++.

Library—A collection of precompiled routines that a program can use.

Linked list—Set of stored data items in which each element or node contains a pointer to the previous or next list element.

List—UI component to display an array of textual or graphical elements.

Locale—The set of information that corresponds to a given language and country.

Log Engine—Used to record events of interest to the user, of which they may not be immediately aware, or of high importance as costs are incurred, such as telephone calls. These events can be retrieved by a viewer application and displayed to the user.

Machine UID—A unique device identification code that should be absolutely unique to a manufacturers product.

Macro—A set of instructions stored in an executable form.

Magic number—A literal number appearing without explanation or obvious meaning in your code.

Makefile—A file containing all the information necessary to specify how to build projects in various environments; generated by the `makmake` tool from an `.mmp` file.

Mask—Bitmap defining transparent regions of another bitmap.

MBM (multibitmap file)—A collection of compressed bitmaps. Used in Symbian OS as an efficient way of storing and retrieving a large number of bitmaps in a single file.

Memory leak—A condition that occurs when applications allocate memory for use but do not free allocated memory when finished.

Menu—A UI control that contains a list of actions the user can perform.

Messaging—A framework for the core messaging functionality, and a framework for providing support for new messaging protocols. The API is used by both client applications and protocol providers.

MIME (Multipurpose Internet Mail Extensions)—A protocol whereby an Internet mail message can be composed of several independent items, including binary and application-specific data.

Mixin—A protocol interface definition designed for "mixing in" in with primary base classes. The basis of the only use of multiple inheritance allowed in Symbian OS. Mixins should contain only pure virtual member functions. This means they describe only the expected behavior of an object.

MMP—Also known as a project file. A hand-edited file whose main purpose is to specify the source files which go to make up a releasable. Used as an input to the `makmake` tool.

MMS (multimedia messaging service)—A mechanism for sending content-rich multimedia messages between mobile phones.

Modal—Describes a type of dialog. A modal dialog must be closed before the user can interact with any other UI control.

Modeless—Describes a dialog that is not modal.

MSDN (Microsoft Developer Network)—A set of services designed to help developers write applications using Microsoft products and technologies (http://msdn.microsoft.com).

MTM (message type module)—Plug-in to the messaging architecture—a group of components that together provide message handling for a particular protocol.

Multi-Field Numeric Editor (MFNE)—Numeric editors that have one or more fields that are separated by data-specific characters. Can be used for the entry of dates, time and so on.

Multitasking—The ability to execute more than one task at once. In computing terms, this means switching from one program to another so quickly that it gives the appearance of executing more than one program at the same time.

Multithreading—The ability some Operating Systems have to execute different parts, or threads, of a program simultaneously.

Namespace—A scope in which declarations and definitions are grouped together.

NIFMAN (Network Interface Manager)—Symbian OS component responsible for network connections.

Note—A UI control used to convey information to the user.

OBEX (Object Exchange)—A set of high-level protocols allowing objects such as vCard contact information and vCalendar schedule entries to be exchanged using either IrDA (IrOBEX) or Bluetooth.

OOD (out of disk)—Condition caused when the current file space is exhausted.

OOM (out of memory)—Condition caused when RAM is exhausted.

Options menu—Menu activated by pressing the Options soft key. (Options is the default value of the left soft key.)

OS (Operating System)—The system software used by computers to schedule tasks and control the use of system resources.

Packet Switched Data (PSD)—A data communications network based on the principles of packet switching.

Panic—A runtime exception caused by a programming error that terminates the current thread of execution.

PCM (Pulse Code Modulation)—The process of taking samples of an analog sound and storing the results as binary data.

PDP (Packet Data Protocol)—A PDP context refers to a set of information such as, security, billing, quality of service and routing, which describes a mobile wireless service call or session.

Personal digital assistant (PDA)—A hand-held computing device.

PIM (personal information manager)—An application that usually includes an address book and organizes unrelated information, such as notes, appointments and names, in a useful way.

Platform UID—A unique platform identification code. Each release of Series 60 will have a different platform ID.

Plug-in—A polymorphic interface DLL used to enhance, or extend the operation of, a parent application.

PNG (Portable Network Graphics)—A graphics format designed as the successor to GIF.

POP3 (Post Office Protocol, v.3)—An open standard for retrieving Internet email from a server.

Port number—The identifier for a logical communications channel between an application and a transport service. Each program or service listens on a particular port for incoming packets, with certain ports permanently assigned to particular protocols by the IANA—for example, port 80 is generally used for HTTP traffic. Port numbers are always used in conjunction with IP addresses when establishing connections to host devices.

POST—HTTP request method for transmitting form data.

PRT—Protocol module.

PUT—An HTTP method for pushing data to a server.

QA (quality assurance)—A system for assuring that commercial products meet certain minimum standards.

Query—UI component used to pose a question to the user.

RAM drive—A temporary file system that is not persisted. typically the `d:` drive.

RAS (Remote Access Service)—A Windows feature that allows remote users to log into a LAN. Can be configured to allow the Series 60 emulator to connect to the network.

Resource files—Files containing data separate from executable code. Their main uses are for defining user interfaces components and for storing localizable data.

RFC 2616—Specification for HTTP 1.1.

Glossary

RFCOMM—An interface that allows an application to treat a Bluetooth link in a similar way as if it were communicating over a serial port.

RGB—Red, green and blue—the primary colors that are mixed to display the color of pixels on a screen. Every color of emitted light can be created by combining these three colors in varying levels.

RISC (Reduced Instruction Set Computer)—A computer processing technology in which a microprocessor understands a few simple instructions, thereby providing fast, predictable instruction flow.

ROM (read-only memory)—Memory used to hold programs and data that must survive when the computer is turned off. It is a permanent memory that can be read but not (easily!) changed. Typically the `z:` drive.

RS-232—The standard for serial data transmission using cables, normally carrying between ±5V and ±12V on both data and control signal lines.

SA (System Agent)—A server that dynamically manages the state of variables whose values reflect the current state of aspects of a number of system components.

SDK (Software Development Kit)—A package that allows software developers to create products to run on a particular platform.

Selection key—The "key" used to confirm a selection. On Series 60 devices this usually means the center position of the navigation controller.

Semaphore—A Kernel object used to synchronize cooperating threads in Symbian OS. Access to the semaphore is through an `RSemaphore` handle.

Serial communication—A system of sending bits of data on a single channel one after the other, rather than simultaneously.

Server—A program that performs services for another program.

Service Discovery Database (SDD)—Database that holds information about a phone's available Bluetooth services.

Service Discovery Protocol (SDP)—Used for locating and describing services provided by, or available through, a Bluetooth device.

Session—The channel of communication between a client and server.

Settings List—A Series 60 UI control for that should be used to present configuration settings to the user.

SIM (Subscriber Identity Module)—The smart card used in mobile phones. It carries the user's identity for accessing the network and receiving calls, and it can also store personal information, such as SMS messages and a phone directory.

SIS file—Symbian installation file, produced by the installation file generator (`makesis.exe`) or the SIS file creator (**Sisar**). Has a suffix of `.sis`.

Glossary

Sisar—SIS file creator.

Skin—An element of the GUI that can be changed to give the interface a different look. Skins are also known as "themes." Supported from Series 60 2.0.

SMIL (Synchronized Multimedia Integration Language)—XML-based language used in the presentation of MMS messages. Pronounced "smile."

SMS (Short Message Service)—Mechanism for sending text messages between mobile phones.

SMTP (Standard Mail Transport Protocol)—Open standard for sending Internet email.

Socket—An abstraction of a communication endpoint between two applications, particularly over TCP or UDP. The Symbian OS ESOCK component provides a generic sockets interface.

Soft key—A button on the device that changes function depending on what you are doing with the phone. Its current function is highlighted using a keyword immediately above the button on the phone's display. Series 60 devices have a left and right soft key—typically the left soft key is for menus and positive responses, and the right soft key is for exiting views and negative responses.

Sprite—An arbitrarily shaped bitmap that may be moved without applications having to redraw the underlying screen. Typically used for pointer cursors and for animated figures in games.

SSL (Secure Socket Layer)—A protocol used for secure Internet communications that transmits data in an encrypted form.

Stack—Area of memory used to implement a data structure that follows the last-in, first-out method of access. The system stack is used to store local program variables, method parameters and return values.

STL (Standard Template Library)—A C++ library of container classes, algorithms and iterators. Note that this is not implemented for Series 60.

Store—A collection of streams.

Stream—The external representation of one or more objects.

Struct—Data structure for an aggregate of elements of arbitrary types.

Symbian—Symbian is a software licensing company, owned by wireless industry leaders, that supplies an Operating System for data-enabled mobile phones: Symbian OS.

Symbian OS—The Operating System on which Series 60 Platform is built. Series 60 1.x is based on Symbian OS 6.1, Series 60 2.x is based on Symbian OS 7.0s.

SyncML (Synchronization Markup Language)—XML-based markup language and open standard for data synchronization, enabling data synchronization between servers and various types of mobile application devices. Optimized for wireless networks.

System watchdogs—Subsystems within Series 60 that monitor performance and may request that your application be closed—for example, the OOM (out-of-memory) watchdog.

TCP (Transmission Control Protocol)—The connection-oriented protocol built on top of IP in TCP/IP. It adds reliable communication and flow-control.

TCP/IP (Transmission Control Protocol/Internet Protocol)—The suite of communications protocols used to make connections on the Internet.

Telnet—Protocol that provides terminal emulation using the TCP/IP protocols. Telnet allows users to log onto and access remote computers.

Themes—See Skin.

Thread—A single unit of execution within a process—threads run concurrently.

THUMB—16-bit instruction set and binary interface for ARM-based processors. If an application is compiled for THUMB, it can only call other functions compiled for THUMB or ARMI. THUMB code runs slower than ARM4 code but takes up less ROM space.

TIFF (Tagged Image File Format)—An image file format.

TinyTP (Tiny Transport Protocol)—An IrDA transport protocol that adds per-channel flow control and SAR (segmentation and reassembly) to the IrDA stack.

TLS (Thread Local Storage)—A machine word of memory that may be used to anchor information in the context of both a DLL and a thread. Used instead of (non-const) static data, which is not supported for Symbian OS DLLs.

TLS (Transport Layer Security)—A protocol used for secure Internet communications that transmits data in an encrypted form. An enhancement of SSL 3.0, defined in RFC-2246.

TRAP—Macro to execute a set of C++ statements under a trap harness. Equivalent to a `catch` statement in C++ exception handling.

Trap harness—Construction associated with the `TRAP` and `TRAPD` macros. Code executed inside a trap harness may leave, returning control to the cleanup part of the harness, and automatically resulting in items being cleaned up from the Cleanup Stack.

TSY—An ETel extension module that handles the interaction between the ETel Server and a particular telephony device or family of devices.

Twip—1/1440th of an inch, or 1/20th of a point. All measurements supported by the GDI are either in pixels for devices or in twips for real-world sizing.

Glossary

UART (Universal Asynchronous Receiver-Transmitter)—Common name for the hardware that drives an RS-232 serial port.

UI (user interface)—The layer of controls, editors and dialogs which allow a user to control a running application.

UID (unique identifier)—A globally unique 32-bit number used in a compound identifier to uniquely identify an object, file type and so on.

UID type—Set of three UIDs which, in combination, identify a Symbian OS object—encapsulated by a `TUidType` object.

UID1—The first UID in a compound identifier (UID type). It identifies the general type of a Symbian OS object and can be thought of as a system level identifier. Executables, DLLs and file stores are all distinguished by UID1.

UID2—The second UID in a compound identifier (UID type). It distinguishes within a general type (defined by UID1) and can be thought of as an interface identifier. Static interface (shared library) DLLs and polymorphic interface (application or plug-in framework) DLLs are distinguished by UID2.

UID3—The third UID in a compound identifier (UID type). It identifies a particular subtype and can be thought of as a project identifier. UID3 may be shared by all objects belonging to a given program, including library DLLs if any, framework DLLs and all documents.

Uikon—Symbian UI layer that provides generic controls. Platform-specific UIs are built on Uikon.

UML (Universal Modeling Language)—A standard notation and modeling technique for object-oriented design.

Unicode—A 16-bit character set that assigns unique character codes to characters in a wide range of languages. Series 60 is a Unicode platform.

URI (Universal Resource Indicator)—The generic set of names and addresses that refer to objects on the Internet. URLs and partial URLs are examples of URIs.

URL (Uniform Resource Locator)—An address of a resource on the Internet.

User Datagram Protocol (UDP)—A connectionless, unreliable transport layer protocol in the TCP/IP protocol suite.

UUID (Universally Unique Identifier)—A unique 128-bit number. UUIDs are used for Bluetooth device addresses.

vCalendar—The open Internet standard for creating, storing and sharing electronic calendar or schedule information. Developed by the Versit consortium and now controlled by the Internet Mail Consortium.

vCards—The open Internet standard for creating, storing and sharing electronic business cards. Developed by the Versit consortium and now controlled by the Internet Mail Consortium.

View—A well-defined representation of user data. Many applications are designed around a set of fundamental data views that form the application's GUI.

WAE (Wireless Application Environment)—Part of the WAP standard, WAE specifies an environment that allows operators and service providers to build applications and services that can reach a wide variety of different platforms.

WAP (Wireless Application Protocol)—Open standard for downloading and presenting browser content for mobile phones.

WBMP (Wireless Bitmap)—Monochrome image format used in WAP.

WCDMA (Wideband Code Division Multiple Access)—Technology for wideband wireless access. Allows very high-speed multimedia services like Internet access and videoconferencing.

WDP—Provides the WAP general datagram transport services above the various data-capable bearer services.

White-box Testing—Software testing that examines the program structure and derives test data from the program logic.

WIN32—Windows 32-bit system libraries to link to in a WINS build.

Window Server—A server which manages the screen, keyboard and (where applicable) pointer on behalf of client applications.

WINS (Windows Single process)—The Emulator build, which runs in a single process environment. (Sometimes refers specifically to Microsoft Visual C++ IDE builds.)

WINSB—WINS platform for Borland C++ Builder emulator builds.

WINSCW—WINS platform for Metrowerks CodeWarrior emulator builds.

WMF (Windows Meta File)—A common Windows graphic file format.

WSP (Wireless Session Protocol)—Provides lightweight WAP Client/Server transactions, with improved reliability over basic datagram services.

WTLS (Wireless Transport Layer Security)—The layer that provides privacy, data integrity and authentication for WAP.

WTP (Wireless Transport Protocol)—Provides lightweight WAP Client/Server transactions, with improved reliability over basic datagram services.

XML (eXtensible Markup Language)—A language specialized for Web documents, enabling the creation of customized tags.

References

Example Applications

The example applications for this book can be downloaded from the following Web sites:

- http://www.emccsoft.com/devzone/
- http://www.forum.nokia.com/books/
- http://www.awprofessional.com/nokia/

The examples are provided as three `.zip` files:

- `EMCCSoft_S60_1x.zip`—the examples for a Series 60 1.x SDK.
- `EMCCSoft_S60_1x_cw.zip`—the examples for a Series 60 1.x CodeWarrior SDK.
- `EMCCSoft_S60_2x.zip`—the examples for a Series 60 2.x SDK.

Each zip should be extracted to the root of the relevant SDK—this will create a directory called `<root-of-SDK>\EMCCSoft` which will contain the examples in a flat structure. Note that, in order for the relative paths used in the creation of the `.sis` files to work, the zip must be extracted to the *root* of the SDK.

`EMCCSoft_S60_1x.zip` is suitable for Series 60 1.x SDKs designed for use with Borland C++ Builder or Microsoft Visual C++. `EMCCSoft_S60_2x.zip` is suitable for *all* Series 60 2.x SDKs.

Symbian OS Books

Here are some other currently or previously available books on Symbian OS programming:

Symbian OS C++ for Mobile Phones
Author: Richard Harrison
Publisher: John Wiley & Sons, Ltd.
ISBN: 0470856114

Programming for the Series 60 Platform and Symbian OS
Author: DIGIA Inc.
Publisher: John Wiley & Sons, Ltd.
ISBN: 0470849487

Symbian OS Communications Programming
Author: J. Jipping
Publisher: John Wiley & Sons, Ltd.
ISBN: 0470844302

Professional Symbian Programming:
Mobile Solutions on the EPOC Platform
Authors: Martin Tasker, Jonathan Allin, Jonathan Dixon, Mark Shackman, Tim Richardson and John Forrest
Publisher: Wrox Press
ISBN: 186100303X

Programming Psion Computers
Author: Leigh Edwards
Publisher: EMCC
ISBN: 0953066304
Out of print, but now available online as a free download in PDF (Adobe Acrobat) format at **http://www.emccsoft.com/devzone/ppc/**.

Other Useful Books

The following are some other useful design, programming and testing books:

Design Patterns: Elements of Reusable Object-Oriented Software
Authors: Erich Gamma, Richard Helm, Ralph Johnson and John Vissides
Publisher: Addison-Wesley
ISBN: 0201633612

UML Distilled: A Brief Guide to the Standard Object Modeling Language
Author: Martin Fowler
Publisher: Addison-Wesley
ISBN: 0321193687

References

The C++ Programming Language, Special Edition
Author: Bjarne Stroustrup
Publisher: Addison-Wesley
ISBN: 0201700735

Effective C++: 50 Specific Ways to Improve Your Programs and Designs
Author: Scott Meyers
Publisher: Addison-Wesley
ISBN: 0201924889

More Effective C++: 35 New Ways to Improve Your Programs and Designs
Author: Scott Meyers
Publisher: Addison-Wesley
ISBN: 020163371X

Exceptional C++
Author: Herb Sutter
Publisher: Addison-Wesley
ISBN: 0201615622

Modern C++ Design: Applied Generic and Design Patterns
Author: Andrei Alexandrescu
Publisher: Addison-Wesley
ISBN: 0201700735

HTTP Pocket Reference
Author: C. Wong
Publisher: O'Reilly UK
ISBN: 1565928628

Software Testing Techniques
Author: Boris Beizer
Publisher: Van Nost. Reinhold
ISBN: 0442206720

Software Engineering: A Practitioner's Approach
Author: Roger S. Pressman
Publisher: McGraw-Hill Education
ISBN: 0073655783

SDKs

Series 60 SDKs are available from Forum Nokia (**http://www.forum.nokia.com**) in the *Tools & SDKs* section. SDKs may also be available from specific licensee and tool manufacturers' Web sites.

IDEs

Information on the latest IDEs that support Series 60 is available from the following Web sites:

- http://www.borland.com/mobile
- http://www.metrowerks.com/MW/Develop/Wireless
- http://msdn.microsoft.com/visualc

Other Web Sites

You may find these other development Web sites useful:

- http://www.forum.nokia.com
- http://www.symbian.com/developer
- http://www.symbian.com/developer/techlib/faq.html
- http://www3.symbian.com/faq.nsf
- http://www.sendo.com/dev
- http://www.siemens-mobile.com/developer
- http://www.newlc.com

And here are some other Web sites providing news, reviews and other information on Series 60 and Symbian OS:

- http://www.wirelessdevnet.com/symbian
- http://symbian.infosyncworld.com
- http://www.allaboutsymbian.com
- http://www.yoursymbian.com
- http://www.symbiandiaries.com

Index

A

`abld.bat`, 6–7, 12, 31, 47–48
  creating, 35
Abstract base classes, 80
Active Objects, 129–131
  checklist, 131–132
  common pitfalls, 142–143
  defined, 129, 705
  implementing, 131–132
  practical example, 132–142
    canceling an outstanding request, 138–139
    construction, 133–134
    destructor, 139
    RunError(), handling errors in, 137–138
    RunL(), 135–137
    starting the Active Object, 134–135
    starting the Active Scheduler, 139–142
  and responsiveness, 220
Active Scheduler, 128–129
  defined, 705
  instantiation, 139–140
  lifecycle of, 140–141
  starting, 139–142
`Add Field`, 304, 306
Advanced application deployment and build guide, 27, 54, 67–72
  ARM targets, building, 70–72
  device identification:
    during execution, 70
    at install time, 69
  device identification UIDs, 68–69
  machine UIDs, 69, 713
  platform UIDs, 67–68
  resource file versions and compression, 70
Advanced communication technologies, 489–551
  HTTP, 490, 491–501
  messaging, 490, 508–543
  telephony, 490, 543–551
  WAP (Wireless Application Protocol), 490, 502–508
Agenda Server, 639–640
Agent APIs, 618
AIF, 33
AIF Builder, 14, 705
`.aif` files, 43–46
  `aifbuilder` tool, 43, 705
  `aiftool` utility, 43
  captions, 45–46
  file localization, 44
  icons, 44
  MIME support, 44–45
`aif` folder, 38
`aifbuilder` tool, 705
`AKN_LAF_COLOR()`, 565
  color palette values for macro, 566
`AknsDrawUtils`, 239
`AknsUtils`, 239
AlarmOrganiser example application, 619
`Alloc()`, 100
`AllocL()`, 95, 100
`AllocLC()`, 100
`AllocReadResourceL()`, 568
Animation, 554–555, 582–615
  architecture, 582–585
  client-side approach to, 586–588
  decoding, 595–597
    Series 60 1.x, 595–596

Index

Animation, *continued*
    Series 60 2.x, 596–597
    direct screen access (DSA), 588–594
        architecture, 589–590
        implementation, 593–594
        key classes for, 590–593
    double buffering, 585–586
    encoding, 597–601
        Series 60 1.x, 598–599
        Series 60 2.x, 599–601
    image conversion, 594–601
    image manipulation, 594–604
    image rotation, 601–602
        Series 60 1.x, 601
        Series 60 2.x, 601–602
    image scaling, 602–604
        Series 60 1.x, 602
        Series 60 2.x, 602–603
    offscreen bitmaps, 585–586
    resource components, 583
Animation example application, 555, 585
AnsPhone example application, 54, 491, 544, 546, 550, 555, 604–606
`.app` filename extension, 36
AppArc (application architecture), 44, 176–177
`Append()`, 101
Application architecture, 174, 178–201
    application initialization, 179–182
    appropriate architecture, choosing, 197–200
    `AppUi` methods, 182–183
    Avkon view-switching architecture, 191–197
        when to use, 197–198
    core application classes, 178–179
    designing an application UI, 183–184
        view terminology, 183–184
    dialog-based architecture, 188–191
        when to use, 199–200
    file handling, 200–201
    traditional Symbian OS control-based architecture, 184–188
        when to use, 198
Application deployment, 26
Application design, 173–221
    application architecture, 174, 178–201
    application framework, 174, 175–177
    ECom, 174, 206–213
    good application behavior, 174
    internationalization, 174
    splitting the application UI and the engine, 174, 201–206
Application engines, 618, 625–649
    calendar engine access, 639–646
        adding, editing and deleting entries, 643–644
        agenda entries, reading, 641–642
        agenda models/concepts, 639
        Agenda Server, 639–640
        anniversaries, 639
        appointments, 639
        calendar alarms, 645–646
        entry model, 639
        events, 639
        indexed model, 639
        instance model, 639
        notification of changes and progress monitoring, 640–641
        searching instances and filtering, 643
        to-dos, 639
    Camera APIs, 627–632
        Camera Server, 628–629
        changes in Series 60 2.x, 630–632
        settings, 630
        taking a picture, 629
        using with an emulator, 628
    compact business cards/vCards, 638
    log engine, 625–627
        log entries, reading, 625–627
    phonebook engine, 632–638
        creating and accessing phonebook items, 632–635
        phonebook searching, 636–637
        receiving notification, 635–636
        speed dialing and common dialogs, 637–638
    photo album engine, 646–649
        photo album folders, accessing, 646–648
        photo album UI components, 648–649
Application Framework, 555, 557, 705
Application Launcher, 33, 40–41, 44–45, 178, 705
Application panics, in Series 60 emulator, 67
Application programming interface (API), 5–6, 76, 78, 97, 705
Application UI components, 223–287
    ContextMenu application, 225
    controls, 224, 225–237
    event handling, 224, 239–244
    forms, 290, 302–310
    menus, 224, 254–264
    NavigationPane application, 225
    panes, 224, 264–286
    resource files, 224, 245–254
    SimpleMenu application, 224
    skins, 224, 237–239
    TitlePane application, 225
Application-initiated redraws, 243
`AppUi()`, 568
`AppUiFactory()`, `iEikonEnv`, 263
Arcs, 576
ARM processor, 20, 706
ARM targets:
  building, 70–72
    building and freezing DLLs, 71–72
    function exports, 70–71
    writable static data in DLLs, 71
ARM4, 6, 20, 706
ARMI, 6, 20, 706
Ascent (metric), 571

Index

ASCII (American Standard Code for Information Interchange), 101, 104, 150, 706
Assertions, 661–662
  side effects of, 662
Associated granularity, `CArray` types, 117
Asynchronous function handling, Series 60 1.x vs. Series 60 2.x, 603
Asynchronous services, 77
  Active Objects, 129–131
  Active Scheduler, 128–129
Audio, 555, 604–614
  audio data, 610–612
  audio device modes, 607–608
  device modes, 607–608
  `.midi` format, 604
  recording, 605–609
  Series 60 audio libraries, 604
  streaming, 612–614
  tones, 609–610
  `.wav` format, 604
Audio example application, 604, 611
AudioPlayer example application, 555
Authorization, 497
`autoexp.dat`, 63
Automatic variables, 33
Avkon, 45, 706
Avkon Skins User's Guide, 239
Avkon view-switching architecture, 191–197
Avkon View-Switching Architecture, 619
  when to use, 197–198
`AVKON_NOTE` control, 314–315
`avkon.hrh`, 41

B

Back key, 304
Baseline (metric), 571
Basic drawing, 554, 559–569
  color and display modes, 565–567
  graphics devices and graphics contexts, 562–565
  pens and brushes, 567
  screen coordinates and geometry, 562
Basic variable types, 76, 82–83
BasicDrawing example application, 555, 560, 563, 565, 567
BC, *See* Binary compatibility (BC)
Bearer, 465, 487, 502–504, 507, 706
Binary compatibility (BC), 71–72, 706
Binary switch setting item reference, 390
`BitBltMasked()`, 580–581
BITGDI library, 559
Bitmap example application, 555, 580
`BITMAP...END` command, 578
Bitmaps, 554, 578–581
  functions, 581
  generating for application use, 578–580
  loading/drawing, 580
  masking, 580–581
  offscreen, 585–586
  START BITMAP commands, 579
Black-box testing, 667, 706
`bld.inf`, 6–8, 13, 30
`bldmake` command, 30
Blit, 586, 706
Bluetooth (BT), 22, 50, 430–431, 434, 438, 446, 471, 706
  baseband, 471
  Bluetooth client, 486
  Bluetooth Security Manager, 477–479
  Bluetooth Server, 485–486
  BluetoothCHAT application, 473–486
  device and service discovery, 479
  DUN (Dial-Up Networking) Profile, 472
  Fax Profile, 472
  File Transfer Profile, 472
  Hands Free Profile, 472
  Headset Profile, 472
  Link Manager Protocol (LMP), 471
  Logical Link Control and Adaptation Protocol (L2CAP), 471
  Object Push Profile, 472
  physical layer, 471
  profiles, 471, 472
  protocols, 471–472
  Radio Frequency Communications (RFCOMM), 472
  `RHostResolver`, 479
  `RNotifier`, device discovery using, 479–480
  security, 477–479
  Serial Port Profile, 472
  Service Advertisement, 472–477
  service discovery, 481–484
  Service Discovery Database (SDD), 472–473, 474
    closing, 477
  Service Discovery Protocol (SDP), 472
  socket communication, 484–486
  stack, 471
BluetoothCHAT application, 431, 473–486
  Service Advertisement, 474–477
  Service Discovery Database (SDD), closing, 477
  service record attributes, creating, 475–476
  service records, removing, 476–477
`bmconv`, 58–60, 472, 578, 706
BMP (bitmap), 559, 579, 706
Body part, resource files, 247, 249–253
Boolean, 82, 83, 507, 706
Borland C++ Builder 6 Mobile Edition, 6, 671
Borland C++ BuilderX Mobile, 6, 8, 671
Borland C++ IDE, 9
Boundary-value analysis, 667
`BUF`, 254
`BUF<n>`, 254
`BUFS`, 254
BullseyeCoverage, 672
`BYTE`, 254

Index

C

C++ namespaces, 81
CA (certificate authority), 706
CActiveScheduler, 87
CActiveScheduler::Start() function, 140–142
CAknAppUi, 258
CAknConfirmationNote, 311
CAknDialog, 291
CAknErrorNote, 311
CAknGlobalNote, 312
CAknInformationNote, 311
CAknNoteDialog, 312
CAknProgressDialog, 312
CAknView, 258
  MenuBar(), 263
CAknWaitDialog, 311, 317
CAknWaitNoteWrapper, 311
CAknWarningNote, 311
Calendar engine access, 639–646
  adding, editing and deleting entries, 643–644
  agenda entries, reading, 641–642
  agenda models/concepts, 639
  Agenda Server, 639–640
  anniversaries, 639
  appointments, 639
  calendar alarms, 645–646
  entry model, 639
  events, 639
  indexed model, 639
  instance model, 639
  notification of changes and progress monitoring, 640–641
  searching instances and filtering, 643
  to-dos, 639
Calendar view switching, 621–622
CallSummary example application, 619
Camera APIs, 627–632
  Camera Server, 628–629
  changes in Series 60 2.x, 630–632
  settings, 630
  taking a picture, 629
  using with an emulator, 628
Camera view switching, 622
Cancel(), 129–130, 142
CancelNoteL(), 325
CArray types, 117–126
  associated granularity, 117
  basic APIs, 118–121
  descriptor arrays, 125–126
  finding, 124–125
  limitations, 126
  sorting, 121–124
CBitmapBitBltMaskContainer::Draw(), 580–581
CBitmapDevice, 563
CBufStore, 158
C-classes, 79

CCoeControl, 567–569
CCoeEnv, 573
CCsvFileLoader() function, CCsvFileLoader, 149
CCustomCtrlDlgCustomControl, 300–302
CDirectFileStore, 158–160
CDirectScreenAccess, 589, 591–592
CEikonEnv, 573
CElementEngine instance, 93–94, 140
CElementList, 92
CElementsEngine, 80–81
CEmbeddedStore, 158
Certificate Generator, 50–52, 706
CFbsBitGc, 563
CFbsBitmap, 578
CFbsBitmapDevice, 563
CFbsDevice, 563
CFileStore, 158–159
CFont, 571
CGraphicsDevice, 563
.chm files, 66
Cipher suites, 453–454, 706
  default order of, 455
  strong cryptography, 454
  weak cryptography, 454
Class variants, 662–663
Cleanup stack, 84, 89–91, 707
  advanced use of, 96–97
    non-CBase-derived heap-allocated objects, 96
    R-classes, 96–97
CleanupClosePushL(), 96–97, 145
CleanupDeletePushL(), 96
CleanupReleasePushL(), 96–97
CleanupStack::PopAndDestroy(), 145
CleanupStack::PushL(), 87
Client, 79, 165, 166, 707
Client MTM API, 512–526
  capabilities, setting, 527–528
  connecting a session to the messaging server, 514–516
  constructing, 516–517
  CSendAppUi class, 531–539
    CreateGeneralMessageL(), 537–538
    creating a message, 535–536
    dynamic menus, 533–534
    getting started, 532–533
    HandleSendAppUiCommandL(), 538–539
    handling commands, 534–535
  MMS MTM:
    adding attachments, 524–526
    creating the message, 523–524
    sending an entry, 526
    using, 523–526
    validating an entry, 526
  MTM, choosing, 528–530
  Send-As API, 526–530
    Send-As object, creating, 527
  SMS MTM, using, 517–523
  UIDs (unique identifiers), 516

Index

using, 514–526
Client pull, 503
ClientAnimation example application, 555, 586–587
Client/server architecture, 77, 165–170
 server sessions, 165–166
  and Inter-Process Communication (IPC), 166–168
 server review, 169
 subsessions, 169–170
Client-side handle, 79
`clindex`, 63
Clipping, 568
Clipping region, 707
`Close()`, 144–145, 148
`CMsvEntry`, 511
`CMsvEntrySelection`, 512
`CMsvOperation`, 512
`CMsvSession`, 511
`CMsvStore`, 511
Codec (coder/decoder), 504, 596, 598, 604, 606, 707
Coding standards, 81, 659–661, 670–671, 683, 696, 707
Collection classes, 76, 109–126
 CArray types, 117–126
  associated granularity, 117
  basic APIs, 118–121
  descriptor arrays, 125–126
  finding, 124–125
  limitations, 126
  sorting, 121–124
 RArray and **RPointerArray** types, 110–117
  basic APIs, 110–113
  finding, 115–117
  sorting, 113–115
Color and display modes, 565–567
COM, 66, 472, 693, 707
Command line building, 9
Command-line tools, IDE vs., 8–9
Comma-separated value (CSV), 77
CommDB, 66, 458–459, 707
Comms module (CSY), 432, 707
Comms Server (C32), 431
 connecting to, 434
 starting, 434
Communications fundamentals, 427–487
 Bluetooth (BT), 431
 infrared, 431
 serial communication, 430, 431–438
 sockets, 430, 438–457
 TCP/IP (Transmission Control Protocol/Internet Protocol), 430, 457–465
Compact business cards, 638, 707
Compact business cards/vCards, 638
Competence centers—Series 60, 6, 707
`ComponentControl()`, 243
Compound controls, 225–228
 creating, 232–236

Compulsory skin-providing controls, 238
CONE (Control Environment), 144, 176, 707
Confirmation note, 311
Confirmation queries, 326, 327
ConfirmationNote application, 313–315
`Connect()`, 144–145
Connected WSP session, 504–505
`Connection`, 497
Connectionless WSP, 505–506
Console applications, 27, 672, 707
`ConstructL()`, 94–95
Container controls, 184, 707
Content-Length, 497
Context pane, 265, 272–274
 basics, 273–274
 defined, 272–274
 displaying an image in, using resources, 274
ContextMenu example application, 225, 262–264
Context-sensitive menus, 262–264, 707
 defining using resources, 262
 displaying, 262–264
Control Environment (CONE), 144, 176, 707
 functions, 568
Control stack, 241
Controls, 224, 225–237, 707
 compound, 225–228
  creating, 232–236
 container, 184, 707
 custom, in dialogs, 300–302
 establishing relationships between, 236–237
 simple, 225–228
  creating, 229–232
 using menus in, 258
 window ownership, 228–229
 and windows, 225
Controls example application, 224, 229, 233, 248
Cooperative multitasking, 127
`Copy()`, 101
`CountComponentControls()`, 243
`CPermanentFileStore`, 158–159
`CPersistentStore`, 158–159
cpp, 9
`CPrinterDevice`, 563
`CreateGraphicsContext()` function, 563
`CreateSession()`, 166
`CreateWindowL()`, 567
Crop, 349, 707
`CSecureStore`, 157, 158
CSendAppUi API, 513–514
`CSendAppUi` class, 531–539
 `CreateGeneralMessageL()`, 537–538
 creating a message, 535–536
 dynamic menus, 533–534
 getting started, 532–533
 `HandleSendAppUiCommandL()`, 538–539
 handling commands, 534–535
`CStreamStore`, 158
CSV (comma-separated value), 77, 707

Index

CSY (serial communications module), 432, 434–435, 439, 708
`CTypefaceStore`, 571
Custom controls:
  creating in a dialog class, 301–302
  in dialogs, 300–302
Custom notes, 311
CustomCtrlDlg example application, 291, 300–302
Customized notes, 313–315
  constructing/executing, 315
`CWaitNoteContainer`, 317
`CWAPBoundCLPushService`, 507–508
`CWAPBoundDatagramService`, 506–507
`CWAPFullySpecCLPushService`, 507–508
`CWAPFullySpecDatagramService`, 507
`CWindowGc`, 563
`CWsBitmap`, 578
`CWsScreenDevice`, 563

D

Dangling pointers, 244
`data` folder, 38–39
Data queries, 328–330
  constructing/executing, 329–330
  defining in resource, 328–329
  types of, 326
`DataAddress()`, 581
Datagrams, 439–440, 457, 507, 708
DataQuery example application, 327–330
Date editor, 425
Daytime, 431, 460, 463, 708
Debugging, 658, 686–696
  applications on target, 692–696
    building the application, 695
    closing a session, 696
    debugging on a device, 695–696
    GDB initialization file, creating, 694–695
    installing and configuring gdbstub, 693–694
    installing the application on a device, 695
    setting up the serial link, 693
  applications on the emulator, 687–689
  defined, 686
  emulator lockup, 690–691
  macros, 689
  memory leaks, finding, 691–692
  resource files, 689–690
  screen flicker, removing, 690
Decoding, 595–597
  Series 60 1.x, 595–596
  Series 60 1.x vs. Series 60 2.x, 603
  Series 60 2.x, 596–597
`def` files, 72
Defensive programming, 661–664
  assertions, 661–662
  class variants, 662–663
  heap testing, 664

`Delete()`, 101
`Delete Field`, 304–305, 306
Descent (metric), 571
Descriptor arrays, 125–126
Descriptors, 76–77, 97–109, 708
  as arguments and return types, 106–108
  classes, 99
  defined, 97
  hierarchy, 98–99
  literals, 101–102
  modifiable API, 100–101
  nonmodifiable API, 99–100
  package, 108–109
  using, 102–106
    `HBufC`, 102–105
    `TBuf`, 105–106
    `TBufC`, 105–106
Development process overview, 4, 6–9
Development reference, 25–73
  multi-bitmaps:
    `bmconv.exe`, 58–60
    colors and palette support, 59–60
Development tools, 27
`devices` command, 29, 36
Dialog class, custom control, creating in, 301–302
Dialog-based architecture, 188–191
  when to use, 199–200
Dialogs, 289–339, 708
  common characteristics of, 291
  constructing/executing, 297–298
  custom controls in, 300–302
  defined, 289
  defining a menu for, 300
  dialog class, writing, 294–295
  dialog data, saving and validating, 295–296
  forms, 290, 302–310
  list, 290, 335–338, *See also* Selection list dialogs
    defined, 335
    markable, 338
  modal, 291
  modeless, 291
  multipage, 298–300
  nonwaiting, 291
  notes, 290, 310–325
  queries, 290, 326–335
  resource, defining, 292–294
  selection list:
    adding icons to, 337–338
    constructing, 337
    defining in resource, 336–337
    executing, 338
  simple, creating, 292–298
  standard, 290, 291–302
    creating a simple dialog, 291–302
    initializing dynamically, 296–297
  waiting, 291
Dial-up networking (DUN), 472, 708
Digia Quality Kit, 673
Direct screen access (DSA), 588–594, 708
  architecture, 589–590

Index

`CDirectScreenAccess`, 589, 591–592
  implementation, 593–594
  key classes for, 590–593
  `MAbortDirectScreenAccess`, 589, 592–593
  `MDirectScreenAccess`, 589, 592–593
  `RDirectScreenAccess`, 589, 590–591
`DisplayMode()`, 581
DLL (Dynamic Link Library), 19, 35–36, 38, 40, 70–72, 128, 708
  polymorphic, 179, 708
`DoCancel()`, 130, 142
Document embedding, 43, 708
`DoExample()` function, 91
Domain name server (DNS), 441, 708
`DOUBLE`, 254
Double buffering, 585–586, 708
`DrawArc()`, 576
`DrawDeferred()`, 243, 567
`DrawEllipse()`, 574–576
`DrawNow()`, 243, 567–568
`DrawPie()`, 576
`DrawPolygon()`, 576–577
`DrawRect()`, 575
`DrawTextVertical()` function, 574
DSA, *See* Direct screen access (DSA)
DTMF (Dual Tone Multi-Frequency), 609, 708
Duration editor, 425
Dynamic menus, 261
Dynamically allocated memory, 33–34
`DynInitMenuPaneL()`, 261

E

`EAikCmdExit`, 259–260
`EAknCmdExit`, 259
`EAknConfirmationNoteFlags` flag, 314
`EAknEditorFlagNoEditIndicators`, 299
`EAknSoftKeyBack`, 260
`EColor64K`, 567
ECom, 174–175, 206–213, 708
  conceptual overview, 208
  DLL, 211–213
  interface, 208–211
EComExample example application, 175, 208
`EDataTypePriorityLow`, 45
`EDataTypePriorityNormal`, 45
Edit Label, 304, 306
Editors, 63, 296, 306, 393–427, 708
  multi-field numeric (MFNEs), 397, 423–427
  numeric, 397, 417–422
  secret, 397, 422–423
  styles, 416–417
  text, 396, 397–417
`EEikDialogFlagCbaButtons`, 293
`EEikDialog-FlagFillAppClientRect` flag, 298
`EEikDialogFlagNoDrag`, 293
`EEikDialogFlagWait`, 293

`EEventKey`, 241
`EEventKeyDown`, 241
`EEventKeyUp`, 241
`EFileRead`, 149
`EFileWrite`, 149
Eikon, 177, 709
Elements application, 77, 80
Elements example application, 77, 80, 92, 95, 103, 111, 118, 122, 132, 200
Ellipses, 575
EMCC Software Ltd., 8
Emulator, defined, 709, *See also* Series 60 emulators
Encoding, 597–601
  Series 60 1.x, 598–599
  Series 60 2.x, 599–601
Engine, 35, 81, 174, 197, 202, 709
Enumerated text setting item reference, 388
EnvironmentSwitch tool, 29, 62
`epoc`, 15, 17–18, 36, 46–47, 709
`EpocCheck` tool, 88, 671
`EPOCHEAPSIZE` statement, 34
`epoc.ini`, 65
`EPOCROC`, 66
EPOCSwitch tool, 28, 62
EPOCToolbar, 62
`EPOSTACKSIZE` statement, 34
`EPriorityStandard`, 130
Error note, 311
`ESimpleMenuCmdNewGame`, 257
ESOCK, 439, 459, 709
ETel, 543–544, 546, 548–549, 709
Event handling, 224, 239–244
  key events, 239–242
    control stack, 241
    focus, 240–241
    offering, 241–242
  observers, 244
  redraw events, 243
Events, 239–242, 709
  offering, 241–242
  redraw, 243
`EVerticalHatchBrush`, 567
Exception handling, 76
Exceptions, 84–88, 709
Execution, Series 60 1.x vs. Series 60 2.x, 603
`EXPORTUNFROZEN` keyword, 72
Externalization, 550, 798
`ExternalizeL()`, 154–156, 581

F

Factory function, 153, 157, 709
FBSCLI library, 559
FFS (Flash filing system), 681, 709
Fields, initializing, 250
File handling, 200–201

Index

File logging, 675–677
File Server, 132, 143–148, 165–166, 170, 175, 177, 433, 568, 709
File system, and good application behavior, 219
Files, 143–151, 709
  `RFile` API, 147–151
  `RFs` API, 144–147
  Symbian OS filenames and pathnames, 143–144
`FilmReel` example application, 619, 628, 630
`FilmReel2` example application, 619, 628
`Find()`, 100
Fixed-period progress notes, 324–325
Flush, 160, 690, 709
`FocusChanged()`, 240
Fold, 709
Font and Bitmap Server, 558
Fonts and text, 554, 569–574
  key classes/functions, 571
  key font classes, using to enumerate through all available fonts, 572–573
  measurements, 570–571
  text effects, 573–574
FontsAndText example application, 555, 572–573
`Format()`, 101
Forms, 290, 302–310, 710
  `Add Field`, 304, 306
  creating in an application, 306–310
  defined, 302
  defining in a resource, 306–308
  `Delete Field`, 304, 306
  `Edit Label`, 304, 306
  edit mode, 302–303
  editing, 305
  executing, 309–310
  form lines, 303–304
  form-derived class, creating, 308–309
  soft keys, 304–306
  view mode, 302–303
Framework, 36–37, 710
Front-end-processor (FEP), 408, 709
`FsSession()`, 144, 568
FTP (File Transfer Protocol), 448, 457, 710
Functional testing, 667–668

G

gcc, 70
GDI (Graphics Device Interface), 559, 710
GDI library, 559
`GET` request, 710
  making, 495–496
`GetDir()`, fetching a directory listing using, 146–147
`GetPixel()`, 581
GIF (Graphics Interchange Format), 559, 710
Global editor, 397

Global message queries:
  creating, 333–334
  handling dismissal of, 334–335
Global notes, 311, 325
Global queries, 327, 332–335
  complexity of, 332–333
  defined, 332
  types of, 335
GlobalQuery example application, 327, 333–335
GMS grid, 367
GNU C++ compiler (gcc), 9, 671, 710
GNU, defined, 710
Good application behavior, 174, 217–220
  Active Objects and responsiveness, 220
  adopting a skeptical/critical approach to development, 217
  checking disk space before saving data, 219
  exiting gracefully, 218–219
  and file system, 219
  handling Window Server-generated events, 218
  hard-coding and magic numbers, 220
  and system watchdogs, 219
  timers, 220
GPRS (General Packet Radio Service), 66
Granularity, 111, 117, 710
Graphical User Interface (GUI), 4–5, 77, 710
  application build process summary, 36
Graphics device classes, 563
Graphics devices and graphics contexts, 562–565
Graphics libraries, 559
Grids, 16–17, 342–343, 364
  basics, 368–375
  concrete grid class, creating, 372–375
  defined, 364, 710
  defining using resources, 368–371
  GMS grid, 367
  markable, 364, 375–376
  menu, 364
  monthly calendar grid, 365–366
  multiselection, 364
  orientation, 365
  pin-up board grid, 366
  selection, 364
  types of, 364
  using, 367–376
`group` folder, 38–39
GSM (Global System for Mobile communication), 502, 651, 710

H

h, 9
HalView example application, 618, 619
Handle, 129, 131–132, 710
`HandleCommandL()`, 259
`HandleControlEventL()`, 244

Index

HandleResourceChange(), 239
Hardware Abstraction Layer (HAL), 618, 649–651, 710
  phone IMEI number, 651
HBufC, 95, 99, 102–105, 106
Header part, resource files, 247–248
  application information resource, 248
  document name buffer, 248
  include statements, 247
  NAME statement, 247
  RSS_SIGNATURE, 248
Heap, 33–34, 78, 710
  allocation, 87
  defined, 78
Heap testing, 664
Height (metric), 571
HelloWorld GUI application, 30–46
  aif files, 43–46
    aifbuilder tool, 43
    aiftool utility, 43
    captions, 45–46
    file localization, 44
    icons, 44
    MIME support, 44–45
  application resource files, 40
  avkon.hrh, 41
  bld.inf, 6–8, 13, 30
  executable and runtime files, 36–38
  HelloWorldCon, 46–47
    building, 47–49
    emulator executable files, 49
    mmp, 48–49
    running, 47–49
    target executable files, 49–50
  localization of applications and resources, 42–43
  mmp, 31–33
  project files and locations, 38–39
  resource compiler, 41–42
  source files, 39–41
  stack and heap sizes, 33–34
  UIDs (unique identifiers), 31–33
HelloWorld project, 4–5
HelloWorld_caption.rsc file, 37, 40, 45–46
HelloWorld.aif file, 37
HelloWorld.app file, 36–37
HelloWorldApp.cpp, 39
HelloWorldApp.h, 39
HelloWorldAppUi.cpp, 39
HelloWorldAppUi.h, 39
HelloWorld.cbx, 7
HelloWorldCon, 46–47
  building, 47–49
  emulator executable files, 49
  mmp, 48–49
  running, 47–49
  target executable files, 49–50
HelloWorldCon example application, 46–48, 50, 55
HelloWorldContainer.cpp, 39
HelloWorldContainer.h, 39

HelloWorldDocument.cpp, 39
HelloWorldDocument.h, 39
HelloWorld.dsp, 7
HelloWorld.dsw, 7
HelloWorldLoc, 27, 53
  .pkg file, 56
helloworld.mmp, 6, 7, 12, 13
HelloWorld.rsc, 36–37
Host, 497
HTML (HyperText Markup Language), 491, 710
HTTP (HyperText Transfer Protocol), 490, 491–501, 711
  data suppliers, 493
  filters, 493
  headers, 492
  HTTPExample application, 493–501
  sessions, 491–492
  transactions, 492
HTTPExample application, 490, 493–501
  GET request, making, 495–496
  implementation, 493–494
  opening a session, 494–495
  POST request, making, 500–501
  request headers, adding, 496–497
  submitting the request/receiving the response, 497–500
  terminating the session, 501
HTTPExample example application, 493
HyperTerminal, 675, 683–684

I

IANA (Internet Assigned Numbers Authority), 460, 711
IAS (IrDA Information Access Service), 466, 711
ICMP (Internet Control Message Protocol), 443, 457–458, 711
ICO, 559, 711
IDE, *See* Integrated Development Environment (IDE)
Image conversion, 594–601
Image manipulation, 555, 594–604
Image rotation, 601–602
  Series 60 1.x, 601
  Series 60 1.x vs. Series 60 2.x, 603
  Series 60 2.x, 601–602
Image scaling, 602–604
  Series 60 1.x, 602
  Series 60 1.x vs. Series 60 2.x, 603
  Series 60 2.x, 602–603
imagefile, 317
imageid, 317
ImageManip example application, 555, 594
imagemask, 317
IMAP4 (Internet Message Access Protocol, v.4), 516

734 Index

IMEI (International Mobile Station Equipment Identification), 651, 711
Import library, Series 60 1.x vs. Series 60 2.x, 603
Import Mobile example application, 13
Inbox, watching, 540–543
`inc` folder, 38
Include header, Series 60 1.x vs. Series 60 2.x, 603
Incoming messages, 539–543
  APIs, 539–540
  watching the inbox, 540–543
Indicators, 284–286
Information note, 311
`Insert()`, 101
`install` folder, 38–39
Installation File Generator, 50–52, 711
Integrated Development Environment (IDE), 4, 7–9, 12–15, 19, 28–29, 40, 47–48, 710
  command-line tools vs., 8–9
  installation tips for, 27
Integration testing, 666–667
Interfaces, 40–41, 46, 80, 82, 111, 118
  defined, 711
Internalize, 156–157, 711
`InternalizeL()`, 156–157, 581
Internationalization, 174, 213–217, 221, 711
  general guidelines for developers, 213–216
  localization, OS support for, 216–217
Internet access point (IAP), 459–461, 711
Internet Protocol (IP), 424, 443, 457, 504, 712
Inter-Process Communication (IPC), 170, 458, 711
IP address, 390, 424, 426, 459, 462, 506, 712
IP address editor, 424
IP address editor setting item reference, 390
IR (infrared), 443, 469, 675, 712
IrCOMM (Infrared Communications), 434, 466, 467, 712
IrDA (Infra Red Data Association), 66, 465, 467, 712
IrDA Serial API, 469
IrDA Sockets API, 467–468
IrLAN (Infrared Local Area Network), 467, 712
IrLAP, 466
IrLMP, 466, 712
IrMUX, 443, 467–468, 712
IrOBEX (Infrared Object EXchange), 467, 470, 712
IrSerial, 431
IrSerial example application, 432, 434–435, 469
IrSockets, 431
IrSockets example application, 439–440, 444–445, 448
IrTranP, 467, 469–470, 712
`IsFocused()`, 240
ISP (Internet Service Provider), 66, 712

J

Java, 500, 557, 712
Jeteye ESI-9680 (Extended Systems, Inc.), 66

JPEG (Joint Photographic Experts Group), 59, 594–595, 712
Just-in-time (JIT), 712
  debugging, 689

K

`KAknsMessageSkinChange`, 239
`KEikMessageColorSchemeChange|`, 565
Kernel, 127, 131, 166–167, 169, 712
`KERN-EXEC` 3, 34
`KErrNoMemory`, 87
Key events, 239–242
  control stack, 241
  focus, 240–241
  offering, 241–242
`KeyMap` keyword, 65
`KUidMsgTypeBt`, 516
`KUidMsgTypeIMAP4`, 516
`KUidMsgTypeIr`, 516
`KUidMsgTypeMultimedia`, 516
`KUidMsgTypePOP3`, 516
`KUidMsgTypeSMS`, 516
`KUidMsgTypeSMTP`, 516

L

L2CAP (Bluetooth Logical Link Control and Adaptation Protocol), 471, 712
`LANG`, 32
Last calling number, retrieving, 550–551
LDRA Testbed, 672
Leave, 84–88, 712
Leavescan, 671
`Leavescan` tool, 88
Leaving, 84
Leaving issues, and the cleanup stack, 88–91
`Left()`, 100
Left adjust (metric), 571
`Length()`, 99
`LIBRARY` statement, 32–34
`LINK`, 254
Linked list, 117, 713
List dialogs, 335–338, *See also* Selection list dialogs
  defined, 335
  markable, 338
List queries, 327, 330–332
  creating/executing, 331–332
  multiselection, 332
  resources, 330–331
Listen queue, 445
Lists, 341–393
  basics, 342, 343

Index

defined, 342
grids, 342–343
settings, 343
vertical, 342–364
Literal search, 66
Literals, 101–102
`LLINK`, 254
`LoadFromCsvFileL()`, 87
Locale, 42, 53, 56, 214, 216–217, 296, 713
Localization of applications, 26
Locally scoped variables, 33
Log engine, 625–627, 713
log entries, reading, 625–627
Logical driver, 432
`LONG`, 254
`LTEXT`, 254

M

`MAbortDirectScreenAccess`, 589, 592–593
Machine UIDs, 69, 713
Macros:
debugging, 689
defined, 713
Magic numbers, 220, 713
Main pane, 11, 286
`makefile`, 14, 713
`makekeys.exe`, 50, 51–52
`makesis.exe`, 50, 51–52
Markable grids, 364, 375–376
creating, 364
drawing, 365
icons, setting, 364–365
Markable list dialogs, 338
Markable lists, 344, 346, 357–360
defining using resources, 358
dynamically changing marked list items, 359–360
marking commands, handling, 358–359
MarkableList example application, 344, 347, 357–359, 376
Mask, 60, 315, 713
Masking color, 60
`MaxLength()`, 101
`MaxSize()`, 101
MBM (multi bitmap file), 59, 274, 279, 349–350, 559, 713
MBMViewer, 62
M-classes, 80
`MCoeControlObserver`, 244
`MCsvFileLoaderObserver` interface, 80–81
`MDirectScreenAccess`, 589, 592–593
MEDIACLIENTIMAGE library, 559
Memory allocation, 33–34
Memory leak, 89, 483, 573, 691–692, 713
Menu grids, 364
Menu lists, 344, 345
`MENU_BAR`, 256–258

`MENU_ITEM`, 257–258
`MENU_PANE`, 257–258
`MENU_TITLE`, 257–258
Menus, 224, 254–264, 713
basics, 256–260
context-sensitive, 262–264
defining using resources, 262
displaying, 262–264
defined, 254
defining using resources, 256–258
dynamic, 261
menu commands, handling, 258–260
submenus, 255
using in a control, 258
Message queries, 327
Messaging, 490, 508–543, 713
APIs, 512–514
Client MTM API, 512–526
CSendAppUi API, 513–514
Send-As API, 513
architecture, 509
entries, 509
entry storage, 510
file system, 510
generic entry handling/unified inbox, 509–510
incoming, 539–543
APIs, 539–540
watching the inbox, 540–543
key classes/data types, 511–512
`CMsvEntry`, 511
`CMsvEntrySelection`, 512
`CMsvOperation`, 512
`CMsvSession`, 511
`CMsvStore`, 511
`TMsvEntry`, 511
`TMsvID`, 511
key concepts, 508–510
Message Index, 510
messaging server and session, 508
Messaging Store, 510
MTMs (message type modules), 509
Messaging example application, 22
Messaging view switching, 623
Metrowerks CodeTEST, 672
Metrowerks CodeWarrior, 6, 8
Metrowerks CodeWarrior C++ IDE, 9
Metrowerks CodeWarrior Development Studio for Symbian OS v.2.5, 671
Metrowerks CodeWarrior IDE, use of devices with, 29
MFNE (Multi-Field Numeric Editor), 423–427
defined, 714
MFNEs, *See* Multi-field numeric editors (MFNEs)
`MGraphicsDeviceMap`, 571
Microsoft Visual C++:
Symbian OS variable expansion, 63
syntax highlighting, 63
Microsoft Visual C++ 6.0, 6
and .NET, 671
Microsoft Visual C++ IDE, 9

736 Index

Microsoft Visual Studio .NET, 64
`Mid()`, 100
`.midi` format, 604
MIME, defined, 713
MIME support, `.aif` files, 44–45
Mixins, 80, 176, 259, 261, 483, 496, 527, 592, 630, 713
`MkDir()`, 145
`MkDirAll()`, 145
`MMdaImageUtilObserver` class, 594
`.mmp` file, 6–8, 12, 46, 72, 251, 649
MMPClick, 63
MMS MTM:
  adding attachments, 524–526
  creating the message, 523–524
  sending an entry, 526
  using, 523–526
  validating an entry, 526
MMS (multimedia messaging service), 401, 490, 508–509, 513–514, 516, 523–524, 537
  defined, 714
Mobile Innovation TRY, 673
`MObjectProvider`, 239
Modal, 142, 189, 291–292, 295, 297, 714
Modal dialogs, 291
Modeless, defined, 189–191, 291, 714
Modeless dialogs, 291
Monthly calendar grid, 365–366
Move (metric), 571
MSDN (Microsoft Developer Network), 19, 714
MsgObserver example application, 491, 539–540
MTM (Message Type Module), 512–513
  defined, 714
MtmsExample example application, 490, 514
Multi Media Server, 558–559
Multi-field numeric editors (MFNEs), 397, 423–427
  date editor, 425
  defining in a resource, 426
  duration editor, 425
  getting/setting a value of, 427
  instantiating, 426
  IP address editor, 424
  number editor, 424
  range editor, 424
  time and date editor, 425
  time editor, 425
  time offset editor, 425
  using an MFNF, 425–427
Multihoming, 459–467
  closing connections, 463–464
  explicit connections, 460–463
  implicit connections, 464–465
  RConnection API, 459–460
  transferring data, 463
Multimedia, 553–615
  animation, 554–555, 582–615
  audio, 555, 604–614
    audio data, 610–612
    audio device modes, 607–608

    `.midi` format, 604
    recording, 605–609
    Series 60 audio libraries, 604
    streaming, 612–614
    tones, 609–610
    `.wav` format, 604
  basic drawing, 554, 559–569
  bitmaps, 554, 578–581
  fonts and text, 554, 569–574
  image manipulation, 555
  Series 60 graphics architecture, 554, 556–559
  shapes, 554, 574–577
Multimedia Messaging Service (MMS), See also MMS MTM
  defined, 523
MultiMediaF example application, 555, 594
Multipage dialogs, 298–300
Multiple SDKs, using, 6
Multiselection grids, 364
Multiselection list queries, 332
Multiselection lists, 344, 346
Multitasking, 77, 127–128, 142, 714
Multithreading, 127, 594, 600, 603

N

`NAME_OF_APP_caption.rss`, 45–46
Namespaces, 81, 113–114
  defined, 714
Naming conventions, 76, 77–81
  C-classes, 79
  M-classes, 80
  namespaces, 81
  R-classes, 79–80
  S-classes, 79
  T-classes, 78–79
Navigation pane, 265–266, 274–286
  displaying a label in, using resources, 281–283
  displaying an image in, using resources, 283–286
  displaying tabs in, 274–276
    using resources, 278–281
  indicators, 284–286
  main purpose of, 274
NavigationPane example application, 225, 278–281, 283
`NBitMapMethods`, 81
`NEikonEnvironment`, 81
New Symbian GUI example application, 61
`NewL()`, overloading to take a read stream, 157
`NewLC()`, 94, 325
  overloading to take a read stream, 157
NIFMAN (Network Interface Manager), 458, 714
Nokia Testing Suite, 673
Non-skin-aware controls, 239
Nonwaiting dialogs, 291

Index

Non-Window-owning controls, 228–229
`NormalFont()`, 568
Notes, 290, 310–325
  customized, 313–315
    constructing/executing, 315
  defined, 714
  global, 325
  predefined, 311–312
  progress, 319–325
    completion of, and user cancellation, 323–324
    creating for a variable-length process, 321
    declaring, 320–321
    defined, 319–320
    executing, 321–322
    fixed-period, 324–325
    updating as a process executes, 322–323
  wait, 316–319
    defining in resource, 316–317
    `MAknBackgroundProcess`-derived class, 318–319
    wrapper object, constructing and executing, 316–318
  wrapped, 310–313
Notification colors, 60
Number editor, 424
Numeric editors, 397, 417–422
  characteristics of, 418
  configuring, 418–422
  defined, 417
  fixed-point editor resource, defining, 419–420
  integer editor resource, defining, 419
  obtaining/validating values in, 420–421
  resources/classes, 418
  setting values in, 421–422
  types of, 417
NumericEditor example application, 397, 418–421, 427

O

OBEX (Object Exchange), 22, 470, 714
Observer Design Pattern, 80
Observers, 244
Offering events, 241–242
`OfferKeyEventL()`, 241, 262
Offscreen bitmaps, 585–586
OOM (out of memory), 664, 685–686, 714
`Open()`, 147–148
Operating system (OS), 15, 34, 126, 477
  defined, 714
`operator()`, 102
`operator[]()`, 100
`operator<()`, 100
`operator=()`, 100
`operator!=()`, 100
`operator==()`, 100
`operator>()`, 100
OpponentForm example application, 305–310, 419
Optionally skin-providing controls, 238
Options menu, 11, 183, 255–256, 262, 266, 269–270, 273, 304–306, 309, 313, 316, 320, 337, 343, 346, 355–359, 378, 383, 714
Out-of-memory testing, 684–686

P

Package descriptors, 108–109
Packet Switched Data (PSD), 459, 714
Panes, 224, 264–286
  context pane, 272–274
  main pane, 286
  navigation pane, 274–286
  soft key pane, 286
  status pane, 264–268
  title pane, 268–272
Panics, 34, 64, 84–88, 91, 100–101, 105, 107, 113, 125–139, 142–143, 152, 248, 294, 423–424, 714
  in Series 60 emulator, 67
Password editor setting item reference, 392
Pathnames, Symbian OS, 143–144
PC-based development options, 9
PC-based platform emulators, 6
PC-Lint, 671
PCM (Pulse Code Modulation), 612, 715
PDP (Packet Data Protocol), 459, 715
Pens and brushes, 567
Performance testing, 668
Petran, 20, 71
Phonebook engine, 632–638
  creating and accessing phonebook items, 632–635
  phonebook searching, 636–637
  receiving notification, 635–636
  speed dialing and common dialogs, 637–638
Phonebook view switching, 620–621
Photo album engine, 646–649
  photo album folders, accessing, 646–648
  photo album UI components, 648–649
Photo Album view switching, 622–623
Physical driver, 432
Pies, 576
Pin-up board grid, 366
`pkg` file format, 52–57
  conditional component installation, 57
  multicomponent installation, 57
  multi-locale installation, 55–56
  running executables during installation, 55
Plain editor, 397
PlainTextEditor example application, 397, 404–406, 410, 427
Platform UID (platform identification code), 53, 715

`PlaySelectedGame()`, 262
Plug-in, 13, 165, 206–207, 715
PNG (Portable Network Graphics), 559, 715
Polygons, 576–577
POP3 (Post Office Protocol, v. 3), 66, 512–513, 516, 715
`PopAndDestroy ()`, 91
Pop-up menu lists, 361–364
 creating, 361–362
 displaying and handling selections, 363–364
 list classes/item definitions, 362–363
 setting the title for, 362–363
PopUpList example application, 344, 347–348, 361–363
Port number, 444–445, 447, 458, 469, 481, 484–485, 504, 715
POST request:
 defined, 715
 making, 500–501
Predefined notes, 311–312
Profiles view switching, 623
Profiling, 670
Program stack, defined, 78
Progress notes, 311, 319–325
 completion of, and user cancellation, 323–324
 creating for a variable-length process, 321
 declaring, 320–321
 defined, 319–320
 executing, 321–322
 fixed-period, 324–325
 updating as a process executes, 322–323
`Project|Properties` menu item, 58
projectname.mmp, 6–7
PRT, 500–501, 504
`Ptr()`, 99

Q

Quality assurance (QA), 658, 659–664
 coding standards, 659–661
 defensive programming, 661–664
  assertions, 661–662
  class variants, 662–663
  heap testing, 664
 defined, 715
Queries, 290, 298, 304–305, 309, 326–335
 confirmation, 326
 data, 328–330
  constructing/executing, 329–330
  defining in resource, 328–329
  types of, 326
 defined, 210, 326, 715
 global, 327, 332–335
  complexity of, 332–333
  defined, 332
  types of, 335

global message queries:
 creating, 333–334
 handling dismissal of, 334–335
list, 327, 330–332
 creating/executing, 331–332
 multiselection, 332
 resources, 330–331
message, 327

R

RAM drive, 326, 715
Range editor, 424
RArray and **RPointerArray** types, 110–117
 basic APIs, 110–113
 finding, 115–117
 sorting, 113–115
RAS (Remote Access Service), 326, 715
R-classes, 79–80, 96–97
`RDirectScreenAccess`, 589, 590–591
`Read()`, 148–149
`ReadResource()`, 568
Recording, 605–609
Recovery testing, 669
Rectangles, 575
Redraw events, 243
`ReportEventL()`, 244
RESOURCE, 32
Resource files, 224, 245–254
 defined, 715
 punctuation, 252
 resource structures, creating, 253–254
 string resources, 250–251
 `STRUCT` field types, 254
 structure, 247–252
  body part, 247, 249–253
  header part, 247–248
 syntax, 245–247
Resource management, 76
RFC 2616, 491, 497–498, 715
RFCOMM, 444, 472, 476, 484, 486, 716
RFile API, 147–151
 opening and closing, 147–148
 reading, 148–149
 seeking, 151
 writing, 149–151
RFs API, 144–147
 `Close()`, 144–145
 `Connect()`, 144–145
 `GetDir()`, fetching a directory listing using, 146–147
RGB, 60, 581, 716
Rich text editor, 398
 character formatting, applying, 413
 configuring, 410–416
 constructing from a resource, 415–416

Index

control:
  creating, 411–412
  using, 416
cursor, positioning, 414–415
defining a resource, 410–411
formatting attributes, setting, 412
setting text in, 413
RichTextEditor example application, 397, 410–413, 415–417, 427
`Right()`, 100
Right adjust (metric), 571
RISC (Reduced Instruction Set Computer), 678, 716
`RmDir()`, 145
ROM (read-only memory), 37, 151, 680, 681, 716
`RReadStream`, 153–154
.rss, 9
RS-232 standard, 472, 675, 716
`RunDlgLD()`, 297–298
`RunError()`, 131
`RunL()`, 130–131
`RWriteStream`, 152–153

S

`sample.app`, 694
`Save()`, 304, 581
Scheme colors, 60
S-classes, 79
Screen coordinates and geometry, 562
`ScreenDevice()`, 568
Scripted testing, 672–673
SDK (Software Development Kit), 4–6, 10, 13–15, 17, 23, 26, 28–29, 31, 41, 45, 47, 50, 52, 59, 61–63, 66–67, 110, 201, 214, 339, 374, 407, 426, 676, 716
Secret editors, 397, 422–423
  defining a resource, 422–423
  instantiating, 423
  resources/classes, 423
Secure sockets, 451–457
  Series 60 1.x, 452–456
  Series 60 2.x, 456–457
`Seek()`, 151
`Selection` button, 36
Selection grids, 364
Selection key, 255, 262–263, 294, 343, 345–346, 378, 383, 686, 716
Selection list dialogs, 336–338
  adding icons to, 337–338
  constructing, 337
  defining in resource, 336–337
  executing, 338
Selection lists, 344
Semaphore, 127–128, 135, 140–141, 169, 716
Send-As API, 513, 526–530

Send-As object, creating, 527
SendAsExample example application, 527–528
Serial communication, 430, 431–438
  Comms Server (C32):
    connecting to, 434
    starting, 434
  CSY, loading, 434
  defined, 430, 716
  serial device drivers:
    loading, 432–433
  serial port:
    closing, 438
    configuring, 435–437
    opening, 434–435
    transferring data over, 437–438
  using, 432–438
Serial Communications Server, 431
Series 60 1.x, differences between Series 60 2.x and, 603
Series 60 1.x WAP APIs, 503–506
  connected WSP session, 504–505
  connectionless WSP, 505–506
  WAP datagrams, sending, 503–504
Series 60 2.x WAP APIs, 506–508
  `CWAPBoundCLPushService`, 507–508
  `CWAPBoundDatagramService`, 506–507
  `CWAPFullySpecCLPushService`, 507–508
  `CWAPFullySpecDatagramService`, 507
  WAP datagrams, sending, 506
Series 60 application wizards, 60–62
  Borland C++BuilderX Mobile, 61
  Metrowerks CodeWarrior, 62
  Microsoft Visual C++ 6.0, 61
Series 60 audio libraries, 604
Series 60 C++ software development kits (SDKs), 4, 5–6
Series 60 emulators, 4, 10–12
  application panics in, 67
  building, 4, 12–14
    from the command line, 12–13
    from an IDE, 13–14
    using Borland C++BuilderX, 14
    using Borland C++IDE Builder 6, 13–14
    using CodeWarrior IDE, 14
    using Microsoft Visual C++ IDE, 13
  configuration, 64–66
    options and information sources, 65–66
  debug mode, 17
  executable locations, 15–17
    debug build emulator, 17
    release build emulator, 15
  hardware color capabilities of, 59–60
  HelloWorld application:
    debugging, 19
    locating/running, 18
  main pane, 11
  Options menu, 11
  running, 4, 15–19

Series 60 emulators, *continued*
  from Borland C++Builder 6 Mobile Edition IDE, 18
  from Borland C++BuilderX IDE, 18
  from the CodeWarrior IDE, 18
  from a command prompt, 17
  from the Visual C++ IDE, 18
  shortcut keys, 699–703
Series 60 graphics architecture, 554, 556–559
  Font and Bitmap Server, 558
  graphics libraries, 559
  Multi Media Server, 558–559
  Window Server, 556–557
Series 60 graphics libraries, 559
Series 60 Software Development Kit (SDK), 4, 10
  installation tips for, 27, 64–66
Series 60-specific coding identification colors-, 60
Series60Tools folder, 62
Server, 53, 66, 79, 127, 165–170, 716
Server push, 503
Server sessions, 165–166
  and Inter-Process Communication (IPC), 166–168
  server review, 169
  subsessions, 169–170
Service Discovery Database (SDD), 472–473, 474, 716
  closing, 477
Service Discovery Protocol (SDP), 472, 477, 716
Session, 716
SessionPath(), 145
SetActive(), 130
SetContainerWindowL(), 567
SetDimmed(), 261
SetFocus(), 240–241
SetLength(), 101
SetMax(), 101
SetObserver(), 244
SetObserver(NULL), 244
SetSessionPath(), 145
Settings lists, 342, 343, 377–379, 381, 383, 385, 716
  basics, 379–392
  binary switch setting item reference, 390
  constructing, 383–384
  defining using resources, 379–381
  deriving, 381–382
  enumerated text setting item reference, 388
  IP address editor setting item reference, 390
  password editor setting item reference, 392
  setting items, 377
    creating, 382
    types of, 384
  settings list values, changing, 383
  slider setting item reference, 384–385
  text editor setting item reference, 387
  time or date editor setting item reference, 389
  using, 378–392
  volume setting item reference, 386
SettingsList example application, 378–384, 391

ShapeDrawer example application, 175, 179, 191, 195, 196, 200, 219
Shapes, 553, 554, 574–577
  arcs, 576
  ellipses, 575
  pies, 576
  polygons, 576–577
  rectangles, 575
Shapes example application, 554, 555, 575, 577
Shortcut keys:
  Series 60 emulators, 699–703
    drawing, 700
    hardware emulation, 702
    miscellaneous, 703
    resource allocation, 701
    Window Server, 700–701
ShowNoteL(), 325
Siemens SX1 smartphone, 6
Simple controls, 225–228
  creating, 229–232
Simple dialogs, creating, 292–298
SimpleDlg example application, 175, 190, 291–294, 296–298, 300
SimpleList example application, 344, 347–352, 354, 358, 368, 372
SimpleMenu example application, 224, 256, 259
SimulateKeyEventL(), 568
.sis file, 50
  building, 21–22, 58
  defined, 716
  installing, 22
sisar utility, 51–52, 57, 717
Size(), 99
Skiing example application, 555, 588, 590–591, 593
Skin-aware controls, defining, 239
Skin-observing controls, 238
Skins, 224, 237–239
  compulsory skin-providing controls, 238
  defined, 237, 717
  non-skin-aware controls, 239
  optionally skin-providing controls, 238
  skin-aware controls, defining, 239
  skin-observing controls, 238
Slider setting item reference, 384–385
SMIL (Synchronized Multimedia Integration Language), 523–526, 717
SMS Inbox, 63
SMS MTM:
  creating an entry, 518–520
  sending an entry, 520–523
  in Series 60, 517–518
  using, 517–523
  validating an entry, 520
SMS OTA (Over the Air), 559
SMS (Short Message Service), 63, 255, 490, 502, 504, 508, 514, 517–520, 625, 633, 638, 668, 717
SMTP (Standard Mail Transport Protocol), 457, 717

Index

Socket communication, Bluetooth (BT), 484–486
Sockets, 84, 430, 438–457, 717
  address family constants, 443
  client socket, connecting to, 446–447
  client/server, 439
  closing, 450–451
  connected sockets, 440–451
  connecting to the Symbian OS Socket Server, 441–442
  connectionless and connected sockets, 439–440
  opening, 442–444
  protocol constants, 443–444
  secure, 451–457
  on Series 60, 438–439
  server socket, connecting to, 444–446
  transferring data over, 448–450
  type constants, 443
Soft key, 10–11, 183, 193, 255–256, 259–260, 304–306, 311, 312, 314, 328, 335, 364, 378, 387
Soft key pane, 286
Software Development Kits (SDKs):
  installation tips for, 27
  versions/selection, 27–29
    Series 60 Version 1.x SDKs, 28–29
    Series 60 Version 2.x SDKs, 29
SOURCE, 32
SOURCEPATH, 32
Special character tables, 403
Sprites, 58, 717
src folder, 38
SRLINK, 254
SSL handshake, 453
SSL (Secure Socket Layer), 451, 453, 456, 487, 717
Stack:
  Bluetooth (BT), 471
  defined, 78, 717
Standard application views:
  using, 619–624
  using resources, 618
Standard dialogs, 290, 291–302
  initializing dynamically, 296–297
  simple dialog, creating, 291–302
Standard Template Library (STL), 76
Start(), 130
StartVibra(), 653–654
Static colors, 60
Status pane, 10, 264–268
  basics, 266–268
  changing the visibility of, 266–267
  context pane, 265
  defined, 264
  navigation pane, 265–266
  position of, 264
  size, handing a change in, 268
  subpanes, 265
  title pane, 265
StatusPane example application, 225, 266–268
STL (Standard Template Library), 76, 109, 717

StopVibra(), 653
Stores, 77, 157–164, 717
  CDirectFileStore, 159–160
  defined, 157, 717
  reading from, 162–163
  steps in using, 163–164
  Stream Dictionary, 161–162
  types of, 158
Stream Dictionary, 161–162
Streaming, 612–614
Streams, 77, 151–157
  defined, 151, 717
  ExternalizeL(), 154–156
  InternalizeL(), 156–157
  overloading NewL() and NewLC() to take a read stream, 157
  RReadStream, 153–154
  RWriteStream, 152–153
Stress testing, 668
String resources, 250–251
Struct, 250, 253, 302, 679
STRUCT, 254
Struct:
  defined, 717
Structural testing, 667–668
Styles, editors, 416–417
Subsessions, 169–170
Symbian, defined, 8, 13, 717
Symbian Installation System (SIS), 27, 50–58
  Certificate Generator, 50–52
  Installation File Generator, 50–52
  makekeys.exe utility, 51–52
  makesis.exe utility, 51–52
  .pkg file format, 52–57
  .sis file:
    build tools, 51–52
    building, 58
  sisar utility, 51–52
Symbian OS, 5–7, 9
  architecture, 71
  as asynchronous operating system, 126
  cleanup stack, advanced use of, 96–97
  defined, 717
  export definitions, 72
  filenames, 143–144
  files, 143–151
    RFile API, 147–151
    RFs API, 144–147
    Symbian OS filenames and pathnames, 143–144
  fundamentals, 75–171
    client-server architecture, 165–170
    collection classes, 109–126
    construction methods, 96
    descriptors, 97–109
  pathnames, 143–144
  stores, 77, 157–164
    CDirectFileStore, 159–160

742 Index

Symbian OS, *continued*
  defined, 157
  reading from, 162–163
  steps in using, 163–164
  Stream Dictionary, 161–162
  types of, 158
  streams, 77, 151–157
    defined, 151
    `ExternalizeL()`, 154–156
    `InternalizeL()`, 156–157
    overloading `NewL()` and `NewLC()` to take a read stream, 157
    `RReadStream`, 153–154
    `RWriteStream`, 152–153
  support for producing devices using non-English languages, 43
Symbian settings tab, 58
SymbianOsUnit, 673
Syntax highlighting, 63
Syntax, resource files, 245–247
System Agent (SA), 651–653, 716
System capabilities, accessing, 618, 649–654
  Hardware Abstraction Layer (HAL), 649–651
    phone IMEI number, 651
  System Agent, 651–653
  vibration API support, 653–654
System testing, 666–667
System watchdogs, 174, 219, 718
  and good application behavior, 219
SystemAgent example application, 619
`SYSTEMINCLUDE`, 32
System-initiated redraws, 243

T

`TAny`, 83
`TARGET`, 32
Target device:
  building, 4
  deploying, 4
Target Series 60 device:
  building for, 19–21
    C++Builder X, 20
    CodeWarrior IDE, 20
  deploying on, 21–22
  running the application on, 22
  .sis file:
    building, 21–22
    installing, 22
`TARGETPATH`, 32
`TBool`, 83
`TBrushStyle`, 567, 574
`TBuf`, 99, 105–106
`TBufBase`, 99
`TBufC`, 99, 105–106
`TBufCBase`, 99

`TChar`, 83
T-classes, 78–79
TCP (Transmission Control Protocol), 465, 471, 492, 718
TCP/IP (Transmission Control Protocol/Internet Protocol), 430, 453, 457–465, 487, 491, 494
  CommDB, 458–459
  IPv6, 458
  multihoming, 459–467
  programming for Series 60, 458
  programming infrared on a Series 60 device, 467–470
    IAS queries, 468–469
    IrDA Serial API, 469
    IrDA Sockets API, 467–468
    IrOBEX (Infrared Object EXchange), 470
    IrTranP, 469–470
TcpipImpEx, 431
TcpipMultihomingEx, 431
`TDes`, 99, 106–108
`TDesC`, 99, 106–108
`TDesC&`, 106
`TDesC8`, 108–109
Telephony, 490, 543–551
  `ETel API`, 544
  getting started, 544–545
  last calling number, retrieving, 550–551
  making a call, 546–548
    checking if caller has hung up, 547
    handling end of call, 548
    opening RCall and dialing, 547
    watching for end of call, 547
  receiving a call, 548–551
    attempting to answer call, 549
    handle end of call, 549–550
    waiting for an incoming call, 549
    waiting for call to end, 549
Telnet, 457, 718
Test harnesses, 681–684
  `RTest`, 682–684
Testing, 658, 664–686
  black-box, 667
  debug output, 674
  file logging, 675–677
  functional, 667–668
  integration, 666–667
  out-of-memory, 684–686
  performance, 668
  recovery, 669
  scripted, 672–673
  serial output, 674–675
  strategies for, 666–669
  stress, 668
  structural, 667–668
  system, 666–667
  on target vs. emulator, 677–681
    directory differences, 680–681
    hardware limits, 678–679
    heap and stack sizes, 678

Index

machine word alignment, 678–679
out-of-disk errors, 679
thread process model, 678
timing differences, 680
writable static data, 680
test executables, running, 686
test harnesses, 681–684
   `RTest`, 682–684
test teams/procedure, 666
tools/techniques for, 669–677
   console applications, 672, 707
   dynamic tools, 670
   resource failure methods, 672–673
   static tools, 670–671
   unit, 666–667
   white-box, 667, 720
Testing example application, 659
TestQuest Pro, 673
Testwell CTC++, 672
`TEventCode`, 263
`TEXT`, 254
Text, *See* Fonts and text
Text editor setting item reference, 387
Text editors, 396, 397–417
   `avkon_flags` properties, 407–408
   configuring, 404–409
   defining a resource, 405–407
   dimensions/input capacity, 399–400
   features of, 398–399
   global editor, 397
   input case, 401–402
   input modes, 401, 408
   instantiating from a resource, 409
   keypad input, filtering, 400–402
   mappings to additional characters, providing, 402–403
   numeric key maps, 402–403, 408
   plain editor, 397
   properties, 404
   resource/classes, 398
   rich text editor, 398
      configuring, 410–416
   special character tables, 403
   types of, 397–398
`TFontSpec`, 571
Themes, *See* Skins
Thread, 34, 64, 67, 127–129, 131, 135, 139–140, 718
Thumb, 6, 20, 718
TIFF (Tagged Image File Format), 559, 718
Time and date editor, 425
Time editor, 425
Time offset editor, 425
Time or date editor setting item reference, 389
Timers, 220
`TInt`, 83
`TInt8`, 83
`TInt12`, 83
`TInt16`, 83
`TInt64`, 83

TinyTP (Tiny Transport Protocol), 470, 718
Title pane, 265, 268–272
   basics, 268–272
   changing the text in, 269–270
      using resources, 270–271
   displaying an image in, 270–271
      using resources, 272
TitlePane example application, 225, 268–271, 273–274, 276, 284–285
`TKeyEvent`, 263
`TLitC`, 102
TLS (Thread Local Storage), 680, 718
TLS (Transport Layer Security), 451–455, 487, 718
`TMsvEntry`, 511
`TMsvId`, 511
Tones, 609–610
`TPckg`, 109
`TPckgBuf`, 109
`TPckgC`, 109
`TPenStyle`, 567, 574
`TPoint`, 562
`TPtr`, 99
`TPtrC`, 99, 108
Traditional Symbian OS control-based architecture, 184–188
   when to use, 198
`Transfer-Encoding`, 497
Transport Secure Layer (TSL), 502
TRAP, 84, 86, 89, 93, 718
Trap harness, 84, 718
Traps, 84–88
`TReal`, 83
`TReal32`, 83
`TReal64`, 83
`TRealX`, 83
`TRect`, 562
`TRequestStatus iStatus` member, 131
`TSize`, 562
TSY, 32–33, 545, 718
`TText`, 83
`TText8`, 83
`TText16`, 83
`TTypefaceSupport`, 571
`TUInt`, 83
`TUInt8`, 83
`TUInt16`, 83
`TUInt32`, 83
Twips, 564, 718

U

UART (Universal Asynchronous Receiver-Transmitter), 675, 719
UI (user interface), 6, 27, 35, 39, 41, 45, 60–61, 76–77, 101, 131, 133, 139, 141–142, 160, 177–178, 719

Index

UID, 32
UID type, 31, 719
UID (unique identifier), 31–32, 44, 49, 53, 68, 160–161, 719
UID1, 31–32, 719
UID2, 31–33, 719
UID3, 31–33, 45, 53, 719
UIDs (unique identifiers), client MTM API, 516
Uikon, 41, 61, 177, 242, 719
UIQ, 5
UML (Universal Modeling Language), 81, 176, 719
Unicode, 13, 20, 76, 83, 98–101, 103–104, 150–153, 247, 254, 438, 448, 674, 688, 718
Unit testing, 666–667
URI (Uniform Resource Indicator), 491–492, 495, 505, 719
URL (Uniform Resource Locator), 397, 403–405, 451, 497, 605, 719
User Datagram Protocol (UDP), 443, 457, 502, 719
User heap macros, 665
User Interface (UI) implementation, 76
User::LeaveIfError(), 86–87
User::LeaveIfNull(), 87
User::LeaveNoMemory(), 87
USERINCLUDE, 32
UUID (Universally Unique Identifier), 474, 719

V

vCards, 638, 717
Vertical lists, 342–364
  basic lists, 348–355
    adding icons to a list, 349–350
    allowing the list to scroll, 352–354
    defining a list using resources, 348
    defining list items using resources, 351–352
    handling list events, 354–355
    instantiating a list using resources, 349
    list item definitions, 352–353
  dynamic lists, 355–357
    changing items dynamically in a list, 356–357
    defining a dynamic list resource, 355
    setting items dynamically in a list, 355–356
  finding items in, 347
  list items/fields, 346–347
  markable lists, 344, 346, 357–360
  menu lists, 344, 345
  multiselection lists, 344, 346
  pop-up menu lists, 361–364
  selection lists, 344
  types of, 344
  using, 347–362
VibraModeStatus(), 654

VibraSettings(), 653–654
Vibration API support, 653–654
View, 510, 550–551
  defined, 720
ViewManager example application, 619
View-switching applications, 619–624
  Calendar view switching, 621–622
  Camera view switching, 622
  Messaging view switching, 623
  nonswitchable applications, 623–625
  Phonebook view switching, 620–621
  Photo Album view switching, 622–623
  Profiles view switching, 623
VirtualKey keyword, 65–66
Volume setting item reference, 386

W

WAE (Wireless Application Environment), 502–503, 720
Wait notes, 311, 316–319
  defining in resource, 316–317
  MAknBackgroundProcess-derived class, 318–319
  wrapper object, constructing and executing, 316–318
Waiting dialogs, 291
WAP datagrams, sending, 503–504
WAP (Wireless Application Protocol), 490, 502–508, 720
  architecture, 502–503
  defined, 502
  Series 60 1.x WAP APIs, 503–506
    connected WSP session, 504–505
    connectionless WSP, 505–506
    WAP datagrams, sending, 503–504
  Series 60 2.x WAP APIs, 506–508
    CWAPBoundCLPushService, 507–508
    CWAPBoundDatagramService, 506–507
    CWAPFullySpecCLPushService, 507–508
    CWAPFullySpecDatagramService, 507
    WAP datagrams, sending, 506
  Series 60 implementation, 503–508
  stack, 502
Warning note, 311
.wav format, 604
WBMP (Wireless Bitmap), 559, 720
WDP, 502–503, 720
Web-safe colors, 60
White-box testing, 667, 720
Width (metric), 571
Win32, 10, 71, 720
Window ownership, 228–229

Index

non-Window-owning controls, 228–229
Window-owning controls, 229
Window Server, 556–557, 720
  shortcut keys, 700–701
Window-owning controls, 229
Windows, controls, 225
WINS (Windows Single process), 64, 678, 720
WINSB, 72, 720
WINSCW, 14, 72, 720
WMF (Windows Meta File), 559, 720
`WORD`, 254
Wrapped notes, 310–313
`Write()`, 149–151
WS32 library, 559
WSP (Wireless Session Protocol), 492, 502–505, 720
`WsSession()`, 568

WTLS (Wireless Transport Layer Security), 502, 720
WTP (Wireless Transport Protocol), 502–503, 505, 720

X

XML (eXtensible Markup Language), 491, 523, 720

Z

`Zero()`, 101

User Interface Creation

- **UI platform development**—Series 60, UIQ and earlier UI platforms
- **Application development**—Messaging, Telephony, Calendar, Contacts
- **UI customization**—Working with licensees to customize Series 60

Java

- **MIDP extensions**—Working with licensees to extend the MIDP APIs
- **JSR implementations**—Implemented for licensees and partner companies
- **Applications**—Including MIDP games and messaging solutions
- **Server components**—Communication systems using HTTP, SMS and WAP Push

Services

We offer a complete end-to-end range of services through our three business groups:

- **Product Creation**—Feasibility studies, specification, base porting, software engineering, project management, systems integration and technical documentation
- **Technical Consulting**—System architecture, workshops, advice clinics and technical support
- **Training**—International reputation for the range and excellence of our standard and customized Symbian-based training courses

Benefits of Working with EMCC Software

Companies work with us for the following reasons:

- **Proven track record**—Delivering successful Symbian OS projects on time and to budget
- **Quality**—Nokia audited QMS system to continually review and improve performance
- **Unrivalled expertise**—Long-standing experience in Symbian OS
- **Reputation**—Widely respected for delivering a first-class service
- **Commitment**—Committed to the success of our customer's projects
- **People**—Carefully selected and trained to the highest standards
- **Focus**—Entirely focused on Symbian OS and related technologies

Wouldn't it be great

if the world's leading technical publishers joined forces to deliver their best tech books in a common digital reference platform?

They have. Introducing
InformIT Online Books
powered by Safari.

■ Specific answers to specific questions.
InformIT Online Books' powerful search engine gives you relevance-ranked results in a matter of seconds.

■ Immediate results.
With InformIT Online Books, you can select the book you want and view the chapter or section you need immediately.

■ Cut, paste and annotate.
Paste code to save time and eliminate typographical errors. Make notes on the material you find useful and choose whether or not to share them with your work group.

■ Customized for your enterprise.
Customize a library for you, your department or your entire organization. You only pay for what you need.

Get your first 14 days FREE!

For a limited time, InformIT Online Books is offering its members a 10 book subscription risk-free for 14 days. Visit **http://www.informit.com/onlinebooks** for details.

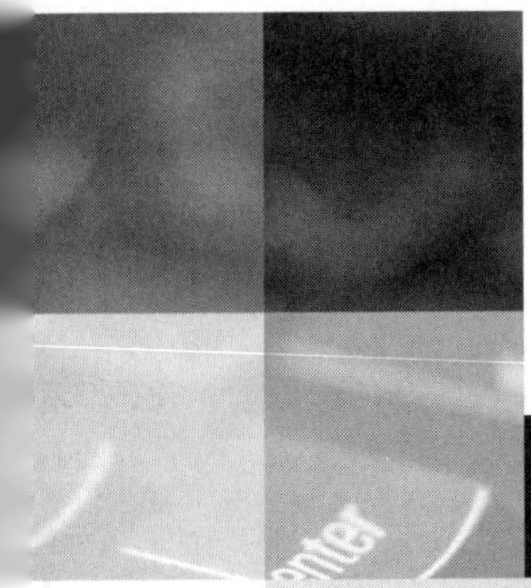

informIT

www.informit.com

YOUR GUIDE TO IT REFERENCE

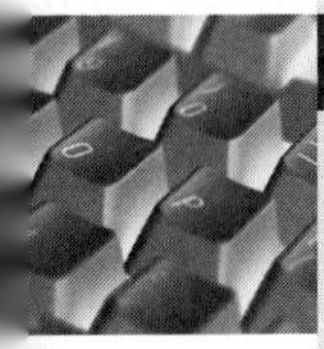

Articles

Keep your edge with thousands of free articles, in-depth features, interviews, and IT reference recommendations – all written by experts you know and trust.

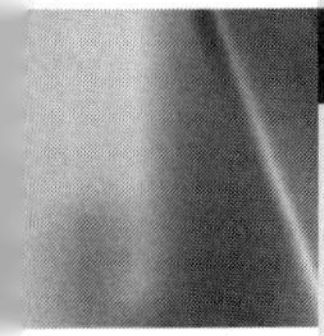

Online Books

Answers in an instant from **InformIT Online Book's** 600+ fully searchable on line books. For a limited time, you can get your first 14 days **free**.

Safari
POWERED BY
TECH BOOKS ONLINE

Catalog

Review online sample chapters, author biographies and customer rankings and choose exactly the right book from a selection of over 5,000 titles.

Register Your Book

at www.awprofessional.com/register

You may be eligible to receive:

- Advance notice of forthcoming editions of the book
- Related book recommendations
- Chapter excerpts and supplements of forthcoming titles
- Information about special contests and promotions throughout the year
- Notices and reminders about author appearances, tradeshows, and online chats with special guests

Contact us

If you are interested in writing a book or reviewing manuscripts prior to publication, please write to us at:

Editorial Department
Addison-Wesley Professional
75 Arlington Street, Suite 300
Boston, MA 02116 USA
Email: AWPro@aw.com

Visit us on the Web: http://www.awprofessional.com